HOMELESSNESS TO HC

HOMELESSNESS TO HOPE

RESEARCH, POLICY AND GLOBAL PERSPECTIVES

Edited by

UDAY CHATTERJEE

Assistant Professor, Department of Geography, Bhatter College, Dantan (Affiliated to Vidyasagar University), Paschim Medinipur, West Bengal, India

RAJIB SHAW

Professor, Graduate School of Media and Governance, Keio University, Fujisawa, Japan

LAKSHMI SIVARAMAKRISHNAN

Professor, Department of Geography, Jadavpur University, Kolkata, India

JENIA MUKHERJEE

Associate Professor, Department of Humanities and Social Science, Indian Institute of Technology Kharagpur, West Bengal, India

RAKTIMA GHOSH

Doctoral Researcher, Indian Institute of Technology Kharagpur, India

ELSEVIER

Elsevier
Radarweg 29, PO Box 211, 1000 AE Amsterdam, Netherlands
125 London Wall, London EC2Y 5AS, United Kingdom
50 Hampshire Street, 5th Floor, Cambridge, MA 02139, United States

ISBN: 978-0-443-14052-5

For information on all Elsevier publications visit our website at https://www.elsevier.com/books-and-journals

Publisher: Mica Haley
Acquisitions Editor: Kathryn Eryilmaz
Editorial Project Manager: Sara Valentino
Production Project Manager: Fahmida Sultana
Cover Designer: Vicky Pearson Esser

Typeset by TNQ Technologies

Contents

5. Migrants and homelessness: Life on the streets in urban India

Rashmi Rai, K. Lakshmypriya, Pallavi Kudal and Roli Raghuvanshi

6. Impact of flood vulnerability on livelihood, health, and homelessness: A case study of flood-affected blocks in lower Shilai watershed, West Bengal, India

Pinki Mandal Sahoo and Lakshmi Sivaramakrishnan

7. Climate-change induced out-migration and resultant women empowerment: Glimpses of resilience from Indian Sundarbans

Anindya Basu, Diotima Chattoraj and Adrija Bhattacharjee

8. Comparing rural and urban homelessness in the context of environmental vulnerability in India's hill states

Nipun Behl and Uttam Roy

9. The price of a roof: How rental stress is fueling hidden homelessness in Khulna City, Bangladesh

Kazi Saiful Islam

Section IV
Homelessness, governancy policy and sustainable solution

17. Exploring the model of Urban Innovation District as a solution for housing need in Neoliberal era: Case of Amravati, India

Sampreeth Inteti and Vibhore Bakshi

18. Urbanization as a remedy for homelessness: An analysis of the positive effect of peri-urban development

Rafia Anjum Rimi and Tasnuva Rahman

19. Meta-analysis of housing policy in India (1990–2022)

Somenath Halder

20. Homelessness in the context of extreme poverty: Social policy from Indonesia

Habibullah Habibullah

21. Housing intervention for people who are homeless in Indonesia: Combining institutional and community-based approaches

Hari Harjanto Setiawan and Yanuar Farida Wismayanti

22. Government policy transformation toward a reformed rental housing ecosystem as a mitigator of homelessness in postpandemic India

Ushnata Datta and Rewati Raman

Contributors

Abdillah Abdillah Universitas Padjadjaran, Gradute Program in Administrative Science, Bandung, Indonesia; Center for Decentralization & Participatory Development Research, Faculty of Social and Political Sciences, Universitas Padjadjaran, Bandung, Indonesia

Addri Attoza Rajshahi University of Engineering & Technology (RUET), Department of Urban and Regional Planning, Rajshahi, Bangladesh

Mondal Avijit Indian Institute of Technology Kharagpur, Department of Architecture and Regional Planning, Kharagpur, West Bengal, India

Roshan Baa St. Xavier's College, Ranchi, Jharkhand, India

Vibhore Bakshi IIT Roorkee, School of Planning and Architecture Bhopal, Department of Urban and Regional Planning, Bhopal, Madhya Pradesh, India

Trisha Barui Council of Architecture (COA), Institute of Town Planners, India (ITPI), Kolkata, West Bengal, India; Indian Institute of Engineering Science and Technology, Shibpur, Government of India, Kolkata, West Bengal, India

Anindya Basu Department of Geography, Diamond Harbour Women's University, West Bengal, India

Nipun Behl IIT Roorkee, Department of Architecture & Planning, Roorkee, Uttarakhand, India

Adrija Bhattacharjee Department of Geography, Diamond Harbour Women's University, West Bengal, India

Rahul Bhushan Centre For Geographical Studies, Aryabhatta Knowledge University, Patna, Bihar, India

Rd Ahmad Buchari Universitas Padjadjaran, Public Administration, Bandung, Indonesia; Center for Decentralization & Participatory Development Research, Faculty of Social and Political Sciences, Universitas Padjadjaran, Bandung, Indonesia

Sweta Chatterjee School of Water Resources Engineering, Jadavpur University, Kolkata, West Bengal, India

Diotima Chattoraj Wee Kim Wee School of Communication and Information, Nanyang Technological University, Singapore, Singapore

Kailash Chandra Das Department of Migration & Urban Studies, International Institute for Population Sciences, Mumbai, Maharashtra, India

Anutosh Das Rajshahi University of Engineering & Technology (RUET), Department of Urban and Regional Planning, Rajshahi, Bangladesh

Ushnata Datta Mcgan's Ooty School of Architecture, Department of Architecture, Ooty, Tamil Nadu, India

Habibullah Habibullah National Research and Innovation Agency (BRIN), Jakarta, Indonesia

Somenath Halder Kaliachak College, Faculty, Department of Geography, Malda, West Bengal, India

Shad Hossain Rajshahi University of Engineering & Technology (RUET), Department of Urban and Regional Planning, Rajshahi, Bangladesh

Sampreeth Inteti Department of Architecture, School of Planning and Architecture, Vijayawada, India

Kazi Saiful Islam Urban and Rural Planning Discipline, Khulna University, Khulna, Bangladesh

Pallavi Kudal Balaji Institute of International Business, Sri Balaji University, Pune, Maharashtra, India

K. Lakshmypriya Christ University, School of Business and Management, Bangalore, Karnataka, India

Alolika Mangal Department of Geography, Vivekananda College, Kolkata, West Bengal, India

Avijit Mondal Council of Architecture (COA), Institute of Town Planners, India (ITPI), Kolkata, West Bengal, India; Indian Institute of Technology Kharagpur, Government of India, Kolkata, West Bengal, India

Ephraim Kabunda Munshifwa The Copperbelt University, Department of Real Estate Studies, Kitwe, Zambia

Chilungu Mwiinde The Copperbelt University, Directorate of Planning, Property and Services, Kitwe, Zambia

Heru Nurasa Universitas Padjadjaran, Public Administration, Bandung, Indonesia

Susmita Paul Council of Architecture (COA), Institute of Town Planners, India (ITPI), Kolkata, West Bengal, India; Architecture and Planning Department, Indian Institute of Technology Roorkee, Government of India, Kolkata, West Bengal, India

Reza Amarta Prayoga Research Center for Social Welfare, Village and Connectivity, National Research and Innovation Agency, Jakarta, Indonesia

Roli Raghuvanshi Shyam Lal College (Evening), University of Delhi, New Delhi, Delhi, India

Margubur Rahaman Department of Migration & Urban Studies, International Institute for Population Sciences, Mumbai, Maharashtra, India

Tasnuva Rahman Rajshahi University of Engineering & Technology (RUET), Department of Urban and Regional Planning, Rajshahi, Bangladesh

Rashmi Rai Christ University, School of Business and Management, Bangalore, Karnataka, India

Rewati Raman Indian Institute of Technology (BHU), Department of Architecture, Planning and Design, Varanasi, Uttar Pradesh, India

Md. Juel Rana Govind Ballabh Pant Social Science Institute, Prayagraj, Uttar Pradesh, India

Rafia Anjum Rimi Rajshahi University of Engineering & Technology (RUET), Department of Urban and Regional Planning, Rajshahi, Bangladesh

Kunaljeet Roy School of Social Sciences and Languages, Vellore Institute of Technology, Chennai, Tamil Nadu, India

Uttam Roy Department of Architecture and Planning, Centre for Transportation Systems (CTRANS), Indian Institute of Technology Roorkee, Roorkee, Uttarakhand, India

Pinki Mandal Sahoo Department of Geography, Mirik College, University of North Bengal, Darjeeling, West Bengal, India

Md. Shakil Ar Salan Rajshahi University of Engineering & Technology (RUET), Department of Urban and Regional Planning, Rajshahi, Bangladesh

Hari Harjanto Setiawan National Research and Innovation Agency (BRIN), Research Center for Social Welfare, Villages and Connectivity, Jakarta, Indonesia

Kasturi Shukla Symbiosis Institute of Health Sciences, Symbiosis International (Deemed University), Pune, Maharashtra, India

Lakshmi Sivaramakrishnan Department of Geography, Jadavpur University, Kolkata, West Bengal, India

Paul Susmita Indian Institute of Technology –Roorkee (IIT–Roorkee), Department of Architecture and Planning, Kolkata, West Bengal, India

Barui Trisha Council of Architecture (COA), Institute of Town Planners, India (ITPI), Kolkata, India

Md. Emaj Uddin Department of Social Work, University of Rajshahi, Rajshahi, Bangladesh

Ida Widianingsih Universitas Padjadjaran, Public Administration, Bandung, Indonesia; Center for Decentralization & Participatory Development Research, Faculty of Social and Political Sciences, Universitas Padjadjaran, Bandung, Indonesia

Yanuar Farida Wismayanti National Research and Innovation Agency (BRIN), Research Center for Public Policy, Jakarta, Indonesia

Vijay Yadav Jawaharlal Nehru University, New Delhi, Delhi, India

Surendra Kumar Yadawa School of Liberal Arts, IMS Unison University, Dehradun, Uttarakhand, India

About the editors

Uday Chatterjee

Assistant Professor, Department of Geography, Bhatter College, Dantan (Affiliated to Vidyasagar University), Paschim Medinipur, West Bengal, India

Dr. Uday Chatterjee, PhD, is an Applied Geographer with a Postgraduate degree in Applied Geography from Utkal University and Doctoral degrees in Applied Geography from Ravenshaw University, Cuttack, Odisha, India. He has contributed many research papers published in various reputed national and international journals and edited book volumes. He has coedited 10 books and is a member of 8 academic societies. His areas of research interest include urban planning, social and human geography, applied geomorphology, hazards and disasters, environmental issues, land use, and rural development. Currently, Dr. Uday Chatterjee is an Assistant Professor in the Department of Geography, Bhatter College, Dantan, India.

Rajib Shaw

Professor in Graduate School of Media and Governance of Keio University, Japan. He is co-founder of a Delhi based social entrepreneur startup, Resilience Innovation Knowledge Academy (RIKA), and chair of the board of two Japanese non-government agencies: SEEDS Asia and CWS Japan. He is the Co-chair of the United Nations Office for Disaster Risk Reduction (UNDRR) Asia Pacific Science Technology Advisory Group (AP-STAG), and CLA for IPCC's 6th Assessment Report. Professor Shaw has 67 books and more than 450 research papers in the field of environment, disaster management and climate change. He is the editor in chief of Progress in Disaster Science. Professor Shaw is the recipient of "Pravasi Bharatiya Samman Award (PBSA)" in 2021 for his contribution in education sector. PBSA is the highest honor conferred on overseas Indian and person of Indian origin from the President of India. He also received United Nations Sasakawa Award in 2022 for his lifetime contributions in the field of disaster risk reduction. More about his work can be found in: www.rajibshaw.org.

Lakshmi Sivaramakrishnan

Professor, Department of Geography, Jadavpur University, Kolkata, India

Prof. Lakshmi Sivaramakrishnan is the Founder of the Geography Department at Jadavpur University. She has taught at North Bengal University and the University of Burdwan as full time Faculty and as Guest Faculty at the University of Tripura, WBSU, Calcutta University, Kalyani University. She is an Urban Geographer who combines the strengths of Regional Planning and Human Geography. She has 60 research papers and 3 books to her credit. She has delivered lectures across universities in India. She has taken up projects under UGC and has been supervising several research fellows. She has guided more than 20 scholars who have obtained their doctoral degree and has also supervised several MPhil scholars. She is on the editorial board of many renowned journals such as *Hill Geographer*, *Regional Science Policy and Practice*, *RSAI*, and *Indian Journal of Regional Science* and at present is the Editor of *Geographical Review of India*, Kolkata.

Jenia Mukherjee

Associate Professor at the Department of Humanities and Social Sciences, Indian Institute of Technology Kharagpur. She pursues transdisciplinary research on volatile delta ecologies of the global South and North. She is currently investigating five large-scale international global partnership projects on coastal livelihoods, climate vulnerabilities, and social resilience. She is the author of the book *Blue Infrastructures* and has published in journals such as *Frontiers in Water, Environment and Planning E, Marine Policy, WIREs Water, Urban Research and Practice*. She is the recipient of several awards and accolades from prestigious organizations such as DAAD and Rachel Carson Center for Environment and Society, Germany, Department of Foreign Trade and Affairs, Australia, the French Embassy.

Raktima Ghosh

Doctoral Researcher, Indian Institute of Technology Kharagpur, India

Raktima Ghosh is a Doctoral Researcher at the Rekhi Centre of Excellence for the Science of Happiness, Indian Institute of Technology Kharagpur. A part of the Global Partnership Project undertaken by the University of Manitoba and funded by SSHRC (Canada), her doctoral work focuses on dried fish social economy and fisheries governance in coastal West Bengal. Specifically, she explores the transdisciplinary possibilities to design solution-focused pathways through applying a participatory approach in dried fish research. Broad areas of her research include environmental governance, fish value chains, social ecological systems, and human geography.

Preface

Emerging challenges such as rapid urbanization processes, unequal economic policies, and political–institutional ideologies combined with perilous ecological and planetary threats are increasingly marginalizing the poor across the world, especially in the global South. Homelessness is the bleak outcome of the structural (policy, sociocultural construct, demographic change, governance mechanisms, neoliberal principles, and geopolitical issues), individual (household conditions, livelihoods), and dynamic social-ecological complexities (climate change impacts including cyclones, flooding, land shrinkage, riverbank erosion, human-induced ecological alterations, conservation, and resource depletion) leading stratification and dispossessions of all sorts to the vulnerable communities. Homelessness is not simply a persistent lack of shelter or house—it extends to outright deprivation of social, economic, and physical security, as well as marginalization of values, hope, purpose, and sense of rootedness in the world. Since colonial times to the present, homelessness has been supported by the intersecting processes of migration, eviction, and ecological displacements. These processes are inextricably attached to the ideas and methods of resource extraction, social transformation, and market logics which suit the interests of a powerful few. In their earliest approaches to considering the life stories of the homeless people, Thomas and Dittmar (1995) rightly place their concerns with locating the "the origins and course of home, not only in the discrete and isolated events of a housing history, but also in the on-going story that is told about this history." This book brings together stories, observations, and critical appraisals that have emerged out of the interdisciplinary studies spanning across North and South. It explores how diverse accounts on homelessness and homeless people are situated within the structural–institutional arrangements of the developing and developed worlds. Through its comparative framework, this book offers a broader understanding of the multiple ways in which homelessness is experienced, perceived, and addressed. On one hand, it lays out the causal issues and implications of homelessness in the "developing" countries of South America, Africa, South and Southeast Asia through delineating how the economies of these countries are highly porous to the global market instruments and transnational agencies that assist the nation-state in crafting "global cities" and conservation prescriptions. Rampant population growth and widening gap between rich and poor contribute to the uneven distribution of resources and self-sustenance policies in the South. Moreover, the biodiversity-rich ecologies of these regions are either degraded or threatened by various anthropogenic activities (hydraulic infrastructures, incessant fish capture, industrial operations), whereas climate change–induced extreme events make marginalized communities more vulnerable. Many countries of the global South have the history of being ripped apart by the colonizer whiteheads instituting perpetual humanitarian crises in borders, migration, and identity. A vast majority of more than 1.1 billion people (including inadequately housed and street sleepers), roughly estimated to be homeless by the United Nations, live in developing countries and the number continues to rise due to socioeconomic uncertainties, climate change, and conflicts. On the other hand, the emerging transboundary issues related to regulatory as well as normative processes in the society have triggered homelessness in some developed countries like the United States. Various studies have examined the connections between racial inequities and homelessness, further specifying the outcomes related to economic mobility, services, societal discrimination, criminalities etc. Indeed, this points to how the mechanisms of homelessness are multipathway and complex in terms of political, economic, and social relationships among and within nations. Indeed, entrenched complexities and issues related to homelessness in fast changing circumstances must be uttered head on and transparent. Much of the conceptualizations and explorations, made on homelessness till date, are based solely on the contexts and findings from the 'developed' world. Despite that the Southern countries encounter wide-ranging dimensions of homelessness and interlinked processes, such realities and driving factors have only received sporadic attention both at the academic and policy levels. This volume will address this gap by developing a holistic, relative, and diverse understanding about homelessness which seeks urgent attention and meaningful actions today.

In this co-edited volume, cross-cutting theoretical framings (such as resilience, well-being, social-ecological systems, sustainability, urban planning, institutions, and gender) and emerging discourses on homelessness will be complemented by empirical findings from the world. It will provide insights into diverse concepts, meanings, perceptions, identities, and values concerning homelessness across rural and urban settings to generate comprehensive understandings. In doing so, this book will critically address the limits of contemporary discussions on homelessness, eviction, and poverty. Broadly, the authors will explore the causations and processes (poverty, migration, eviction, displacements, disability) of homelessness to shed light on physical, social, ontological, territorial, and cognitive facets of homelessness in both local and regional contexts across the world. In making meanings and space for the voices of the people in the street, the essays will draw in the accounts of hardships, lived experiences, involvements, and trauma while also recognizing the influences of class, caste, gender, race, and ethnicity. Furthermore, this book foregrounds emerging interdisciplinary frameworks, placing priorities on social justice, rights-based, and collaborative approaches. It will lay a strong and inclusive focus on viable transitions through identifying, comparing, and advocating for inclusive, actionable measures and policies. This book will embrace contributions from interdisciplinary researchers who are involved with ethnographic, historical, and sustainability research across the plane of social sciences—sociology, human geography, history, economics, psychology, development studies, population studies, South Asian studies, and political science. It will be a useful guide to the students, researchers, practitioners, and policymakers, interested in expanding the concerns on homelessness as well as formulating effective pathways for improvements or change. State-of-the-art methodologies and unbiased arguments about relevant issues will explicitly characterize the volume.

This book is a comprehensive collection of emerging discourses which inform and get informed by the field-based observations on homelessness. What is unique about the volume is that it accommodates a variety of key themes, methods, and critical outlooks, sharply manifesting the interdisciplinary nature of the authorship. Added to that, it builds upon the current scholarship on homelessness focusing on high-, medium-, and low-income countries of the world, and tracing out the commonalities, variabilities, and interconnections within the processes and contexts of homelessness across nations. The novelty of the proposed book also relates to that it adheres to a solution-focused approach, which emphasizes collaboration among practitioners, activists, grass-roots organizations, and researchers in designing action-oriented pathways. Through carving out the situational complexities, comparative understandings, meanings, and stories which cohesively paint the scenario of homelessness around the world, this volume calls for all-important academic and policy discussions. Indeed, it has the impulse to draw the attention of academicians, practitioners, and policymakers alike to reimagining, retrospecting, and redesigning housing with multiple layers of homelessness.

This book has been made possible with the dedicated efforts of academicians, social scientist experts, urban planners, policymakers, and local activists around the world. Heartfelt appreciation goes to the exceptional contributors of each chapter, whose hard work played a crucial role in bringing this book with rich content. Special gratitude is extended to the anonymous reviewers, whose insightful comments and suggestions significantly enhanced the depth of research and the overall quality of the book. As perpetual learners, we are deeply thankful for the invaluable support received from students, parents, family members, teachers, and collaborators. Their collective assistance has been instrumental in streamlining the rigorous editing process. Last, but not least, we express our sincere admiration for our publisher and the publishing editor, Katy Eryilmaz, Senior Acquisitions Editor, Science, Technology, and Society, book series Elsevier, Sara Valentino Senior Editorial Project Manager and her team, Elsevier, whose unwavering support and encouragement have been indispensable throughout this endeavor.

West Bengal, India **Uday Chatterjee**

Japan **Rajib Shaw**

West Bengal, India
Lakshmi Sivaramakrishnan

West Bengal, India **Jenia Mukherjee**

West Bengal, India **Raktima Ghosh**

SECTION I

Introduction

CHAPTER

1

Home, homeless, and homelessness

Paul Susmita[1], *Barui Trisha*[2] *and Mondal Avijit*[3]

[1]Indian Institute of Technology—Roorkee (IIT—Roorkee), Department of Architecture and Planning, Kolkata, West Bengal, India; [2]Council of Architecture (COA), Institute of Town Planners, India (ITPI), Kolkata, India; [3]Indian Institute of Technology Kharagpur, Department of Architecture and Regional Planning, Kharagpur, West Bengal, India

1. Introduction

Homelessness represents an acute manifestation of societal marginalization, subjecting an extensive cohort of individuals to the perils of malnutrition and acute impoverishment. The formidable challenges that underpin this condition encompass a glaring dearth of accessible healthcare provisions, compounded by the burden of exorbitant costs, thereby engendering deleterious repercussions for mental well-being. The confluence of these adversities establishes an incubatory environment conducive to the propagation of substance abuse, a phenomenon further exacerbated by the prevailing inadequate living conditions. Within this milieu, the susceptibility of women and children to instances of violence is markedly amplified, a reality further accentuated by the attendant stigmatization and social ostracization. Ultimately, the scourge of homelessness transgresses the boundaries of a mere socioeconomic conundrum, constituting an erosion of fundamental human rights. At its crux, the genesis of homelessness is rooted in the bedrock of extreme destitution, insufficiency in the availability of affordable housing, glaring socioeconomic disparities, lack of ownership, lack of identity, instances of prejudiced practices. Urban locales, characterized by the juxtaposition of paltry wages, soaring rental expenditures, and a relentless inflationary surge in the cost of living, precipitate a harrowing dilemma for individuals, compelling them to make the stark choice between subsistence and secure lodging. This inexorable confluence impels a considerable contingent to gravitate toward urban centers, albeit involuntarily, culminating in a significant stratum compelled to eke out an existence on the unforgiving streets. The Universal Declaration of Human Rights (UDHR), adopted by the United Nations General Assembly in 1948, proclaims in Article 25(1) that "Everyone has the right to a standard of living adequate for the health and well-being of himself and of his family, including food, clothing, housing and medical care." It is noteworthy that domestic strife, intrafamilial turbulence, and the specter of substance-induced abuse, together with the ominous specter of physical and

Homelessness to Hope
https://doi.org/10.1016/B978-0-443-14052-5.00001-X

psychological violence, collectively impel a substantial cohort to extricate themselves from their familial moorings in pursuit of sanctuary elsewhere. This intricate tapestry not only underscores the somber ramifications attendant to homelessness but also serves as a stark reminder of its inextricable linkage with the broader ambit of societal marginalization.

A home or shelter constitutes an intrinsic imperative that embraces the facets of habitation, security, solace, and entitlement. Moreover, it harbors a profound psychological import interwoven with self-regard, affiliation, and individuation. In contradistinction, destitution emerges as a worldwide phenomenon of greater ubiquity than distinctiveness. Political, economic, cultural, societal, and environmental variables collectively orchestrate the existence of those bereft of domiciles, exerting influence upon prospective enhancements. As stipulated by the Census of India, a census house is delineated as a habitation possessing the following attributes: a roofing structure, a distinct primary entrance accessible from a thoroughfare, and formal acknowledgment as an autonomous residential entity. Consequently, abodes characteristic of slum or squatter settlements, despite their marginalized status, frequently elude categorizations denoting homelessness by virtue of their adherence to these prescribed criteria. Conversely, pavement-dwellings are customarily precluded from census house classification due to the impermanent nature of their roofing configurations and structural compositions (Census of India, 2011). Homelessness is apprehended through the prism of a continuum of material habitation conditions that bear the imprint of emotional underpinnings. Understanding homelessness involves considering the range of living conditions shaped by emotions. The concept of "home" highlights how people adapt over time, while the economic factors influence the physical aspects of homelessness. Identity and culture help us grasp the complexities of real experiences, which can inform solutions for homelessness. In simpler terms, to grasp homelessness, we must look at the various living conditions influenced by emotions, how people adapt to their surroundings over time, and the economic factors that affect the physical aspects of homelessness. Additionally, understanding people's identity and culture can help us address the complexities of their real experiences, which can guide us in finding solutions to homelessness. As stipulated by the United Nations Human Rights "Homelessness is a profound assault on dignity, social inclusion and the right to life. It is a prima facie violation of the right to housing and violates a number of other human rights in addition to the right to life, including non-discrimination, health, water and sanitation, security of the person and freedom from cruel, degrading and inhuman treatment."

2. Significance of addressing homelessness and global scenario

The Universal Declaration of Human Rights (UDHR) of 1948 explicitly articulates in Article 25(1) that every individual possesses the entitlement to a standard of living that ensures their health and well-being, along with that of their family. This encompasses essential provisions such as food, clothing, housing, medical care, and essential social services. Furthermore, this declaration underscores the right to security in situations such as unemployment, illness, disability, widowhood, advanced age, or other instances of livelihood deficiency that are beyond an individual's control. Considering the gravity of these principles, it is noteworthy that the global issue of homelessness and inadequate housing persists. In 2005, the United Nations estimated that around 100 million people were homeless worldwide, with an additional

1.6 billion lacking proper housing. A more recent development, in 2021, saw the World Economic Forum reporting a figure of 150 million homeless individuals globally. Another study conducted in urban centers revealed that over 60% of homeless individuals faced challenges in accessing basic healthcare services. In urban areas characterized by high living costs, nearly 40% of individuals experiencing homelessness reported instances of substance abuse as a coping mechanism. An alarming 75% of homeless women and children encountered incidents of violence or exploitation, underscoring the heightened vulnerability within this demographic (Global Homelessness Statistics, 2023). The lack of affordable housing is evident in the fact that over 20% of individuals in urban environments faced a distressing choice between rent and food expenditure, contributing to their precarious circumstances (Fig. 1.1).

Significant advancements have been made in ameliorating slum-related living conditions, the intersection of internal population growth within these areas and rural–urban migrations necessitates a holistic and adaptive approach in tackling this multifaceted global issue. Between 1990 and 2014, there was a noteworthy reduction in the proportion of the global urban population residing in slums, declining from 46% to 23%. This positive trend was a reflection of concerted efforts to address this issue. However, the complex landscape of urbanization and development dynamics in numerous developing

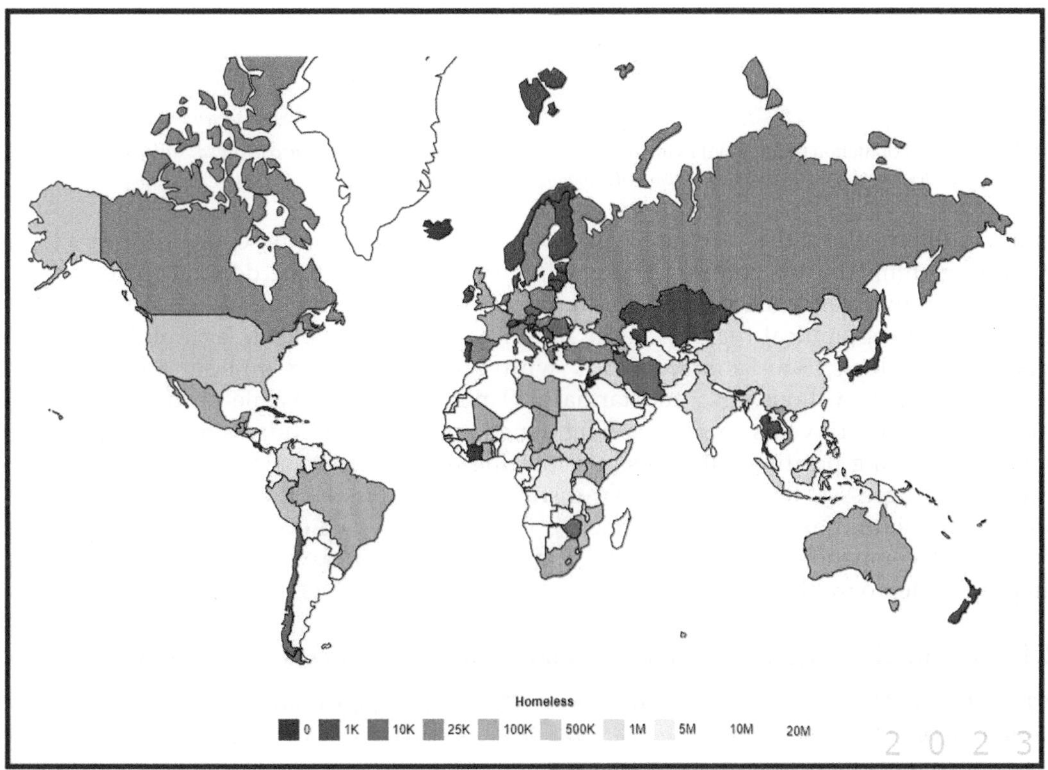

FIGURE 1.1 Country wise homelessness as per 2023. *Data Source: World Population Review. https://worldpopulationreview.com/*

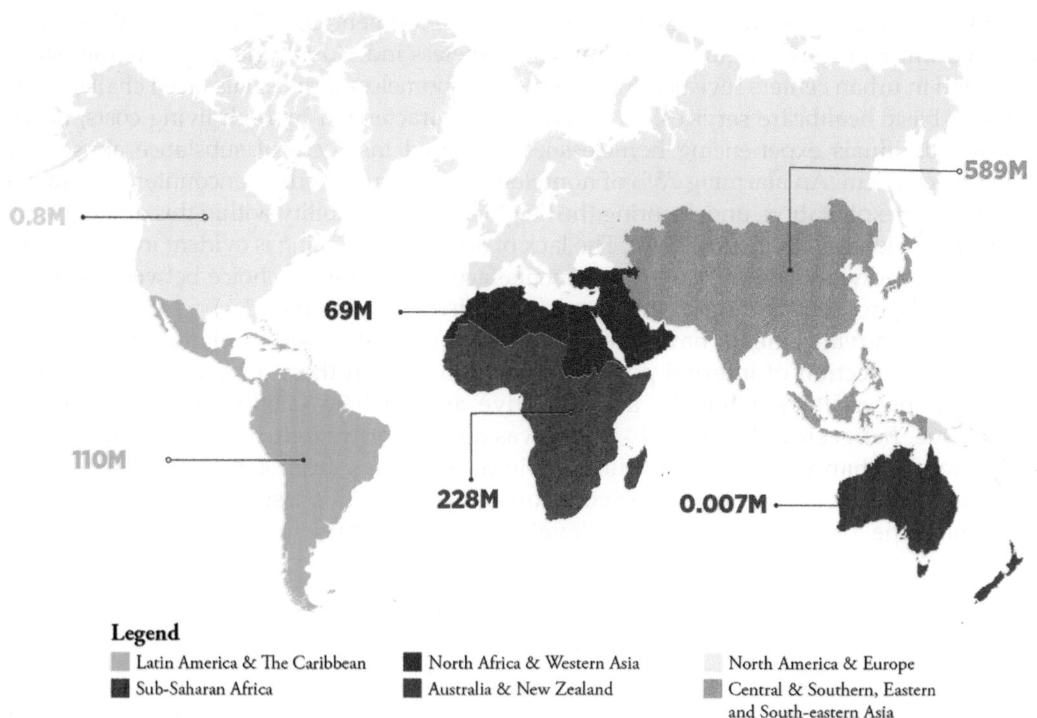

FIGURE 1.2 Population (per region) living in slums and informal settlements. *Source: United Nations, 2023. https://unhabitat.org/sites/default/files/2020/06/the_urban_sdg_monitoring_series_monitoring_sdg_indicator_11.1.1.pdf*

countries inherently led to an inevitable escalation in the absolute count of individuals inhabiting slums. In this context, it is crucial to recognize that the demographic concentration of individuals residing in slum-like conditions is predominantly observed within three major geographical regions: Latin America and the Caribbean, accounting for 110 million individuals; Sub-Saharan Africa, with a staggering 228 million; and East and South-eastern Asia, housing a substantial 589 million (2020). While strides have been taken in the past 15 years to place a heightened emphasis on addressing the challenges of slums, yielding positive outcomes in relocating millions from such conditions, the trajectory of progress is tempered by two distinct factors. Firstly, the persistent growth of populations within existing slums offsets a portion of the achieved gains. Secondly, the intricate phenomenon of rural–urban migrations introduces a dynamic that further shapes the demographic landscape of slum populations (Fig. 1.2).

2.1 Examination of global initiatives and policies to address homelessness

2.1.1 United Nations housing policies and SDG 11: "Sustainable Cities and Communities"

The synergy between UN housing policies and "Sustainable Cities and Communities" SDG 11's Indicator 11.1.1 is ***Global monitoring of slums remains a key concern for achieving***

the right to adequate housing. It is underscored by their shared emphasis on data-driven methodologies, collaborative frameworks, and robust accountability mechanisms to effectively confront homelessness as an integral facet of sustainable urbanization. The recent proliferation of statistics accentuates the magnitude of the homelessness predicament, thereby galvanizing concerted endeavors aimed at engendering a more all-encompassing and inclusive urban milieu. The rate of provisioning and availability of suitable and affordable housing on the global market significantly lags behind the pace of urban population expansion. This striking disparity underscores a pertinent concern that prompts inquiries into policy strategies aimed at enhancing the quality of life for individuals residing in slum environments. Concurrently, this issue necessitates a strategic approach to mobilize substantial investments in order to realize the objective of providing adequate housing. This endeavor is intricately linked to the broader goal of attaining Sustainable Development Goal (SDG) target 11.1 by the year 2030.

2.1.1.1 United Nations housing policies and SDG 11: "Sustainable Cities and Communities"—indicator 11.1.1 in relation to homelessness

1. **Monitoring Homelessness:** Aligned cohesively with the tenets of SDG 11, UN housing policies astutely acknowledge Indicator 11.1.1 as a seminal gauge to meticulously monitor the urban habitat's living conditions, inclusive of homelessness. The metric's focal ambit encompasses the quantum of urban inhabitants ensconced within substandard domiciles, notably encompassing slums or informal enclaves.
2. **Intrinsic Urban Predicament:** The essence of Indicator 11.1.1 is underpinned by a clarion call to confront head-on the formidable challenge of homelessness within the broader tapestry of sustainable urban advancement. The metric not only spotlights the exigent imperative to grapple with homelessness but also enunciates the imperative of assiduously quantifying the proportion of urban denizens bereft of adequate sanctuary.
3. **Data-Informed Decision Architecture:** Central to UN housing policies is an accentuated thrust on harnessing the potency of data collation as a linchpin for informed policy determinations. Indicator 11.1.1's prominence accentuates the pronounced need to methodically accrue data pertaining to homelessness, a pivotal toolkit that empowers governance structures in crafting laser-focused interventions while concurrently charting the trajectory of progress toward ensuring judicious habitation.
4. **Precisely Targeted Interventions:** The cogency of UN housing policies is exemplified by a cogent emphasis on redressing the domiciliary exigencies of marginalized strata. Indicator 11.1.1, in harmonious synergy, fortifies this narrative by its calibrated spotlight on enclaves marked by subpar accommodations, rudimentary settlements, and habitations of deficiency—all of which commonly intersect with the contours of homelessness.
5. **Holistic Urban Schema:** The lexicon of UN housing policies resonates with the clarion call for holistic urban schematics that encompass the specter of homelessness. SDG 11's quintessential blueprint envisions urban enclaves that are hinged on the paradigms of safety, inclusivity, resilience, and sustainability—tenets seamlessly aligned with the antecedent campaign to mitigate homelessness.

6. **Consortium and Collaborative Leverage:** UN-driven initiatives fervently trumpet the clarion of collaborative impetus, entailing symbiotic alliances encompassing governance paradigms, nongovernmental edifices, and stakeholder assemblages. The solemn precept of addressing homelessness requisites collaborative endeavors—an overarching dynamic that proffers immediate sanctuaries, transitional abodes, and enduring antidotes, emblematic of SDG 11's ethos of synergistic cooperation.
7. **Gender Prism and Homelessness:** The canvass of Indicator 11.1.1's canvass accentuates the undercurrent of gender-mediated inflections inherent in homelessness. The axiom posits a trenchant acknowledgment that women and girls often grapple with exacerbated susceptibility within precincts typified by informal domiciles, casting a resonance with the overarching objectives of SDG 11 encompassing gender parity and empowerment.
8. **Locus of Population Dynamics:** Recent statistical inferences, resonant with the 2021 census, expound upon the perilous liaison between precipitous urbanization and an augmenting specter of homelessness. As per empirical accounts, the urban demographic's transient fringe teems with approximately two million destitute individuals bereft of permanent domiciliary shelters, further accentuating the prescient impetus to fold homelessness within the SDG 11 ambit.
9. **COVID-19's Pernicious Reverberations:** The traumatic reverberations of the pandemic, predicated on the COVID-19 maelstrom, have exacted a disproportionate toll on homeless contingents. Indicator 11.1's compass resonates with the exigencies of emergencies, evoking a consonance with the UN's overarching campaign to imbue the rubric of homelessness mitigation within the domain of disaster preparedness.
10. **Oversight and Pacing Progress:** The sentinel import of SDG 11's Indicator 11.1 lies in its potency as an emblem of accountability germane to the realm of inadequate sheltering, inclusive of homelessness. UN housing policies, in harmonious unison, foment a climate of oversight, stimulating governance cadres to chart, appraise, and recalibrate their trajectories in tandem with the unassailable imperatives enshrined within the UN's housing policy panorama.

2.1.2 UN Guidelines for the implementation of the right to adequate housing

The **16 Guidelines for the Implementation on the Right to Adequate Housing** contain under each guideline specific implementation measures for States, public authorities, and regional and local governments. As per UN "The Guidelines provide States with a set of implementation measures in key areas of concern, including homelessness and the unaffordability of housing, migration, evictions, climate change, the upgrading of informal settlements, inequality and the regulation of businesses. All of the implementation measures are informed by the urgent need to reclaim housing as a fundamental human right. Implementation of the Guidelines will substantially alter how States treat housing, creating a new landscape where housing can be secured as a human right for all" (Box 1.1).

BOX 1.1

The 16 guidelines for the implementation on the right to adequate housing

Guideline No. 1. Guarantee the right to housing as a fundamental human right linked to dignity and the right to life;

Guideline No. 2. Take immediate steps to ensure the progressive realization of the right to adequate housing in compliance with the standard of reasonableness;

Guideline No. 3. Ensure meaningful participation in the design, implementation and monitoring of housing policies and decisions;

Guideline No. 4. Implement comprehensive strategies for the realization of the right to housing;

Guideline No. 5. Eliminate homelessness in the shortest possible time and stop the criminalization of persons living in homelessness;

Guideline No. 6. Prohibit forced evictions and prevent evictions whenever possible;

Guideline No. 7. Upgrade informal settlements incorporating a human rights-based approach;

Guideline No. 8. Address discrimination and ensure equality;

Guideline No. 9. Ensure gender equality in housing and land;

Guideline No. 10. Ensure the right to adequate housing for migrants and internally displaced persons;

Guideline No. 11. Ensure the capacity and accountability of local and regional governments for the realization of the right to adequate housing;

Guideline No. 12. Ensure the regulation of businesses in a manner consistent with State obligations and address the financialization of housing;

Guideline No. 13. Ensure that the right to housing informs and is responsive to climate change and address the effects of the climate crisis on the right to housing;

Guideline No. 14. Engage in international cooperation to ensure the realization of the right to adequate housing;

Guideline No. 15. Ensure effective monitoring and accountability mechanisms;

Guideline No. 16. Ensure access to justice for all aspects of the right to housing;

https://documents-dds-ny.un.org/doc/UNDOC/GEN/G19/353/90/PDF/G1935390.pdf?OpenElement

3. Indian scenario of home, homeless, and homelessness

3.1 Census of India

The empirical evidence culled from the 2011 census portrays a portrayal wherein an estimated cohort of nearly 17.7 lakhs individuals are situated in a state of homelessness.

Regrettably, the inherent limitations of this census exercise preclude a comprehensive enumeration of the entirety of the homeless populace. Adding to this complexity, the decennial rhythm of these surveys renders the acquired data temporally stagnant, harking back a decade. Contrastingly, perspicacious insights emanating from the Commissioners of the Supreme Court reveal that approximately 1% of the urban citizenry, constituting an approximate aggregate of ~37 lakhs, grapples with the stark reality of homelessness. Demonstrative of a conscientious endeavor, the Ministry of Housing and Urban Affairs conducted an auxiliary survey under third-party auspices in 2019 to meticulously unravel the contours of the urban homeless stratum, culminating in an estimated tally of approximately 23.93 lakh individuals ensnared within the throes of homelessness. This confluence of statistics reflects a disconcerting trend, with the proliferation of homelessness being propelled by the twin catalysts of population augmentation and the convoluted ramifications of the COVID-19 pandemic. Inextricably interwoven, these agents of change have jointly conspired to augment the ranks of the homeless demographic. This demographic dynamic assumes a geographical facet of significance, with Rajasthan, Gujarat, Uttar Pradesh, Maharashtra, and Haryana standing forth as the quintet of states harboring the most substantial concentrations of homeless individuals. This stark delineation underscores the imperative for calibrated policy initiatives and concerted interventions to ameliorate the predicament of this marginalized stratum (Box 1.2).

BOX 1.2

Statistical insight

The 2011 census underscores an alarming count of nearly 17.7 lakhs people grappling with houselessness, marking a critical juncture in the evaluation of urban vulnerability. The underpinnings of the census procedure's inherent limitations unfurl, obstructing the holistic encapsulation of the entire homeless populace within its ambit.

The temporal constraints of the decennial census modality are pronounced, rendering the acquired data a decade-old artifact, reminiscent of a bygone era. The discerning pronouncement by the Commissioners of the Supreme Court situates the percentage of homeless individuals within the urban populace at 1%, culminating in a voluminous aggregate exceeding ~37 lakhs.

A tangible testament to conscientious governance, the Ministry of Housing and Urban Affairs, through a judicious third-party survey, has approximated the figure of urban homeless at a staggering 23.93 lakhs.

The confluence of population upsurge and the unprecedented disruptions triggered by the COVID-19 pandemic has catalyzed the precipitous ascent in the ranks of the homeless stratum, accentuating the urgency of a proactive response.

The stark delineation of Rajasthan, Gujarat, Uttar Pradesh, Maharashtra, and Haryana as the states grappling with the highest homeless population underscores the regional disparities in addressing this socio-economic conundrum.

3.2 Indian government policies

The jurisprudential and policy dimensions unveiled herein underscore the imperative for a comprehensive and equitable housing paradigm, one that not only acknowledges the constitutional sanctity of the right to dignified shelter but also transmutes policy rhetoric into substantive and transformative change for the entirety of India's homeless populace. **Legal Entitlements of Homeless Individuals Article 21 of the Indian Constitution** serves as a lodestar for the foundational entitlement to life and personal liberty. The august Supreme Court, through its jurisprudential pronouncements, has unequivocally affirmed that the right to dignified shelter resides harmoniously within the ambit of the right to life. This legal edifice confides the solemn duty of ensuring housing provisions unto the state, thereby crystallizing a mantle of responsibility in this domain. The mitigation of destitution borne from homelessness could be substantially alleviated through the establishment of accessible shelters, thereby casting a harsh spotlight on the state's conspicuous lapses in extending a commensurate social safety net to its citizenry. Striving for Comprehensive Housing Solutions In pursuit of its overarching vision, the Government of India has articulated a laudable aspiration to furnish housing for all by the culmination of 2022. This formidable endeavor has engendered the formulation of an intricate web of policies and subsidiary frameworks under the aegis of the Pradhan Mantri Awas Yojana. However, the fruition of the governmental imperative to bestow housing for the entire populace remains regrettably incomplete in the absence of a concerted focus on encompassing the segments ensnared within the throes of homelessness.

3.3 Examination of Indian government initiatives and policies to address homelessness

The Indian government has embarked upon a series of initiatives and policy measures designed to tackle the issue of homelessness and enhance housing conditions for marginalized populations. These endeavors are integral components of broader strategic frameworks aimed at fostering sustainable urban development, reducing poverty, and fostering social inclusivity. Below are pivotal initiatives and policies in this context:

1. **Pradhan Mantri Awas Yojana (PMAY)—Housing for All:** Commenced in 2015, PMAY stands as a flagship program with the paramount objective of rendering affordable housing accessible to both urban and rural populations categorized as economically disadvantaged by the year 2022. The program encompasses two distinct facets: Pradhan Mantri Awas Yojana (Urban) and Pradhan Mantri Awas Yojana (Gramin). Its overarching objectives encompass the construction of dwellings, the enhancement of existing housing infrastructure, and the provision of financial assistance to eligible beneficiaries.
2. **Deendayal Antyodaya Yojana—National Urban Livelihoods Mission (DAY-NULM):** This initiative is centered around poverty alleviation and the augmentation of livelihood prospects for the urban underprivileged. DAY-NULM endeavors to orchestrate the organization of urban disadvantaged communities into self-help groups, accompanied by endowing them with skill development training and financial backing for endeavors

that generate income. The ultimate aspiration is to contribute substantively to the amelioration of living standards.

3. **National Urban Housing Fund (NUHF):** Conceived as a vehicle for resource generation via bond issuance, NUHF is meticulously tailored to channel funds into the construction of affordable housing solutions for the urban economically challenged demographic. Its primary objective lies in enabling the establishment of housing projects attuned to the requirements of society's vulnerable segments (2015).
4. **Rajiv Awas Yojana (RAY):** While subsequently succeeded by PMAY, RAY's initiation aimed to establish a slum-free urban landscape in India. This initiative was meticulously oriented toward heightening the living conditions within slums and conferring property rights to those inhabiting these areas. RAY also sought to provide support for in-situ slum redevelopment and the realization of affordable housing solutions.
5. **Housing Rights for Urban Homeless (HRUH) scheme:** This scheme is singularly targeted at providing shelter and indispensable services to the urban homeless populace. Central to this endeavor is the establishment of shelters and interim housing solutions within urban locales. The scheme's central tenet is ensuring that homeless individuals are accorded a secure and conducive abode.
6. **National Urban Rental Housing Policy:** Currently in the developmental phase, this policy is poised to address the intricacies associated with rental housing. Such an initiative is particularly pivotal for individuals with modest means who may lack the resources to acquire property of their own (National Urban Rental Housing Policy (Draft), 2015).
7. **Scheme of Shelter for Urban Homeless (SUH) under the National Urban Livelihoods Mission (NULM)**: The National Urban Livelihoods Mission (NULM) has set forth a focused objective: the provision of permanent shelter accompanied by essential services for urban homeless individuals. This objective is slated to unfold in a phased manner through the Shelter for Urban Homeless (SUH) Scheme. The mission's underlying motivation resides in the aspiration to rectify the predicament faced by those lacking a stable abode, contributing to their socio-economic well-being and engendering a more inclusive urban landscape. The fundamental objective of the 2007 National Urban Housing & Habitat Policy (NUHHP) is to foster the sustainable advancement of living environments throughout the nation. This policy endeavors to establish an impartial distribution of land, housing, and amenities at affordable rates, catering to all segments of society.

In India as the most populated country, the issue of housing and homelessness is a multifaceted challenge that encompasses a diverse range of living conditions and social. Although there are national schemes and policies to resolve the issues of homelessness and housing crisis but the primary concern in this context is the lack of critical insights on policymaking regarding homelessness and housing. While there are data and facts available, there is often a lack of in-depth analysis and understanding of the policy decisions made in this regard. This absence of comprehensive evaluation hinders the effectiveness of policies meant to address homelessness. Indian housing and homelessness policies often encounter significant drawbacks in both formulation and implementation. These drawbacks can include inadequacies in budget allocation, bureaucratic inefficiencies, and a disconnect between policy intent and on-ground execution. Another crucial issue in the Indian scenario is the lack of policies that consider all societal strata. Homelessness affects individuals from various socioeconomic backgrounds, and policies must be comprehensive enough to address the specific needs of different groups. Policies with significant gaps have a noticeable impact on daily life as they contribute to the proliferation of slums in cities and towns.

4. Challenges and opportunities in addressing homelessness

The issue of homelessness is complex and multifaceted, encompassing both urban and rural environments. The dynamics of homelessness here are intricately linked with a myriad of socio-economic factors and policy challenges. In this perspective, two distinct categories of homeless populations emerge, each influenced by diverse factors and requiring tailored interventions. The first category pertains to urban homelessness. This phenomenon is predominantly attributed to a convergence of factors such as destitution, the influx of migrant workers seeking employment opportunities, climate-induced migration, individuals displaced due to political reasons, lack of secure land ownership, and the vulnerability of specific communities. Urban centers, often seen as magnets of economic opportunity, attract a steady stream of migrants in search of better prospects. However, the challenges of housing affordability, limited social safety nets, and inadequate urban planning contribute to the prevalence of urban homelessness. Vulnerable groups, including orphans and individuals from marginalized communities, often find themselves without a stable place of residence and lacking access to basic services. The larger cities and urban agglomerations worldwide have been facing the major issues of urban migrants and formation of slums in the middle of the cities as well as in the peripheral surroundings. Indian metropolitan cities including Mumbai, Delhi, Kolkata, Chennai, Bangalore, Hyderabad are no exceptional. In the present decade the scenario has become scarier due to the presence of poor infrastructure and impacts of climate change. The second category pertains to rural homelessness, which unfolds within a distinct context. The rural homelessness is prominent in developing nations of Asia and Africa majorly. Nigeria has the highest homeless people around 24.1 million in the world, which is 10% of their total populations (2021). Communal conflicts, often arising from socio-political tensions, can result in the displacement of rural populations, rendering them homeless. Political factors, driven by local power dynamics, can lead to the expulsion of individuals from their homes. Syria still holds the unfortunate record for hosting the most significant number of forcibly displaced individuals globally, encompassing over 6.6 million refugees on the international stage and an additional six million people who have been internally displaced within the nation's borders (2020). Syria stands as the nation with the highest homeless rate globally, with approximately one-third, or around 29.6%, of its population of 22.1 million individuals experiencing homelessness. Syria's homeless crisis serves as an alarming testament to the multifaceted challenges endured by its population. The dire consequences of the ongoing war, coupled with staggering deficits in basic amenities like water and food, cast a long shadow on the nation's socio-economic fabric, exacerbating the plight of its homeless and vulnerable citizens (2022a). Moreover, climate-related impacts such as natural disasters, which are not uncommon in the Indian subcontinent, can displace rural communities

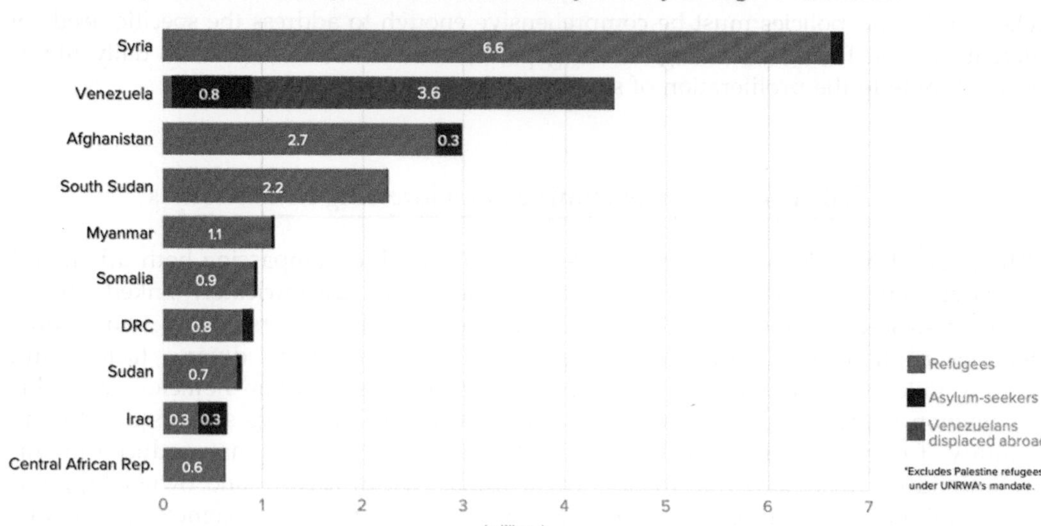

FIGURE 1.3 Top international displacement situations by country of origin. *Source: UNHCR.*

and render them homeless. Rapid industrialization and deforestation also contribute to the dislocation of rural populations who depend on traditional livelihoods, pushing them into homelessness (Fig. 1.3). Homelessness has persisted globally since the early 20th century, attributed to a convergence of factors such as rapid development, globalization, the establishment of urban hubs, exponential population growth, political and social tensions, among others. While these conventional factors have contributed to the issue, contemporary challenges have further exacerbated the situation. Notably, modern crises like climate-induced homelessness and land scarcity have compounded the complexity of the problem, presenting additional hurdles that demand attention and resolution.

4.1 Housing and land rights

Housing and land rights represent a remarkably intricate challenge, encompassing dimensions beyond the scope of homelessness alone. Nevertheless, it's imperative to recognize that homelessness is frequently rooted in the absence of secure land rights and identity-related issues. In the Indian context, this issue disproportionately affects segments such as slum dwellers, the urban poor, and tribal communities. Notably, the Manipur land rights and housing dispute between local communities has garnered international attention. India's complex landscape has witnessed instances where land rights intricately intersect with issues of identity and livelihood. One such poignant example is the Manipur case, where a dispute over land rights has taken on global significance. Back in 1990, clashes erupted due to land disputes. The Kuki community contended that as many as 350 of their villages had been uprooted, leading to tragic outcomes including the loss of over 1000 lives and the displacement of 10,000 individuals. Additionally, in 1993, tensions flared between the Meitei Pangal

(Muslim) and Meitei communities. The unfortunate culmination saw a bus carrying Muslim passengers being set ablaze, resulting in the tragic loss of over 100 lives (2017). This underscores the profound interplay between land rights, housing, and intricate social dynamics that can culminate in critical consequences. In India and beyond, addressing these issues necessitates not only policy measures but also fostering a deeper understanding of the historical, social, and economic underpinnings that shape the complexities of housing and land rights disputes.

4.2 Climate migration and homelessness

Climate change stands as the predominant catalyst behind the burgeoning crisis of homelessness that the world grapples with daily. This global phenomenon is not confined to transitory disruptions but perpetuates across generations through climate-induced dislocation, affecting millions of people each year. As per UN IOM (International Organization for Migration) stated that "**Climate migration** is the movement of a person or groups of persons who, predominantly for reasons of sudden or progressive change in the environment due to climate change, are obliged to leave their habitual place of residence, or choose to do so, either temporarily or permanently, within a State or across an international border." The forthcoming COP27 summit is poised to center its deliberations on the issue of climate-driven homelessness, recognizing the urgency of this matter. At this conference, nations and stakeholders will immerse themselves in extensive dialogs with the aim of crafting effective strategies to tackle the multifaceted challenges faced by climate refugees. Among the foremost objectives is the exploration of avenues for providing shelter, vital assistance, and essential resources to those who experience displacement due to climate-related factors (Fig. 1.4).

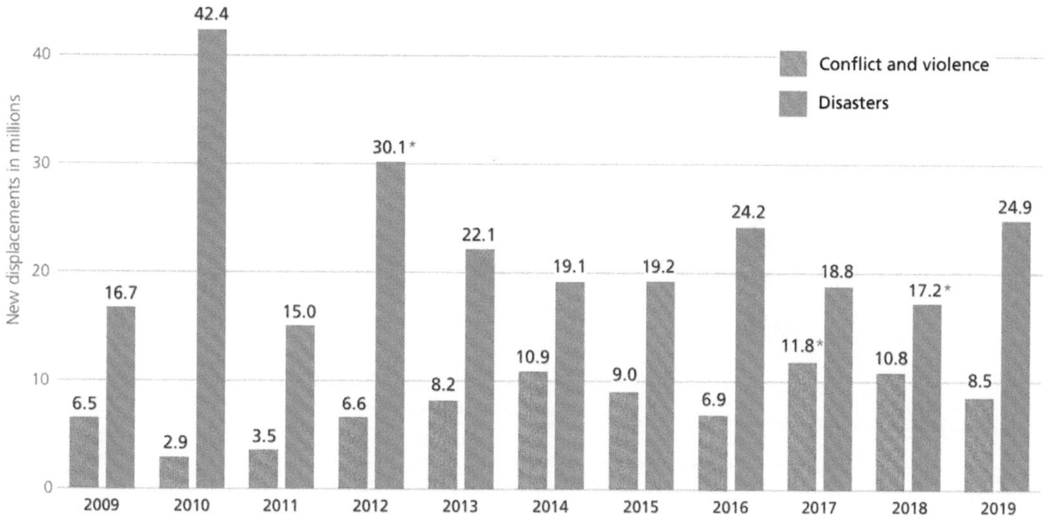

FIGURE 1.4 New displacements by conflict, violence, and disasters worldwide (2009–2019). *Source: Global Report on Internal Displacement Report 2020. https://www.internal-displacement.org/sites/default/files/publications/documents/2020-IDMC-GRID-executive-summary.pdf*

COP27 serves as a pivotal juncture for fostering international cooperation among nations. Collaborative endeavors will be directed toward the creation of comprehensive policies and proactive initiatives designed to preempt and ameliorate the displacement triggered by the vicissitudes of climate change. By doing so, a concerted effort will be made to curtail the incidence of individuals finding themselves without homes owing to the far-reaching ramifications of environmental factors. As per the Global Report on Internal Displacement report 2020 by NRC, it has been mentioned that "Conflict and disasters triggered 33.4 million new internal displacements across 145 countries and territories in 2019. Most of the disaster displacements were the result of tropical storms and monsoon rains in South Asia and East Asia and Pacific. Bangladesh, China, India and the Philippines each recorded more than four million, many of them pre-emptive evacuations led by governments. Many evacuees, however, had their displacement prolonged because their homes were damaged or destroyed" (Global Report on Internal Displacement, 2020) (Fig. 1.5).

In case of India, climate homelessness has become a major issue in last 2 decade. Cyclones are impacting the people in the coastal regions. The home and land of coastal communities are getting submerged. As per the Climate Action Network South Asia (CANSA) reports, an estimated 45 million individuals within India alone will be compelled to undertake migration by 2050 as a result of climate-related disasters, reflecting a threefold surge compared to the present statistics. The IDMC 2020 Report has published that in the realm of disaster-induced displacement, India stands as the epicenter of

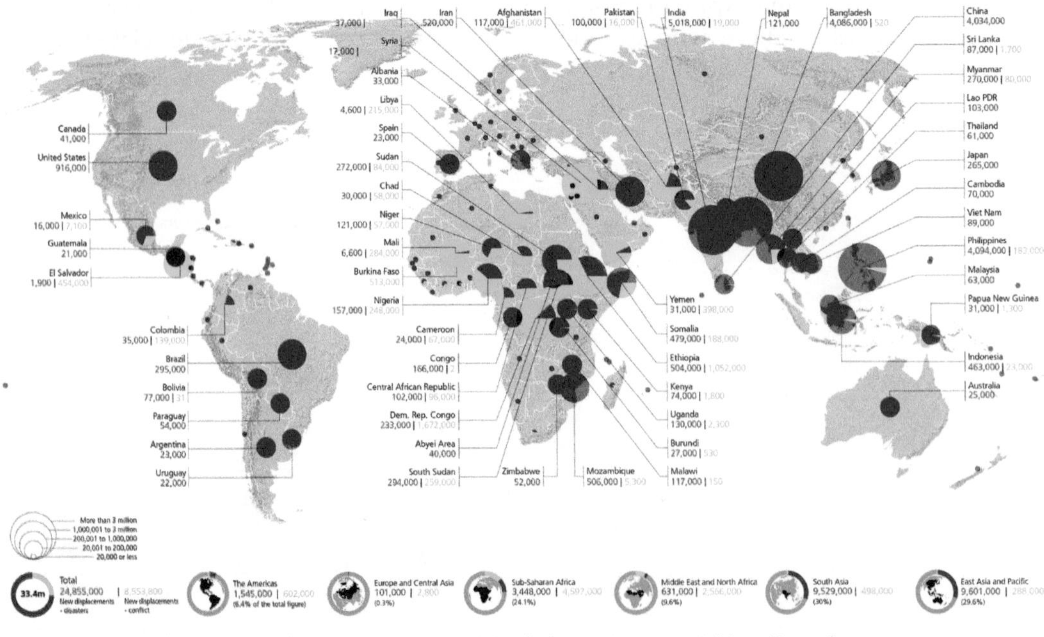

FIGURE 1.5 New displacements by conflict and disasters in 2019. *Source: Global Report on Internal Displacement Report 2020. https://www.internal-displacement.org/sites/default/files/publications/documents/2020-IDMC-GRID.pdf*

extensive occurrences within South Asia and ranks among the countries with the most elevated global rates. Over the duration of a 10-year period, spanning from 2008 to 2019, an average of around 3.6 million individuals faced displacement on an annual basis. Notably, a substantial proportion of these instances unfolded during the monsoon season. The climate homeless from the neighboring countries like Bangladesh, Myanmar are also approaching the east coast border of India due to lack of home and basic amenities (2022b, 2020).

5. Conclusion

Homelessness and the challenges it poses are universal in scope, constituting an intricate and evolving issue that spawns novel causative factors with the passage of time. This predicament transcends its immediate impact, exerting influences that resonate through generations to come. It stands as a multifaceted challenge, intertwining social, economic, and environmental dimensions within the contemporary global landscape. In today's world, the issue of homelessness is no longer solely attributed to traditional factors like supply and demand imbalances or migration. Instead, we are witnessing a shift where socio-political factors and climate-related issues are becoming dominant contributors to homelessness statistics. This evolving landscape highlights the need for a more comprehensive approach to understanding and addressing the problem. This chapter endeavors to elucidate the intricate panorama of homelessness at both the global and national levels. However, it is imperative to acknowledge that the discourse surrounding the concepts of home, homelessness, and the myriad factors that contribute to this phenomenon is far-reaching and inexhaustible. The subject encompasses a wealth of dimensions that continue to shape our societies, necessitating continued dialog and exploration. Socio-political impacts, such as conflicts, economic disparities, and policy decisions, play a significant role in driving individuals and families into homelessness. For example, political instability and economic downturns can lead to job loss and financial insecurity, ultimately resulting in people losing their homes. Therefore, it's crucial to acknowledge the social and political dimensions of homelessness to create effective policies and interventions. Climate-related issues also increasingly contribute to homelessness. Natural disasters, such as hurricanes, floods, and wildfires, can displace large populations, leaving them without shelter. The frequency and intensity of these events are on the rise due to climate change, making it imperative to incorporate climate considerations into homelessness prevention and relief strategies. It's important to recognize that homelessness can be temporary and often occurs due to unforeseen external issues. Sudden health crises, family emergencies, or other unexpected events can push people into homelessness. As a result, it is vital for policymakers to address not only chronic homelessness but also provide solutions for those experiencing temporary homelessness. Allocating emergency funds for housing and support services becomes essential in these cases. In summary, the contemporary understanding of homelessness must go beyond supply and demand dynamics and migration. The influence of socio-political factors and climate issues on homelessness statistics is becoming more pronounced. To effectively address this multifaceted challenge, policies and funding allocation should take into account the evolving nature of homelessness, including its temporary aspects and the various underlying causes that drive individuals and families into homelessness.

AI Disclosure

During the preparation of this work the author(s) used Paraphrasing tools to improve readability and language of the work. After using this tool/service, the author(s) reviewed and edited the content as needed and take(s) full responsibility for the content of the publication.

References

2011. Census of India. https://censusindia.gov.in/nada/index.php/home, https://censusindia.gov.in/census.website/.

2015. *National Urban Rental Housing Policy (Draft).* Delhi: Government of India Ministry of Housing and Urban Poverty Alleviation. https://mohua.gov.in/upload/uploadfiles/files/National_Urban_Rental_Housing_Policy_Draft_2015.pdf.

2020. Global Report on Internal Displacement. Norwegian Refugee Council: World. https://www.internal-displacement.org/sites/default/files/publications/documents/2020-IDMC-GRID.pdf.

Brigadier Sushil Kumar Sharma Deputy General Officer. (2017). *The complexities of tribal land rights and conflict in Manipur: Issues and recommendations.* Manipur: Vivekananda International Foundation. https://www.vifindia.org/sites/default/files/the-complexities-of-tribal-land-rights-and-conflict-in-manipur.pdf.

Global homelessness statistics. (2023). https://www.homelessworldcup.org/homelessness-statistics. (Accessed 5 August 2023).

Further reading

Obasi, C. O. (2021). *The homeless-poor and the COVID-19 stay-at-home policy of government: Rethinking the plight of homelessness in Nigeria.* SAGE. https://doi.org/10.1177/21582440211045078

Govt. of India. (2015). *Refinance scheme for urban housing fund.* India: Govt. of India. https://mohua.gov.in/upload/uploadfiles/files/11RFUrbanHousingFundUCBs.pdf.

Greater Change. (2022). *Which country has the highest rate of homelessness?.* https://www.greaterchange.co.uk/post/which-country-has-the-highest-rate-of-homelessness. (Accessed 9 August 2023).

United Nations Human Rights. (2022). *Homelessness and human rights Special Rapporteur on the right to adequate housing.* https://www.ohchr.org/en/special-procedures/sr-housing/homelessness-and-human-rights. (Accessed 27 July 2023).

World Economic Forum. (2020). *This is how many people are forcibly displaced worldwide.* https://www.weforum.org/agenda/2020/06/displacement-numbers-world-refugee-day/. (Accessed 1 August 2023).

CHAPTER

2

Conceptualizing homelessness in context to global south

Kunaljeet Roy[1], Alolika Mangal[2] and Sweta Chatterjee[3]

[1]School of Social Sciences and Languages, Vellore Institute of Technology, Chennai, Tamil Nadu, India; [2]Department of Geography, Vivekananda College, Kolkata, West Bengal, India; [3]School of Water Resources Engineering, Jadavpur University, Kolkata, West Bengal, India

1. Introduction: Conceptualizing "homelessness"

Understanding "homelessness" from theoretical and practical perspective requires critical discussion on the idea of "home." Home actually articulates a "critical geography of home" in which home is understood as an emotive place and spatial imaginary that encompasses lived experiences of everyday, domestic life alongside a wider, and often contested, sense of being and belonging in the world (Blunt & Dowling, 2022). The authors marked about the arguments of the Humanistic Geographers who placed "home" at the Center of their work. With a prime focus on human association and creativity, humanistic geographers, explored the avenues in which places are meaningful and found significant from people's perspectives. The theoretical lineages of humanistic notion of geography are multiple and include domains of phenomenology, humanism, and existentialism (Bachelard, 1994 cited by Blunt & Dowling, (2022). In respect of the idea of home, we actually find a focus on the meaning of "home," i.e., how the people relate to and experience their respective dwellings as well as how they formulate a sense of "home" in terms of comfort and belonging. Home is not simply considered as a house or living space for the humanistic geographers but is regarded as a very special place which is irreplaceable center of significance from the perspective of belongingness and emotional attachment. For Dovey (1985) as cited by Blunt & Dowling, (2022) the term "home" signified a unique relationship between people and their environment; a connection through which they create strong sense of their own world. Home is regarded as "hearth," the pivotal point through which humans ere centered in this earth.

UN-HABITAT has acknowledged the notion of "homelessness" as one of the natural concerns since its inception. The concern was further confirmed by the habitat agenda which includes seven articles directly directing toward the idea of homelessness. During the

https://doi.org/10.1016/B978-0-443-14052-5.00002-1

Millennium Summit of September 2000, global leaders agreed to recognize the living conditions of the slum dwellers as one of the striking issues related to human rights. They also agreed to find out fruitful way to improve the living conditions of at least 100 million slum dwellers to a satisfactory level by 2020. Though setting the goal is considered as a very positive step taken toward alleviating survival crisis of the urban poor in global scale. The role of UN-HABITAT is to work jointly with the Governments, local civic bodies and civil society to find out fruitful ways and means of providing inexpensive and suitably located land for the urban poor. This includes technical and logistic support, implementation of innovative financing mechanisms, community based microcredit schemes and private sector investments in pro-poor urban land development. Furthermore, intervention to facilitate access by the urban poor toward basic infrastructure and civic amenities including safe drinking water and improved sanitary facilities has been included in the agenda. The global think tank actively promoted the idea of "inclusive city" where the urban poor are able to have an effective voice in decisions affecting their everyday survival. This initiative has also focused in incorporating voices from all the stakeholders especially women, minorities, and the marginal sections. The very idea of "Right to City," theoretically marked by Henri Lefebvre and critically conceptualized by David Harvey (2008) includes not only the right to secure livelihood and "home" inside a city but also to a wide range of actions including access to credit and the review of building standards, affordability of public housing, access to safe living environment, availability of basic amenities and so on. Bulk of the issues related to this are found in global south. In the cities of developing countries, due to population growth and diminishing land availability, the need of the hour is to promote and apply pro-poor and socially inclusive urban policies and housing strategies to deal with these emerging challenges of homelessness. Fowler et al. (2019) from the context of public health complexities defined "homelessness" as an enduring threat to public health across the globe. Mostly the children, families and marginal sections are facing housing insecurity; they struggle to deal with their utmost needs because of their inability to achieve shelter. The authors computed the complexity of homelessness by using a simple mathematical formula:

$$d\,homelessness(t)\,/\,dt = entries(t) - exists(t);$$

where "d" represents change; "homelessness" represents total number of people remaining homeless; "t" represents time, "entries" represent people entering into the situation of homelessness at a given point of time; "exists" entries' represent people exiting from the situation of homelessness at a given point of time. The authors further highlighted the complex interplay between the individuals, interpersonal and other socio-economic and political factors associated with homelessness. Apart from that, they also highlighted psychological and personal struggles, relationship between the family members, friends, partner creating a vicious cycle. These complexities resulting into housing insecurity and are associated with the systematic responses to homelessness. Individuals with mental illness or people with special needs face immense psychological setback due to this complex interplay of factors related to "homelessness" (Fig. 2.1).

For the authors, these complex causes seek complex solutions. They require multifaceted support network and assistance to adapt in response to changing housing requirements and often include residential and nonresidential backing. In case of the global south, poverty and lack of affordable public housing has remained two of the biggest challenges in quest to

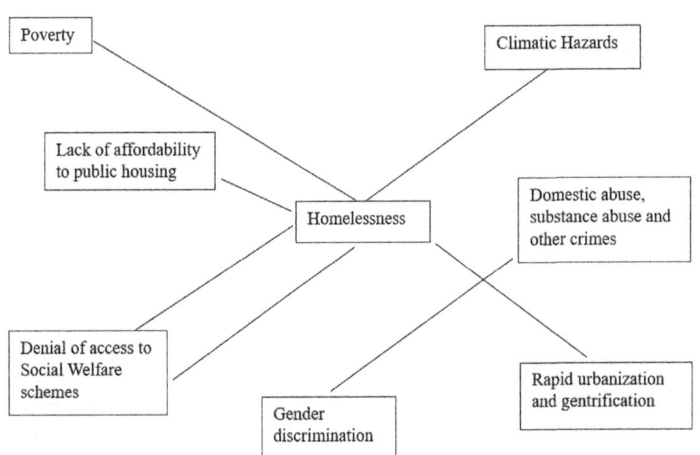

FIGURE 2.1 Complex interplay of factors associated with "homelessness." *Source: Modified after Fowler et al. (2019).*

deal with the issue of "homelessness." Apart from that, we have also included the quotient of climate change, the potential threat to adapt to the hazard outcomes and the surmounting crisis related with the growing numbers of the "environmental/climatic refugees" worldwide, especially in global south. This study aims to identify the complexity of factors associated with the idea of "homelessness" and augment detailed analysis of spatial variation of the same across global south. We have considered different genres of literature and official sources which provided the database to analyze the global scenario of "homelessness," especially in the study area (i.e., global south). The study has two connected segments of physical/natural and socio-political and economic sources of "homelessness" for providing an overall picture of complexities and vulnerabilities associated. Graphical illustration and associated data analysis has been performed to justify the textual arguments.

2. Complex factors associated with "homelessness" in global south

2.1 Climate change, recurring hazards, and vulnerability: the climatic refugees

Climatic Refugee or the Environmental Migrant is the one who is forced to leave their native place due to the adverse impacts of climate change belonging to several climatic hazards such as sea level rise, desertification, drought conditions, and extreme flood conditions. These severe environmental conditions make their place of residence uninhabitable for living and other amenities. Clubbed with climate change issues and conflicts and political instability conditions can also be exacerbated for the Climatic refugees. Myers (1993), Rashad (2020) advocated proper causes for the displacement that will lead to reach by 200 million climatic refugees by 2050 that are followed: Gradual environmental changes such as land degradation attributed to lack of agricultural practices, lack of rural infrastructures for agriculture, population growth and poverty issues; natural disasters are the second cause of displacement. Environmental refugees are trapped in the graticules of disaster issues, poverty issues and demographic pressures. The other causes are "environmental accidents" and "crises" that are the results of infrastructure projects (Table 2.1).

TABLE 2.1 Tabulation on migration currents into three possible categories.

	Disasters		Exportation		Deterioration	
Subcategory	Natural	Technological	Development	Ecocide	Pollution	Depletion
Origin	Natural	Anthropogenic	Anthropogenic	Anthropogenic	Anthropogenic	Anthropogenic
Reason for migration	Unintentional	Unintentional	Intentional	Intentional	Unintentional	Unintentional
Duration	Acute	Acute	Acute	Acute	Gradual	Gradual
Estimated displaced/ homeless persons	7000	144,000	1.3 million	7 million	15 million	115,000
Case specific	Montserrat	US-TMI	China-3G	Vietnam	Bangladesh	Ecuador-Amazon

Modified after Bates (2002).

2.1.1 *African context*

According to Addaney et al. (2019), though climatic refugees got international attentions however, there is no explicit legal framework for the same. African Regional Law does not cover the legal protection for the Climatic Refugees where, Climate induced Displacement is referred for the protection and assistance of the displaced persons to the 2009 African Union Convention however, it does not cover the migration beyond the borders. As per the Climate change scenario over Africa, it can be asserted that the Food and Agriculture Organization (FAO) observes that 319 million hectares of Africa's land is vulnerable to desertification issues. As per United Nations, this inhabitability will lead to migration of about 50 million Sub-Saharan migrants elsewhere by 2020 along with 700 million Africans will be forced to be displaced across the continents due to land degradation by 2050. FAO already revealed that displacement of about 26.4 million people occurred due to climate related disasters in Africa between 2008 and 2015. Castles and Rajah (2010) highlighted two scenarios where, in the case of slow-paced events (such as decline in rainfall intensity or changes in crop fertility), the inhabitants tried to develop possible adaptation strategies to get rid of this situation such as planting new crops with the help of possible irrigation systems, changing existing agricultural practices and income sources. Adesina et al., (2021) accounted the numbers of refugees from the different regions in Africa from the span of 1990–2017 where, the highest one from Somalia (19,217,481) and lowest one from Lesotho (238). Total 66.3% of the refugee population in Africa is from seven countries out of 54 countries in Africa within the same span. Here for the span of 1990–2017, number of refugees over African countries has been incorporated to show the trends of variation of refugees. Primary drivers of African migration are climate change issues, natural disasters, economic problems and violence, social tensions. Eastern Africa shows up the highest number of refugees and lowest for the North Africa (Fig. 2.2).

2.1.2 *Asia and Pacific context*

Edes et al. (2018) briefed that every year displacement is occurring due to floods, droughts, land degradation, soil degradation, typhoons, and cyclones. Deaths, displacement, and

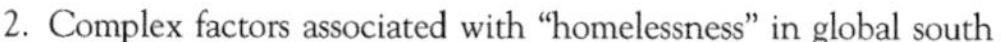

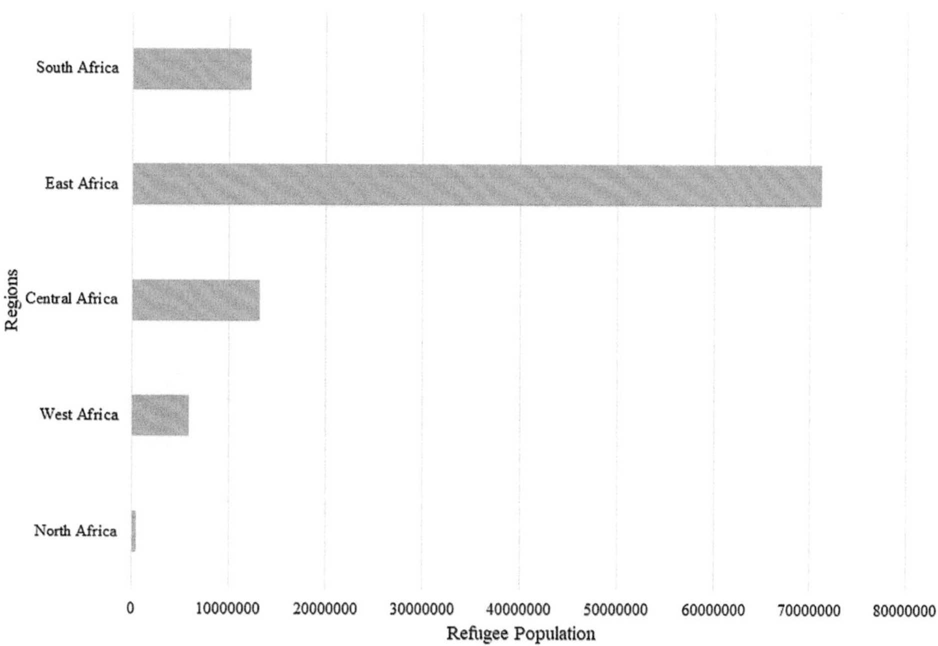

FIGURE 2.2 Estimation of the refugees in African continent estimation of the refugees in African continent. *Source: Data from Adesina et al., (2021).*

property loss are the commons as the aftermath. As per the statistics, between 2009 and 2011, 13.2 million people displaced in 2009, 31.8 million people displaced in 2010 and in 2011, 10.7 million people got displaced. In 2011, regional statistics revealed that 3,800,000 people of East Asia got displaced, 3,400,000 people of South East Asia, 3,500,000 people of South Asia and 9000 people of Central and West Asia got displaced for the possible climate related and extreme weather events (Fig. 2.3).

Hugo and Bardsley (2013) asserted that several states of South East Asia are exposed to Typhoons those results in havoc displacement where, *Haiyan typhoon* hit the regions of Philippines on 2013 that killed more than 6000 people and four million people got displaced. Typhoon of 1912 killed an estimated 15,000 people. Ayazi and Elsheikh (2019) described the correlation between sea level rise and forced migration over the Asian countries where, *Thailand* being 5 feet above the sea level and its sinking at the rate of 0.8 inches every year. It has been estimated by a 2015 report from Thailand's National Reform Council, the city could be completely submerged by 2100 that will cause 14 million residents' displacement. As per the report by *Indonesia's* Ministry of Marine Affairs and Fisheries, already 24 islands have been lost due to sea level rise between 2005 and 2007. *London* is also in the list of future displacement due to melting of polar ice sheets and mountain glaciers that can increase sea levels over the coming decades. *Maldives* will experience a possible sea level rise of half a meter by about 2100. If the sea levels rise by 1 m, Maldives could be completely inundated by 2100. Ober (2019) deciphered that there is an uneven distribution of displacement in the selected countries of Asia due to several cumulative factors

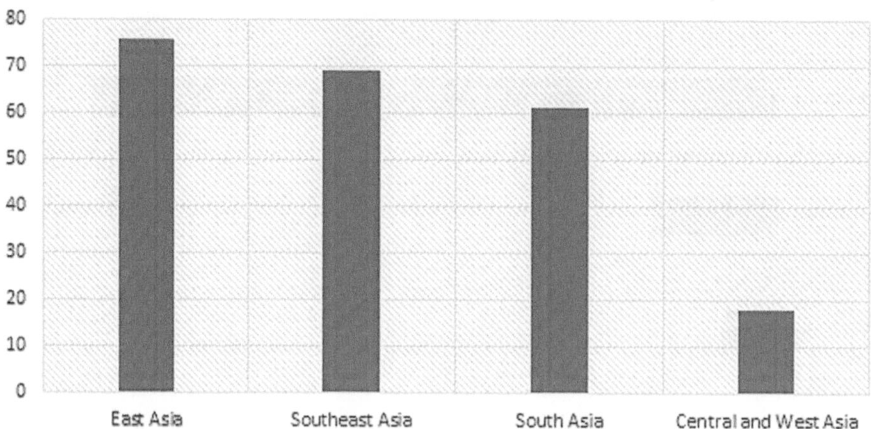

FIGURE 2.3 Displaced population in Asia–Pacific region due to hazards (2010–21). *Source: Data from Disaster Displacement in Asia and the Pacific, IDMC.*

such as population size, hazards exposure and associated vulnerability and resilience against those. Tropical storms of 2011 in South East Asia resulted the worst flooding in 50 years that affected Cambodia, Myanmar, Thailand, and Vietnam and displacement of about 214,000 people occurred in Cambodia. Due to these hazards over South East Asia, primarily the seasonal movement or climatic displacement tends to be local based or rural-urban based. Duration of movement is also temporary as after displacement due to the onset of hazard exposure, they immediately attempt to get back to their native place after the disasters. Along with these factors of movement, socio-economic status, gender and education and other demographic factors are also responsible for the timing and onset of mass movement due to climatic issues. Women's social norms and behaviors, limited education expects to be there for their family or kin during the onset of disasters instead of movement rather than men. Mbaye (2017) made a connection with climate change issues and international migration that would result in a way where, most people migrate internally in search of work however, some proportion of them migrate internationally if they connect with new networks and get more opportunities in the urban areas. Villaseca (2017) found that migration corridors between Asia and the Middle East and within East and South East Asia especially toward Malaysia and the Republic of Korea are condemning. Nishimura (2018) links between slow onset of the aftermaths of climate change, human rights, and the cross-border people's movement.

2.1.3 *Latin American context*

Rigaud et al. (2021) also projected through the report Groundswell: Preparing for Internal Climate Migration (2018) that 17 million or 2.6% of the total population will move from their native places to escape the aftermath of climate change. On an average account, it can be said that rather than Sub-Saharan Africa and South Asia, Latin America will be impacted less by

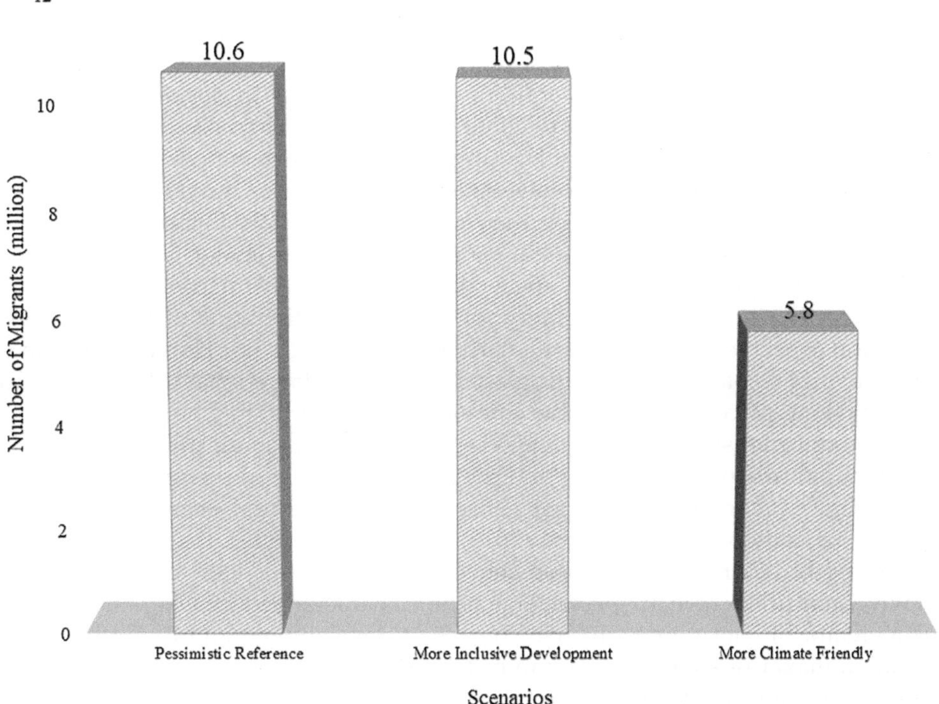

FIGURE 2.4 Number of migrants (million) under the "three scenarios" over Latin America. *Source: Data from Rigaud et al. (2021).*

climate change and the severity of internal migration. Two reasons followed up for this statement that includes agricultural employment is quite lower in this region and Latin America ha the great potentialities in adaptive capacity, stronger economies, strong financial resources that can cope up the situation (Fig. 2.4).

Delavelle (2013) asserted that environmental migration is used as the key strategy being a risk management strategy irrespective of droughts and rainfall variability conditions over Meso-America for mitigating the aftermath of household consumption and agricultural revenues. *"Agricultural Squeeze"* refers to Mexicans those migrates in response to poor agricultural outputs against the land degradation situation.

2.2 Chronicles of poverty and challenges for affordable public housing

Johnsen and Watts (2014) linked critically the prominence of poverty in context of homelessness which has varied over time. Poverty has been acknowledged globally as the most important and significant factor for homelessness. Empirical evidences indicated that experience of poverty is a common denominator found among the majority of the homeless people across the world. The impact of poverty in causing homelessness is determined partly by macro-level structural conditions such as welfare regimes, housing and labor markets, and the remaining by complex exchanges between the aforesaid macro- and microlevel components including the level of state protection, support channels and welfare policies, access to social, cultural and human resources etc. The occurrence of "homelessness" in global south, especially in the neo-liberal era of city planning in developing nations has been viewed as an outcome of failure to combat with acute urban poverty and lack of inclusive urban design to cater the urgent housing needs of the urban poor. For Ghosh (2019), in the developing world, the idea of homelessness refers to those who reside in open spaces without having basic lodging regarding their possessions; mostly found in the slums and shanty squatters. The term "homelessness" is persistent across major global cities and its not a contemporary phenomenon. In the wake of the neo-liberal command along with the rapid "gentrification" and regeneration of urban areas the issue of homelessness becomes a critical aspect to deal with, particularly in global south. For Harvey (2008) in order to convert the urban spaces as a site for global investment large-scale urban restructuring activities are taking place throughout the cities across the world; in global south it is taking shape even more furiously. The author expressed his concerns about the relationship exists between the rapid urban restructuring and the marginalization of the urban poor. This act of marginalization occurs on different axes of spatial, socio-cultural, economic, and political scale of marginalization. Banerjee-Guha (2019) critically addressed the neo-liberal agenda of restructuring of older city spaces and the struggle of urban poor to find passage into these newly "gentrified" urban spaces. The author characterized neoliberalism in present times as a creator of enormously diverse courses of development and underdevelopment; concurrently a maker of a changing urban order. She severely criticized the neoliberal city processing as an entity in absorbing the surplus capital at the price of a burgeoning intensity of dispossession of the urban masses rejecting the "right to the city." For her, the cities are emerging as hyperactive sites of "creative destruction," to pull apart the traditional, the old and giving birth to the new, causing havoc on the city space and lives of the urban poor. Cities are getting transformed rapidly, thus pushing out the poor, the marginals from the core of public space, leading to further critical extreme. In the global south, since the advent of the neo-liberal regime, this has become an overpowering procedure in most of the cities. The author cited David Harvey's (2008) notion of "annihilation of space" which leads to the state involvement into cleanse the city spaces of those who are left behind in the wake of the globalization. Here the idea of "homelessness" is discussed from a critical perspective where the changing socioeconomic restructuring of urban spaces after the neo-liberal regime began post 1990s. The radical scholars like Harvey criticized the post-modern "space" making where the policy framing remained biased in some way, especially in the developing areas of global south. Aspects of forced gentrification, urban regeneration and annihilation of urban "ghettos" and enclaves like immigrant colonies, slums, etc., are developing passage of "selective urbanism," leading toward more segregation and polarization of marginal groups.

We surveyed literature to get further insights on the homelessness and its relation to poverty and lack of affordable housing in major cities of global south including South Asian cities of Mumbai, Delhi, Kolkata, Dhaka, Lahore; Southeast Asian cities of Singapore, Hong Kong, etc., and some selected cities of Africa and of Latin America. The critical and complex components of homelessness have been studied from the following conceptual domains: Table 2.2.

TABLE 2.2 Bases of homelessness in global south.

Domain	International standpoint	Global south
Urbanization process	Process of population shift from rural to urban cities motivated by several factors.	Booming economy, availability, and access to better resources in cities experience an influx of population in the form of migration.
Formation of slums	Imperative problem associated with rapid urban growth in specific regions of the world.	The most important aspect of urbanization in the developing countries of the global south region.
Poverty	Poverty is pronounced deprivation in wellbeing and comprises many dimensions. It includes low incomes and inability to acquire the basic goods and services necessary for survival with dignity.	Major reason of homeless in southern region and a social deprivation of a large number of people living in developing and underdeveloped country.
Housing	Housing is more appropriately defined as a process involving the interaction between an organism and the environment.	Housing refers to pavement dwelling, construction of small house to modern villa relate to rural to urban migration, to unemployment to employment, informal to formal sectors depending on the needs.
Migration	The movement of people across a specified boundary to establish a new permanent or semipermanent residence triggered by several factors.	Economic and social development of urban areas and cities attract the movement of human resources changing urban geographical space.
Public policies	A public policy is a deliberate and careful decision that provides guidance for addressing selected public concerns. It is a decision-making process that addresses identified goals, problems, and concern.	Countries of this region are framing policies without evaluating actual ground scenario properly. Implementation of schemes and policies are not properly working in the favor of homeless people lacking in addressing the issue properly.

Source: Various sources of literature Review.

2.2.1 "Homelessness" in South Asia

2.2.1.1 Indian scenario

Sattar (2014) stated a comparative study of Indian urban homeless while describing the role of HUDCO for urban development. While describing homelessness the definition has broadly described under social economic conditions stating the various dimensions of homelessness. Several causes like poverty, failure of the housing system, domestic violence, lack of support, unrest social political scenario, natural calamities, mental weakness, and disabilities of various forms are identified as root causes of homeless. Decade long data from 2001 to 2011 has been used to show the magnitude of growing homelessness in the country state-wise and sector wise along with government role in handling the problem is also mentioned. It is also known that though from 1961 to 1981 homeless population rate has shown an increasing trend post that it has fallen but in case of urban areas the number of houseless populations has upward trend since 1961 to 2011. In terms of rural urban homelessness to total population urban homeless is showing a marginal rise since the beginning of 20th century. A comparative study between mega cities show that Kolkata is experiencing major growth of homeless population due to minority influx and apartheid while greater Mumbai stands second. In Mumbai majority of homeless people are daily wagers, vendors, taxi drivers, beggars while in Kolkata homeless are rick saw pullers rag pickers, domestic helps, and some who are facing hardships becoming professional organ blood donors and engaged themselves in prostitution even specially women. In Delhi mostly homeless are the permanent and seasonal migrants, unskilled laborers domestic servants beggars. Though government has several national policies for housing for economic weaker sections and LIG. Schemes of JNNURM cities, IHSDP (Integrated Housing and Slum Development Program), Indira Awas Yojana (IAY) are working to settle this rising urban problem.

Bhattacharya (2019) talked about various dimensions of homelessness in the Indian Cities. As a resultant phenomenon of globalization, rapid urbanization, booming economy in certain parts of the world leads to homelessness in Global South. As a meticulous result Indian cities are facing a rapid growth of squatters, ghettos, pavement dwellers who are often termed as homeless population. 20th century being the leading year of massive high-rise bloom which simultaneously creating a space of rising homeless population. The real estate sparks, gentrification process, land acquisition, displacement and urban hazard added homeless population to the cities like Mumbai, Delhi, and Kolkata. Global South has marked urban growth in the form of displacement as residential capitalism. The author has focused her work mainly on the homeless situation of Kolkata. "The city of joy" is considered to produce space for "Bastis" which are slums, depicting the cities uncanny decaying of urban nature. The encroaching stalls spill over the footpaths limiting the usage of free moving space in the city generating new homelessness called "pavement dwellers." Though authorities are actively evicting those sites still it seems to be post-colonial urbanism. With those glitzy skyscrapers the city is experiencing space shortage of housing for that informal human being. Speak (2019) showed the concept of homelessness in developing world which is considered to be the part of Global South. As the developing world is very divergent in terms of social, political, cultural, economic, and environmental it came up with certain facets of homelessness in general. While describing homelessness he used terms like rough sleeping, dwelling in pavements, shanties, living in rehabilitation camps have evolved with the passage of time. But in a broader sense homelessness is referred to as slums for the developing world. While describing major push factors or driving forces of growing homeless population, economic reasons specially poverty is considered among the biggest. It can be major economic failures which push rural to urban migration like from suburban areas to Delhi, Gurugram, or climatic change like Saharan African or seeking for job opportunities in a greater world. Displacement is major concerned with homelessness. Global climate change and associated with environmental disaster forcing people to experience new ways of deprivation. Cities with larger population and its over growing nature in a planned or unplanned manner facing the worst situation. Some violent evictions or some prior informed removals of informal settlements are adding the same taste to the insecure people. Sociopolitical reasons are working in a similar manner. Khorakiwala (2018) discussed in his work how several government policies are working to deal with urban homelessness in Indian cities. India is under the influence of neo-liberal macroeconomic policies which are creating a major segregation between have and have-nots. The urban policies generated higher land values denounced the presence of marginal groups indirectly creating peripheral spaces for people to live. The government policies of National Urban Housing and Habitat (2004) has addressed the issue huge number of migrations by creating integrated townships but it has turned out detrimental for the weaker section ultimately unaffordable to urban homeless people. Still the policy to some extent is able to move the slum population to better housing conditions and providing all the basic amenities of living. But to improve the way of life there is a need of refurbishment of existing shelter with furthermore demand for zones of separate residential, commercial, public areas for a sustainable healthy environment. The other scheme for urban homelessness in 2013 is put forwarded to give a specific space for urban homeless people as certain laws are obstructing the presence of urban homeless people. These are specifically for the people migrating from other places as a major source of labor force to meet the city's need itself. The scheme came up with special shelters for the deprived people who are trying hard to thrive in the city. It was a step to provide shelter beyond slums but the shelters are indeed nonfunctional. These places are small to live and grown up in an unplanned manner. Lacks of unavailability of lands is the due cause of failure of the policy. However, while addressing the issue of urban homelessness the thinking should be done in creating opportunity for all and create a sense of equity. Banerjee-Guha (2019) took the example of Mumbai (Bombay) the largest cosmopolitan city in India. Considering the basic necessities of the urban poor such as shelter, transport, water and sanitation, employment openings, etc. Following the wave of globalization, major Indian cities like Mumbai, Bangalore, Hyderabad emerged as favorite destination for land developers wooing the real estate business and redevelopment of slums and desolated, dilapidated spaces began. Ghosh (2019) in this context highlighted the critical case of Delhi where he focused on the making of "homelessness." It revealed that the large-scale destruction in the last 3 decades which rendered thousands of people homeless. They are deprived of their basic rights and requirements. The issue of night-shelters and the situation of shelter-less population is also growing as a critical issue in major cities.

2.2.1.2 Scenario of Bangladesh

Biswas (2013) has stated in his work about the homeless situation in Dhaka, Bangladesh as an alarming situation in developing countries in Global South. The study not only focused on the problems of street dwellers but also aimed to recognize the alleyways leading to such degenerating situation. The daily life problems, social factors, health risk, and socio demographic nature of this kind of urban issues is growing in Dhaka city. Like all other glittering cities of the developing nations Dhaka showed a tremendous growth over the decades while pushing marginalized people to a state of utter despair. The homeless people are poverty stricken, having a shifting nature, following the leading occurrence of natural disasters. Even those people are facing a constant threat of eviction, devoid of civic amenities and basic needs. However, the demographic characteristics of this homeless population indicate higher female ratio beckoning Islamic background striking between age group of 30–40 years. The capital city of Bangladesh is not only experiencing rapid growth of urban population but also a major hub of economic sources, job opportunities, thriving livelihood. Analyzing the basic causes for growing homeless population in the city demographic structure is considered to be the main reason. Apart from these physical ailments, growing chronic diseases, respiratory illness, infectious disease, living in unhealthy conditions leading to higher morbidity rate as they lack in proper health care. The livelihood problem associated with constant struggle for space made homeless people to sleep on footpaths, rail stations, bus stops, market places, dilapidated houses. In addition to this, several other problems including lack of availability of food, drinking water, and toilets also forcing them to thrive under vulnerable situations. The entire scenario of those people generates other social problems of safety security, protection which directly made them associated with illegal activities, substance abuse, criminal offenses. The constant sufferings, fear of eviction apprehended by the law enforcing authority, lack of opportunities make them feel vulnerable.

2.2.1.3 Scenario of Pakistan

Younes et al. (2021) analyzed the situation prevailing in Lahore from a specific home. As being one of the oldest city Lahore is experiencing population pressure where people are found in distressful situation as a reason of broken families, stepparents, domestic violence, insecurities, physical abuse, etc. The increased number of immigrants caused a huge pressure on existing land. The homeless people are mainly found in parks, slums, roads, railways depicting city's lower living standard. Bus stops and Shahdara are hot spots of these homeless people (Shah & Butt, 2011). As the government is reluctant in its part of work mostly nonprofit organizations are working in favor of these people. As results there are several shelter homes formed to protect these homeless people. The study focused on the socioeconomic factors which are responsible for growing such shelters. Data showed that shelter is a home for people aged between 10 and 24 years. While focusing on the factors, it is found that mostly domestic violence, economic condition, mental condition, delinquent behavior, broken homes are reasons behind such occurrence. The following table reveals the figures of the homeless persons in major South Asian nations which is found very alarming. Delaying in administrative and civic society intervention in this context will further aggravate the present scenario (Table 2.3).

2.2.2 *"Homelessness" in Southeast Asia*

Banerjee-Guha (2019) asserted that the contemporary cities have emerged as highly combative spaces, both within the academic discussions and in the everyday experience in the city. Cities are becoming crucially inter-connected to the process of materializing neoliberalism as a philosophy, especially in the context of the global south. In Hong

TABLE 2.3 Scenario of homelessness in South Asia.

Country	Population (in millions)	Slum-dwellers (in millions)	Homeless population (in millions)
India	1378	65	1.77
Bangladesh	146	5.3	5
Pakistan	204	32	20
Sri Lanka	21.8	0.65	N. A
Myanmar	52	3	1
Afghanistan	40	2.6	0.5
Nepal	28	2.8	0.25

World Bank and National Govt. figures, 2014–20. Data from Sekhri (2021).

Kong, as she cited about the homeless used to live in "cage homes," which she critically pointed as inhuman arrangement of iron beds tied up on all sides, placed one above the other. While stating the major cause of homelessness poverty stride as the main cause in South Asian countries like Indonesia, Malaysia, Thailand, and Singapore. Homelessness in terms of poverty has been addressed in this work. While analyzing the overgrowing homeless population in these selected countries of South Asia. Different factors like social, demographic, cultural, political, geographical landscape are taken into account. While poverty has been polarized urban population folks migrating from Trans regional spaces are the worst victims of the problem. The chronic nature of poverty associated with several demographic factors promotes episodic homelessness, whereas temporary homeless, secondary homelessness, tertiary homelessness is rapidly growing in developing countries of south-east Asia. Malaysia from a socio demographic aspect shows most homeless individuals are men and also the ethnic composition are root causes of growing phenomena. Thailand accounts a large number of homeless people whose job insecurities and aging of population marked as the major cause of homelessness. Though Singapore is considered as developed nation still research suggested a part of population seems to be homeless under certain urban issues which pushed them to such distressful situation. A variety of common problems like living in unhygienic condition to drug abuse, illness, social and political reasons push forward a major part of population into poverty-stricken situation which leads to homelessness in these countries.

2.2.3 *"Homelessness" in Africa*

2.2.3.1 South African scenario

Tenai and Mbewu (2020) highlighted in their work about increasing street homelessness in South Africa and religious factor playing a major role for that. While describing social economic causes of increasing homelessness among black Africans they stated multiple facts like broken families, dysfunctional families, and lack of shelter, food, basic amenities, and access to proper housing. Though the South African Housing Policy (SAHP) attempts to respond to homelessness in South Africa, the churches are working and acting properly to address the problem of street homelessness. From the biblical perspective the church asks for moral rights and want to create a utopian society to help the homeless and criticize authorities for discrimination. However, the ministry of the MCSA helped internal and external migrants in the time of various incidences of xenophobic attacks in South Africa. Displaced and destitute migrants were taken to safe shelter and they are provided with required food and clothing. Policy implementation while in terms of taking care of homeless people Methodist role was certain and widely popularized. Social meetings addressed issues like poverty is discussed and solutions were given to fight against. To attract people toward holiness telling spiritual stories indeed act as an important tool while helping the homeless people. The Methodist church of South Africa with the help of government initiatives, a better platform is given to homeless people. Policy makers applied the formula of engaging street people to policy framing as without their presence the mission of eradication of poverty will not be successful. The focus was not to form any homelessness situation and church should create more space for the policy makers and the homeless people to act for the betterment of the situation.

3. Conclusion

A growing debate exists within the scientific community about the consequences of global climate change which posed several bigger threats to the survival and well-being of the common folk. A report released in the recent past by the Global Humanitarian Forum, estimated that globally every year climate change takes a toll of over 300,000 lives and more than 325 million are affected severely. It is also estimated that nearly 04 billion people throughout the world are in a vulnerable stage due to severe adverse effects of climate change. These impacts are not homogeneously distributed. It is the poorest of the population, who are the worst hit as per Rajendra Pachauri, the ex-chairman of the IPCC. Vulnerability in this regard is determined by two key factors: exposure-sensitivity and adaptive capacity. The former one is the nature of a climate or weather condition (exposure) attached with whether or not the individual encounters the sensitivity. On the other hand, the adaptive capacity is the capacity to react successfully to the variety of exposure sensitivities or how well an individual can deal with a specific occurrence. Hence, we may infer from their arguments that the essential factors associated with the aforesaid exposure sensitivity and adaptive capacity that the people who are living in the territorial and littoral margins of the ecologically sensitive areas emerged as the most vulnerable population in this regard. We have covered the literature and reports published in recent past about the climate induced forced migration and the growing numbers of the environmental refugees in context to global "homelessness." The present challenges which the climatically vulnerable population of the global south, especially in the eastern Africa or the parts of central Asia, south Asia and Latin America are facing is not only due to climate changes in general. Here we can say that the decision makers can plan accordingly and assess whether current strategies are adequate enough in addressing possible future vulnerabilities to global climate change. Future vulnerabilities can be possibly prevented by planning combating strategies well in advance. To implement this in reality, all the key stakeholders need to have access to the created knowledge and to device the effective knowledge transfer and to ensure community involvement for the dissemination of the research and policy findings. On the other hand, the influence of poverty in causing homelessness is determined by both the macro-level and micro-level structural conditions such as welfare regimes, public housing conditions, labor-market scenario with individual vulnerabilities like ill-health or substance misuse etc. Remarkably, the effects of poverty can be arrested by applying a number of factors such as the magnitude of the welfare policies and schemes, individual access to social, economic and cultural human capital. Homeless population face tremendous barrier to avail basic civic amenities, ensuring minimum wages and income, affordability to decent housing and several other stages of discrimination is also observed in their case. The scholars believed that the link between poverty and homelessness can be broken if the structural intervention from the policy makers and the administrative interventions are ensured for the micro- to macroscale of action. In short, the complexity of the phenomenon of homelessness in respect to crisis situations evolved due to poverty and scarcity of affordable housing can be solved by the administration by adopting holistic and welfare approach toward the civic requirements and ensuring the accessibility of the urban poor toward the basic amenities.

This work is primarily based on the available literature, outcomes of conferences, research articles, and published reports by scholars belonging to globally recognized institutions and think tanks. We have tried to cover the geographical and territorial extension of the "global south" and utmost emphasize have been given on extracting the complexity of interconnected factors related to the central idea of "homelessness." Valid examples and conceptual insights have been furnished in this chapter for critical evaluation of the scholarly point of view and justified on the basis of our understanding of this issue. However, due to lack of secondary evidences and limited amount of access available, we are not been able to amplify the whole argument in context to the geographical vastness.

References

Addaney, M., Jegede, A. O., & Matinda, M. Z. (2019). The protection of climate refugees under the African human rights system: Proposing a value-driven approach. *African Human Rights Yearbook, 3*, 242.

Adesina, M., Adeel, M., Omigbile, O., Abiodun, A., Adehunoluwa, E., Oladele, R., Olufadewa, I., Abudu, F., Onathoja, O., & Adeyelu, N. (2021). Trends and drivers of refugees in Africa. *European Journal of Environment and Public Health, 6*(1), 1–8. https://doi.org/10.21601/ejeph/11379

Ayazi, H., & Elsheikh, E. (2019). *Climate refugees: The climate crisis and rights denied*. UC Berkeley: The Othering & Belonging Institute.

Banerjee-Guha, S. (2019). *Homeless in neoliberal cities: View from Mumbai, housing* (pp. 62–75).

Bates, D. C. (2002). Environmental refugees? Classifying human migrations caused by environmental change. *Population and Environment, 23*(5), 465–477.

Bhattacharya, D. (2019). Speculation and homelessness in Indian cities, the JMC review. *An Interdisciplinary Social Science Journal of Criticism, Practice and Theory, 3*, 22–36.

Biswas, T. (2013). *Pathways into homeless and problems of homeless people in Dhaka city. M.Phil. Thesis*. National Institute of Preventive and Social Medicine. Bangabandhu Sheikh Mujib Medical University.

Blunt, A., & Dowling, R. (2022). *Home* (2nd). Oxon and New York: Routledge.

Castles, S., & Rajah, C. (2010). *Environmental Degradation, Climate Change, Migration and Development. 5th Accion Global de los Pueblos sobre Migracion.*

Delavelle, F. (2013). *Climate induced migration and displacement in Mesoamerica*. Discussion Paper. Nansen Initiative Central America Consultation.

Edes, B. W., Hill, J., Reckien, D., Dobias, R. J., Ito, Y., & Manzano-Guerzon, H. M. (2018). *Addressing climate change and migration in Asia and the Pacific*. Asian Development Bank.

Fowler, P. J., Hovmand, P. S., Marcal, K. E., & Das, S. (2019). Solving homelessness from a complex systems perspective: Insights for prevention responses. *Annual Review of Public Health, 40*, 465–486. https://doi.org/10.1146/annurev-publhealth-040617-013553

Ghosh, S. (2019). Understanding homelessness in neoliberal city: A study from Delhi. *Journal of Asian and African Studies*, 1–13.

Harvey, D. (2008). The right to the city. *New Left Review, 53*, 23–40.

Hugo, G., & Bardsley, D. K. (2013). *Migration and environmental change in Asia* (pp. 21–48). Springer Science and Business Media LLC. https://doi.org/10.1007/978-94-007-6985-4_2

Johnsen, S., & Watts, B. (2014). *Homelessness and poverty: Reviewing the links*. Heriot-Watt University.

Khorakiwala, Z. (2018). Urban homelessness in India: A. Policy review. *International Research Journal of Social Sciences, 7*(12), 17–20.

Mbaye, L. (2017). Climate change, natural disasters, and migration. *IZA World of Labor*. https://doi.org/10.15185/izawol.346

Myers, N. (1993). Environmental refugees in a globally warmed world. *BioScience, 43*(11), 752–761. https://doi.org/10.2307/1312319

Nishimura, L. (2018). *The slow onset effects of climate change and human rights protection for cross-border migrants*. Office of the United Nations High Commissioner for Human Rights.

Ober, K. (2019). The links between climate change, disasters, migration, and social resilience in Asia: A literature review. *SSRN Electronic Journal*. https://doi.org/10.2139/ssrn.3590184

Rashad, S. M. (2020). African climate refugees: Environmental injustice and recognition. *Open Journal of Political Science, 10*(03), 546–567. https://doi.org/10.4236/ojps.2020.103033

Rigaud, K.,K., Sherbinin, A., Jones, B., Arora, A., & Adamo, S. (2021). *Groundswell Africa: A deep dive on internal climate migration in Tanzania*. World Bank.

Sattar, S. (2014). Homelessness in India. *Shelter, 15*(1), 9–15.

Sekhri, N. (2021). South Asia monitor: A perspective on from and of interest to the region. In *South Asia's housing crisis: Yawning gap between slogans and ground realities*. https://www.southasiamonitor.org/contributors/564/Nirupama%20Sekhri. (Accessed 17 June 2023).

Shah, T. H., & Butt, H. (2011). Sleep comes all the way : A study of homeless people in Lahore. *Pakistan, 1*(3), 207–217.

Speak, S. (2019). *Affordable housing and social protection systems for all to address homelessness, the State of Homelessness in Developing Countries*. United Nations.

Tenai, N. K., & Mbewu, G. N. (2020). Street homelessness in South Africa: A perspective from the Methodist Church of Southern Africa. *HTS Teologiese Studies/Theological Studies, 76*(1). https://doi.org/10.4102/hts.v76i1.5591

Villaseca, E. (2017). *Addressing governance challenges in a changing labour migration landscape*. Conference Paper. International Labour Organization..

Younes, I., Jabbar, R., Iqbal, A., & Liaqut, A. (2021). Factors affecting homelessness in Pakistan: A case study of shelter homes in Lahore, Pakistan. *Journal of the Punjab University Historical Society, 34*(1).

Further reading

IDMC. (2022). *Disaster displacement in Asia and the pacific: A business case for investment in prevention and solution*. https://www.internal-displacement.org/sites/default/files/publications/documents/220919_IDMC_Disaster-Displacement-in-Asia-and-the-Pacific.pdf. (Accessed 23 May 2023).

Jasni, M. A., Hassan, N., Ibrahim, F., Kamaluddin, M. R., & Che Mohd Nasir, N. (2022). The interdependence between poverty and homelessness in Southeast Asia: The case of Malaysia, Indonesia, Thailand, and Singapore. *International Journal of Law, Government and Communication, 7*(29), 205–222. https://doi.org/10.35631/ijlgc.729015

Mayer, B. (2015). Climate change, migration, and international law in Southeast Asia. *Adaptation to climate change: Asean and comparative experiences* (pp. 337–358). World Scientific Publishing Co.. https://www.worldscientific.com/worldscibooks/10.1142/9642.

UN Habitat. (2006). *Homelessness: Global perspective. Report of the international conference on homelessness, India Habitat Centre*.

Wandel, J., Riemer, M., Gómez, W., Klein, K., Schutter, J., Randall, L., Morrison, M., Poirier, S., & Singleton, C. (2010). *Homelessness and global climate change: Are we ready? A report from the study on the vulnerability to global climate change of people experiencing homelessness in waterloo region*.

SECTION II

Rural urban homelessness and socioecological drivers

CHAPTER

3

Urbanization, poverty, and homelessness in Zambian cities: A threat to the achievement of sustainable development goals

Chilungu Mwiinde[1] *and Ephraim Kabunda Munshifwa*[2]

[1]**The Copperbelt University, Directorate of Planning, Property and Services, Kitwe, Zambia;**
[2]**The Copperbelt University, Department of Real Estate Studies, Kitwe, Zambia**

1. Introduction

Globally, studies such as Refs. Adetokunbo and Emeka (2015), Baumann et al. (2013), Kennedy (2015), Satterthwaite and Mitlin (2013), Susser (1996), Timmer et al. (2019) show a direct link between urbanization, poverty, and homelessness. The formulation of Sustainable Development Goals (SDGs) is partly a response to these global challenges. Specifically, SDGs 1 and 11, "No poverty" and "Sustainable Cities and Communities," respectively, then become important global calls for action to deal with these challenges. Urbanization, as a trigger, can have both positive and negative consequences. For instance, urbanization provides greater opportunities for employment, social services, and modern facilities which are essential in improving living conditions and sustaining social progress (Cohen, 2006). However, unmanaged and rapid urbanization remains a global challenge, especially in developing countries, as it negatively affects social equity, public health, and the environment (Gu, 2019). Despite rapid urbanization and population growth, evidence shows a decline in new housing stock and inadequate urban housing infrastructure to support the increasing population. This has greatly contributed to homelessness, especially among the urban poor (Akinluyi & Adedokun, 2014; Bodo, 2019; Chiluba et al., 2021). Homelessness then becomes one of the major challenges to economic development and the welfare of citizens, as housing is universally acknowledged as one of the most basic human needs for life (Akinluyi & Adedokun, 2014). With the current global focus on SDGs, urbanization and cities play a greater

Homelessness to Hope
https://doi.org/10.1016/B978-0-443-14052-5.00003-3

role in attainment of such goals (Chen et al., 2022; Kroll et al., 2019) and homelessness endangers the ultimate success of all the 17 SDGs (Casey & Stazen, 2021). Studies, though, show that the specific linkages have to be contextualized within different environments. This chapter thus investigates the link between urbanization, poverty, and homelessness in the Zambian context. It focuses on four major cities of Zambia (Lusaka, Kitwe, Ndola, and Livingstone). It further links the findings to strategies toward the achievement of SDGs asserting that the current conditions are a threat to the achievement of these goals. After this opening section, this chapter then highlights the rationale of the study, which is followed by the materials and methods used. The section on results and discussions is then given, followed by the limitations of the study. Sections on recommendations and conclusions are then given before ending with acknowledgments.

2. Limitations of the study

This study was mainly limited to the data collection process and focused largely on literature review. Nonetheless, this chapter still provides insights into urbanization, poverty, and homelessness in major Zambian cities in lights of the SDGs.

3. Materials and methods

This chapter was mainly based on a case study approach, literature review, and interviews with key informants. The understanding of urbanization, poverty, and homelessness in relation to SDGs was based on a combination of peer-reviewed published literature and expert knowledge. The Google Scholar search engine was used to filter published research in peer-reviewed journals or authoritative public reports from official organizations (such as United Nations). The review process involved searching the literature based on terms associated with "urbanization," "homelessness," "poverty," and SDGs. The arguments and evidence found in the literature and reports, together with the expertise of planners in the specified cities, guided the discussion on the interaction among urbanization, homelessness, poverty, and SDGs. The findings are helpful in improving understanding of urbanization and homelessness, especially in relation to SDGs as well as strategies for sustainable urbanization and housing. This chapter then focused on Zambia as a case study.

4. Rationale of the study

Sub-Saharan Africa, with an urban population estimated at 42%, is arguably one of the most urbanized regions in the world; with serious implications on the provision of housing and other urban services for the majority of urban dwellers (Bodo, 2019; Chigudu & Chavunduka, 2021; Herner et al., 2018). The result is that 56% of the urban population is said to be living in informal settlements which lack clean water and sanitation facilities for residents, consequently threatening the attainment of sustainable development goals in developing countries (Casey & Stazen, 2021; Chen et al., 2022; Chiluba et al., 2021). Sustainable urbanization and housing concerns influence the global direction toward sustainable development in terms of economic, social, resource, environmental, and ecological aspects and provide focused perspectives for implementing the SDGs (Chen et al., 2019; Chen et al., 2022). This calls for an urgent need to understand the interaction between urbanization, poverty, homelessness, and SDGs, hence the justification for this study.

4.1 Urbanization as a global phenomenon

Urbanization is currently a global phenomenon affecting all nations. Among various definitions, urbanization is perceived as an increase in the global population proportion living in cities (Akinluyi & Adedokun, 2014; Bodo, 2019). It involves transforming human production and lives from rural to cities including population concentration, drastic landscape changes and nonagricultural activities which affect the social and natural environments (Chen et al., 2022; Tonne et al., 2021). Most literature has referred the continued rural–urban migration (mass movement of people from rural settlements to cities) and rate of natural increase (more births than deaths) to be major sources of population increase, especially with advanced technology and developmental projects. The majority of studies consider rural–urban migration to have been the greatest source of rapid urbanization between 1950s and 1970s, while natural increase has been the greatest source in the past 2 decades (Bodo, 2019). Individual and corporate efforts to reduce commuting time and transportation expenses while improving jobs, education, housing, marketplace, and competition opportunities in cities have been other natural causes of urbanization (Akinluyi & Adedokun, 2014; Chiluba et al., 2021). The current trajectory shows an unabating rise in the urban population.

In most developed countries, urbanization contributes to improving economic conditions and living standards of people, especially when complemented by industrialization. Unfortunately, rapid urbanization in most developing countries has led to rapid population growth, resulting in observable adverse effects such as poverty, housing crisis, and inadequate land for development (Akinluyi & Adedokun, 2014; Bodo, 2019; Chiluba et al., 2021). Poverty is evidenced through lack of well-paying jobs and growth in informal employment opportunities. This has contributed to homelessness resulting in declining quality of life and increases in diseases, infant mortality, malnutrition, insecurity, and breakdown of family cohesiveness and community spirit (Bodo, 2019). Housing crisis is evidenced through inadequate housing which remains a global concern and has greatly contributed to homelessness which persists among the majority, poor urban dwellers. This has resulted in housing informality for millions and growth of informal settlements and acute service delivery challenges. Inadequate land is evidenced through growth of settlements in flood prone areas, consequently increasing the intensity and frequency of harmful flood events and environmental degradation (World Bank, 2022).

4.2 Demographic trends in the urbanization process

Cities in developing countries, have usually served as sources of job opportunities (Bodo, 2019; World Bank, 2022) which has greatly contributed to rapid urban population growth as the majority of the people see opportunities in urban areas. For instance, the global urban population was 34% of the total in 1960 which increased to 43% by 1990. By 2014, this percentage had grown to 54% globally (Bodo, 2019; UN-DESA, 2014). The World's population residing in cities is currently estimated at 55% and is anticipated to reach 60% by 2030 and 65% by 2050 (Bodo, 2019). Regardless of the scale and speed of urbanization globally, it is anticipated that about 90% of future growth would occur in developing countries across Asia and Africa (UN-ESC, 2019); with serious implications given the economic capacities of countries in these regions. Ensuring sustainable global prosperity will require greater understanding of the unsustainable conditions (including homelessness) resulting from the current urbanization conditions (Bodo, 2019; Chen et al., 2022; Fu et al., 2019).

4.3 Homelessness as a human and global challenge

Homelessness still remains a global challenge affecting all facets of mankind. It is one of the most acute forms of material deprivation as it makes people incapable of enjoying permanent accommodation (Casey & Stazen, 2021; Chiluba et al., 2021). Among various definitions by different jurisdictions, countries, and regions, homelessness is seen as "living in severely inadequate accommodation without security of tenure and access to basic services" (Casey & Stazen, 2021, p. 67) and is categorized (Speak, 2004). It thus includes people living in buildings below the minimum standards, temporary accommodation, and on the streets. It can further be categorized as (Chiluba et al., 2021): *primary* for those living on the streets; *secondary* for those moving between temporary shelters and emergency accommodation; and *tertiary* for those living in private boarding houses lacking private bathroom or security of tenure. The homeless mostly are poor and lack steady income which makes them incapable of acquiring and maintaining regular, safe, secure, and adequate housing. Despite having the universal basic human rights to adequate housing, inadequacy has been one of the greatest causes of homelessness in cities (Bodo, 2019). For instance, it is estimated that globally, the homeless people who require access to housing and basic infrastructure services from 25 years and above are about 50 million; those evicted forcefully are about 15 million; those within the European Union 4.1 million; those displaced annually due to climatic events are 22 million; those living in adequate housing conditions are 1.6 billion and those living in informal settlements are 883 million (Chiluba et al., 2021). These statistics reveal a growing problem globally, although more acutely in Sub-Saharan Africa. Homelessness implications include lacking some basic human rights such as health care, education, privacy (decent housing), clothing, employment, and social security (Casey & Stazen, 2021; Chiluba et al., 2021). The homeless mainly suffer from poor health (mortality), poverty (inability to earn basic income and other elements for survival), hunger, lacking access to education, clean water, sanitation, and connectedness. This also contributes to rising inequalities and inhibiting development of sustainable and inclusive cities (Akinluyi & Adedokun, 2014; Herner et al., 2018) which in turn greatly jeopardizes attainment of the SDGs (Chen et al., 2022). For sustainable development, governments should therefore ensure addressing homelessness in most SDG areas.

4.4 Urbanization, poverty, homelessness, and SDGs

SDGs are a universal call to action with intensions of eliminating poverty, protecting the planet and attaining shared global peace and prosperity and providing liveable cities (Chen et al., 2022; UN, 2018a, 2018b). This chapter mainly focuses on SDG 11, which specifically concentrates on urban areas and settings and covers all dimensions of the other 16 SDGs (UN, 2018a, 2018b). SDG 11 is an urban goal aimed at making cities safe, inclusive, resilient, and sustainable. This makes urbanization go beyond a demographic phenomenon, but also a transformative process for stimulating many global development aspects and a key for international development policy (UN, 2018a, 2018b). Globally, cities play a key role in attaining all SDGs and that requires promoting sustainable social, economic, and ecological dimensions for guiding the international community toward future sustainable development (Chen et al., 2022; Fu et al., 2019). They are involved in promoting sustainable planning, innovation, local governance of infrastructure, resource management, and economies (Appio et al., 2019), which are key for global sustainable development (Chen et al., 2022; Scoones et al., 2020). Cities impact directly and indirectly on land and energy use, resource consumption and climate change as they are key in driving industrialization and economic growth and are recognized as innovation and investment hubs. Urban areas remain the strings connecting all SDGs as most of the SDG targets comprise an urban component. For instance, SDG 11.4 recognizes the connection between cities and culture as key in driving and enabling attainment of numerous city related SDG targets. This needs to be mainstreamed across numerous SDG indicators. Urbanization therefore becomes key in enabling rural–urban connections that provide for a balanced territorial development and addressing numerous persistent global challenges including homelessness, climate change, pollution, resilience, and environmental degradation (UN, 2018a, 2018b; UN-ESC, 2019). Since homelessness is one of the crudest manifestations of poverty and inequality (UN-ESC, 2019), addressing it greatly contributes to eradicating poverty, ensuring healthy lives and sanitation for all and making cities and human settlements inclusive, safe, and resilient. Paradoxically, homelessness is absent from SDGs which greatly jeopardizes the attainment of almost all the 17 Goals. Considering that what gets measured is what gets done, not mentioning homelessness in the SDGs would make the homeless to continually remain ignored, unattended to and left behind in pushing for progress (Casey & Stazen, 2021; UN, 2018a, 2018b; UNDP, 2018).

SDG 11 focusing on sustainable cities and communities has usually been considered to be vital for sustainable urbanization (Akuraju et al., 2020; Ghani et al., 2021, pp. 1–28). Numerous studies (Anwar et al., 2022; He et al., 2022; Xu et al., 2022; Zhao et al., 2022) have confirmed that urbanization connotation and scope of action go beyond SDG11 and have interactions with all other SDGs (Chen et al., 2022; Fu et al., 2019). Furthermore, several studies have examined the interactions between specific areas and SDGs (Fu et al., 2019), including environmental, hygiene, energy systems (Parikh et al., 2021), food security (Giannetti et al., 2020), climate action, and ecological security (Maes et al., 2019). As a complex systemic process of urban-rural transformation, urbanization, involves various components including, humans, environment, technology, land, and management (Chen et al., 2022; Chua et al., 2019; Wang, 2020). A cursory glance at the current state of cities in Sub-Saharan Africa clearly show that the inspirations of SDG 11 are far from being met. As noted above, many lack basic infrastructural facilities, have high incidence of poverty, and are growing more uncoordinated and informally. Changing the current trajectory therefore requires a serious relook at how African governments are domesticating the provisions of SDG 11.

5. Urbanization, poverty, and homelessness in Zambia

This chapter uses Zambia as a case study in order to contextualize the above discussion. It thus shows that Zambia has been grappling with the twin problems of urbanization and poverty resulting in challenges of homelessness for the majority in urban areas, such as in the cities of Lusaka, Kitwe, Ndola, and Livingstone. This homelessness is particularly evidenced in the rapid growth of unplanned settlements in the last 2 decades. In order to achieve SDGs 1 and 11, these trends need to be reversed.

5.1 Status of Zambian cities and population growth trends

Zambian cities serve as key market locations and service hubs. They play diverse economic roles that add value in specific sectors and have been greatly impacting the economic landscape, fostering economic growth, structural transformation, and poverty reduction. For instance, cities contributed about 83% to the Gross Domestic Product (GDP) in 2015. These economic activities are mainly concentrated in cities, with the Copperbelt Province contributing 30% of GDP, Lusaka Province 27%, and Southern Province 10% (World Bank, 2022). The key economic activities in the four cities include tourism (Livingstone), real estate and construction (Lusaka), and copper mining (Kitwe and Ndola). Cities thus assist in improving household welfare and lowering poverty by providing market access and service hubs for their neighboring towns. For instance, the level of consumption for Zambian households living beyond 2 hours' distance from the closest city (with a population size of 100,000) is 40% lower than those living less than 30 min away (World Bank, 2022). The level of welfare in highly dense areas is also 21% higher than that of less populated areas. For example, densely populated areas with better large markets connectivity have better access to core public services, lower poverty levels (World Bank, 2022), human capital, higher labor productivity, higher portion of employees in nonfarm and higher wage jobs (60% compared to 20% in least populated areas) contrary to those distant low-density and poorly connected areas (World Bank, 2022). Furthermore, cities have been drivers of Zambia's economic growth and consequently poverty reduction. For instance, there has been a significant drop in poverty levels in cities between 2010 and 2015, with the largest drop in Lusaka (by 7.1%), followed by other urban areas (4.6%) and a modest decrease in rural areas (by 1.6%) (World Bank, 2022). Clearly, cities are an important part of the country's economic survival; and so are its linkages to the surrounding rural districts. In terms of population, Zambia is among the most urbanized countries in Sub-Saharan African (Chiluba et al., 2021). Based on the observed and projected population, as well as spatial growth of cities, Zambia's future is anticipated to be urban. For instance, over the past 2 decades, Zambia's urban population has grown from 9.89 million in 2000 to the current estimate of 19.42 million (Data from: https://datacatalog.worldbank.org/dataset/world-development-indicators/). It is thus estimated that by 2030, nearly half (48%) of the Zambian population will live in cities; this is anticipated to increase to 60% by 2050 (UN-DESA, 2014; World Bank, 2022). The implication is that the challenges of poverty and homelessness have an upward trajectory, if no major actions are taken. This again speaks to the need for the Zambian government to relook at the implementation of SDGs in the urban areas, particularly number 11.

5.2 Urbanization and poverty

Zambia, with a GNI per Capita of US$1,348World Economic Outlook database: April 2022 IMF Zambia—Country Data from: https://IMF.org, is classified as one of the poor countries in the World. Thus, despite rapid growth of cities, regional spatial disparities in poverty and population growth have remained evident across the country. Copperbelt and Lusaka regions have lowest poverty rates while Luapula and Western have highest rates (World Bank, 2022). It is therefore clear that urbanization and economic activities have not sufficiently translated into overall national economic development and poverty reduction. Thus, regardless of numerous anticipated benefits of urbanization, current evidence shows that if unmanaged, it can lead to a number of negative consequences such as increased poverty and homelessness. It is worth noting that the current status is that of increased urban population with no corresponding increase in gainful employment; with many finding their livelihoods in informal business activities. The results are poverty and homelessness. In other words, the urban poor start failing to pay for necessities including housing, health care, education, food, and childcare. The issues of underemployment and unemployment are argued to be central causes of homelessness in developing countries including Zambia (Chiluba et al., 2021; Herner et al., 2018). This in turn greatly negatively affects the attainment of SDGs (Chen et al., 2022).

5.3 Housing informality and service delivery challenges

Seventy percent of Zambia's population is estimated to live in informal settlements and 90% of the labor force is employed in the informal sector (ILO, 2019). This makes housing informality and service delivery challenges important facets of homelessness. Despite the fact that Zambian cities are growing and attracting rural population, this growth has mainly been informal, which exacerbates vulnerability to disasters and hamper sustainable development (UN-DESA, 2014). Rapid, unmanaged, and underfinanced urbanization has resulted in housing crises in cities and consequently high levels of housing informality and severe service delivery challenges in that infrastructure growth is unable to meet the population growth and increased services demand. This is evident in Lusaka, Kitwe, Ndola, and Livingstone, with a similar trend in some smaller cities (World Bank, 2022). The challenges of inadequate investment in housing and infrastructure are clearly visible in the major cities of Zambia where land for development is often allocated without support services (Mushinge et al., 2022). In many instances, the construction of houses commences without full services on sites (Munshifwa et al., 2017). Thus, newly opened up development areas are often characterized with acute service delivery challenges resulting in lack of basic services including water, solid waste management, sewerage and electricity, as well as increased vulnerability and insecurity of tenure. In many such cases, construction is undertaken largely through individual initiatives which at times so not even meet the urban development standards, hence defeating the inspirations of creating "liveable communities" as envisaged under SDG 11. It is argued that lacking affordable, safe, and stable housing has been the greatest direct cause of homelessness countrywide. However, discrimination also inhibits access to housing, employment, and justice among other helpful services (UN-Habitat, 2020). This greatly negatively affects the attainment of the SDGs. Based on SDG11, addressing this requires promoting prosperous, inclusive and resilient urban settlements, as indicated in the draft National Urbanization Policy (2021–30) and territorial development which is more balanced (World Bank, 2022).

5.4 Vulnerability to effects of climate change

One of the concomitant effects of homelessness is increased vulnerability to the effects of climate change such as flooding. In other words, rapid urban growth has been spreading to unsafe areas. Over time, unplanned settlements have expanded into flood prone area, increasing the risk for settlers. For instance, Ref. Nchito (2007) reported two settlements, Mutendere and Kalikiliki which at the time were separated by a stream and prone to flooding. In most cities, entire areas previously seen as marshlands and catchment areas for the stream are now by both formal and informal housing development. Following heavy rains during the current 2022/23 season (December to April), most media reports carried pictures of overflooded settlements, washed away bridges, and displacement of people. The result has been increased vulnerability to flood including increased prevalence of diarrheal diseases (Kabinga, 2016).

6. Government strategies for sustainable urbanization and housing

The preceding sections show an immediate need for government to devise strategies to manage urbanization, which should include addressing homelessness. Regardless of the extension of social programs and numerous government strides aimed at addressing urbanization and associated homelessness, it still remains rampant in Zambia due to high poverty levels, inadequate housing, and discrimination (Herner et al., 2018; UN-Habitat, 2020). In line with the SDGs, the Zambian Government has outlined some urban sector priorities in the eighth National Development Plan which are aimed at addressing national housing issues including land policy and urban planning; decentralization and service delivery; climate change and disaster risk management; private sector development; job creation and micro-small and medium enterprises (MSMEs) (GRZ, 2022). These are planned to be achieved through the following three main urban policies which are aimed at providing a clear understanding of the role urbanization plays in inclusive, sustainable and resilient national growth and specify urban sector aspirations for the nation (World Bank, 2022), these include

1. **Draft National Urbanization Policy (2021–30)**—aimed at promoting prosperous, inclusive, and resilient urban settlements. The draft policy recognizes the link between urbanization and its negative effects, hence the need for the development of liveable and "resilient" urban settlements.
2. **National Lands Policy**—visualizing a transparent land administration and management system for inclusive sustainable development by the year 2035. As noted earlier, the development of housing and related infrastructure is often in ecologically sensitive areas because of the land allocation systems. This policy therefore aims for a more transparent system which recognizes sustainability in urban development.
3. **National Housing Policy (2020–24)**—aimed at facilitating the provision of sustainable, decent, and affordable housing for all socioeconomic groups in Zambia. This chapter

also pointed to the serious challenges in the provision of decent housing, resulting in the rapid growth of informal settlement. This Housing Policy recognizes these challenges and propose some policy measures to deal with them.

These three examples provide direction on the thinking of government in dealing with urbanization, poverty, and housing. They also reveal intentions of incorporating sustainability aspects in policies with emphasis on prosperity, inclusivity, resilience, transparency, and decency; all important facets of "sustainable cities and communities." It is, however, important to note that experience across Africa shows that formulation of policies is often not the problem but their implementation. This chapter thus asserts that it is a coordinated and multisectoral implementation approach which will help in reducing homelessness by improving the quality of life through greater opportunities for employment, social services modern facilities and provision of sustainable and inclusive housing infrastructure. However, rapid and unmanaged urbanization greatly contributes to homelessness mainly through its negative effects including poverty and housing crisis. On the other hand, high levels of homelessness might escalate poverty levels (Chiluba et al., 2021; Herner et al., 2018; UN-Habitat, 2020; World Bank, 2022; ZSA, 2020; ZSA et al., 2019).

With regards to the interaction among urbanization, homelessness and SDGs, studies suggest that urbanization is one of the greatest contributors of homelessness, which in turn greatly affects the attainment of SDGs. Urbanization has been found to be greatly linked with all the 17 SDGs and about 151 (89%) of their targets respectively. Such relations include topics such as: poverty eradication, infrastructure, energy consumption, international cooperation and environmental protection based on SDGs 11 (sustainable cities and communities); 15 (life on land); and 12 (responsible consumption and production). Although there is no direct interaction between urbanization and SDGs 4 (Quality education) and 5 (Gender equality), homelessness affects all SDGs targets. This suggests that urbanization and its associated homelessness greatly contribute to and might worsen the achievement of SDGs targets which determine the sustainability of the actual development path of cities/regions respectively (Zhao et al., 2022). The stronger relational effects on SDGs suggest urbanization to be one of the greatest effective controls in attainment of SDGs (Chen et al., 2022) provides a useful framing of the interaction between urbanization and SDGs which are discussed under four dimensions, namely: public health and social welfare; economic growth and energy consumption; natural resource use and ecological/environmental impacts; and international cooperation for development. This chapter adopts this topography and extends it to include poverty and homelessness within the Zambian environment (see Fig. 3.1).

This framework posits that there is a link between urbanization, poverty, homelessness, and the attainment of most SDGs, as discussed in the following subsections.

6.1 Public health and social welfare equality

Zambia is reported to be one of the most urbanized countries at 43%. Rapid and unmanaged urbanization generates numerous problems including high urban poverty, inadequate housing, medical and public hygiene infrastructure which has greatly contributed to high homelessness levels among the urban poor (Chen et al., 2022). This has consequently led to growth of informal settlements which lack safe drinking water, public sanitation, and

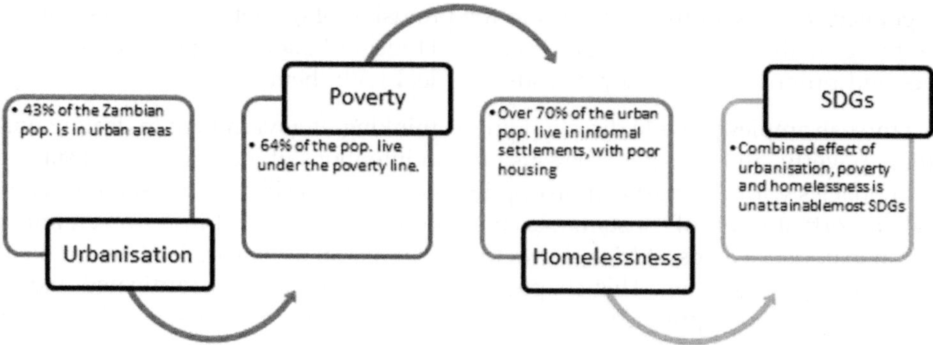

FIGURE 3.1 Urbanization, poverty, homelessness and the attainment of SDGs.

face high poverty and environmental pollution levels thereby threatening public health in cities (Aliu et al., 2021). Such settlements in Zambian cities currently house about 70% of national urban population (World Bank, 2022); with challenges of poor health and unequal social facilities. Globally, informal settlements dwellers exceeded one billion in 2020 and about 85% of them were mostly in Central and South Asia (359 million), East and Southeast Asia (306 million), and Sub-Saharan Africa (230 million) (Sachs et al., 2022). In line with the SDGs, managed and sustainable urbanization, as well as effective governance, provides better opportunities for employment, income, resources and improved access to health and education facilities and investment in infrastructure including housing. This helps in reducing levels of homelessness (Chiluba et al., 2021; Herner et al., 2018) and improves public health and life expectancy of urban dwellers (Liu et al., 2017). Globally, apart from Sub-Saharan Africa, there is a negative correlation between urbanization and adult mortality.

6.2 Economic growth and energy consumption

One of the clear indicators of unmanaged urbanization is the high rate of unemployment in urban areas. For instance, Zambia's unemployment level stands at 64% with most of the available jobs being of low-productivity and informal (World Bank, 2022). Globally, 6.2% of the population lacked employment in 2021 due to Covid-19 compared to 5.4% before the pandemic (Sachs et al., 2022). Pressing emphasis on sustainable urbanization and housing would help in realizing green economic development and a clean energy (Chen et al., 2022). Managed urbanization also accelerates economic development through increasing energy consumption scale and growth as well as addressing energy inefficiencies through dense spatial forms and population (Chen et al., 2022). SDGs also assert that managed urbanization helps in provision of clean and green energy, decent work as well as industrialization, innovation and adequate related infrastructure. However, unmanaged urbanization negatively affects energy consumption due to overcrowding especially in informal settlements (resulting from homelessness) and greater demand for energy. For instance, only about 21% of Lusaka households have access to electricity and 16% of households in other urban areas (World Bank, 2022) and over 700 million of the global population lacked access to electricity in 2020, out of which over three-quarters lived in Sub-Saharan Africa (Sachs et al., 2022).

6.3 Natural resource use and ecological/environmental impacts

Unmanaged and rapid urbanization and its associated homelessness and unemployment has resulted in growth of informal settlements extending into unsafe, flood prone areas thereby increasing the intensity and frequency of harmful flood events which exacerbates the effects of climate change. Ref. Munshifwa et al. (2021) show allocation of land in ecologically sensitive areas in the cities of Kitwe and Ndola. This results in environmental and ecological degradation and climate change, consequently increasing disaster risks, especially flood related (He et al., 2022; Liu et al., 2017). Urban development planning should emphasize sustainable urbanization which is inclusive of functional ecosystem services and biodiversity value for urban resilience (Chen et al., 2022).

6.4 International cooperation for development

SDGs consider urbanization as a key developmental tool aligned with 2030 agenda. Hence, it supports international cooperation for development as evidenced in the past in squatter upgrading projects. Since urbanization-associated homelessness is one of the crudest manifestations of poverty and inequality (UN-ESC, 2019), international cooperation might help in bridging high inequalities and adaptive capacity among developing countries, like Zambia, consequently aiding attainment of SDGs (Zhao et al., 2022). It also helps in addressing specific urbanization barriers in attainment of SDGs, eliminating and mitigating urban adaptation risk factors and linking global and local perspectives. Sustainable urbanization and housing development requires international collaborative governance in reconciling environmental, social and needs for economic development issues (Chen et al., 2022).

7. Recommendations

This chapter has shown that urbanization is one of greatest contributors of homelessness in cities. Furthermore, urbanization and its associated homelessness greatly affect the attainment of most of the SDGs. Transforming Zambian cities and attaining SDGs require focusing attention on sustainable urbanization and housing for improved public welfare through policy reforms and targeted investments. This entails prioritization and collaboration among various key government ministries as well as international cooperation (Chen et al., 2022; World Bank, 2022). This chapter therefore recommends four reform areas and their respective priority areas as follows:

1. **Urban planning:** Planning provides a framework within which urbanization and economic growth can be managed. For instance, the Urban and Regional Planning Act of 2015 stipulates that all Local Authorities should prepare Integrated Development Plans (IDPs), which are broad instruments for capturing issues related to social, economic, and environmental. The state of Zambian cities currently shows failure in this aspect.
2. **Land service delivery and housing:** The rapid growth of informal settlements is a manifestation of failure of the formal system to provide services and housing for the growing urban population (Ephraim Kabunda Munshifwa, 2019). This is acknowledged

in both the National Land Policy and the National Housing Policy, however implementation of remedial measures is lacking.

3. **Urban land administration:** Just like planning, an effective land administration system supports economic development in many ways, including by clearly defining property rights, increased land asset accumulation and improved real estate and land markets operation. Challenges with the computerized land management system at the Ministry of Lands has resulted in numerous problems on the land including corruption, illegal allocations, etc. (Mushinge et al., 2022).
4. **Sustainable urban practices:** Integration of sustainable urban practices will greatly assist cities in the management of current challenges. For instance, this calls for the halting of allocations in ecologically sensitive areas and encouraging more nature-based solutions to the treats of climate change.

8. Conclusions

This chapter was aimed at investigating urbanization, poverty, and homelessness in lights of the SDGs. The chapter focused on the major cities in Zambia (Lusaka, Kitwe, Ndola, and Livingstone). The study found a strong relationship between urbanization, poverty, and homelessness. It thus argued that sustainable urbanization improves economic conditions and living standards for urban dwellers, consequently reducing homelessness while unmanaged and rapid urbanization generates numerous problems including high poverty levels and inadequate housing which greatly contribute to homelessness and the concomitant effects of environmental degradation, poor health and safety standards, vulnerability to disasters, etc.

Acknowledgments (if any)

The authors wish to thank the Planners in the four cities of focus for their helpfulness comments during interviews.

References

Adetokunbo, I., & Emeka, M. (2015). Urbanisation, housing, homelessness and climate change. *Journal of Design and Built Environment*, (2).

Akinluyi, M. L., & Adedokun, A. (2014). Urbanization, environment and homelessness in the developing world: The sustainable housing development. *Mediterranean Journal of Social Sciences, 5*(2), 261–271. https://doi.org/10.5901/mjss.2014.v5n2p261

Akuraju, Vamsidhar, Pradhan, Prajal, Haase, Dagmar, Kropp, Jürgen P., & Rybski, Diego (2020). Relating SDG11 indicators and urban scaling – an exploratory study. *Sustainable Cities and Society, 52*, 101853. https://doi.org/10.1016/j.scs.2019.101853

Aliu, I. R., Akoteyon, I. S., & Soladoye, O. (2021). Living on the margins: Socio-spatial characterization of residential and water deprivations in Lagos informal settlements, Nigeria. *Habitat International, 107*, 102293. https://doi.org/10.1016/j.habitatint.2020.102293

Anwar, A., Sinha, A., Sharif, A., Siddique, M., Irshad, S., Anwar, W., & Malik, S. (2022). The nexus between urbanization, renewable energy consumption, financial development, and CO2 emissions: Evidence from selected Asian countries. *Environment, Development and Sustainability, 24*(5), 6556–6576. https://doi.org/10.1007/s10668-021-01716-2

Appio, F. P., Lima, M., & Paroutis, S. (2019). Understanding Smart Cities: Innovation ecosystems, technological advancements, and societal challenges. *Technological Forecasting and Social Change, 142*, 1–14. https://doi.org/10.1016/j.techfore.2018.12.018

Baumann, T., Bolnick, J., & Mitlin, D. (2013). The age of cities and organizations of the urban poor: The work of the South African homeless people's federation. *Empowering Squatter Citizen: Local Government, Civil Society and Urban Poverty Reduction*, 193–215. https://doi.org/10.4324/9781849771108

Bodo, T. (2019). Rapid urbanisation: Theories, causes, consequences and coping strategies. *Geographical Research, 2*, 32–35.

Casey, L., & Stazen, L. (2021). Seeing homelessness through the sustainable development goals 15. *European Journal of Homelessness, 15*(3), 63–71. https://www.feantsaresearch.org/public/user/Observatory/2021/EJH_15-3/EJH_15-3_A4_v02.pdf.

Chen, M., Chen, L., Cheng, J., & Yu, J. (2022). Identifying interlinkages between urbanization and sustainable development goals. *Geography and Sustainability, 3*(4), 339–346. https://doi.org/10.1016/j.geosus.2022.10.001

Chen, M., Gong, Y., Lu, D., & Ye, C. (2019). Build a people-oriented urbanization: China's new-type urbanization dream and Anhui model. *Land Use Policy, 80*, 1–9. https://doi.org/10.1016/j.landusepol.2018.09.031

Chigudu, A., & Chavunduka, C. (2021). The Tale of two capital cities: The effects of urbanisation and spatial planning heritage in Zimbabwe and Zambia. *Urban Forum, 32*(1), 33–47. https://doi.org/10.1007/s12132-020-09410-8

Chiluba, L., Kabwe, M., & Chibesa, F. (2021). *Causes of homelessness in Zambia*. https://www.researchgate.net/publication/351036259.

Chua, Roy Y. J., Huang, Kenneth G., & Jin, Mengzi (2019). Mapping cultural tightness and its links to innovation, urbanization, and happiness across 31 provinces in China. *Proceedings of the National Academy of Sciences, 116*(14), 6720–6725. https://doi.org/10.1073/pnas.1815723116

Cohen, B. (2006). Urbanization in developing countries: Current trends, future projections, and key challenges for sustainability. *Technology in Society, 28*(1–2), 63–80. https://doi.org/10.1016/j.techsoc.2005.10.005

Fu, B., Wang, S., Zhang, J., Hou, Z., & Li, J. (2019). Unravelling the complexity in achieving the 17 sustainable-development goals. *National Science Review, 6*(3), 386–388. https://doi.org/10.1093/nsr/nwz038

Ghani, Fatima, Tsekleves, Emmanuel, & Thomas, Yonette Felicity (2021). *Urbanization and cities as drivers of global health*. Springer Science and Business Media LLC. https://doi.org/10.1007/978-3-030-05325-3_3-1

Giannetti, B. F., Agostinho, F., Eras, J. J. Cabello, Yang, Zhifeng, & Almeida, C. M. V. B. (2020). Cleaner production for achieving the sustainable development goals. *Journal of Cleaner Production, 271*, 122127. https://doi.org/10.1016/j.jclepro.2020.122127

GRZ. (2022). 8th national development plan 2022 - 2026. In *Socio-economic transformation for improved livelihoods*. GRZ. http://www.zda.org.zm/wp-content/uploads/2022/09/8th-NDP-2022-2026.pdf.

Gu, C. (2019). Urbanization: Positive and negative effects. *Science Bulletin, 64*(5), 281–283. https://doi.org/10.1016/j.scib.2019.01.023

He, Jianjian, Yang, Yi, Liao, Zhongju, Xu, Anqi, & Fang, Kai (2022). Linking SDG 7 to assess the renewable energy footprint of nations by 2030. *Applied Energy, 317*, 119167. https://doi.org/10.1016/j.apenergy.2022.119167

Herner, M., Parker, C., McClellan, N., Muscato, L., & Guenther, E. (2018). Unmet needs of homeless at a shelter in an area undergoing urbanization. *Journal of Rural and Community Development, 13*.

ILO. (2019). *Informality and poverty in Zambia: Findings from the 2015 living conditions and monitoring survey*. October 2018. Geneva: ILO: International Labour Office https://www.ilo.org/wcmsp5/groups/public/—africa/—ro-abidjan/—ilo-lusaka/documents/publication/wcms_697953.pdf.

Kabinga, F. M. (2016). *Exploratory study of enviromental health factors associtaed with the prevalence of diarrhoea diseases in mtendere township-lusaka district Zambia (doctoral dissertation, university of Zambia)*.

Kennedy, M. (2015). Urban poverty and homelessness in the international postcolonial world. *Reworking Postcolonialism: Globalization, Labour and Rights*, 57–71. https://doi.org/10.1057/9781137435934_4

Kroll, C., Warchold, A., & Pradhan, P. (2019). Sustainable Development Goals (SDGs): Are we successful in turning trade-offs into synergies? *Palgrave Communications, 5*(1). https://doi.org/10.1057/s41599-019-0335-5

Liu, Y., Xiao, H., Lv, Y., & Zhang, N. (2017). The effect of new-type urbanization on energy consumption in China: A spatial econometric analysis. *Journal of Cleaner Production, 163*, S299–S305. https://doi.org/10.1016/j.jclepro.2015.10.044

Maes, M. J. A., Jones, K. E., Toledano, M. B., & Milligan, B. (2019). Mapping synergies and trade-offs between urban ecosystems and the sustainable development goals. *Environmental Science & Policy, 93*, 181–188. https://doi.org/10.1016/j.envsci.2018.12.010

Munshifwa, E. K., Mwenya, C. M., & Mushinge, A. (2021). Urban development, land use changes and environmental impacts in Zambia's major cities: A case study of Ndola. In *Sustainable real estate in the developing world* (pp. 63–81). Emerald Publishing Limited.

Munshifwa, Ephraim Kabunda (2019). Adaptive resistance amidst planning and administrative failure: The story of an informal settlement in the city of Kitwe, Zambia. *Town and Regional Planning, 75*(1), 66–76. https://doi.org/10.18820/2415-0495/trp75i1.8

Munshifwa, Ephraim Kabunda, Ngoma, Wilson, & Makenja, Ikugile (2017). Major determinant of physical development on urban residential land: The case of Kalulushi Municipality in Zambia. *International Journal of Social Science Studies, 5*(6), 79. https://doi.org/10.11114/ijsss.v5i6.2434

Mushinge, A., Munshifwa, E. K., Jain, N., Chileshe, R. A., & Ngosa, M. (2022). Effectiveness of public institutions in addressing illegal acquisition of state land in Zambia: Case of Ministry of Lands and Kitwe Municipality. *African Journal on Land Policy and Geospatial Sciences, 5*(1), 72–080.

Nchito, W. S. (2007). Flood risk in unplanned settlements in Lusaka. *Environment and Urbanization, 19*(2), 539–551. https://doi.org/10.1177/0956247807082835

Parikh, Priti, Diep, Loan, Hofmann, Pascale, Tomei, Julia, Campos, Luiza C., Teh, Tse-Hui, Mulugetta, Yacob, Milligan, Ben, & Lakhanpaul, Monica (2021). Synergies and trade-offs between sanitation and the sustainable development goals. *UCL Open Environment, 2*. https://doi.org/10.14324/111.444/ucloe.000016

Sachs, Jeffrey, Kroll, Christian, Lafortune, Guillame, Fuller, Grayson, & Woelm, Finn (2022). *Sustainable development report 2022*. Cambridge University Press. https://doi.org/10.1017/9781009210058

Satterthwaite, D., & Mitlin, D. (2013). *Empowering squatter citizen: "local government, civil society and urban poverty reduction"*.

Scoones, I., Stirling, A., Abrol, D., Atela, J., Charli-Joseph, L., Eakin, H., Ely, A., Olsson, P., Pereira, L., Priya, R., van Zwanenberg, P., & Yang, L. (2020). Transformations to sustainability: Combining structural, systemic and enabling approaches. *Current Opinion in Environmental Sustainability, 42*, 65–75. https://doi.org/10.1016/j.cosust.2019.12.004

Speak, S. (2004). Degrees of destitution: A typology of homelessness in developing countries. *Housing Studies, 19*(3), 465–482. https://doi.org/10.1080/0267303042000204331

Susser, I. (1996). The construction of poverty and homelessness in us cities. *Annual Review of Anthropology, 25*(1), 411–435. https://doi.org/10.1146/annurev.anthro.25.1.411

Timmer, D. A., Stanley Eitzen, D., & Talley, K. D. (2019). Paths to homelessness: Extreme poverty and the urban housing crisis. *Paths to Homelessness: Extreme Poverty and the Urban Housing Crisis*, 1–210. https://doi.org/10.4324/9780429301124

Tonne, Cathryn, Adair, Linda, Adlakha, Deepti, Anguelovski, Isabelle, Belesova, Kristine, Berger, Maximilian, Brelsford, Christa, Dadvand, Payam, Dimitrova, Asya, Giles-Corti, Billie, Heinz, Andreas, Mehran, Nassim, Nieuwenhuijsen, Mark, Pelletier, François, Ranzani, Otavio, Rodenstein, Marianne, Rybski, Diego, Samavati, Sahar, Satterthwaite, David, … Adli, Mazda (2021). Defining pathways to healthy sustainable urban development. *Environment International, 146*, 106236. https://doi.org/10.1016/j.envint.2020.106236

UN. (2018a). *Sustainable cities, human mobility and international migration*. A Concise Report. UN. https://doi.org/10.18356/a11581d8-en

UN. (2018b). Tracking progress towards inclusive, safe, resilient and sustainable cities and human settlements. In *SDG 11 synthesis report. High level political forum 2018*. http://uis.unesco.org/sites/default/files/documents/sdg11-synthesis-report-2018-en.pdf.

UN-DESA. (2014). *Revision of the world urbanization prospects*.

UNDP. (2018). *What does it mean to leave no one behind? United Nations Development Programme*. UNDP. https://www.undp.org/publications/what-does-it-mean-leave-no-one-behind.

UN-ESC. (2019). *Affordable housing and social protection systems for all to address homelessness : Report of the secretary-general*.

UN-Habitat. (2020). *World cities report-the value of sustainable urbanization*.

Wang, Yan (2020). Urban land and sustainable resource use: Unpacking the countervailing effects of urbanization on water use in China, 1990–2014. *Land Use Policy, 90*, 104307. https://doi.org/10.1016/j.landusepol.2019.104307

World Bank. (2022). *Zambia urbanization review policy note*. Washington D.C: World Bank. https://documents.worldbank.org/en/publication/documents-reports/documentdetail/099410012022220621/P1777290c949c70c9082a008455f0bdd8a9.

Xu, Zihan, Peng, Jian, Qiu, Sijing, Liu, Yanxu, Dong, Jianquan, & Zhang, Hanbing (2022). Responses of spatial relationships between ecosystem services and the Sustainable Development Goals to urbanization. *Science of the Total Environment, 850*, 157868. https://doi.org/10.1016/j.scitotenv.2022.157868

Zhao, Zhongxu, Pan, Ying, Zhu, Jing, Wu, Junxi, & Zhu, Ran (2022). The impact of urbanization on the delivery of public service–related SDGs in China. *Sustainable Cities and Society, 80*, 103776. https://doi.org/10.1016/j.scs.2022.103776

ZSA. (2020). *Labour force survey report*. GRZ. https://www.mlss.gov.zm/wp-content/uploads/2022/03/2020-Labour-Force-Survey.pdf.

ZSA, (MOH) Zambia, ICF. (2019). *Zambia demographic and health survey 2018*. Lusaka, Zambia, and Rockville, Maryland, USA: Zambia Statistics Agency, Ministry of Health, and ICF. https://dhsprogram.com/pubs/pdf/FR361/FR361.pdf.

CHAPTER

4

Strengthening rural–urban linkage through growth center to reduce homelessness

Tasnuva Rahman, Addri Attoza, Rafia Anjum Rimi, Shad Hossain, Anutosh Das and Md. Shakil Ar Salan

Rajshahi University of Engineering & Technology (RUET), Department of Urban and Regional Planning, Rajshahi, Bangladesh

1. Introduction

Bangladesh is a country where over 60% of the total population lives in rural areas (Rural Population (% of Total Population) - Bangladesh, 2021) where most of the rural areas lack connection between rural areas and urban areas, and rural people cannot sell their products outside. Due to this, they do not get the proper price for their product which has a bad effect on the economic condition of the rural people. Inadequate income, high rates of poverty, and unemployment all contribute to homelessness. Homelessness is most visible in places with rapid economic growth, which drives up housing costs, as well as those with high unemployment due to decreasing sectors such as farming, wood, mining, or fishing in rural areas (Richter et al., 2019, pp. 197–213). Due to a lack of viable occupations and consistent income, rural family income is much lower (Cloke et al., 2002). As a result, most people leave the city and shift to urban areas for their earnings. When they leave the village and come to the city, the first bad situation they face is homelessness. Poverty rates are greater in rural regions than in metropolitan areas and it's another problem of homelessness, although rural areas are the main territory for raw materials and are more related to primary activities and lead a simple life (DESA, 2021). These raw materials are prime needs of urban people, and rural people's lives improve when they expand their product beyond their boundaries (Gebre & Gebremedhin, 2019). In this way, both rural and urban people get benefitted. Their socioeconomic condition develops through the rural–urban linkage, which means a link between rural and urban locations, and in the same way connections also exist between people and their

Homelessness to Hope
https://doi.org/10.1016/B978-0-443-14052-5.00004-5

activities as well as in their lifestyle. The rural–urban linkage works through the growth center of the rural area (Barua & Akter, 2015). Rural growth centers are built on selectivity, while on the other hand, social amenities are decentralized. It recommends where amenities should be positioned at various levels of the hierarchy (Sharma, 2021). Growth center is basically the CBD (Central Business District) of rural area. Mainly it is the rural market which is responsible for the area's social and economic activities. It's basically a focal point and directly related to the rural development (Mondal & Das, 2010). As a result, the entire growth of all villages within a region may be assured in an integrated manner at a low cost. Rural growth centers are entities that provide spatial connectedness to services for adjacent rural regions as well as their own population (Moseley, 1973). These growth centers, notably rural growth centers, also serve as a decentralizing unit for larger urban centers, helping to disperse the population concentration in metropolitan regions by offering services in rural areas (Sharma, 2021).

Development in rural areas supports socioeconomic developments (Dudek & Wrzochalska, 2017). Socioeconomic development can be identified through indicators like life expectancy, literacy rates, and employment rates, more specifically, the growth of commerce, transportation, and communications are the indicators of economic activity as well as it promotes extended economic development. When people lack these facilities in rural areas, they migrate to urban areas, where they struggle to have proper occupation and places to live (Rural to Urban Migration, 2012). Income disparity, as opposed to low earnings, has a role in increasing the occurrence of homelessness. Lower salaries make it more difficult to purchase homes (O'Regan et al., 2021). In most cases, they stay in illegal establishments and locations. This also decorates the condition of a city, making it overpopulated with migrated people who cannot afford standard living. Ultimately, this creates a severe homelessness problem. So, strengthening rural–urban linkage through growth centers can simulate socioeconomic development and eventually reduce the homelessness problem (Tacoli, 1998). The growth center approach has become an integral part of the formulation and implementation of regional policies in developing and reducing the homelessness problem (Shah, 1974). Mundumala, a growth center situated near Rajshahi, has been developing and continuously enriching in its economic context. Rural people of this area can contribute directly to the economy by producing food and other necessary elements for living (Khan, 2001). Their life, occupation, and occasion move around agriculture, and most of the people are employed through agriculture. In this area, there are many agricultural infrastructures to meet the food requirements of rural people. This area represents how the rural–urban linkage through growth center is strengthened and simulates their socioeconomic development.

This socioeconomic development, catalyzed by the rural–urban linkage, has created income opportunities and substantially reduced the issue of low income in rural areas. As a result, rural residents can have an affordable standard of living, which acts as a potent pull factor, dissuading them from migrating to overcrowded urban areas. It is the lack of employment opportunities and subpar living conditions in rural regions that often drive people toward cities, exacerbating the homelessness problem. In rural areas, the lack of proper housing options further contributes to the challenge, whereas in urban centers, the issue of homelessness is already prevalent due to overpopulation and limited resources. Thus, the strengthening of the rural–urban linkage, primarily through the establishment of growth centers, plays a crucial role in stimulating socioeconomic development and, in turn, significantly reducing the homelessness problem. Considering these facts, the objective of this chapter was to express the linkage of urban and rural areas through the growth center of Mundumala and highlight the impact of the socioeconomic of the rural area and the reduction of the homelessness problem. The study covers all the aspects like agriculture and urbanization to show the linkage between the urban and rural areas. The study was conducted to thorough PRA tools in the Mundumala, Tanore area, which will bring out much other information about the area and the impacts of rural–urban linkage. The study can be further used to develop the agriculture, market area, and other rural infrastructure and understand the community approach of the rural people and develop the life of rural people. The study is limited in the sense that the selected study area is small and can be less impactful in terms of assuming the situation of all rural areas of Bangladesh. The study's focus and observation is confined to PRA tools and questionnaire survey. The result represents the aspects where rural–urban lineage is created through growth centers that simulated socioeconomic development, and eventually reduction of the homelessness problem is seen in the area. Relevant studies show that rural–urban links begin because urban and rural life depends on one another for survival, and this support takes many forms. Bangladesh's rural areas are inherently defined by their agriculture. Growth centers are the primary means through which agricultural goods are transported to metropolitan areas. Growth centers are an essential component of Bangladesh's rural–urban connectivity (Kafy et al., 2019). But urban planners typically focus on metropolitan hubs in cities while ignoring rural areas. Planners for rural regions mostly disregard metropolitan areas and limit their definition of rural areas to farming communities and their surrounding terrain. However, urban and rural communities depend on a mix of agricultural and nonagricultural economic sources. Agricultural and other commodities from rural-based producers are transported to urban markets for local customers and national and international markets related to growth centers (Tacoli, 2003). The interaction between urban and rural regions is crucial for the social and economic growth of both rural and urban populations. Instead of focusing on urban and rural areas individually, it would be essential to consider how they interact (Gebre & Gebremedhin, 2019). A study discovered that the PRA tool has been effective in facilitating genuine community participation in problem-solving, planning, institutional building, and the implementation of development activities and ensuring the long-term viability of grassroots organizations (Shah & Baporikar, 2010).

2. Methodology

In this study, data were collected through PRA survey and questionnaire survey. Participatory Rural Appraisal (PRA) is a systematic, semistructured approach conducted on-site. It embodies a bottom-to-top methodology that involves rural inhabitants in understanding and analyzing their way of life. This technique is beneficial for community members to assess their livelihoods and associated challenges. In this chapter, PRA's emphasis on active community involvement ensures that the voices and perspectives of all stakeholders are considered, especially those of marginalized groups like the homeless population. By incorporating their insights, the research process can be directed toward developing targeted interventions that address their specific needs and concerns. Furthermore, PRA enables the collection of valuable community insights, contributing significantly to the formulation of a comprehensive strategy aimed at reinforcing the connection between rural and urban areas. This method supported by focus group discussion highlights power dynamics in society; it is valuable for evaluating how local people portray homelessness and the reason behind homelessness and the condition of rate of homelessness people in the study area. The methodology of this chapter includes a theoretical framework, selection of study area and focus group selection, conduction of PRA techniques, preparation of questionnaire, and a collection of data and analytical techniques. The flow chart of the methodology steps is shown in Fig. 4.1.

2.1 Theoretical framework

The theoretical framework of the study describes theoretical terms that have helped to identify the current problems of the sites. Theoretical framework has guided the research and analysis on-site, determining what are the suggestions and recommendations needed to be proposed, and the statistical relationships are also represented. Due to growth center development, product consumption and selling development are increased. Because of product consumption, different types of agricultural sectors are developed frequently.

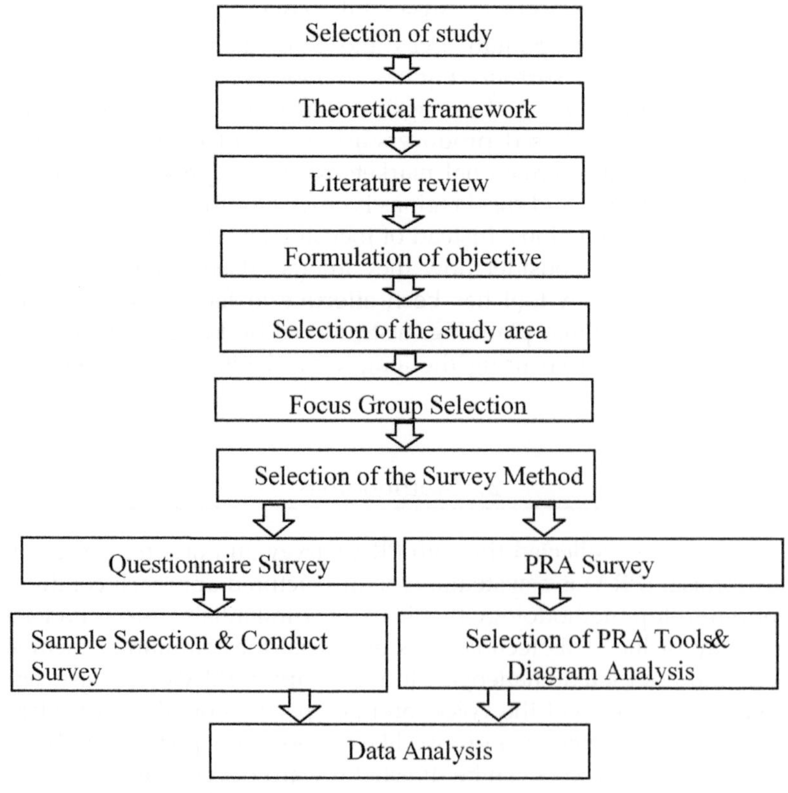

FIGURE 4.1 Methodology of the study.

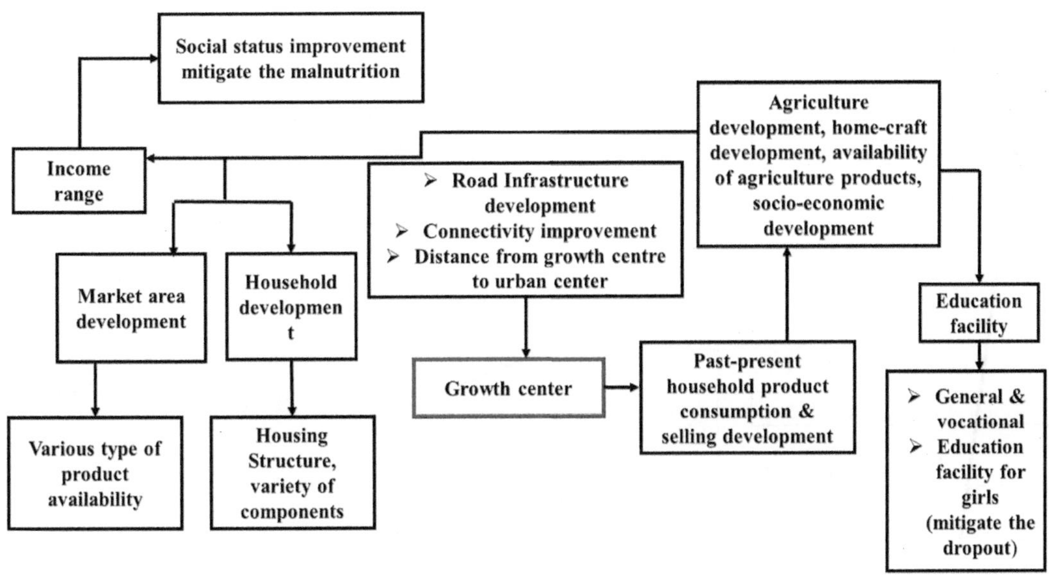

FIGURE 4.2 Theoretical framework.

Agricultural product produce, agriculture-related technology, and home craft development rise very easily because of growth center development. Besides agricultural development, the economic condition of people who lived in this area also enhanced and that is why they pursue their education facilities frequently without any economic-related problems. Because of this agricultural development, market and household development also increased. Through market area development, different types of product availability increased surrounding that area. In this area, housing is developed based on housing structure, which develops the homelessness condition. Since all over sectors are developed, income range is increased. As a result, socioeconomic development of Mundumala Hat improved (Fig. 4.2).

2.2 Study area profile

Study area Mundumala is selected for this study purpose. Because a well-developed growth center (Mundumala Hat) is situated here, it plays a vital role to develop rural–urban linkage. Mundumala is a major village of Tanore Upazila in the Rajshahi District. Tanore Upazila is located between 24°29′ and 24°43′ north latitudes and between 88°24′ and 88°38′ east longitudes. Nichole and Niamatpur Upazilas bound it on the north, Paba and Godagari Upazilas on the south, Mohanpur and Manda Upazilas on the east, Nichole, Nawabganj Sadar, and Godagari Upazilas on the west (Fig. 4.3).

The growth center of the study area is named "Mundumala Hat." Mundumala Hat is situated in the center of the Mundumala village. Adjacent to nearly 12 villages are farmers producing agro-products (especially paddy) that directly contribute to the rural economy, an important characteristic of a growth center. Besides, agricultural production inputs (seeds, fertilizers, pesticides) are easily available, increasing agricultural productivity. Not only the

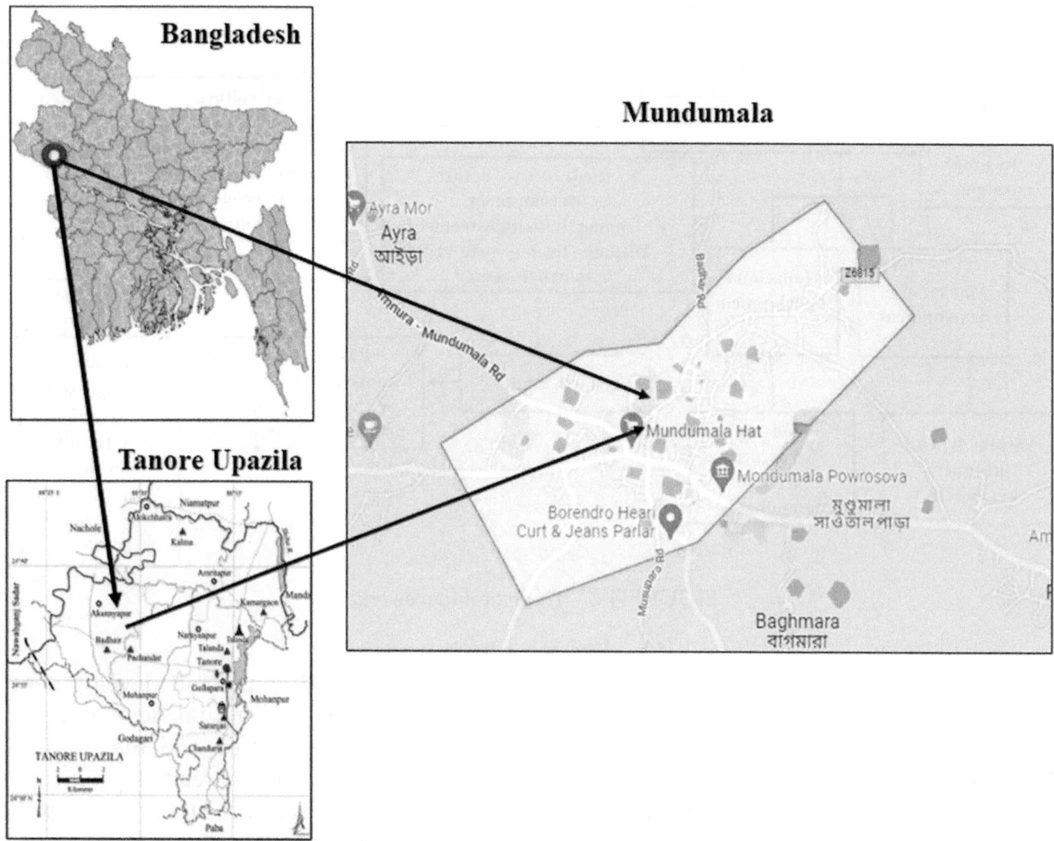

FIGURE 4.3 Study area map.

agricultural products but also many other products (like fruits, fish) and different types of traders are involved in marketing those products in Mundumala Hat. It is also a place of social gathering and interactions, another important target of growth centers. On the hat day (Monday), many traders and consumers gather here and exchange products. This increases social relationships among people of different characteristics and also different areas. Besides the commercial and social aspects, this growth center also creates temporary nonfarm employment opportunities in hat days and at the same time increases communication between traders, farmers, and also with consumers.

2.3 Habitational background of Mundumala Hat

Social map describes the habitational pattern of Mundumala Hat. Social map seeks to explore the spatial dimensions of peoples' realities. The focus here is on depicting habitation patterns and the nature of housing and social infrastructure: roads, drainage systems, schools, and water facilities. Focus groups were formed to create a social map and conduct other PRA techniques. All members were local residents and had been living in the study

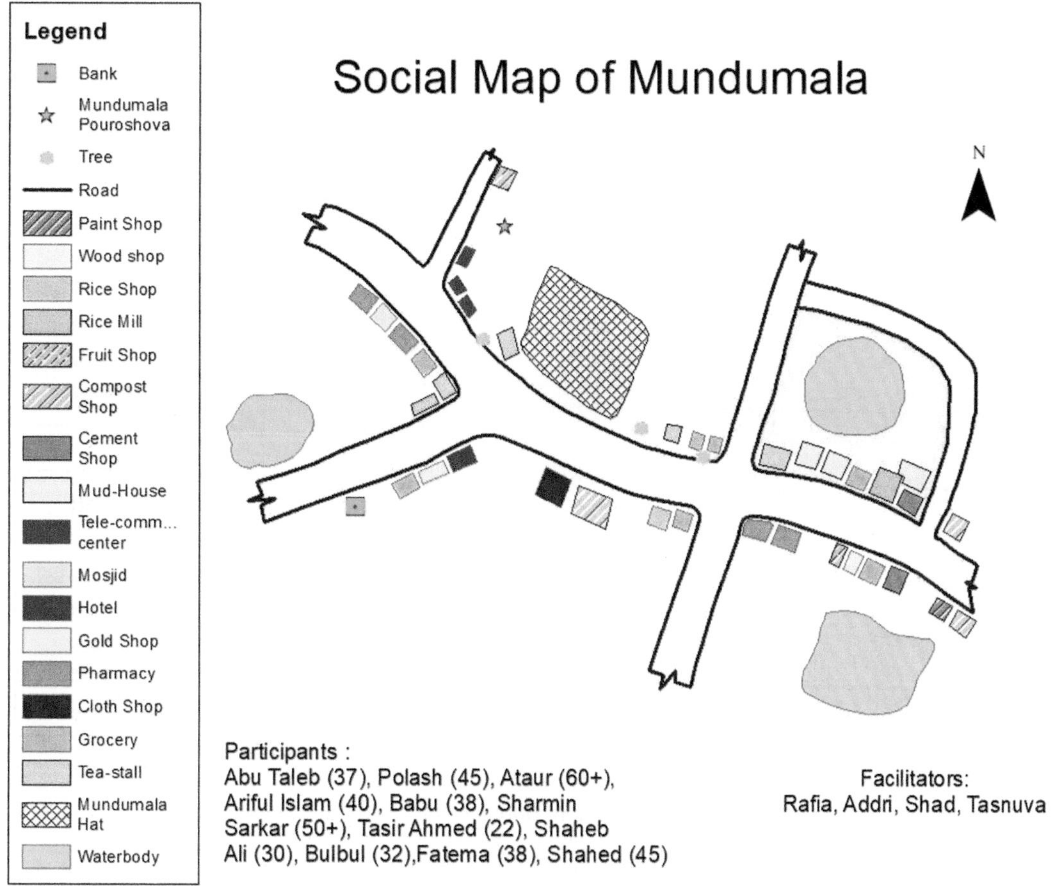

FIGURE 4.4 Social map of Mundumala Hat.

area for a considerable amount of time. They identified the significant feature such as school, college, hat area, important shops, and also relative other features (Fig. 4.4).

2.4 Natural resource of the study area

The natural resource of the study area is represented by the resource map which is the most used PRA technique. The social map focuses on habitation; the resource map focuses on the land, water body, fields, and vegetation. The map is done by the focus group. It is considered to have a depth analysis of the surroundings. In Mundumala, the most used agriculture area is paddy cultivation land and water body is used for fishing (Fig. 4.5).

2.5 Focus group Selection

To highlight the socioeconomic condition of the Mundumala Bazar, basically two types of surveys such as questioner and PRA survey were conducted, where three types of focus

Resource Map of Mundhumala

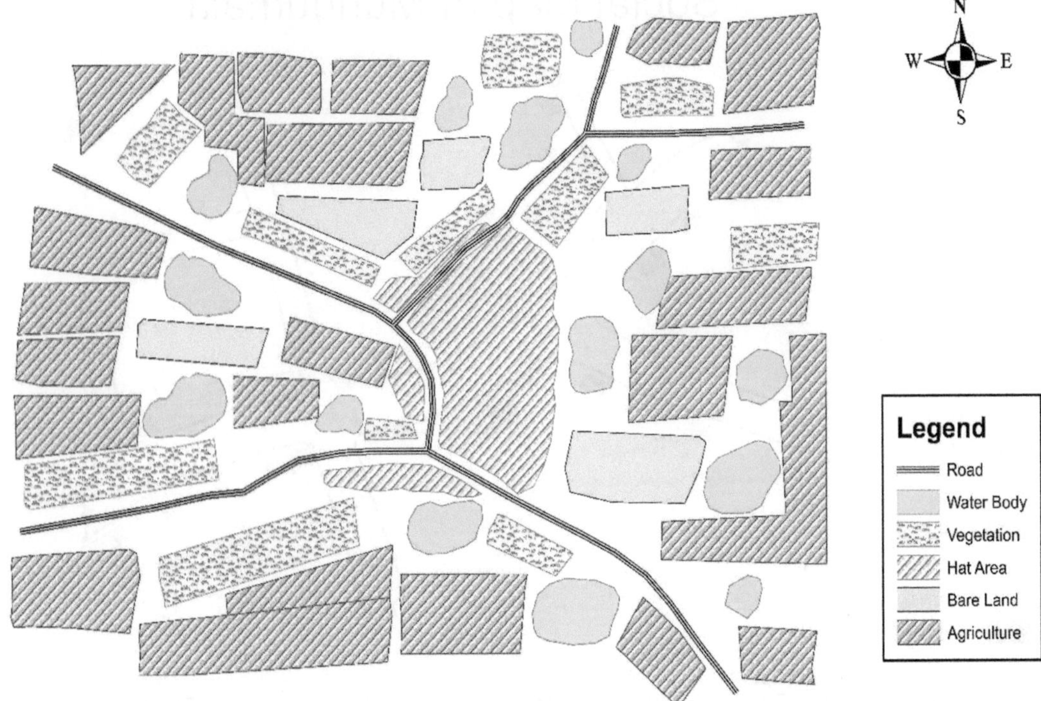

Participants: Abu Taleb , Babu ,Sattar , Asif, Roton, Babul, Robi

Faciliators: Rafia, Addri, Shad, Tasnuva

FIGURE 4.5 Resource map of Mundumala.

groups (growth center, agriculture, and household) were picked that are suitable for the study purposes. For the first two types of group people, focus groups were formed where each group consisted of 13 members. All members were residents or related to the activity and have been living in the study area for a considerable amount of time. Their livelihood is directly or indirectly dependent on Mundumala Hat. The age range of these people is 22–70 years. So, they can provide original problem or situation about the study area.

2.6 Selection of the survey method

Two types of survey are selected to analyze the rural–urban linkage through growth center to reduce homelessness which are PRA survey and questioner survey. PRA surveys are conducted through some PRA techniques such as social map, resource map, mobility map, service and opportunity map, trend analysis, seasonal diagram, cause–effect diagram, impact diagram, and network diagram. For questioner survey, a well-designed questionnaire survey has been conducted with 68 samples from Mundumala with 90% confidence level. The survey was done through field survey. Cluster sampling was selected. There is one cluster group those people who are living in the village. The analysis mainly focuses on the effect of rural–urban linkage on the rural people's socioeconomic condition that are influenced by growth center and transportation facilities. So the survey included the issues to cover the present scenario of transportation facilities, growth center, and present socioeconomic condition of rural people that involve accessibility toward linked growth center, mobility ensured by the condition of the road, transport system, distance and travel time, etc.

3. Analysis and discussion

3.1 Agricultural aspect on Mundumala Hat

Production of agricultural goods within a year: Seasonal diagram is created based on agriculture and economic condition by the focus groups. In Mundumala area, paddy, vegetables, fruits, and fishing are common agricultural things. In this diagram, agricultural production ranges are represented according to six seasons. People are happier in late autumn, autumn, and summer season, and their happiness bleaks in winter, spring, and rainy season, respectively. In Mundumala, fishing, vegetables are yielded every season, but these production rates are increased in the rainy season, autumn, and late autumn. The average production rate of paddy in autumn to winter and the paddy production increased intensively in autumn and late autumn season. For fruit production, summer and rainy seasons are preferable, but fruit production increased in summer. The impacts of these agricultural conditions show that most of the people of Mundumala benefited economically from agriculture. Because of this production range of agriculture, there was an increase in employment opportunity and an increase in product selling capacity and seasonal sells capacity. This also brings positive and fruitful changes in people's lifestyle and economic condition. The reason for these developments is the flourishment of technologies (Fig. 4.6).

Agricultural changes over the era: Significant agricultural products are paddy, vegetables, fruits, fishing, and others. In the following diagram, it is seen that the production of paddy, vegetables, and fruits has increased sequentially from 2008 to 2022 consecutive years. But fish production dropped after 2013. And other products have decreased and also increased (Fig. 4.7).

In 2008–10, all agricultural resources' production rate was at minimal range. **In 2011–16,** paddy, vegetables, and fruits production increased, but fishing and other production rates decreased in 2014–16 period. **In 2017–22,** all production rate increased at the same level. Paddy, vegetables, and fruit production have increased as technology improved over these years. Agricultural research developed. Farmers now have better access to the agriculture officers and information; better seed, low irrigation variety of paddy, vegetables, and fruits have been researched and introduced in the market. Many vegetables were seasonal in the past but now they are cultivated throughout the year. This is helpful to satisfy the need of present demand of increased population. As a result, their economic development occurred around agriculture. Employment opportunities have been created and people are able to export their cultivated products after meeting up their own requirements.

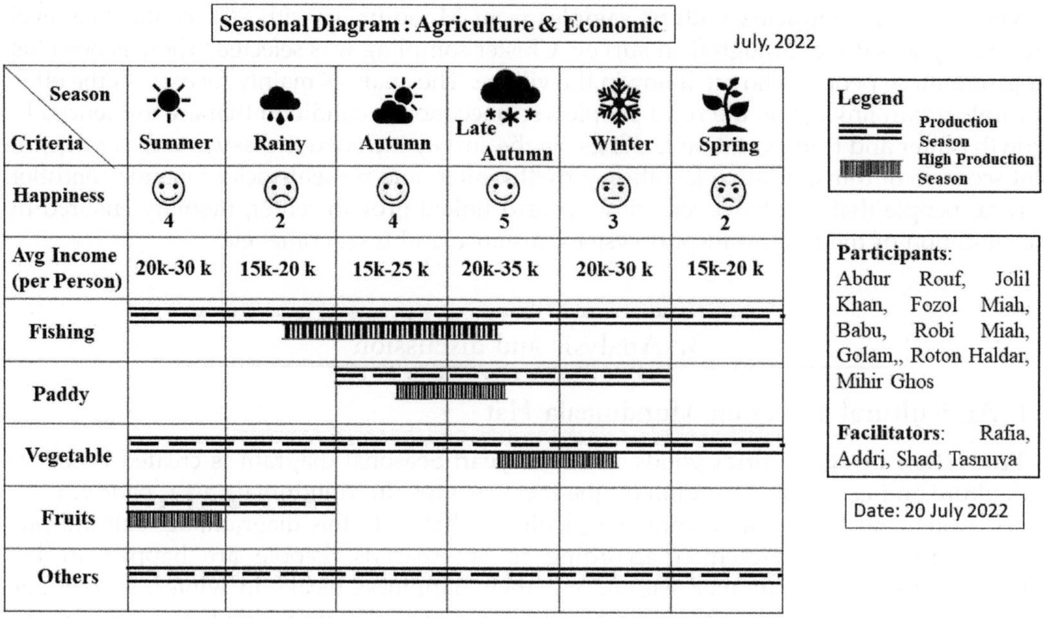

Seasonal Diagram : Agriculture & Economic

July, 2022

Criteria \ Season	Summer	Rainy	Autumn	Late Autumn	Winter	Spring
Happiness	4	2	4	5	3	2
Avg Income (per Person)	20k-30 k	15k-20 k	15k-25 k	20k-35 k	20k-30 k	15k-20 k
Fishing						
Paddy						
Vegetable						
Fruits						
Others						

Legend
Production Season
High Production Season

Participants: Abdur Rouf, Jolil Khan, Fozol Miah, Babu, Robi Miah, Golam,, Roton Haldar, Mihir Ghos

Facilitators: Rafia, Addri, Shad, Tasnuva

Date: 20 July 2022

FIGURE 4.6 Seasonal diagram of agriculture and economy.

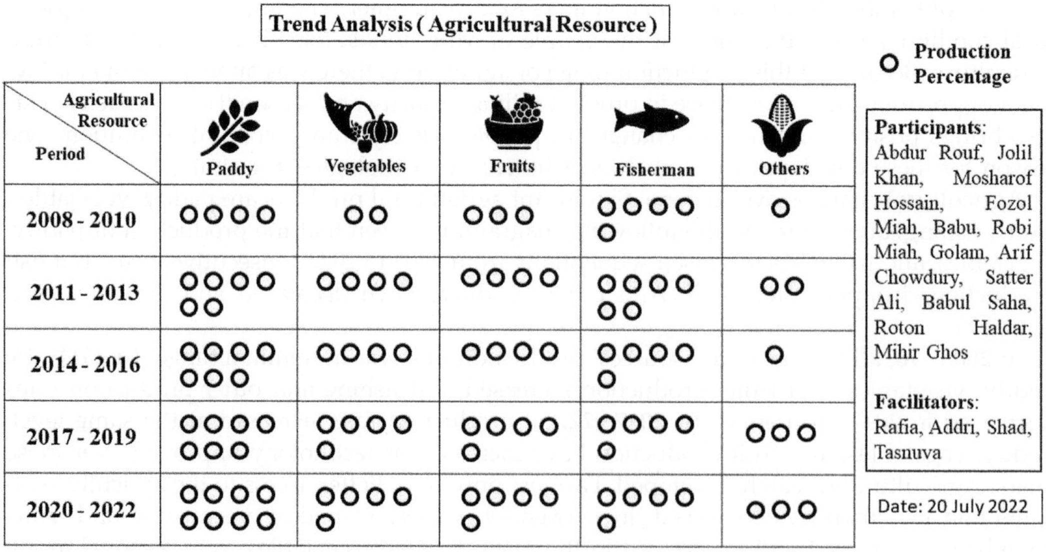

Trend Analysis (Agricultural Resource)

O Production Percentage

Period \ Agricultural Resource	Paddy	Vegetables	Fruits	Fisherman	Others
2008 - 2010	OOOO	OO	OOO	OOOO O	O
2011 - 2013	OOOO OO	OOOO	OOOO	OOOO OO	OO
2014 - 2016	OOOO OOO	OOOO	OOOO	OOOO O	O
2017 - 2019	OOOO OOO	OOOO O	OOOO O	OOOO O	OOO
2020 - 2022	OOOO OOOO	OOOO O	OOOO O	OOOO O	OOO

Participants: Abdur Rouf, Jolil Khan, Mosharof Hossain, Fozol Miah, Babu, Robi Miah, Golam, Arif Chowdury, Satter Ali, Babul Saha, Roton Haldar, Mihir Ghos

Facilitators: Rafia, Addri, Shad, Tasnuva

Date: 20 July 2022

FIGURE 4.7 Trend analysis of agricultural resource.

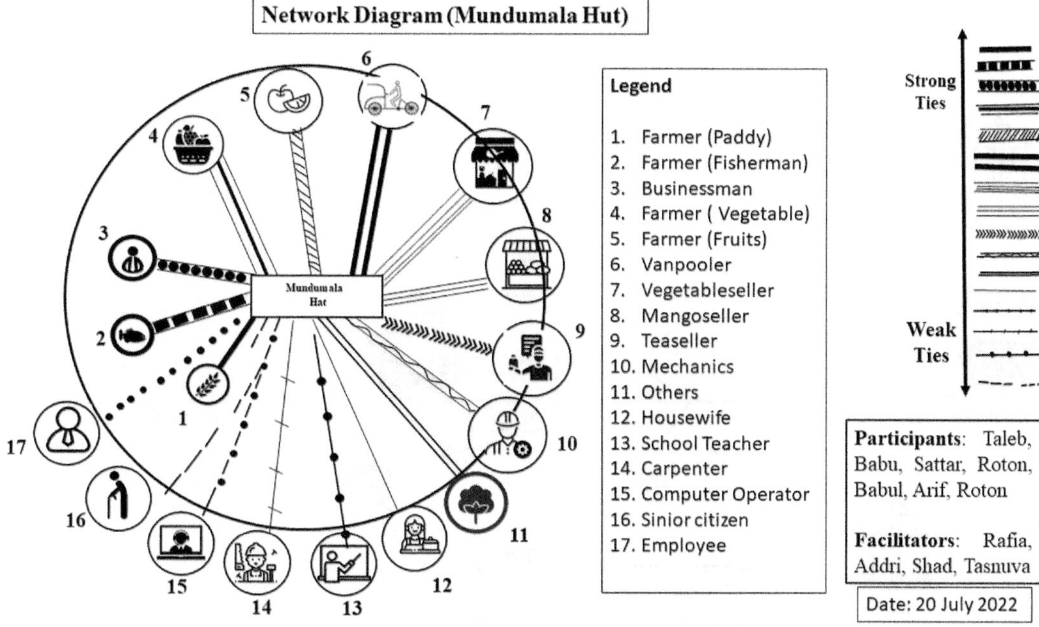

FIGURE 4.8 Network diagram.

3.2 Effect of rural–urban linkage on Mundumala Hat

Relationship between different occupations with Mundumala Hat: This relationship is represented by network diagram. The following diagram represents the relationship between the Mundumala Hat and people of different occupations. The diagram also interprets the strong or weak relationship between them. Here, the strongest relationship is with the paddy farmer; then the relation gets weaker with every step. The sequence of the relation is fishing farmer, business, vegetable farmer, fruit farmer, van pooler, vegetable seller, mango seller, tea seller, mechanics, others, housewife, school teachers, carpenter, computer operator, senior citizens, employee, etc (Fig. 4.8).

Strong ties sector: Farming sector has strong relationship with the growth center of Mundumala Hat. Agriculture-dependent occupations have significantly stronger relation as it is a rural growth center and most of its areas are used for agriculture purposes. Thus, most of their economic activities move around farming. Farming is the primary activity of rural area, so the Mundumala Hat is more connected to the farmers. Agriculture is the main driving force of rural area. The development of the area moves around the agriculture, so they have closer relation to it and their social, cultural, and economic life moves around agriculture. After producing good they trade their product in the market where their product goes outside the area. Their activities expand from primary activities.

Medium ties sector: Vanpooler, vegetable seller, tea seller, and mango seller have medium types of connection with the growth center of Mundumala Hat.

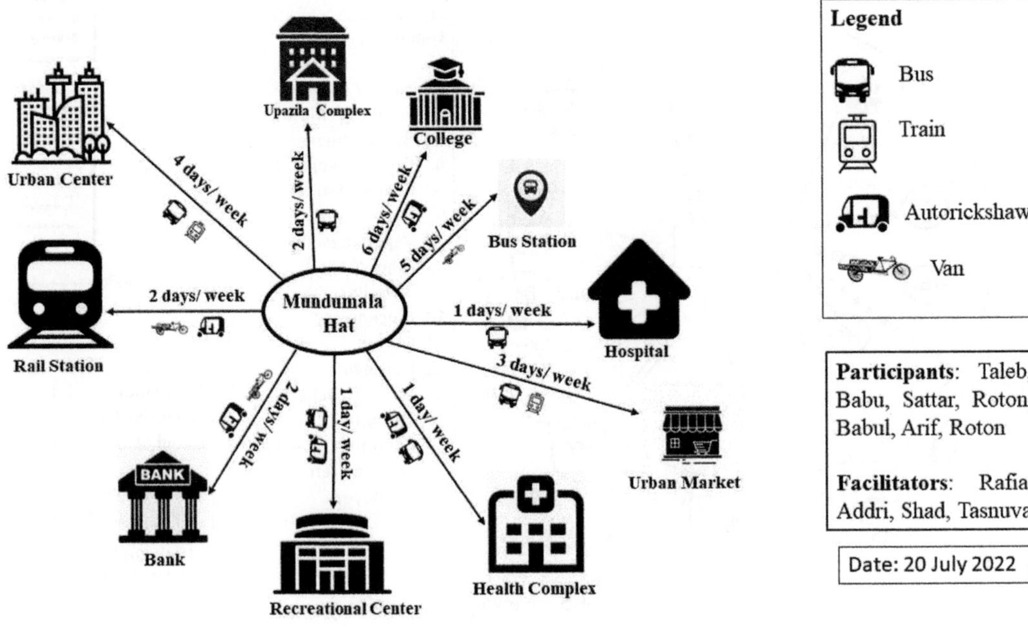

FIGURE 4.9 Mobility map.

Weak ties sector: School teacher, carpenter, computer operator, senior citizen, and employee have weak relationship with the growth center of Mundumala Hat, as this sector contributes to the socioeconomic development in very low range.

Movement pattern of the community of Mundumala: The following diagram illustrates the movement pattern of the community of Mundumala and its surrounding area by mobility map PRA techniques. People of Mundumala move for different types of services and opportunities (Fig. 4.9).

3.2.1 *Frequency of visit in different places*

- **Visit each day:** Boys and girls visit for their education purpose 6 days per week.
- **Visit 5 days per week:** Community has to visit 5 days per week at bus station to travel to the urban center and market.
- **Visit 4 and 3 days per week:** Community has to visit urban center and urban market for their needed purpose.
- **Visit 2 days per week:** Community has to visit 2 days at bank for transaction purpose and at rail station for traveling and delivering their selling product outside of their region.
- **Visit 1 day per week:** Community visiting the recreational center, health complex, and hospital for mental and physical purposes.

3.2.2 *Mode of communication*

- **Auto rickshaw and van:** They visit bank, rail station, recreational place, health complex, bus station by auto rickshaw and van. These places are within Mundumala area and not so far from the study area.
- **Bus:** They travel urban center, urban market, hospital, and urban complex by bus. These places are far distance from the Mundumala Hat.
- **Train:** They sometimes travel to the urban center and market by train.

Very flexible movement of Mundumala communities are urban, college, bus station, and market area based. Because of this flexible movement, there is an increase in the connection of Mundumala growth center with the other necessary facilities and service opportunity. This connection also increases their economic conditions, employment sector, and literacy sectors.

3.3 Availability of service and opportunities surround Mundumala Hat

The availability of services and opportunities is represented by services and opportunity map. This diagram represents the distance of various services from the Mundumala Hat. Service facilities are bus stations, hospital, town market, health complexes, residential center, bank, rail station, urban center, Upazila complex, and college. It shows not only the available importance services but also people's feelings of deprivation and their needs (Fig. 4.10).

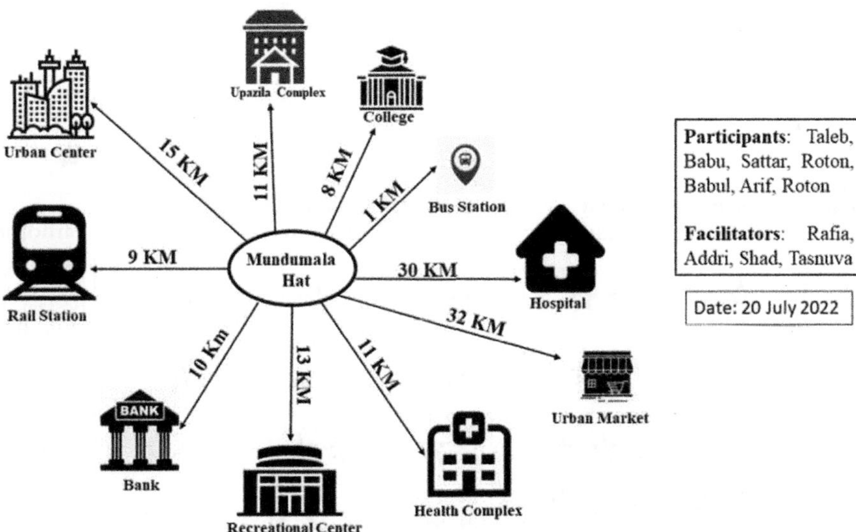

FIGURE 4.10 Service and opportunity map.

3.4 Available services

Bank is the most important service of any area. In Mundumala Hat, two or three different banks are situated. For different monetary transactions, the people of Mundumala area go to the nearest bank, which is about 10 km from the study area. Another important service is the hospital that is situated far from Mundumala Hat, and the distance range of this place is 30 km. Maximum people go there because of their serious condition of health. Otherwise, they get their treatment from a health complex, which is about 11 km distance from Mundumala Hat. For going to the urban center and market, people mostly use bus and train. The bus and train station are close to the Hat area which is 1 km and 9 km far from the study area, respectively. To pursue their education facility, they travel 8 km of distance every day.

Opportunities for Mundumala growth center: To sell their agricultural product and other economic product outer of the rural area for economic growth, they traveled to the urban center and urban market, which are about 15 and 32 km from Mundumala area, respectively. These two centers are their opportunity. It is observed that hospital, town market, and urban center are the most distant; these reflect that they have good reasons to be more dependent on the Hat area. Their small-scale requirements are fulfilled, so they do not feel the urge to go far. Other day-to-day needs like college, bank, rails station, and bus station are at a moderate distance.

3.5 Changes in road network facilities over the period

The road network facilities are segregated into various facilities like road surface, road width intersection, pedestrian facility, parking, etc (Fig. 4.11).

Trend Analysis (Road network)

☆ Development percentage

Road Surface / Period	Road Surface	Road Width	Intersection	Pedestrian Facility	parking
2008 - 2010	☆	☆☆	☆	☆	☆
2011 - 2013	☆☆	☆☆	☆	☆☆	☆
2014 - 2016	☆☆	☆☆☆	☆☆	☆☆	☆☆
2017 - 2019	☆☆☆☆	☆☆☆☆	☆☆	☆☆☆	☆☆
2020 - 2022	☆☆☆☆	☆☆☆☆	☆☆	☆☆☆	☆☆☆

Participants: Abdur Rouf, Jolil Khan, Mosharof Hossain, Fozol Miah, Babu, Robi Miah, Golam, Arif Chowdury, Satter Ali, Babul Saha, Roton Haldar, Mihir Ghos

Facilitators: Rafia, Addri, Shad, Tasnuva

Date: 20 July 2022

FIGURE 4.11 Trend analysis of road network.

The changes over the era from the year 2008–22 are presented in the diagram. Road surface increased over the year, and significant change was observed after 2017. Road widths are expanded year by year to cover increasing traffic demand. Road development followed by the pedestrian facility and parking has been developed every forwarding year. Road connectivity and transportation development are musts. Bangladesh, as a developing country, has focused on road infrastructure development in recent years. And road connectivity can circulate and accelerate economic growth. It also helps people to seek better employment opportunities and to look for better services and facilities.

3.6 Effect of growth center on household

Market area development: At rural area, 43% of the shop are kaccha bazar. People usually sell their household (vegetables, fish, rice) products to the markets to earn their livelihood.

After that, 26% of shops are groceries and gradually the percentage decreases. Eventually, we observed a less number in the wholesale shop (8%) as fewer people have that sufficient product to sell in the market (Fig. 4.12).

As Mundumala Hat is a rural market, agricultural products like paddy and vegetables are sold in the market, so kaccha bazar is established having most of that area. Eventually, the vegetables are transported to nearby villages.

3.7 Agriculture development

Due to urban and rural linkage, most of the agricultural land in the rural area has changed. For example, due to the linkage, they are able to collect any other materials from the urban

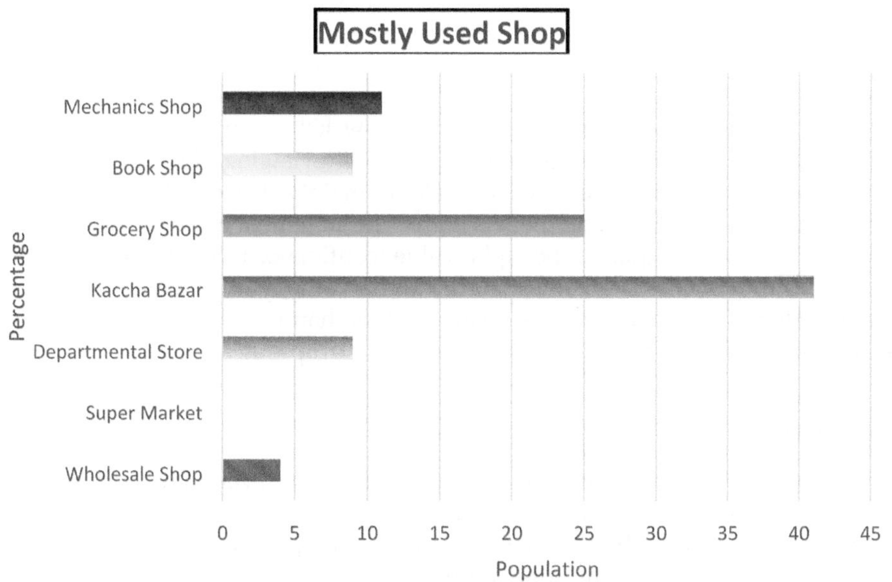

FIGURE 4.12 Mostly used shop of Mundumala area.

area, including fertilizers required for land, so that there is no obstacle in improving the quality of land, and agricultural land is also increasing. With the increase in agricultural land, they are now selling agricultural products to local customers and outside areas. In this way, economic development is also increasing. From cross-tabulation chi-square analysis, the P-value (probability) is less than the alpha value (confidence level), $P = .05844715$. In that case, it can be said that changes are taking place in most agricultural lands, resulting in increased production. As a result, agricultural products can now be sold to outside and local consumers.

3.8 Household development

The homelessness problem has been reduced with rural economic development. Improvement in market regulation, transportation, and road network has affected rural–urban linkage, leading to changes in people's living. The relation between the house construction material like tin, bamboo, brick, wood, etc., and how many years ago the people of that area reconstructed their houses were analyzed by chi-square. The P-value of the chi-square test represents the relationship between the selected variables. The two variables are associated if the probability value is less than or equal to the significant level (alpha value $= 0.5$) according to the decided significance level. If the P-value is greater than the alpha value (0.5), you conclude the variables are independent. The P-value is 0.62, which means the time frame and house reconstruction material are not associated.

3.9 Education development

Most people prefer General Education System. Only 6% of people prefer Vocational Education System, and the rest prefer General Education System.

At present, in most of the rural areas of Bangladesh, girls are given in marriage at a young age. Due to this problem, the dropout rate of female students is higher in the field of education. According to 2011 data, 54.10% of boys in Bangladesh get educational opportunities, whereas the rate of girls is comparatively less, i.e., 49.40%. In particular, girls have fewer educational opportunities than boys in rural areas. Reviewing the cross-tabulation chi-square analysis, it is seen that the level of students mainly drops out at higher secondary level, and the number of girls among them is more. The statement of P-value is "If the P-value (probability) is less than or equal to the alpha value (confidence level), then the two variables are associated. If the P-value is greater than the alpha value, you conclude the variables are independent." Here, in the analysis, the P-value is less than the alpha value, $P = .442287$. In that case, the higher secondary level girls are likely to drop out.

3.10 Satisfaction

The household people were asked about their satisfaction level with the overall development of the area. About 86% of respondents' reactions were average about the change which indicates the reduction of the homelessness problem, 10% people are satisfied and 4% reaction is bad about the development, they are not satisfied. The graph shows that the people of the area require more involvement from the concerned authority about solving their problems. Though they feel their situations have improved from the past. This has an impact on their life and community behavior, when the user of the area is satisfied, they are more concerned about protecting the resources and working for the society. So, the satisfaction level of the community is also important in terms of assessing the condition.

3.11 Effect of Mundumala Hat

This chapter includes the effect of the Mundumala Hat on the socioeconomic development of the Mundumala area people and expresses the linkage between rural and urban areas. Socioeconomic development reduces the homelessness problem. For findings on this effect and the causes of these effects on the Mundumala area cause–effect diagram, impact diagram was used.

Cause–effect diagram: The socioeconomic condition of the Mundumala area is increasing, and the homelessness problem is reducing day by day. The causes behind the development are the development of the growth center. The following figure presents the cause–effect diagram of Mundumala Hat on socioeconomic development done by local people, i.e., farmers, buyers, sellers, and dealers (Fig. 4.13).

3.12 Causes behind the development

- **Infrastructure**: Infrastructural development that occurred from the road network, market area, waste management, and households. The infrastructure development of the

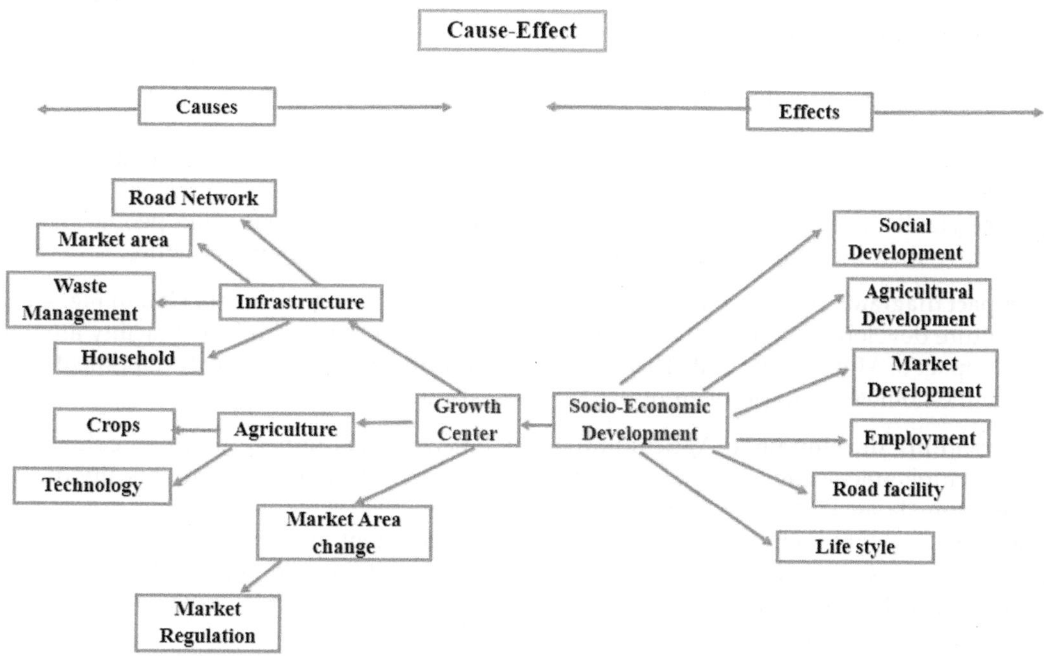

FIGURE 4.13 Cause–effect diagram.

Mundumala Hat increased rapidly from 2009. As a result, people's transportation links create interlinkage trade and develop their economic condition.

- **Agriculture**: Agriculture is classified into crops and technology. After increasing technological availability farmers are introduced to advanced equipment and agricultural method which are the reason for increasing crop production. As a result, farmers can export their crops and develop their economic conditions.
- **Market area change:** Another factor related to the growth center is market area change which is caused by market area regulation. Supply and demand maintenance are an important factor to develop a growth center. In Mundumala Hat, supply and demand of production was maintained, as a result growth center developed day by day. Also, government market regulation is another reason for growth center development.

3.12.1 *Effect of the development*

People's income ranges have increased because of the growth center development. Also, facilities and agricultural change has occurred for this reason. Some important effects are

- **Social development:** Because of the development of social infrastructure, educational facilities, the literacy rate has increased. Also, people used to communicate in urban areas and create a linkage between rural and urban areas.
- **Agricultural development:** Crop production increased, and people moved to secondary activity, i.e., the rice industry are grown, and the processed product can be exported outside the area, especially in urban areas.
- **Market development:** Because of the growth center development, market area is expanding. Also, the activity of the market area has increased in different sectors. As a result, polarization effect is created.
- **Employment:** Mundumala area people moved from primary activity to secondary activity, and a small-scale industry is established. As a result, employment opportunity is created.
- **Road facility:** Road infrastructure development, street light, and parking facilities are developed.
- **Lifestyle:** Because of growth center development, employment generation people improve their lifestyle, living facilities also improved household infrastructure.

Impact diagram: The villagers create the major impact of the Mundumala Hat. In Fig. 4.14, agriculture development, market development, social development, and road facility are the main impact of Mundumala Hat. These impacts enforce the socioeconomic development and reduce the homelessness problem of the Mundumala area.

- **Agriculture development:** Agricultural development–based impact increased cultivation, employment rate, economic condition, and proper lifestyle in the community of Mundumala area.
- **Market development:** Because of market area development, the unemployment problem decreased. As a result, economic condition increased and lifestyle improved.
- **Social development:** Both market and social development has equal impact on Mundumala growth center.

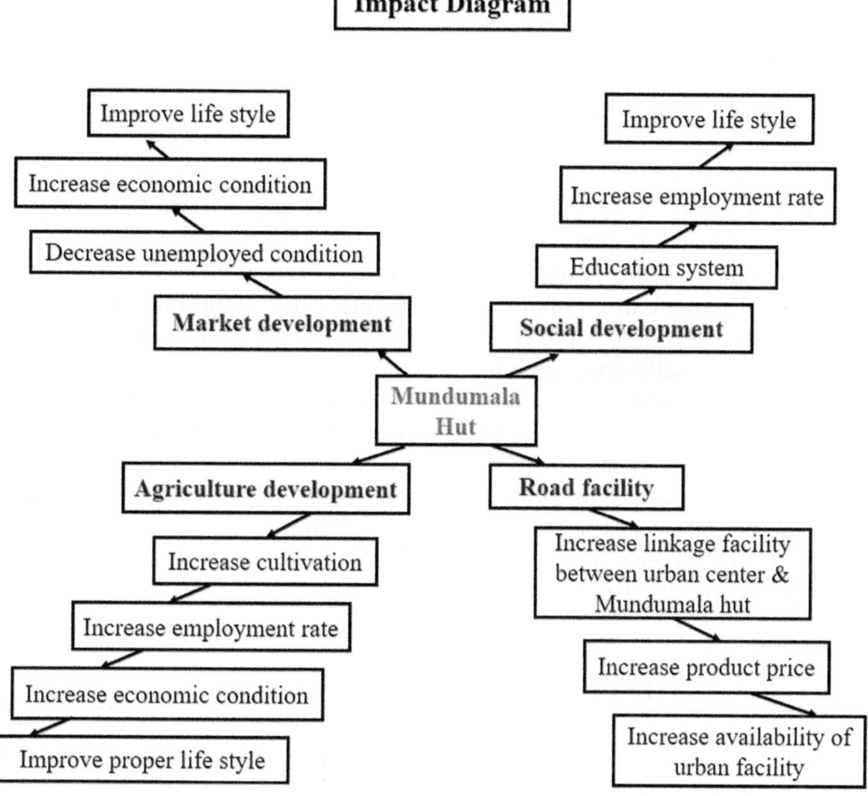

FIGURE 4.14 Impact diagram.

- **Road facility:** Road facility impacts other things like linkage facility between urban center and Mundumala; product price and availability of urban facilities are increased. Later, creating urban center and Mundumala linkage facility increased their service and opportunity facilities that help to reduce unemployment conditions. Technological things are easily accessible for agricultural purposes after this road facility impact (Fig. 4.14).

3.12.2 *Effects on homelessness*

It is seen that the issue of poverty in Bangladesh is closely tied to a significant number of landless households, resulting from the transition away from an agricultural, rice-centric system. This shift has directly endangered rural household incomes, leading to a rise in rural-to-urban migration and contributing to the growing rural and urban homeless population.

The establishment of the Mundumala Hat, a growth center, has played a pivotal role in addressing these challenges. According to discussions with local groups, the growth center has generated substantial economic opportunities, resulting in a noticeable decrease in the homeless population. Several key factors have driven this positive change. From the analysis of PRA tools, it is found that there is an emergence of job opportunities in secondary activities, and small-scale industries have provided a stable income source for local residents.

Moreover, developing essential infrastructure, including improved road networks and market facilities, has enhanced access to resources and markets, thereby fostering trade, increasing income levels, and improving the socioeconomic condition. The significant improvements resulting from the rural–urban linkages have not only alleviated the issue of low income but have also improved living standards in rural areas, discouraging the departure of residents to densely populated urban centers.

The case of Mundumala Hat illustrates the vital role of local economic development and infrastructure enhancement in combating homelessness. By integrating the growth center with regional and global markets, the rural community has diversified its livelihood options, reducing reliance on unstable income sources.

Furthermore, the improved connectivity facilitated by upgraded road infrastructure has strengthened the local economy by enabling smoother transportation of goods to broader markets. This emphasis on sustainable and inclusive development in Mundumala Hat serves as a robust model for addressing global economic uncertainties and vulnerabilities at the community level. The success of the growth center highlights the potential of localized initiatives in mitigating the adverse impacts of global economic shifts and drivers on vulnerable communities, thereby curbing the risk of homelessness.

4. Findings and recommendations

After analyzing the growth center with PRA methods, the importance of growth centers in driving socioeconomic development in rural areas is found. The Mundumala Hat has been identified as one such growth center due to its instrumental role in creating employment opportunities and establishing rural–urban linkages. This has contributed significantly toward reducing migration and homelessness in the region. Moreover, by providing necessary facilities in the area, such as markets, healthcare centers, and educational institutions, people are less inclined to move toward cities. Therefore, the development of growth centers is essential in addressing the challenges faced by rural areas and promoting sustainable and inclusive development. The PRA method has been used to identify the local needs and preferences of the community. It is a useful tool for policymakers to consider the people's opinions and requirements while formulating policies. However, despite the positive aspects of growth centers, there are certain challenges that need to be addressed. For instance, the lack of waste disposal and drainage facilities in Mundumala area poses a significant challenge to the region's sustainable development. The authorities need to address this issue by providing drainage facilities that meet the area's requirements. Furthermore, the increase in secondary activities in the region has led to a rise in vehicular movement, making the current road infrastructure insufficient for total transportation. Therefore, there is a need to improve the road conditions and provide proper road facilities to meet the transport demand. Overall, growth centers like Mundumala Hat are crucial for driving socioeconomic development in rural areas and reducing homelessness. However, policymakers need to address the challenges faced by these growth centers, such as waste disposal, drainage facilities, and road infrastructure, to promote sustainable and inclusive development in the region. The PRA method can be used to identify the local needs and preferences of the community, and policy-makers should consider these opinions while formulating policies.

Findings from this chapter underscore the critical importance of prioritizing local needs, particularly the establishment and development of growth centers, as a pivotal strategy in effectively reducing homelessness. The analysis indicates that a deliberate focus on local initiatives, such as growth centers, is not only beneficial but also imperative for combating the homelessness problem. Consequently, it is recommended a strategic emphasis on the expansion and enhancement of growth centers as an approach to address homelessness within studies context and potentially extend this strategy to other regions facing similar challenges. This approach not only bridges the socioeconomic divide between rural and urban areas but also serves as a potent driver for creating sustainable, stable communities that are less susceptible to homelessness.

5. Conclusion

Bangladesh's rural areas determine greater rates of population increase and a lower rate of economic growth. In terms of infrastructure and economic activity, Bangladeshi rural areas remained undeveloped after independence. Especially rural economic conditions are underdeveloped, and the main reason is the lack of a proper growth center. Due to this it can be seen that rural people migrate to urban areas for employment and economic development and they face homelessness problems. Rural–urban linkage plays an important role to establish a growth center. In this research study, growth centers that serve as a hub for connecting rural and urban areas can be crucial to the rural economy and help foster socioeconomic growth. Homelessness may be reduced by concentrating on rural economic development. Both rural and urban residents benefited from this way of life and low-income rural people also get relief from the homelessness problem. On the other hand, it is hardly possible for us to know the impact of the growth center on socioeconomic conditions without knowing the socioeconomic condition of that area. For this reason, PRA seems to have a strong tool to provide the opinion of the local people and give the proper value to their notion. The socioeconomic situation of the inhabitants of Mundumala, which appears to have a growth center, primarily depends on that. PRA allows for gathering community input to understand better the effects of socioeconomic conditions on the rural population and the connections between urban and rural hubs. The rise of trade, transportation, and communications are indications of economic activity and support further economic development. Indicators like life expectancy, literacy, and employment represent socioeconomic development which reduced local people's migration to urban areas and reduces homelessness. So, strengthening the links between rural and urban areas through growth centers is very essential. It is also found that the overall income of the Mundumala area flourished which helped them to improve their social status as well as the nutrition facility. Besides, it has provided some crucial concepts such as most of them are dependent on agriculture, most of them are lower middle class, and they face various obstacles which have been reduced due to the growth center of that area and it has strengthened through linkage; moreover, the networking problem with their urban area is eliminated and their socioeconomic conditions have developed. So, this study of Mundumala represents a growth center that is an important hub of rural areas and also identified issues that required to be solved to improve more. Furthermore, the study can be used for more ways to develop the growth center policy developments and look for other measures for strengthening rural–urban linkage. To solve the homelessness issue, the growth center should be given enough focus to develop.

References

Barua, U., & Akter, R. (2015). Rural-urban linkage through growth centers in Bangladesh. *Bangladesh Research Publications Journal, 10*(4), 314–320.

Cloke, P., Milbourne, P., & Widdowfield, R. (2002). *JSTOR: Rural homelessness* (pp. 1–26). https://doi.org/10.2307/j.ctt1t89cd7.4

DESA, UN. (2021). World social report: Reconsidering rural development. In *World social report*.

Dudek, M., & Wrzochalska, A. (2017). Making development more sustainable? The EU cohesion policy and SOCIO-economic growth of rural regions in Poland. *European Journal of Sustainable Development, 6*(3). https://doi.org/10.14207/ejsd.2017.v6n3p189

Gebre, T., & Gebremedhin, B. (2019). The mutual benefits of promoting rural-urban interdependence through linked ecosystem services. *Global Ecology and Conservation, 20*, e00707. https://doi.org/10.1016/j.gecco.2019.e00707

Kafy, A.-A., Mohiuddin, H., Hossain, N., & Roy, S. (2019). Identifying the growth centers connectivity based on road network and market accessibility in the north-west part of Bangladesh. *Journal of Bangladesh Institute of Planners, 10*, 105–115.

Khan, M. (2001). *Rural poverty in developing countries: Implications for public policy. Economic issues*. International Monetary Fund, ISBN 978-1-58906-006-7. https://doi.org/10.5089/9781589060067.051

Mondal, B. K., & Das, K. (2010). Role of growth center: A rural development perspective. *Journal of Bangladesh Institute of Planners, 3*, 129–141.

Moseley, M. J. (1973). The impact of growth centres in rural regions—I. An analysis of spatial "patterns" in Brittany. *Regional Studies, 7*(1), 57–75. https://doi.org/10.1080/09595237300185051

O'Regan, K. M., Ellen, I. G., & House, S. (2021). How to address homelessness: Reflections from research. *The Annals of the American Academy of Political and Social Science, 693*(1), 322–332. https://doi.org/10.1177/0002716221995158

Richter, R., Fink, M., Lang, R., & Maresch, D. (2019). *Key takeaways*. Informa UK Limited, ISBN 9781351038461. https://doi.org/10.4324/9781351038461-6

(2021). Rural population (% of total population)—Bangladesh. World Bank.

Rural to urban migration.(2012). https://www.studysmarter.co.uk/explanations/human-geography/population-geography/rural-to-urban-migration/.

Shah, I. A., & Baporikar, N. (2010). Participatory rural development program and local culture: A case study of Mardan, Pakistan. *International Journal of Sustainable Development and Planning, 5*(1), 31–42. https://doi.org/10.2495/SDP-V5-N1-31-42

Shah, S. M. (1974). Growth centers as a strategy for rural development: India Experience. *Economic Development and Cultural Change, 22*(2), 215–228. https://doi.org/10.1086/450706

Sharma, P. K. (2021). Rural growth centres for regional development in Bundelkhand region of Madhya Pradesh, India. *Transactions of the Institute of Indian Geographers, 43*(1), 31–42.

Tacoli, C. (1998). Rural-urban interactions: A guide to the literature. *Environment and Urbanization, 10*(1), 147–166. https://doi.org/10.1630/095624798101284356

Tacoli, C. (2003). The links between urban and rural development. *Environment and Urbanization, 15*(1), 3–12. https://doi.org/10.1177/095624780301500111

CHAPTER

5

Migrants and homelessness: Life on the streets in urban India

Rashmi Rai[1], *K. Lakshmypriya*[1], *Pallavi Kudal*[2] *and Roli Raghuvanshi*[3]

[1]**Christ University, School of Business and Management, Bangalore, Karnataka, India;** [2]**Balaji Institute of International Business, Sri Balaji University, Pune, Maharashtra, India;** [3]**Shyam Lal College (Evening), University of Delhi, New Delhi, Delhi, India**

1. Introduction

India has a sizable and diversified migrant population, essential in determining the nation's social, cultural, and economic environment. India has a long migration history, with records of people traveling between locations for trade, exploration, conquering, and other purposes. People or entire families are forced to migrate to cities for a means of life due to structural and social inequalities, expanding poverty, declining livelihoods, and other issues. In the course of this process, some of them end up homeless. To provide people with basic facilities, safety, and security, the government is becoming increasingly concerned about the rapidly rising rate of urban homelessness. Homelessness has several social, emotional, and psychological components besides the apparent lack of a home. According to the Census of Census India (2011), there are 139 million internal migrants in India overall, including those moving between and within states. The Economic Survey of India 2017 estimates that between 2011 and 2016, there were about nine million interstate migrants in India (Government of India, 2017). According to Census 2011 data, 37% of all interstate migrants, or 20.9 million persons, are from the states of Bihar and Uttar Pradesh (Government of India, 2017). The two major Indian cities of Delhi and Mumbai, which together have a combined population of 9.9 million, are the ones that draw the most migrants (Hindustan Times, 2016). Academically, there are still open debates about the reasons for and effects of Migration and, subsequently, as how mobility relates to poverty and inequality. There are still disciplinary disparities in methods for comprehending Migration. In this article, the authors present their view of (labor) migration processes based on the worldwide literature, emphasizing

Homelessness to Hope
https://doi.org/10.1016/B978-0-443-14052-5.00005-7

that although there are lessons to be learned about Migration in general, the connections between migration and poverty, is highly context-specific. Indian migrants confront a variety of difficulties and dangers. The informal economy employs many migrants, who frequently deal with exploitation, low pay, hazardous working conditions, and a lack of social security benefits. Additionally, they have trouble accessing crucial services like healthcare, education, and others. Migrants frequently reside in slums or unofficial colonies, lacking essential utilities, and are socially excluded. In addition, migrants experience discrimination in their new communities due to linguistic and cultural limitations. The aim and approach of this study are to study the nature of migration and the motives for such migration, as well as whether present policies are suitable to solve the growing problem of migrant exploitation in India. A more holistic approach to migration and its issues has also been thoroughly examined. In addition, policy framework revision to minimize the negative aspects of migration is briefly discussed.

1.1 Migration and homelessness

Urban homelessness is the absence of a stable, reliable, and suitable place to spend the night for a person or household. It includes a variety of living situations, such as squatting, using homemade shelters, residing in vacant buildings, or staying in short-term lodgings like emergency shelters or hostels. Poverty, unemployment, a lack of affordable housing, mental health problems, substance addiction, domestic violence, and social exclusion are just a few of the complicated causes of homelessness. Homelessness is a term used to describe a person who does not have a place to live or has spent the night on the streets. According to the Census of India, "houseless people" are individuals who do not reside in "census houses." The latter is used to describe "a structure with a roof." According to the United Nations, "homeless" in 1999 included "those sleeping without shelter, in constructions not meant for habitation, and in welfare institutions." In India, most people without homes are found in locations like roadside ditches, sidewalks, drainage pipes, stairwells, platforms, etc. The concept of homelessness in India is multifaceted and encompasses individuals who do not live in regular residences. Universal Declaration of Human Rights refers to homelessness as people without a regular dwelling due to their inability to acquire safe and adequate housing (Jha & Kumar, 2016). However, the definition of homelessness in India, as per the Census India (2011), includes those who do not live in Census houses but stay in various open spaces such as pavements, roadsides, railway platforms, and temples(Jha & Kumar, 2016). *One of the narratives while studying the state of homelessness was that She was swinging her water bottle, holding the hands of her elder brother and walking down to her school in an upscale locality. Mira is one of the faces of a happy child who has adapted to the home away from her native in the foothills of Siliguri. Her father works as a security guard in one of the upscale apartment complexes in Bangalore and her mother works in a garment manufacturing unit. They are one of the faces of joblessness and poor living conditions in their native land and had to move out to make both ends meet. Though they survived out of the pandemic life was harsh during those times but they had nothing to go back to in their native.*

The worst part of being homeless isn't just finding a place to live; it's losing one's sense of value in life, losing one's dignity, and constantly being exploited by the public, the police, and the laws that place them in a vulnerable position. According to Waldron (1991), homelessness in contemporary entrepreneurial, neoliberal cities is merely a matter of not having a place to call one's own. Because of the division and segregation of urban environments, homeless migrants highlight the contradiction of urbanization. Despite being unseen, they comprise a sizable portion of the Metropolitan population. Being lost in the city is a case of minimum citizenship, devoid of the right to the city, and is subject to ongoing violence. The expanding claim for the "citizen's right" to various public areas like walkways, pavements, and parks challenges their existence on the streets and pavements (where they live). Many academics have demonstrated how underpaid and exploitative work in the informal sector can be, frequently involving a network of numerous contractors or agents (Mosse et al., 2005). Most migrants start and remain homeless, leading lives of misery and dislocation, as a result of which they are denied rights and privileges. They are arguably the most anonymous, silent, and unseen segment of a city's population. According to Veness (1993), middle-class worldviews delegitimize lower-class lifestyles, making "the poor" seem weird and distant. A fascinating perspective regarding how people who are homeless are perceived in modern society is covered by Parsell (2010). He contends that the earlier literature on homelessness had been focusing too much on describing the qualities of persons who are homeless in his piece titled "Homeless is what I am, not who I am?" They are not only labeled as the "other" due to what they lack but also because "they have lost their identity." There is a lot of emphasis on describing how of homeless people are, they are unfortunately also identified as the "other" not only because of what they lack, but also because "they have become depersonalized" (Datta, 2020). One of the examples that can be cited is the one of Delhi. It has one of the greatest proportions of interstate migrants in its population: 40% (or more than 63 lakh) of Delhi residents are migrants, with overhalf hailing from Uttar Pradesh. A huge proportion of Delhi migrants are poor, unskilled, or semiskilled artisanal rural residents. The effects of homelessness can be devastating for both individuals and society. Homeless people frequently face considerable obstacles when accessing needs like food, clean water, sanitizers, healthcare, and education. They are more susceptible to problems with their physical and mental health, drug usage, aggression, and exploitation. Public services and resources, such as emergency shelters, healthcare facilities, and law enforcement, are also strained by homelessness. Internal migration has far-reaching repercussions for migrants and their families. Typically, migrant labor lives in poor conditions, with insufficient access to drinking water and essential utilities. Many labor migrants have not registered and hence do not have access to the PDS system or ration cards.

The National Statistical Office started a survey in April 2017 to estimate employment and unemployment indicators in urban areas and rural areas. The Periodic labor force survey conducted from July 2020 to June 2021 collected additional information on migration particulars and temporary visitors. However, due to the COVID-19 pandemic, fieldwork faced disruptions, and data collection was completed by September 2021. The sample design used a rotational panel scheme in urban areas and random sampling in rural areas. The total number of surveyed migrants during this period is shown in Fig. 5.1 (PIG, 2022). Total migration percentage was found to be 28.9% from rural and urban areas and male and female. The survey includes migrants with different usual places of residence (UPR) and temporary visitors staying temporarily in a location different from their UPR.

Fig. 5.2 provides a detailed breakdown of the reasons for migration among males and females. As per this survey 22.8% of male migrants were in search of employment. This refers to individuals who migrate to urban or other regions for better job opportunities or to improve their current employment prospects. Further 20.1% males migrate in search of better

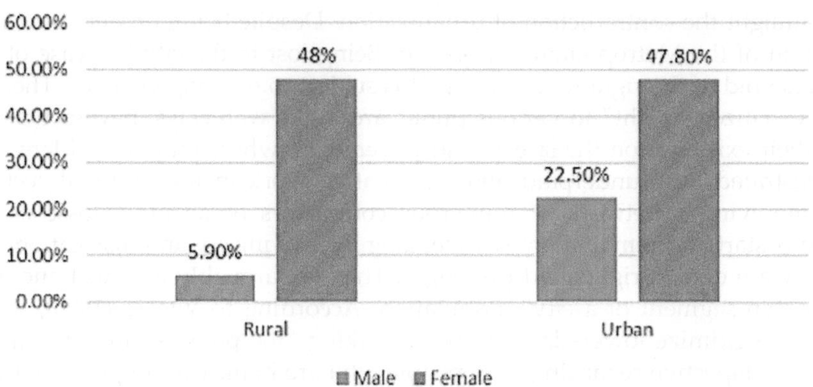

FIGURE 5.1 Migration rate in percentage from PLFS in July 2020–June 2021.

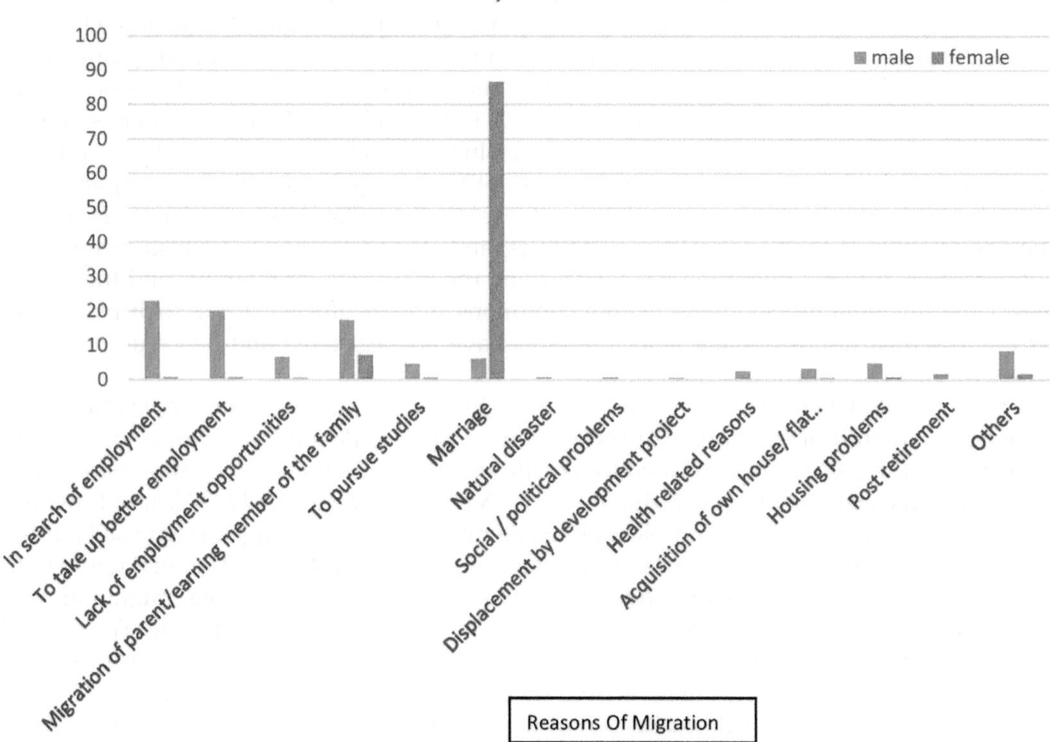

FIGURE 5.2 Percentage distribution of male and female migrants for various reasons.

employment opportunities. This category includes individuals, particularly males, who migrate for specific employment, work, or business opportunities. It encompasses factors like proximity to the workplace, better job prospects, or job transfers. 6.7% males migrate due to job losses or the closure of businesses or industries, primarily driven by a lack of employment opportunities in their place of origin. 17.5% males and 7.3% females migrate due to migration of parent/earning members of the family. Only 4.7% males migrate to pursue educational opportunities in different regions, often for higher studies or specialized courses. Maximum, i.e., 86.8% females migrate due to marriage, as they move to their husband's residence after marriage. Another important reason for migration highlights in the figure above is the housing problem. 4.8% of males migrate due to housing problems.

1.2 Factors leading to urban migration

Internal migration is more widespread in modern India than international migration. Internal migrants make up the majority of migrants in India; they cross internal boundaries to find work in industries, including construction, manufacturing, domestic work, agriculture, and services. These migrants generally relocate from rural to urban regions. Uttar Pradesh, Bihar, Rajasthan, and Madhya Pradesh are among India's main migration routes due to their high outmigration rates. Better employment prospects, housing circumstances, educational chances, and other causes. Both push and pull variables influence internal Migration. Both motivation and removal of variables affect internal migration. In contrast to pull factors like the promise of better jobs, education, and living standards in metropolitan areas, push elements in rural areas include poverty, a lack of work possibilities, natural disasters, political unrest, and social disputes. India has a massive internal migration problem. There were over 450 million internal migrants in India in 2011, which made up about 37% of the entire population, according to the 2011 Census of the nation. The issue of migration is still up for debate for various reasons. There are at least as strong voices (made stronger in times of crisis) calling for a decrease in the number of immigrants, even though there is usually a preference for or acceptance of people moving when there is a need for labor.

1.3 Magnitude of homelessness

Estimating the number of homeless people in India is difficult due to differences in data collection methodologies. According to the 2011 Census, 1.77 million persons were homeless, accounting for 0.15% of the population (Jha & Kumar, 2016). Alternatively, alternative organizations say that these figures are unrealistic, with some suggesting that the homeless population in Delhi alone is substantially more significant than the official Census. India has a diverse population of homeless people, including older adults, disabled people, unmarried men and women, and women. The percentage of homeless who are also mentally ill and have children on the streets is likewise significant (Jha & Kumar, 2016).

1.4 Literary approach and research gaps

Structural change or the reallocation of labor among economic sectors with varying levels of labor productivity contributes to an economy's growth process. The pattern of growth determines the way of workforce reallocation. However, GDP is positively impacted when labor is reallocated from lower-productivity sectors to higher-productivity ones. It has long been acknowledged that this contribution might be rather significant in low-income countries because productivity gaps between industries are frequently wide (Ghose, 2021). India changed to a higher growth path trajectory in the 1990s based on the strength of its economic reforms 1991 and the acceleration of subsequent economic reforms in the 2000s. Since the adjustments, annual growth has frequently surpassed 5%. Economic reforms led to structural changes in the Indian economy, including a declining agricultural sector, a growing services sector, and widening regional imbalances. There has been a growth in interstate migration throughout time, particularly in urban India, according to the spatial characteristics of migration, such as distance and stream-wise migration. Long-distance migration from northern parts of the country to southern India is also on the rise. It is only natural that the interstate migration pattern differs significantly between states, given the spatial heterogeneity underlying the various levels of development. However, there has recently been an increase in the migration of low-skilled, uneducated individuals from rural parts of developing states to metropolitan cities in pursuit of employment opportunities, revealing the interstate gaps in development (ILO, 2020). Numerous studies conducted in the rural states show high rates of emigration from impoverished and drought-prone regions of backward states like Andhra Pradesh, Bihar, Orissa, Rajasthan, and Jharkhand to developed conditions that provide a variety of job prospects in the unorganized sector. Around the 1980s, the trend of southward migration began, significantly growing throughout the ensuing decades. The movement has previously been for south Indians to go north, searching for employment in cities like Mumbai and Delhi (Parida, 2019). Kerala is also a haven for migrant workers; Kerala has drawn thousands of migrant workers from other states through the years. These migrant laborers are crucial to Kerala's industrial and commercial sectors. The Kerala State Planning Board said 2021 that migrant workers who work in Kerala remit Rs 7.5 billion to their families in other states (Shahina, 2023).

According to the IOM (2020), natural disasters accounted for 61% (17.2 million) of all new displacements, while war and violence accounted for 39% (10.8 million). Most internal migration is caused due to extreme drought and other climatic change-infused disasters than internal violence (Joshi, 2020). India produces 150,000 tonnes of municipal solid trash daily, of which more than half is deposited in landfills or goes unmanaged, and over 30,000 water bodies have been infringed upon. Compared to metropolitan regions, India's rural areas have a 35% greater loss of life from air pollution-related illnesses. Circular Migration is widespread in drought-prone parts from remote rural areas. These areas have shallow agroecological potential (Bellampalli & Kaushik, 2020). Environmental crimes persist unabatedly, and India's position has fallen nine places in the world in terms of achieving the sustainable development targets set by the UN. Extreme events like drought, floods, heat waves, wildfires, and slower-onset effects like altered rainfall patterns, rising sea levels, increased salinization, and decreased soil fertility pose severe risks for environmental and human settlements as a result of the destruction or loss of land and property, large-scale migration of people cause population displacement. India suffered extreme weather occurrences on 271 days between January and October 2022, which resulted in nearly 2900 fatalities. According to the research, "disasters accounted for the majority of displacement in Southern Asia in 2018," except Afghanistan, where conflict and violence had a more significant impact on displacing people from their homes. According to the state of India's environment report, 2023 only reported 9 of Asia's 93 disasters in 2019, yet it was responsible for approximately 48% of the fatalities.

2. Recent trends in migration

Studies on migration and development are generally optimistic about the possibility of transitory internal labor migration to aid the eradication of poverty and the advancement of humankind in low-resource environments. Temporary labor migration, when compared to permanent migration, entails shorter cycles of movement in which migrant workers return to their village and typical residence after a term of employment elsewhere. Because input costs are typically low and economic returns are less variable and more evenly distributed within the foreigner base than those from international labor migration, temporary labor migration is believed to benefit the poorest households the most in India and other contexts. However, local realities such as exploitative employment practices, a domestic legislative climate that primarily disregards migrant workers and their rights, and distress-induced movement patterns temper this optimism in India. Instead of being a free decision, this labor movement is driven by the need to provide for one's family (Deshingkar & Akter, 2009; Deshingkar, 2006; Deshingkar & Start, 2003; Mosse et al., 2005; Rogaly et al., 2001; Rogaly & Rafique, 2003). Aiming to maximize the apparent benefits of this workforce and their households, proponents of temporary migrant labor strive to enhance support for their movement by changing policy and specific development initiatives (Deshingkar & Akter, 2009). The demographic change that is occurring in various parts of the nation and the effects it is having on the workforce is becoming more and more apparent. Migration has become an essential component in determining the population makeup of different regions. Some of the nation's longest labor migration corridors connecting eastern and northeastern India with the southern Indian states have developed over the past few decades. While some regions of the country benefit from the demographic dividend, others have a severe shortage of workers in occupations requiring strenuous physical labor. More and more attention is being drawn to the demographic change that is taking place in various parts of the nation and how it is impacting the labor force. Migration is now a serious issue.

In contrast to Kerala, which is experiencing a population loss, where salaries in the informal sector are high in India, the tribal areas of Odisha, the wages in the agricultural sector are still as low as 120 rupees per day (Gramvikas, 2019). Behind insights from Thuamul Rampur, India, 2019 migration is not viewed positively context in various countries and this is also the case in India. Even though the UNDP's 2009 Human Development Report emphasizes how important Migration is to human Development, governments typically try to prevent movement from rural to urban regions, and there is a lack of integration between Migration and the development process. Additionally, many people have claimed that restricting immigration would be inappropriate because it is crucial to human Development and fulfilling dreams. While efforts have been made to address the concerns of migrant populations in India, there is still a need for comprehensive policy frameworks that provide adequate protection, social security, and better living conditions for migrants. Identifying and solving the challenges faced by migrants is crucial for nurturing inclusive and Sustainable Development in India.

2.1 Case studies: Successful interventions and best practices

This section discusses a few examples of successful interventions and best practices implemented by state governments in the past for urban homeless people. These case studies highlight initiatives to assist homeless people through collaborations with NGOs. Here are a few notable examples.

2.1.1 Tamil Nadu: "Amma Unavagam" (mother's kitchen)

The Tamil Nadu government initiated the "Amma Unavagam" program in 2013. It is a subsidization program conceived by the Ministry of Food and Civil Supplies. This scheme aims to provide low-cost meals to the urban poor and homeless individuals. The state government collaborated with NGOs and self-help groups to establish and operate these canteens, ensuring affordable and nutritious food for those in need. Amma Unavagam, which translates to "Mother's Kitchen," is a noteworthy drive introduced by the government of Tamil Nadu, India. Launched in 2013, this initiative was a collaboration between the state government and various stakeholders, including NGOs and self-help groups. The primary objective of Amma Unavagam was to tackle food insecurity and address the hunger issue prevalent among vulnerable populations in Tamil Nadu. The program aimed to ensure that even those with limited financial resources could access quality food by offering subsidized meals at low prices. The initiative drew inspiration from the concept of inclusive governance, where the state plays a vital role in providing essential services to its people (Alagumalai, 2018). Under the Amma Unavagam scheme, a network of canteens was established throughout Tamil Nadu, especially in urban areas where homelessness and poverty are more prevalent. These canteens provided hygienic and wholesome meals at highly affordable prices. The menu typically consisted of a balanced mix of rice, sambar (lentil-based stew), curry, and buttermilk. Amma Unavagam aimed to alleviate the financial burden on the urban poor, including people experiencing homelessness, by setting prices significantly below the market rates. The low-cost meals ensured that people had access to regular and nutritious food and reduced their dependence on unhealthy and unhygienic alternatives.

Additionally, the initiative generated employment opportunities as the state government collaborated with NGOs and self-help groups to operate and manage the canteens. This created a positive socioeconomic impact by supporting local communities and empowering individuals to become self-reliant. Amma Unavagam received widespread appreciation for its effectiveness in addressing food insecurity and supporting people experiencing poverty in urban areas, including homeless individuals. The initiative demonstrated the government's commitment to promoting social welfare and inclusive growth. It also served as a model for similar subsidized meal programs implemented in other states across India. Focusing on affordable and nutritious meals, the initiative played a crucial role in alleviating hunger and improving the overall well-being of people with low incomes in urban areas, including homeless individuals, in Tamil Nadu.

2.1.2 *Rajiv Awas Yojana*

RAY was a centrally financed scheme for slum rehabilitation and housing the urban poor, including homeless individuals. Several states, such as Madhya Pradesh, Odisha, and Kerala, successfully implemented RAY by adopting inclusive and participatory approaches to ensure sustainable and affordable housing solutions (Ministry of Housing and Urban Affairs, 2023).

The RAY started in 2011 to achieve a "Slum free India." The program was implemented in two phases: a preparatory phase and an implementation phase. The implementation phase was till 2022. RAY followed a performance strategy, which included the preparation of a Slum-free City Plan of Action (SFCPoA) and the Development of houses for chosen areas. The scheme gave exhaustive guidelines to achieve targets as soon as possible. The scheme is presently under review.

2.1.3 *Karnataka: "Basava Vasati Yojana"*

Basava Vasati Yojana was launched by the Government of Karnataka. The scheme is named after Basavanna, a revered philosopher and social reformer from Karnataka. The Basava Vasati Yojana aims to provide cheap houses to homeless individuals in rural areas of Karnataka. The scheme aims at providing eligible individuals with financial assistance for the construction of houses or for making necessary improvements to their existing homes. The scheme focuses on improving the living conditions of marginalized communities and people from Scheduled Castes (SCs), Scheduled Tribes (STs), and other backward classes (OBCs). This housing scheme plays a crucial part in providing a dignified life and reasonable living standards to vulnerable sections of society, fostering inclusive development in rural and urban areas of Karnataka (Rajiv Gandhi Housing Corporation Pvt, n.d.). These case studies represent a snapshot of successful interventions and best practices introduced by state governments in India. However, the landscape of government initiatives and NGO collaborations is ever-changing. With new programs and interventions being introduced regularly, it is challenging to track all the existing schemes catering to the needs of homeless people.

3. Methodology

The research methodology adopted for studying the complex issue of migrant homelessness in urban India is majorly qualitative. First of all, a detailed review of the literature is done to get a comprehensive understanding of the problem and generate valuable insights that can inform evidence-based policies and effective interventions. Qualitative interviews were conducted with homeless migrants to gain in-depth insights into their experiences, challenges, and perceptions. These interviews were semistructured and narrative-based. One of the narratives is also quoted in detail in one of the sections. Researchers also visited the sites where homeless migrants try to settle down. These areas are on the footpaths or small dwellings under a subway or a flyover bridge in an urban city. Fieldwork involved direct observation and immersion in the environments where homeless migrants reside. Researchers tried to witness the living conditions, interactions, and daily struggles of the migrants, which provided a contextual understanding of the issue. A narrative analysis was done to identify recurring themes, patterns, and qualitative insights into the challenges faced by homeless migrants. These patterns, challenges, and issues are discussed in various sections of the chapter. The cases of state government interventions were also studied in detail to understand how various state governments are dealing with this problem of homelessness. The major limitation of the research is nonavailability of recent census data. The census was the last census data available which was in 2011. Henceforth the exact homeless individual data for recent years is not available. Hence a quantitative analysis was not done to understand the trend.

4. Results and discussion

While the Indian Constitution guarantees the work, residence, and liberty of movement to all its citizens, the experiences of migrant workers often contradict the egalitarian principles enshrined in the Constitution. It is imperative to address the challenges faced by migrant workers and ensure their rights and dignity are protected. Only through concerted efforts and effective policies can we create an inclusive society where the rights of all citizens, including migrant workers, are upheld. The recent pandemic COVID-19 and the resulting lockdown measures have highlighted the existing vulnerabilities within migrant worker communities. It has become evident that their lack of legal protections, limited access to healthcare, and precarious living conditions make them especially susceptible to the adverse effects of such crises. The pandemic has exposed the urgent need for comprehensive and inclusive policies that address the specific needs of migrant workers.

4.1 Dealing with informality

Addressing the issue of informality in the Indian economy is crucial for sustainable and inclusive growth. The informal economy in India has been a tenacious encounter, even with the sturdy levels of economic development taking place over the past 2 decades. It continues to be a priority for governments to address this issue. Formalizing recruitment processes is an essential step toward reducing informality. Ensuring that wages are fixed relatively and transparently helps prevent the exploitation of workers and encourages compliance with labor regulations (Brief, 2023). Additionally, promoting the use of formal payment systems, such as direct bank transfers, can help track wages and ensure timely and accurate payments to workers. Many workers in the informal economy do not have access to social security benefits, such as healthcare, pension, and insurance (Srivastava, 2019). By expanding social security coverage to informal workers, governments can provide them with a safety net and improve their overall well-being.

4.2 Ensuring migrant workers' access to justice

Assuring migrant workers' admittance to impartiality as an unalienable right will enable them to overcome unreported pay loss, fight fraud and denial, and resist slavery and other forms of abuse, such as violence and harassment. The legal obligations of trade unions, employers, labor departments labor courts, and law implementing authorities in reacting to desecrations of migrant workers' labor rights must be clearly defined. For the advantage of migrant employees, labor departments must ensure that the labor codes are instigated and play a substantial role in grievance redress. An adequate legal assistance system must be able to process scant proper evidence and ply migrant populations who may come across challenges. One of the significant issues faced by migrant workers is the unaccounted loss of wages. Many employers exploit their vulnerable situation by withholding or manipulating their salaries, making it challenging for them to support themselves and their families. Strict regulations and monitoring mechanisms should be implemented to ensure employers comply with fair wage practices. Additionally, creating awareness among migrant workers about their rights and providing them access to legal remedies can help them overcome this loss and seek appropriate compensation. Migrant workers often fall victim to fraud and denials of their rights, such as contract substitution, deceptive recruitment practices, and document confiscation. These practices further exacerbate their vulnerability and hinder their ability to seek justice (ILO, 2022). Governments should establish robust regulatory frameworks to combat fraudulent activities in the recruitment process and strengthen oversight mechanisms. Furthermore, ensuring transparency and accountability in recruitment is crucial to protecting migrant workers' rights and preventing fraudulent practices. One of the most distressing issues faced by migrant workers is bondage and abuse, including violence and harassment (Hastie, 2017). Many workers find themselves trapped in exploitative and abusive working conditions, unable to escape due to threats, debt bondage, or fear of reprisals. To counter this, it is imperative to establish effective mechanisms to promptly identify and address cases of abuse (Harkins & Åhlberg, 2017). Strengthening labor inspections, providing confidential reporting channels, and creating safe spaces for victims to seek assistance is essential to protecting migrant workers from bondage and abuse.

4.3 Establishing unorganized workers' social security boards for protection and welfares

In the background of elevated informality in the Indian economy, there is a pressing need to strengthen social fortification bases at the national and subnational levels. This will help ensure that migrant workers receive the minimum required levels of social protection. Additionally, extending universal social protection coverage is crucial to address the vulnerabilities faced by these workers. To address these challenges, it is essential to strengthen social protection floors. This involves establishing a solid framework that guarantees minimum benefits and safeguards for all workers, regardless of their employment status (Hirose et al., 2011). By implementing social protection floors, the government can ensure that migrant workers have access to essential services, such as healthcare, maternity benefits, unemployment benefits, and pensions. Establishing Social Security Boards for workers in the unorganized sector, with simple and effective modes of worker registration, including self-registration processes, is paramount to providing social security benefits and protection to unorganized workers (van Ginneken, 2013). Furthermore, digitizing registration records enables interstate portability, ensuring that workers can access their entitlements regardless of location. All states must prioritize implementing these measures to uplift the lives of unorganized workers and create a more inclusive and equitable society.

4.4 Safe living and working environment

The migrant workers must have access to noble living spaces that provide them privacy, security, and comfort. These living spaces should meet basic standards of hygiene and sanitation to ensure the well-being of migrants. In addition to dignified living conditions, it is equally essential to provide migrants with safe and nonprecarious work environments (Prakasam, 2014). Migrants should be protected from hazardous working conditions, such as exposure to harmful substances, dangerous machinery, or excessive physical exertion. Employers should adhere to strict safety regulations and provide appropriate protective equipment to guarantee the welfare of migrant workers. Furthermore, it is essential to prioritize the physical and mental health of migrants (John, 2021). Healthcare services should be readily available, including preventive care, treatment for illnesses or injuries, and mental health support. Migrants should not face any discrimination or barriers when seeking medical assistance.

4.5 Supporting worker organizations and collectivization

Enabling workers' collectivization and organization is crucial to achieving a more equitable and empowered workforce. Encouraging workers to come together and form collective organizations can create a stronger sense of solidarity and collective bargaining power. One of the critical benefits of collectivization is the ability to negotiate better working conditions, wages, and benefits. When workers are organized, they can collectively bargain with employers to secure fair compensation and improved workplace policies. This can lead to higher job satisfaction, increased productivity, and a more harmonious working environment. Beyond the immediate benefits for workers, collectivization also has broader societal implications (John, 2021). We can reduce income inequality and create a more just society by empowering workers and giving them more control over their work lives. Collectivization promotes a more inclusive economy where wealth and power are more evenly distributed.

4.6 The role of significant players and the policy framework

While the specific roles and responsibilities of each stakeholder are substantial, it is essential to remember that an integrated strategy that considers the intersectionality of the various types of vulnerabilities experienced by migrants will be crucial in ensuring that the responses are not limited to specific silos. To fully address the numerous matters that have been recognized, the road map for developing a policy framework is essential, which will ensure the inclusion of internal migrant workers in India (ILO, 2020). For this purpose, a wide array of stakeholders, from the Government of India to subnational governments, from CSOs to development aid agencies, and the UN, from employers to trade unions, should work in coherence and unison (Tiwari, 2022). Driving significant and lasting change will require a comprehensive and systemic strategy that properly considers the contribution of migrant workers to the Development of the nation and acknowledges the importance of their voice and agency.

4.7 Managing stressed migration through rural development

The National Commission on Rural Labor (NCRL) acknowledges the significance of prioritizing rural Development to reduce distress migration and promote safe Migration. The NCRL has recognized the centrality of this policy focus and has recommended its implementation. By strengthening rural development initiatives, we can create better opportunities and living conditions in rural areas, thereby reducing the need for individuals to migrate out of distress (Rhoda, 1983). This approach addresses the root causes of Migration and empowers individuals and communities to make informed decisions about their future (Simran, 2018).

4.8 Strengthening state-level labor departments to address migrant worker issues

To effectively address the challenges faced by migrant workers, the Ministry of Labor and Employment (MoLE) must take the lead in strengthening and equipping the departments of labor at the state level. This document outlines the importance of enhancing staffing and budgetary allocations for labor departments to ensure the welfare and protection of migrant workers. By taking the lead and prioritizing the strengthening of state-level labor departments, the MoLE can effectively address the issues faced by migrant workers. Enhancing staffing and budgetary allocations will empower these departments to provide comprehensive support, enforce labor laws, facilitate skill development, and establish effective grievance redressal mechanisms. These measures must be implemented to ensure the welfare and protection of migrant workers across the country.

4.9 Including migrants in India's urban policy formulation

India's urban landscape is undergoing rapid transformation, with cities experiencing a significant influx of migrant workers. These migrants contribute immensely to the urban economy and play a crucial role in developing cities. However, their needs and concerns often go unnoticed in formulating urban development programs and policies. To address this issue, it is imperative that the Ministry of Housing and Urban Affairs takes the initiative and ensures that the needs of migrant workers are purposefully included in India's urban policy formulation, which will lead to the creation of inclusive and sustainable cities. The MoHUA must take the lead in ensuring their participation, addressing their unique needs, and recognizing their contributions. By doing so, India can harness the full potential of its migrant workforce and create cities that are genuinely inclusive and prosperous for all.

4.10 Foster social dialog to formulate policies and programs

The National Commission for Enterprises in the Unorganized Sector (NCEUS) has proposed the establishment of conciliation committees at the district level and dispute resolution councils at lower levels. These committees and councils aim to ensure the inclusion of unorganized workers, including migrant workers, in the dialog and decision-making processes. The conciliation committees would serve as platforms for resolving conflicts and disputes between employers and workers, mainly focusing on the concerns of migrant workers (2022). These committees would provide a space for open discussions and negotiations, ultimately leading to fair resolutions and the formulation of appropriate policies.

4.11 Facilitation centers for migrant workers

The primary purpose of these facilitation centers is to assist migrant workers with their registration process and provide them with relevant information regarding their social security entitlements. By facilitating the registration process, migrant workers can ensure their legal documentation is in order, protecting them from potential exploitation or mistreatment. Additionally, these centers serve as an essential source of information, providing that migrants are aware of their rights and entitlements (2022). One of the key advantages of these facilitation centers is the continuity of contact they provide. Migrant workers often face isolation and are disconnected from their families and support networks. The centers serve as a bridge, offering a platform for communication and support. Through regular contact, the centers can address any concerns or issues faced by migrant workers, offering guidance and counseling services.

4.12 Skilling and education for migrants

Regarding migration, education, and skilling can significantly impact the choices and opportunities available to individuals. By equipping them with relevant knowledge and practical skills, these programs enable young people to pursue better prospects in their communities rather than being forced to migrate in search of employment elsewhere. One key aspect is the range of destinations that migrants from these areas typically choose (2022). Understanding these patterns makes it possible to develop educational and skilling programs tailored to these specific destinations' needs. By comprehending the range of occupations, goals, and markets precise to migration areas, it is advisable to implement high-impact skilling and vocational training programs for the youth in these source areas. By doing so, individuals can find better opportunities within their communities, reducing the need for migration in search of employment.

4.13 Purposeful enumeration in capturing heterogeneous living arrangements

The heterogeneity in living arrangements reflects the diverse socioeconomic settings within urban areas. Rented rooms in city centers and industrial peripheries cater to individuals seeking affordable accommodation close to employment opportunities. Open settlements, on the other hand, are often characterized by informal housing structures and a lack of basic amenities. These settlements are home to marginalized communities facing challenges accessing essential services. In addition, addressing the healthcare needs of migrant workers requires a multifaceted approach. Incentivizing ASHAs, offering services in languages that migrant workers understand, and establishing mobile health clinics are essential measures to meet their healthcare needs. By implementing these measures, we can contribute to refining the comprehensive health and well-being of migrant workers and their families.

4.14 Priority for funding internal labor migration

Development aid agencies should allocate resources to support policy development and advocacy efforts that protect the rights of internal labor migrants. This includes promoting inclusive labor market policies, ensuring fair wages, and providing social protection mechanisms (2022). Funding should be directed toward programs that enhance the skills and capacities of internal labor migrants, enabling them to access better employment opportunities and improve their socioeconomic conditions. Development aid agencies should prioritize funding for projects that enhance access to essential services for internal labor migrants, such as healthcare, education, housing, and sanitation. Adequate provision of these services is necessary to ensure the well-being and dignity of migrants.

5. Internal migration—the push and pull factors

Though the migrant population makes up a minority group in a developing country's urban population, their impact on economic policies, poverty reduction, housing, health, and jobs is tremendous. Haan and Yaqub (2010) points out the challenge and reasons for the Migration of the rural population who keep moving between rural and urban spaces. Though few early research opinions referred to the migrants from third-world countries as needy, who had to migrate as they did not have any option to survive in their rural hamlets. In a study conducted by Dyson (1996) on population and food prospects, he anticipated the movement of the peasantry community in India to towns. Numerous issues stem from the large-scale, destitute Migration of people from rural to urban regions in India, which results in uneven urbanization and severe urban deterioration. Mass migration causes an increase in poverty, unemployment, and underdevelopment (Tyagi & Siddiqui, 2013). From India's rural areas, about five million people migrate annually, with three million of them moving to cities (Parida, 2019). In the majority of the underdeveloped and agrarian states, the rate of rural outmigration is significant. Rural—urban migration in India is being fueled by agricultural mechanization. More and more people are moving to the more developed states, especially the larger metropolises and medium-sized cities, which are increasingly modern. Issues relating to social protection, the depletion of natural resources, and the adverse effects of environmental degradation and climate change (2018). Migration and forced displacement have a variety of primary causes, including conflicts, violence, and natural catastrophes. Socioeconomic causes, such as poverty, food insecurity, lack of employment prospects, limited access to social protection, the depletion of natural resources, and the adverse effects of environmental degradation and climate change, force many migrants to relocate. Much of the pull-and-push factor theory is analogous to the migration theory. The pull and push theories suggest both beneficial and detrimental aspects of migration. Better work, transportation, health, and educational opportunities are pull factors. War, discord, and inadequate facilities are all push forces in many areas. Aside from these causes, migration may also be prompted by psychological or personal factors (Lee, 1966).

6. Migrant's life in urban India

Even though India has a history of bloodshed and internal disturbances due to political unrest and inter-ethnic violence, there haven't been many refugees. However, a high amount of internal displacement has been brought on by war, conflict, violations of human rights, and forced relocation. Internal displacement takes on various shapes, much like internal economic migration. These are influenced by the pursuit of livelihood strategies and the use of mobility as a means of self-defense in the face of the specific risk dynamics of that conflict. As a result, internal displacement patterns may be highly context-specific, posing unique issues for those impacted (and their hosts) regarding aid, protection, and finding solutions. However, in areas of armed conflict, both those who remain in place and IDPs who return face far greater threats to their safety or means of subsistence, particularly in situations where "immobility" is mandated as a tactic of war or control. Politically induced displacements have surfaced since independence; north-east India has experienced two significant violent battles since independence: the Assam movement, the Naga movement, and the Riots for Gorkhaland in Darjeeling are a few of these. A constant stream of displaced individuals is being produced by the violence and retaliatory actions taken by the government and other forces opposed to the secessionists (2023). In the last few months, India has witnessed riots in the northeastern state of Manipur; on May 3, a violent uprizing broke out in Manipur as thousands of tribal members opposed proposals to grant the majority Meitei tribe protected status as a Scheduled Tribe, claiming the group already had advantages in the state. The categorization system sets quotas for government employment and college admissions to address historical and structural injustice and prejudice. According to media estimates, violent battles, primarily between the ethnic Meitei and Kuki tribes, have resulted in at least 70 fatalities, 35,000 displaced persons, and the destruction of over 1700 homes. According to estimates, 250,000 Kashmiri Pandits have moved to Jammu and other cities like Delhi as a result of the widespread anarchy brought on by political instability, the killing of Kashmiri Pandits by fundamentalist secessionist organizations, and ongoing violations of fundamental human rights by both the state and militant groups. People forced to flee Kashmir have been unable to do so because of the constant reality of periodic atrocities, despite the election and restoration of civil rights.

6.1 Government policies and interventions

Rapid urbanization in India has exacerbated the issue of homelessness and contributed to the migrant crisis. As cities expand and attract more economic opportunities, rural individuals move to cities for better earnings (Shelter Homes, 2023). However, the shortage of affordable urban housing has left many vulnerable to homelessness, particularly the poorest urban dwellers. Marginalized groups face social exclusion and discrimination, making them more vulnerable to homelessness. Natural disasters and forced evictions further contribute to homelessness as communities are displaced without adequate resettlement plans. Addressing these multifaceted causes is crucial to tackling the problem of homelessness in India (Goel et al., 2017). The 2011 Census revealed a lack of 7.1 million urban housing units in India, a significant increase from 5.1 million units recorded in 1991. This shortage, coupled with inadequate resettlement policies and the informal working of labor, has resulted in limited availability of cheap housing amenities and good working conditions for migrants (Jha & Kumar, 2016). To alleviate the difficulties experienced by migrants, the Indian government has launched several programs. These include the National Rural Employment Guarantee Act (NREGA), which provides 100 days of guaranteed employment to poor people in rural areas. The Inter-State Migrant Workmen (Regulation of Employment and Conditions of Service) Act, 1979, aimed at protecting the rights of interstate migrant workers. To further promote migrant welfare, the government has introduced programs like the One Nation, and One Ration Card which will provide major commodities at subsidized rates. Addressing homelessness in India requires a multifaceted approach that includes affordable housing programs, social welfare initiatives, skill development, employment opportunities, and comprehensive support services. The national and state governments have implemented policies and programs to address homelessness, but significant challenges remain in providing sustainable solutions for the homeless population. The following section offers a brief on various national- and state-level policies designed to address this issue.

6.1.1 National policies

Census conducted in 2011 revealed that there were 17,73,040 homeless individuals in India, including 938,348 in cities and 834,692 in villages. With the government's vision of "Housing for All," the Ministry of Housing and Urban Affairs (MoHUA) implemented the Pradhan Mantri Awas Yojana-Urban (PMAY-U) on June 25, 2015. This initiative aims to provide proper brick-and-mortar houses to the highly vulnerable population without homes. States and Union Territories (UTs) have conducted demand surveys under PMAY-U to assess housing demand. Under this scheme, 114 lakh houses were sanctioned till December 2021 in various states and Union Territories. As per the Economic times daily newspaper 1.15 cr houses sanctioned under PMAY-Urban, 56.20 lakh units already built, while the remaining are at multiple stages of construction (Economic Times, 2022). Further to tackle housing issues in rural areas Pradhan Mantri Awaas Yojana-Gramin (PMAY-G) was launched in rural areas by the Ministry of Rural Development on April 1, 2016. A target of building 2.95 crore houses is set in this scheme. Under PMAY-G, beneficiaries receive unit assistance of Rs. 1.20 lakhs in plain areas and Rs. 1.30 lakhs under challenging areas for house construction. Beneficiary selection under PMAY-G is primarily based on the Socio-Economic and Caste Census (SECC) 2011 data. Landless beneficiaries are given top priority. However, providing land for construction is the responsibility of the respective states. Of the 446,058 people with no land or house, 202,719 have been provided land, and 3128 have received financial assistance for land purchase. The status of this provision varies across states and UTs.

(i) National Urban Housing and Habitat Policy (NUHHP), 2007 seeks to foster affordable housing across the nation, ensuring fair and affordable access to land, shelter, and services for all segments of society. However, the urban homeless face the most significant vulnerability among these groups. Urban homeless are essential to the ecosystem, and modern cities run due to their labor and efforts. Despite their efforts and contribution, they are the ones who suffer maximum and lack all sorts of essential things to live a life of dignity. They face numerous challenges with respect to basic amenities and clean and dignified living spaces (Nambiar & Mathew, 2022). Scheme of Shelter for Urban Homeless (SUH) aims to provide permanent shelters with all basic amenities to urban homeless population.

(ii) National Urban Livelihoods Mission (NULM): NULM aims to reduce poverty and homelessness in urban areas. It provides shelter, skill development training, and livelihood support for the urban poor, including homeless individuals (TNULM, 2023).

6.1.2 State-level policies

(i) Tamil Nadu Urban Livelihoods Mission (TNULM): The Tamil Nadu National Urban Livelihood Mission aims to provide permanent community shelters for the most vulnerable homeless groups, i.e., single women with children, old aged, differently abled, and mentally challenged people. Objectives include ensuring shelter access with basic infrastructure, water supply, sanitation, safety, and security. The mission also focuses on providing access to government entitlements like social security pension, credit linkage, education, and affordable housing for the homeless population (TNULM, 2023).

(ii) Delhi Urban Shelter Improvement Board (DUSIB): The DUSIB is incorporated to provide shelter and support services to homeless individuals in Delhi. It operates various night shelters and works toward enhancing the living conditions and well-being of those experiencing homelessness (Delhi Urban Shelter Improvement Board, 2023).

(iii) Kerala Apna Ghar project: The Apna Ghar Project was designed to provide interstate migrant (ISM) workers with rental-based, secure, and clean hotel accommodations. Over the last 2 decades, Kerala has seen a steady influx of ISM workers from other Indian States, mainly Bihar, Uttar Pradesh, Orissa, West Bengal, and Assam. Due to rising levels of Education and emigration, notably to the middle east, Kerala's labor force experienced a substantial gap due to the loss of Keralites from the basic labor force. The ISM workers are now filling that void and have established themselves as a vital and fundamental component of Kerala's economy in a variety of industries, including agriculture, construction, hospitality, etc. (Peter et al., 2020). Under the project, a four-story facility with more than 50 rooms, 32 kitchens, 96 bathrooms, eight dining halls, and laundry facilities was constructed in Kanjikode, Palakkad, as part of the pilot project. The structure has a fire suppression system, rainwater collection systems, diesel generators, 24-h security, and CCTV cameras.

Several states in India have established Urban Shelter Improvement Boards or similar bodies to address homelessness and provide shelter for the urban homeless population. However, these boards' specific names and structures may vary from state to state. A few of the Urban Shelter Improvement Boards are listed below:

- Delhi: Delhi Urban Shelter Improvement Board (DUSIB)
- Maharashtra: Maharashtra Housing and Area Development Authority (MHADA)

- Karnataka: Karnataka Slum Development Board (KSDB)
- Tamil Nadu: Tamil Nadu Slum Clearance Board (TNSCB)
- West Bengal: West Bengal Housing and Infrastructure Development Corporation (WBHIDCO)
- Kerala: Kerala State Housing Board (KSHB)

6.1.3 Effectiveness and gaps in policy implementation

Despite the presence of these policies, their effectiveness in addressing the needs of homeless migrants remains limited. Several gaps in policy implementation contribute to this situation.

- Inadequate Shelter Facilities: The existing shelter infrastructure is insufficient to accommodate the growing number of homeless migrants. Many migrants are still deprived of basic shelter and live in deplorable conditions.
- Lack of Awareness and Accessibility: Homeless migrants often face challenges accessing information about government schemes and facilities. The lack of awareness hampers their ability to benefit from these programs.
- Limited Coverage and Targeting: The policies often fail to address the unique needs of homeless migrants. Some migrants, particularly those living in informal settlements, remain excluded from the benefits of these policies.
- Inadequate Support Services: While shelter provision is crucial, comprehensive support services such as healthcare, skill development, and social integration are often lacking. This hinders the long-term stability and well-being of homeless migrants.
- Weak Monitoring and Evaluation Mechanisms: Insufficient monitoring and evaluation mechanisms make it difficult to assess the impact and effectiveness of these policies. Without regular assessments, necessary improvements and modifications cannot be made.

While policies for homeless migrants in India have been implemented, their effectiveness is hindered by gaps in policy implementation. Adequate shelter facilities, improved awareness and accessibility, targeted coverage, comprehensive support services, and robust monitoring and evaluation mechanisms are crucial for addressing the needs of homeless migrants effectively. By addressing these gaps, policymakers can create an inclusive and supportive environment that uplifts and empowers homeless migrants in India.

7. Recommendations and way forward

The paradox of urbanization is featured by the homeless conditions in which rural migrants inhabit the urban landscapes in search of livelihood. This is the reality of metropolitan cities in India and many other developing countries (Jha & Kumar, 2016). The government has been going down heavily with civic authorities and the police dismantling the urban street dwellers polythene covered shambles on the footpaths, outer area of railway lines destroying their minimal household materials due to encroachments. Homelessness is the result of Government's inability to provide housing facilities in urban areas to displaced population, though resettlement policies and agreements are made at state level and metropolis levels, the number of squatter settlements have been increasing in urban areas with no proper accountability on ownership of land (Cooper, 1995). The increasing mobility of migrant workers has necessitated the establishment of structures and institutions to address the unique needs of this population (Speak, 2022). These structures play a crucial role in ensuring the well-being and protection of migrant workers and their families. The existing state machinery serves as a foundation for supporting migrant workers within the host country. By integrating structures within this machinery, governments can better address the specific challenges faced by migrant workers. This includes the provision of essential services such as healthcare, education, and housing. Additionally, state-run agencies can collaborate with nongovernmental organizations (NGOs) to offer legal aid, counseling, and job placement services to migrants. As we encounter high mobility nature of the migrant workforce it is imperative to involve various structures and institutions to address their rights and requirements. Embedding these structures within the existing state machinery and interstate mechanisms ensures a comprehensive and coordinated approach to supporting migrant workers and their families. By prioritizing their well-being, societies can harness the valuable contributions of migrant workers while also upholding their fundamental rights. Between 2015 and 2022, 20 million households, including 18 million slum and 2 million nonslum households, will face a housing shortage, according to the Pradhan Mantri Awas Yojana-Housing for All (Urban) scheme of the Ministry of Housing and Urban Affairs, Government of India (Ministry of Housing and Urban Affairs, 2023).

Nevertheless, the interventions in certain cases do not yield results due to the fact that, in order to meet the needs of housing and infrastructure, it is implied by the state that significant dwelling must be constructed. Housing construction also runs into challenges pertaining to construction quality and occupant comfort and safety. However, because of a lack of technical capability on the part of the implementing agencies or the lack of novel construction materials, resettlement homes are frequently built with conventional, inefficient materials. Further, families may be more likely to buy energy-inefficient gadgets like bulbs, fans, etc., since they are inexpensive than their more expensive, energy-efficient counterparts as a result energy use and carbon emissions increase with additional burden of waste generation and poor segregation and disposal from these resettlement areas Aggarwal (2018). Resettlement projects in Coimbatore is a testimony of how effectively indigenous supplies can be used effectively in resettlement projects. Traditional technologies which can be locally sourced to meet climatic conditions can be an effective way in ensuring involvement of local sources to address the issue of homelessness.

The establishment of effective response systems necessitates a strong political will at the central level. Political leaders must recognize the significance of investing resources, time, and effort into these systems to address the diverse needs of the population. This commitment ensures that the necessary policies and initiatives are put in place to support the functioning and sustainability of the response systems. Furthermore, development in high out-migration regions requires a multi-faceted approach that addresses the protection of local wage markets, the provision of social security, creation of local employment, reduction of vulnerabilities to bonded-labor-like situations and the elimination of labor trafficking. The Working Group on Migration recommends including information regarding workers' status surveys by the NSSO, alongside a separate survey about migration. In order to facilitate policy decision-making, regular data collection and synthesis should be carried out in a way that allows early dissemination of macro and small data. A national nodal agency with an overarching responsibility for the integration and protection of migrants in a range of sectors, such as employment rights and welfare entitlements, should be established at the federal level to ensure the systematic gathering of data on flows of migration, ensuring portability of various entitlements related to food supplies, education, health, and early childcare. In addition to fiscal allocations, establishing a national helpline for migrant workers can play a significant role in safeguarding their rights and addressing their grievances. Such a helpline would provide a platform for workers to call in, register their concerns, and seek resolution to any violations they may have encountered. This helpline could be staffed by trained professionals who are well-versed in the rights and protections afforded to migrant workers. By establishing a fair migration governance system, rural communities can thrive, ensuring the well-being and prosperity of their residents. These efforts will contribute to the overall development of migrants and create a more equitable and inclusive society.

8. Conclusion

In the course of India's future, it is imperative that we strive to "Build Back Better" by addressing the challenges faced by migrant workers and their families. The notion of "Building Back Better" which first surfaced in conversations on disaster recovery a few years ago, has grown to be the one that has gained the most traction in terms of post-COVID-19 restoration. The approaches being created and discussed at the international level by decision-makers, policy advisers, and civil society emphasize collaborative global action and sustainable reconstruction. A call for optimism is Building Back Better. The pandemic that has ravaged every region of this planet is still difficult for the globe to comprehend. Resources and strength are needed to rebuild and restore a society. To achieve this, we must embark on a transformative journey that prioritizes decent work and social justice for these marginalized individuals. By doing so, we can create a more inclusive and equitable society that leaves no one behind. To truly "Build Back Better," we must prioritize the concept of decent work for migrant workers. This means providing them with fair wages and benefits, safe working environments, and social protection. By promoting decent work, we can enhance their economic security, improve their living standards, and foster social cohesion. This encompasses addressing the complete barriers and inequitable practices that deter their access to education, healthcare, and other essential services. By ensuring equal opportunities and rights for all, we can create a more inclusive society where migrant workers can thrive. By addressing the challenges they face, implementing policy reforms, and strengthening social safety nets, we can create a more equitable society that ensures the well-being and dignity of all its members. Let us embark on this transformative journey and strive for a future where no one is left behind.

References

Aggarwal, D. (2018). *Resettling India's urban poor sustainably*. The Energy and Resources Institute. https://www.teriin.org/article/resettling-indias-urban-poor-sustainably. (Accessed 12 November 2023).

Alagumalai, A. (2018). Food security and amma unavagam in Tamil Nadu – with special reference to madurai city. *Shanlax International Journal of Arts, 5*.

Bellampalli, P. N., & Kaushik, R. (2020). Identification of the determinants of rural workforce migration: A study of construction segments in Udupi district, Karnataka, India. *Review of Development and Change, 25*(2), 256–270. https://doi.org/10.1177/0972266120980187

Brief, I. L. O. (2023). *Intervention model: For extending social protection to migrant workers in the informal economy*.

Census India. (2011). https://censusindia.gov.in/census.website/data/census-tables.

Cooper, B. (1995). *Shadow people: The reality of homelessness in the 90's*. http://gopher://csf.colorado.edu:70/00/hac/homeless/GeographicalArchive/reality-australia. (Accessed 16 September 2023).

Datta, A. (2020). Circular migration and precarity: Perspectives from rural Bihar. *Indian Journal of Labour Economics, 63*, 1143–1163. https://doi.org/10.1007/s41027-020-00290-x

Delhi Urban Shelter Improvement Board. (2023). https://delhishelterboard.in/main/?page_id=3346. (Accessed 6 May 2023).

Deshingkar, & Akter, S. (2009). *Migration and human development in India*. Human Development Research Paper, 13.

Deshingkar, P., & Start, D. (2003). *Seasonal Migration for livelihoods in India: Coping, accumulation and exclusion*. Overseas Development Institute.

Deshingkar, P. (2006). *Internal migration, poverty, and development in Asia. London*. Retrieved from ODI Briefing Paper 11.

Dyson, T. (1996). *Population and food. Global trends and future prospects*. https://doi.org/10.4324/9780203977156

Economic Times. (2022). *1.15 cr houses sanctioned under PMAY-Urban, 56.20 lakh units already built: Govt to LS*. Economic Times. https://economictimes.indiatimes.com/industry/services/property-/-cstruction/1-15-cr-houses-sanctioned-under-pmay-urban-56-20-lakh-units-already-built-govt-to-ls/articleshow/90418065.cms.

Ghose, A. K. (2021). Structural change and development in India. *Indian Journal of Human Development, 15*(1), 7–29. https://doi.org/10.1177/09737030211005496

Goel, G., Ojha, M. K., & Ghosh, P. (2017). Urban homeless shelters in India: Miseries untold and promises unmet. *Cities, 71*, 88–96. https://doi.org/10.1016/j.cities.2017.07.006

Government of India. (2017). *Economic survey 2016–17*. Ministry of Finance, Government of India. https://www.indiabudget.gov.in/budget2017-2018/es2016-17/echapter.pdf.

Gramvikas. (2019). *Challenges of migrants and families left behind insights from Thuamul Rampur, India*. https://www.gramvikas.org/wp-content/uploads/2019/06/Gram-Vikas-final-_-web-1.pdf. (Accessed 29 July 2023).

Haan, de A.,, & Yaqub, S. (2010). Migration and poverty: Linkages, knowledge gaps and policy implications. In K. Hujo, & N. Piper (Eds.), *South-South migration. Social policy in a development context*. London: Palgrave Macmillan. https://doi.org/10.1057/9780230283374_6

Harkins, B., & Åhlberg, M. (2017). *Access to justice for migrant workers in South-East Asia*. International Labour Organization.

Hastie, B. (2017). The inaccessibility of justice for migrant workers: A capabilities-based perspective. *The Windsor Yearbook of Access to Justice, 34*. https://id.erudit.org/iderudit/1057056ar.

Hindustan Times. (2016). *What the 2011 census data on migration tells us*. Hindustan Times. https://www.hindustantimes.com/delhi-news/migration-from-up-bihar-disproportionately-high/story-K3WAio8TrrvBhd22VbAPLN. html.

Hirose, K., Nikac, M., & Tamagno, E. (2011). *Social security for the migrant worker: A rights-based approach*. International Labour Organization.

ILO. (2020). *Road map for developing a policy framework for the inclusion of internal migrant workers in India*. International Labour Organization. https://www.ilo.org/newdelhi/whatwedo/publications/WCMS_763352/lang–en/index.htm. (Accessed 10 June 2023).

ILO. (2022). *Road map for developing a policy framework for the inclusion of internal migrant workers in India*. https://www.ilo.org/wcmsp5/groups/public/—asia/—ro-bangkok/—sro-new_delhi/documents/publication/wcms_763352.pdf.

IOM. (2020). *World migration report*. International Organization for Migration. https://publications.iom.int/system/files/pdf/wmr_2020.pdf. (Accessed 29 July 2023).

Jha, M. K., & Kumar, P. (2016). Homeless migrants in Mumbai: Life and labour in urban space. *Economic and Political Weekly, 51*(26–27), 69–77. http://www.epw.in/system/files/pdf/2016_51/26-27/Homeless_Migrants_in_Mumbai_0.pdf.

John, J. (2021). *A study on social security and health rights of migrant workers in India.* Kerala Development Society.

Joshi, H. (2020). Extreme weather events causing most migration. *Monogabay.* https://india.mongabay.com/2020/02/extreme-weather-events-causing-most-migration/#:~:text=climate%2Dinduced%20disasters.-,Photo%20by%20Hridayesh%20Joshi.,to%20tropical%20storms%20and%20floods. (Accessed 24 November 2023).

Lee, E. S. (1966). A theory of migration. *Demography, 3*(1), 47–57. https://doi.org/10.2307/2060063

Ministry of Housing and Urban Affairs. (2023). https://mohua.gov.in/. (Accessed 20 June 2023).

Mosse, D., Gupta, S., & Shah, V. (2005). On the margins in the city: Adivasi seasonal labour migration in western India. *Economic and Political Weekly, 40*(28), 3025–3038.

Nambiar, D., & Mathew, B. (2022). Roles played by civil society organisations in supporting homeless people with health care-seeking and accessing the social determinants of health in Delhi, India: Perspectives of support providers and receivers. *SSM - Qualitative Research in Health, 2.* https://doi.org/10.1016/j.ssmqr.2022.100157

Parida, J. K. (2019). Rural-urban migration, urbanization, and wage differentials in urban India. In *Internal migration, urbanization, and poverty in Asia: Dynamics and interrelationships* (pp. 189–218). India: Springer Singapore. https://doi.org/10.1007/978-981-13-1537-4_8

Parsell, C. (2010). Homeless is what i am, not who i am: Insights from an inner-city brisbane study. *Urban Policy and Research, 28*(2), 181–194. https://doi.org/10.1080/08111141003793966

Peter, B., Sanghvi, S., & Narendran, V. (2020). Inclusion of interstate migrant workers in Kerala and lessons for India. *Indian Journal of Labour Economics, 63*(4), 1065–1086. https://doi.org/10.1007/s41027-020-00292-9

PIG. (2022). *Ministry of Statistics and Programme Implementation.*

Prakasam, S. (2014). Living conditions of migrant service workers in urban India. Case study of chandigarh. *Journal of Sociology and Social Work, 2*(1), 99–121.

Rajiv Gandhi Housing Corporation Pvt. https://ashraya.karnataka.gov.in/. (Accessed 29 7 2023).

Rhoda, R. (1983). Rural development and urban migration: Can we keep them down on the farm? *International Migration Review, 17*, 34–64. https://doi.org/10.2307/2545923

Rogaly, B., & Rafique, A. (2003). Struggling to save cash: Seasonal migration and vulnerability in West Bengal, India. *Development and Change, 34*(4), 659–681. https://doi.org/10.1111/1467-7660.00323

Rogaly, B., Biswas, J., Coppard, D., Rafique, A., Rana, K., & Sengupta, A. (2001). Seasonal migration,social change and migrants rights: Lessons from West Bengal. *Economic and Political Weekly, 36*(49), 4547–4559.

Shahina, K. K. (2023). *Migrant crisis in Kerala.* https://www.outlookindia.com/national/the -guest-worker-crisis-magazine-274725. (Accessed 12 March 2023).

Shelter Homes. (2023). *India's migrant crisis pointed to another problem – its lack of shelter homes.* https://scroll.in/article/968374/indias-migrant-crisis-pointed-to-another-problem-its-lack-of-shelter-homes. (Accessed 26 July 2023).

Simran. (2018). Assessing the inter-relationship between rural–urban migration and rural development: A case study of Delhi. *OIDA International Journal of Sustainable Development, 11*, 11–24.

Speak, S. (2022). *The state of homelessness in developing countries. In presentated at: United Nations expert group meeting on "affordable housing and social protection systems for all to address homelessness".* https://www.un.org/development/desa/dspd/wp-content/uploads/sites/22/2019/05/SPEAK_Suzanne_Paper.pdf. (Accessed 18 September 2023).

Srivastava, R. (2019). Emerging dynamics of labour market inequality in India: Migration, informality, segmentation and social discrimination. *Indian Journal of Labour Economics, 62*(2), 147–171. https://doi.org/10.1007/s41027-019-00178-5

Tiwari, P. (2022). *Skilling is key to curbing unemployment, migration in rural India.*

TNULM. (2023). *Shelter for urban homeless (Suh).* TNULM. Available at:.

Tyagi, R. C., & Siddiqui, T. (2013). *Causes of rural-urban migration in India: Challenges and policy issues* (Vol 5, pp. 173–176). United Nations Educational, Scientific and Cultural Organization, 6.

van Ginneken, W. (2013). Social protection for migrant workers: National and international policy challenges. *European Journal of Social Security, 15*(2), 209–221. https://doi.org/10.1177/138826271301500206

Veness, A. R. (1993). Neither homed nor homeless: Contested definitions and the personal worlds of the poor. *Political Geography, 12*(4), 319–340. https://doi.org/10.1016/0962-6298(93)90044-8

Waldron, J. (1991). Homelessness and the issue of freedom. *UCLA Law Review, 39*, 295–324.

Further reading

FAO IFAD IOM WFP. (2018). *The linkages between migration, agriculture, food security and rural development.* https://www.ifad.org/documents/38714170/40721506/The+Linkages+between+Migration%2C+Agriculture%2C+Food+Security+and+Rural+Development.pdf/85c3c0c5-d803-4966-a2bc-13b58c772b50?t=1533219475000. (Accessed 3 February 2023).

International Labour Organization. (2022). *Fair recruitment and access to justice for migrant workers.* Switzerland: International Labour Organization. https://www.ilo.org/wcmsp5/groups/public/—ed_protect/—protrav/—migrant/documents/publication/wcms_850615.pdf.

Lama, M. P. (2023). Internal displacement in India: Causes, protection and dilemmas. *Forced Migration Review, 8.*

CHAPTER

6

Impact of flood vulnerability on livelihood, health, and homelessness: A case study of flood-affected blocks in lower Shilai watershed, West Bengal, India

Pinki Mandal Sahoo[1] *and Lakshmi Sivaramakrishnan*[2]

[1]Department of Geography, Mirik College, University of North Bengal, Darjeeling, West Bengal, India; [2]Department of Geography, Jadavpur University, Kolkata, West Bengal, India

1. Introduction

Destructive natural events occur regularly across the world, although most do not cause enough damage to be considered as a natural disaster. Most common among those are floods. Flooding is one of the serious, common, and costly natural disasters that affect the communities around the world. Flood vulnerability is usually affected by the land use characteristics of the areas under the influence of flood. That is to say, a flood will have different levels of vulnerability according to the landuse characteristics, potential for damage, number of people living in the region as well as availability of resources. Floods cause damage to houses, industries, public utilities, and property resulting in huge economic losses, apart from loss of lives. Though it is not possible to control the flood disaster totally, by adopting suitable structural and nonstructural measures the flood damages can be minimized.

The significance of the study on this area are summed up as follows:

- It seems that high intensity, regular, devastating flood is a common characteristic feature of a large stream and its network. But it is found that, though the Shilai River is a much smaller stream with lesser number of tributaries and distributaries, it is one of

Homelessness to Hope
https://doi.org/10.1016/B978-0-443-14052-5.00006-9

the most vulnerable streams in Paschim Medinipur, responsible for recurrent destructive flood in every year in Ghatal subdivision.

- The study area, i.e., the lower Shilai watershed is a hydrological unit, where the tributaries of this river, have huge impact on the watershed as these rivers contribute huge amount of water and sediment to the Shilai river especially during the monsoon season of the year, for which lower reaches of Shilai river is subjected to be influenced by both sedimentation and flood.
- One of the most important characteristics of this Shilai drainage system is the excessive bank erosion in Shilai and all of its tributaries. Such a situation develops the possibility of flooding as the banks are unable to hold the excessive water especially during monsoon.
- The majority of the watershed areas is inhabited by the scheduled caste, scheduled tribe, and other backward caste people, who face vulnerability each year. Therefore, immediate measures are needed to save them from the flood situation and awareness campaign is needed to develop their own resilience strategies, otherwise the homelessness situation continues to prevail and take a grave shape.

A compact study of Shilai river flood and its impact has been made by Sukul (2008), in his paper, *Annual Floods of the Silabati and the Surrounding Ecosystem*, in the book *River Floods: A Socio-Technical Approach*. He summarized his research identifying the impacts due to Shilai river flooding. People living with floods besides Shilabati river, usually habituated with annual floods. They build their houses on mounds and use small boats for transport during flood and make embankments in the northern part of the river to protect the rice crop from flooding. His detail study also reveals that the Shilabati river at present become highly polluted. The industrial effluent of Damodar river is poured into Rupnarayan and brought upstream up to Shilabati River. Moreover, the overfishing has reduced the large fishes, affecting the breeding system. This compelled the people to migrate to an extent specially the present generation.

The book, *River Floods: A Socio-Technical Approach* published another important paper of Chaudhury (2008) on *Floods—Impact on Human Health*, which minutely researched the frequency and intensity of flood and its potential health impact. The effects are broadly divided into two groups. The direct health effects can be categorized as those resulting at the time of flood event or immediately after and those longer term effects which may appear and lasts for months or even year after flood. The number of deaths associated with flooding, i.e., mortality from drowning, heart attack and injuries are closely to the life-threatening characteristics of flood falls under direct impact of flood.

In the journal *Water 2011*, Spaliviero, M., Dapper, M. D., Mannaerts, C.M., and Yachan, A. (2011) in their article *Participatory Approach for Integrated Basin Planning with Focus on Disaster Risk Reduction: The Case of the Limpopo River* have suggested that a participatory approach is a suitable method for basin planning integrating both water and land aspects. A qualitative research approach is adopted in this paper. In particular, the "*Living with Floods*" experience in the lower Limpopo River, in Mozambique, is described as a concrete example of a disaster adaptation measure resulting from a participatory planning exercise. The main focus of this article is on participatory approach for a more integrated basin planning at the local, national and subregional scales, bringing together different stakeholders in the planning process to

(i) reduce the "distance" between decision-makers and the local population; (ii) maximize the use of local knowledge and resources, hence empowering local communities; and (iii) positively influence decision-making at higher (national and even subregional) levels. This gives the opportunity to think about the present scenario of the Shilai river flooding and its management by the locals.

2. Study area

The lower Shilai watershed lies between the latitudinal extension of 22° 25′40″N to 22°52′30″N and that of the longitudinal extension is 87°27′ E to 87°50′40″E. The study area is the lower reaches of Shilai river, regionally consists of lower Shilai watershed and administratively includes three blocks of Ghatal subdivision and one municipality within it. This was chosen as study area because of some distinguishing features like, Shilai river is a much smaller stream with lesser number of tributaries and distributaries, but this is an annually flood prone area, with highly sedimented riverbed. Simultaneously the region is attributed with extreme fertile land which has the capacity to support double even triple cropping resulting huge crop production possibility and also provides the economic support to the farmers, cultivators, agricultural laborers to maintain their livelihood. But due to regular flood and consequent water inundation and deterioration in their living condition, health and to some extent their homelessness (Figs. 6.1 and 6.2) (Murty, 1998, pp. 51–64).

3. Objectives of the present study

- To analyze the flood proneness of lower reaches of Shilai river.
- To assess the flood impact on livelihood, health, and homelessness condition of the inhabitants.
- To identify the flood resilience strategies of the local community to withstand flooding situation.

4. Materials and methods

Multiple databases, both primary and secondary, were required for the research work and all are processed to develop as database and maps through SPSS, Excel, ARC GIS 10.3, and ERDAS 9.5. The primary data has been obtained on the basis of cross-sectional questionnaire survey, focused-group interview, group discussion and observation method which is based on flood situations, particularly related to flood prone areas in the lower Shilai watershed, people's socioeconomic and health condition, effect on their livelihood pattern during floods, their response to the disaster, condition of medical, transport and power sector during flooding and government's plans for overall development of the region. The secondary database mainly includes the Climatic data like water discharge data, Height of water level at Shilai pick up barrage data, Agricultural Data, Livestock Census, Topographical sheets (73N/5,

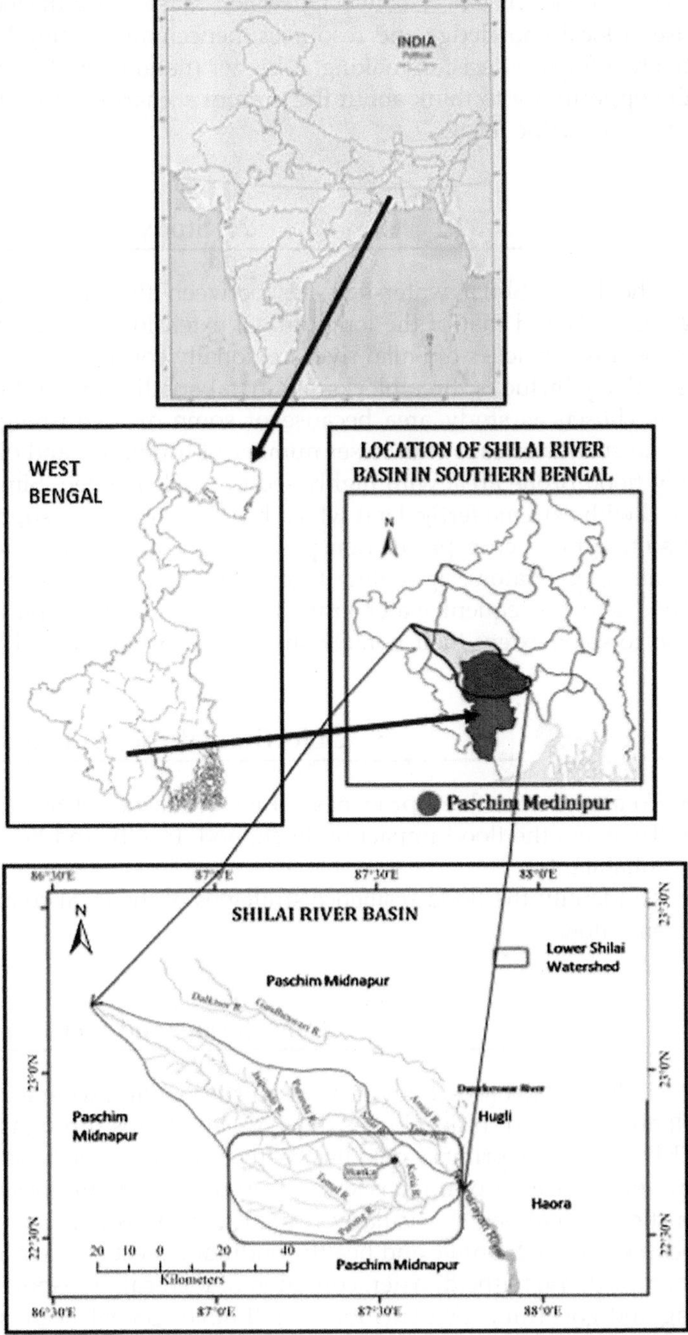

FIGURE 6.1 Location map of lower Shilai river watershed. *Prepared by Researcher.*

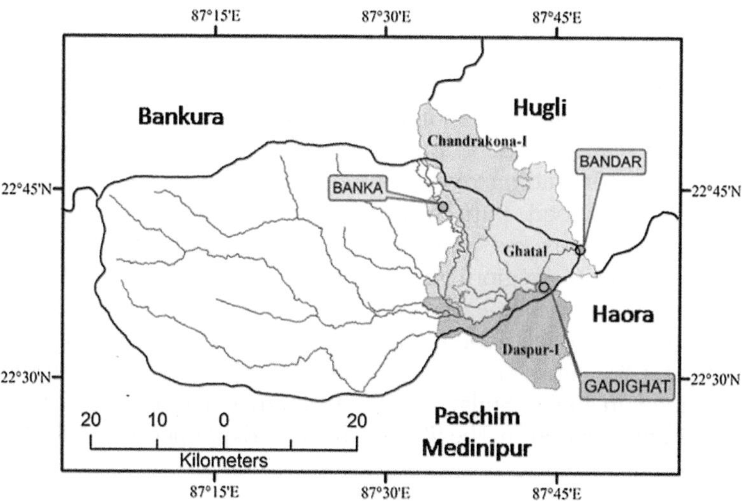

FIGURE 6.2 Study area blocks affected by flood in lower Shilai watershed. *Prepared by Researcher.*

73N/9, 73N/10, 73N/13, and 73N/14), Satellite Images [IRS AWIFS (2015), IRS LISS III(2010), LANDSAT ETM (2001), and LANDSAT TM(1990)], Mouza maps and Police station maps, District Planning series map, Watershed Atlas, Unpublished Reports, Departmental Project Reports and innumerable books, online and hardcopy journal papers, reports, numerous PhD thesis, etc.

Both quantitative and qualitative analyses, in addition to the application of Remote Sensing and GIS techniques are the major tools to drive the present research work. Several arithmetic, statistical analyses as well as quantitative analyses of qualitative data have been followed, which helps to identify problematic scenario and also show the path of proper decision making and planning for future. To identify the extent of flood exposure to local people, through purposive random sampling, cross-sectional household survey of 662 households covering almost 3338 people of three blocks are surveyed in the study area conforming both flood prone mouzas of the said three blocks and the flood prone town area of Ghatal block, along both sides of Shilai River, ranging about 100 mt from the river bank to beyond 2 km reach. The detailed survey includes querying the male population of 68%, 64%, 85%, and 77% of Daspur I, Ghatal, Ghatal municipality, and Chandrakona I block, respectively, and the rest querying of the female population. Vulnerability to flood hazards is much influenced by age, gender, caste, permanency of stay in flood prone areas, Literacy rate, household income and assets, the quality of housing and basic services, environmental health risks within the workplace, all round human development status and multidimensional poverty situations of the flood hazard prone people. So, the most vulnerable groups like Individuals/households living in poor quality homes and neighborhoods that lack adequate provision for water, sanitation, and drainage, Income-earners particularly intervened by flood hazard, people living in remote villages, scheduled caste or scheduled tribe people with little prosperity who can be discriminated in obtaining adequate incomes, housing and basic services in many cases, in respect of particular ethnic groups or castes basis, senior citizen, women, and young children who are the most vulnerable to flood hazards were being surveyed.

The primary data has been obtained on the basis of cross-sectional questionnaire survey, focused-group interview, group discussion and observation method which is based on flood situations, particularly related to flood prone areas in the lower Shilai watershed, environmental factors which are degrading river health, people's socioeconomic and health condition, effect on their livelihood pattern during floods, their response to the disaster, condition of medical, transport and power sector during flooding and government's plans for overall development of the region. For perception analysis, the developmental status of the flood affected population are expressed by calculating Human Development Index (HDI), the livelihood status of the flood affected population are expressed by calculating Human Poverty Index (HPI), Thurston Scaling Technique is being applied to identify the triggering factors of flood in lower Shilai reaches through perception analysis; the extent of impact assessment of Such Flood In The Study Area Are Explained By Multidimensional Factor Analysis.

5. Shilai river and drainage system characteristics of lower Shilai watershed

The study area is the lower Shilai reaches which is incorporated within lower Shilai watershed whose total area is 1930.15 square km. Shilai is the main river of the study area which causes recurrent floods. Shilai river originates from Boro Gram village of Maguria-Lalpur panchayat under the jurisdiction of Hura Police station of eastern part of Purulia District. From Purulia district or earlier known as Manbhum district, Shilai river enters into Paschim Medinipur through Garbeta-II block. It runs first in an easterly direction. After entering into Pashchim Medinipur, near Kharkusum the river is bisected into two parts. The eastern part of the flow of the river came to be known as Ketia Khal whereas the western part of the flow keeps its name as Shilai River. Both of these rivers flow almost in parallel fashion and turn to the south-east and then south and ultimately enter into the Ghatal subdivision through Chandrakona II block. Near Narajol, it takes a sharp turn to the north and meet with Dwarkeswar river at Bandar. The main tributaries of Shilai river during its flow are Parong Nadi, Tamal Nadi, Buriganga Nadi, Donai Nadi, Kubai Nadi, etc. Due to regular alluvial deposition from the main river and its tributaries, the riverbed gradually grows and in rainy season it causes floods in Ghatal, and parts of Chandrakona I and Daspur-I blocks of Ghatal subdivision which area is physiographically known as lower reaches of Shilai river. From the geographical point of view, it is a depressed area (Sukul, 2008).

6. Reasons for flooding in the lower reaches of Shilai river

Flood can be in the form of local inundation due to heavy rain or due to accumulation of run off/discharge after heavy rain in the catchments and or due to breach of embankment. Major contributing factors for flood can be categorized both as natural and artificial (Chowdhury, 2008).

6.1 Natural causes identified for flood

- Heavy local rainfall, high intensity rainfall within a short span of time, variability of rainfall in the study area is the main reason of flood in Shilai river.
- Heavy flow of water in the drainage channels within the basin boundaries during the peak flood period, the river rules high.
- During monsoon the traditional earthen embankments get partially damaged due to over topping at reaches of low elevation, percolation of flood water, resulting to breaches at weak reaches.
- Flood in Shilai river is totally landscape related, the Shilai river catchment area especially in the south eastern portion comprising the Ghatal block and Daspur I block are saucer shaped areas.
- Lower part of the Shilai river has been jacketed by putting up and raising of ex zamindari bundhs. Now it is a place of flooding.
- The downstream of the river gets silted up constantly due to tidal effect which causes flooding.
- Due to gentle longitudinal slope of Shilai riverbed, it has lost drainage efficiency which causes flood in the study area.
- Bank slides or bank erosion causes blocking in the course of stream and then sudden release of blockage over the stream during monsoon, results excess water to flow over the riverbed.
- There is inadequate drainage to carry away surface water quickly, the Shilai river has a few number of distributaries to regulate the excessive water during monsoon, therefore causing flood (Das et al., 2011).

6.2 Anthropogenic reasons of flood

- During monsoon, the area of lower Shilai reaches faces flood hazards almost every year due to the combined effect of high spillway discharge from the Kangsabati dam at Mukutmonipur in the district of Bankura. The contribution of accumulated water from the uncontrolled catchments below the dam may be very high when there is rainfall in the district of Bankura or Paschim Medinipur. The situation further worsens when it synchronizes with the release from Kangsabati Dam. Consequent high run off and flood discharge from D. V. C. system and tidal lockage in the river Rupnarayan worsen the flood situation.
- Shilai river in the entire course of the stream flow from the time of independence till date, has resulted the river to be sedimented and silted, has reduced the water holding capacity of the main river and its tributaries, therefore any climatic fluctuation and consequent human induced activities has worsened the situation as flood intensity increases with each passing year.
- The land-use characteristic of majority of the catchment area is either riverbed cultivation or pisciculture. Such practices results in flood mostly during rainy season, as the river bed which is attributed to carry the river water is being utilized as cropland. This practice is considered as environmentally destructive in nature.

- Construction of boro-bundhs across the river for Rabi and Boro irrigation are also causing siltation of the riverbed. This is an irrational practice which may reduce the irrigation costs of the farmers but on other hand degraded the riverbed condition reducing river thalweg, hampering the river flow especially during rainy season.
- Lack of proper control of land use following irrational method like tree cutting, absence of tree plantation, agricultural activities like tremendous usage of pesticides, fertilizers deteriorate the quality of the soil which does not help in percolation of water, moreover soil erosion is taking place, resulting in uncontrolled surface flow, which during the monsoon period appear to be destructive and causes recurrent flood.
- The sluice gates/lock gates are being closed except Bakshi and Ranichak, as this have its impact on the river flow during monsoon as silted up rivers have become unbearable to carry excess water and as the sluice gates are closed, therefore the river has no option other than to flood the two flanks of its course.
- The height of barrage affecting Shilai river like Shilai pick-up barrage, Kangshabati barrage is getting lower as it is also sedimented because of accumulation of water for years, lacking in artificial cleaning. Therefore, if high intensity rainfall occurs within a short span of time, as the water holding capacity of the barrages is reduced, so they release more water frequently causing devastating flood in the catchment areas of Shilai river.
- Deforestation in the upper catchment areas reducing retaining capacity of water and holding soil and consequent soil erosion resulting in silting of riverbeds, causing the Shilai river non-navigable along with its other tributaries and distributaries. This results in shifting course of the river which may cause flood in the entire riverine region.
- Beside these, numerous other triggering factors are responsible for flood in the lower reaches of Shilai river. Among the prime reasons of flood, huge flood flow from upstream locations, embankment breaching both due to erosion and high-water level, outflow from Shilai pick up barrage or Kangshabati barrage are important.

7. Downstream effects of Shilai pick-up barrage on rivers of lower Shilai watershed

The study on the impact of pick-up barrages across the flow of alluvial rivers in the lower Shilai watershed shows that discharge, sediment, channel, and floodplain morphology is interrelated to the flooding situation. If the discharge of the Shilai river is studied for last 39 years, it is found that the discharge has been varied depending on the amount and intensity of rainfall, height of outflow from the barrage etc. The highest discharge is found in the year 1986 which is about 2439 cusec. Beside this year, 1989, 1993, 1998, 1999, 2001, 2006, 2008, 2011, and 2012 are the years of greater discharge of water in Shilai river which is more than 2000 cusec. While the year 1978, 1984, 1985, 1995, 2003, 2004, 2005, 2009, and 2014 has lowest amount of river flow discharge, i.e., below 1500 cusec. The minimum amount of discharge of water however is found in the years 1979, 1980, 1981, 1982, 1994 when it almost reached below 1000 cusecs. Disparity in the amount of rainfall is the prime cause of such a great variability of discharge of water. Such a variability of flow upon the stream has its impact over the navigability as huge amount of siltation takes place. There is another close relation between discharge and frequency of flooding. The years 1981, 1984, 1986, 1987, 1989, 1993, 1998, 1999, 2001, 2006, 2007, 2008, 2011, and 2012 faced flooding about twice in the lower reaches of Shilai river. Rest of the years however faced the flooding only once if we consider for last 39 years. Therefore, the Discharge from the Shilai Pick-Up Barrage at Kadamdeulihas a direct impact on the flooding at lower Shilai watershed region.

8. The extent of flood damage in lower reaches of Shilai river

It is found that large population of most of the Gram Panchayats of the three blocks are facing this recurrent devastating flood and its consequent occurrence. It is found that among them 62 mouzas of 4 g panchayats that is almost 50% of the total areas are regularly flood affected and this flood creates severe problems to the regular livelihood of the people. In Ghatal block there are 156 mouzas under 12 g panchayats (2011 census) covering 229.91 square km area, and among them about 125 mouzas that is almost 80% of the total are flood affected. Similarly, in Daspur I block covering 168.3 square km. area and among 10 g panchayats, 5 are fully and 7 are partly being flood damaged on regular basis and constituting about 157 mouzas, 93 are regularly affected due to Shilai river flood that is almost 59% of the total are flood affected along with its infrastructure, cropland as well as population. Such a great extent of flooding makes the region distinguishable as area of people living with floods in South Bengal. Floods are the most common natural calamities that West Bengal has to face almost every year in varying degrees. Flood is an inevitable natural phenomenon occurring from time to time in all rivers and natural drainage systems, which not only damages the lives, natural resources and environment, but also causes the loss of economy and health (Kolay, 2007, pp. 210–278) (Figs. 6.3 and 6.4).

9. Impact assessment of Shilai river flood

From the primary survey over 662 households in the study area covering 3338 persons, as well as from the secondary survey, it is evident that the occurrence of the flood in the lower reaches of Shilai river areas has aggravated the situation especially due to slow but continuous change-over of the hydro geomorphic factors and its characteristics for past a few years. In addition, various other dimensions have identified by the local people as well as district planners as the causes of local flood (Messner & Meyer, 2006).

9.1 Impact on homelessness

One of the striking issues to be identified during the flood event in lower Shilai is the homelessness of the people, and getting a temporary family accommodation with mass in a school premise or in a government relief center. It has been found that when such annual, devastating, recurring flood event takes place in extreme, people become temporarily homeless even for a few months of stretch. It implies that people left homeless with no available drinking water, lack of food, medicine, electric power, and life of them

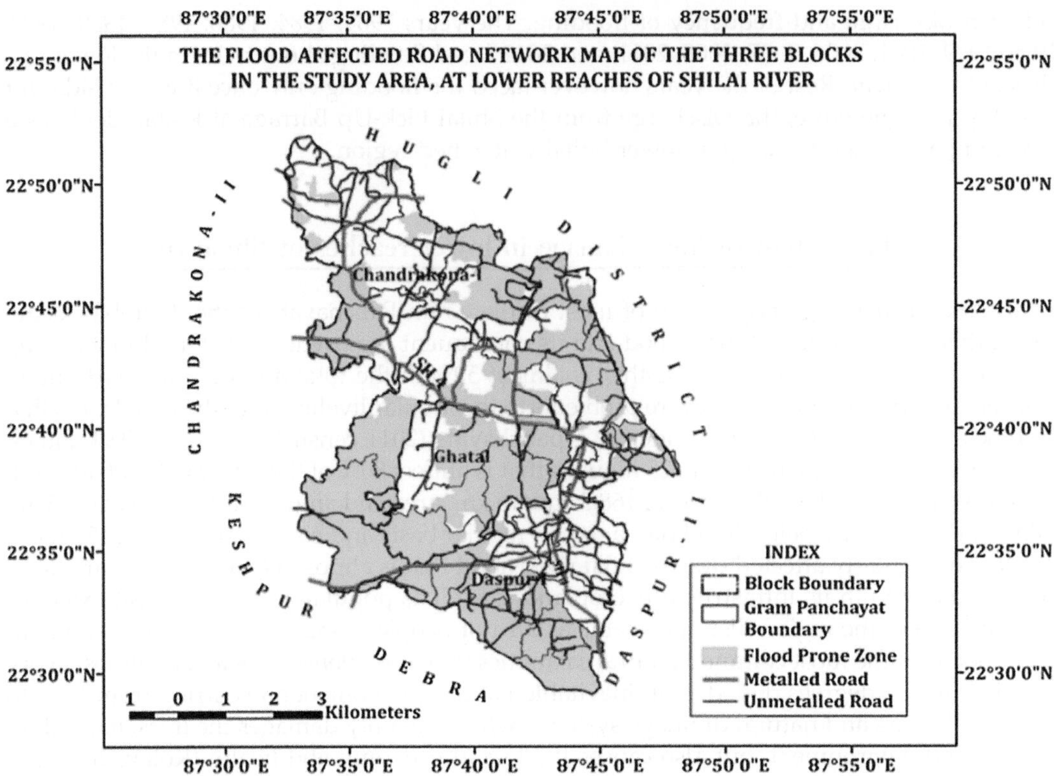

FIGURE 6.3 The flood prone mouzas at lower Shilai river reaches. *Prepared by Researcher.*

almost become standstill as during the water stagnation day, both public and private means of transport becomes completely unavailable. People's lives become confined into the shelter house with the government help in terms of food, clothing, essentials etc. within government school premises. But the homeless people have almost no access to the important documents, home assets, etc. Students may lose their school books, copies, examination admit cards, their parents may lose their deeds, certificate, credentials, any written records when they lose their home due to extreme sudden flood, which they have mere hope to recover in near future. Therefore, a person's lifelong earning goes in vain with the immediate and recurrent occurrence of flood, and make people shelter less on frequent interval. Moreover, the recurrent annual flood compels the family to think about their present and future condition whether living with floods or to shift permanently to some other places. The decision is quite difficult as most people have their forefather's land to cultivate, or may be engaged in either pisciculture or in any other kind of business work. To shift to a new place actually means to earning a new livelihood, maintaining their new house, new school for their children, along with that also maintaining their previous land for cultivation, house, etc. After reaching to a certain age, most of the people do not get interested to take the new risk in life. Therefore, people of lower Shilai watershed are left with only one option, that, during flooding time they have to

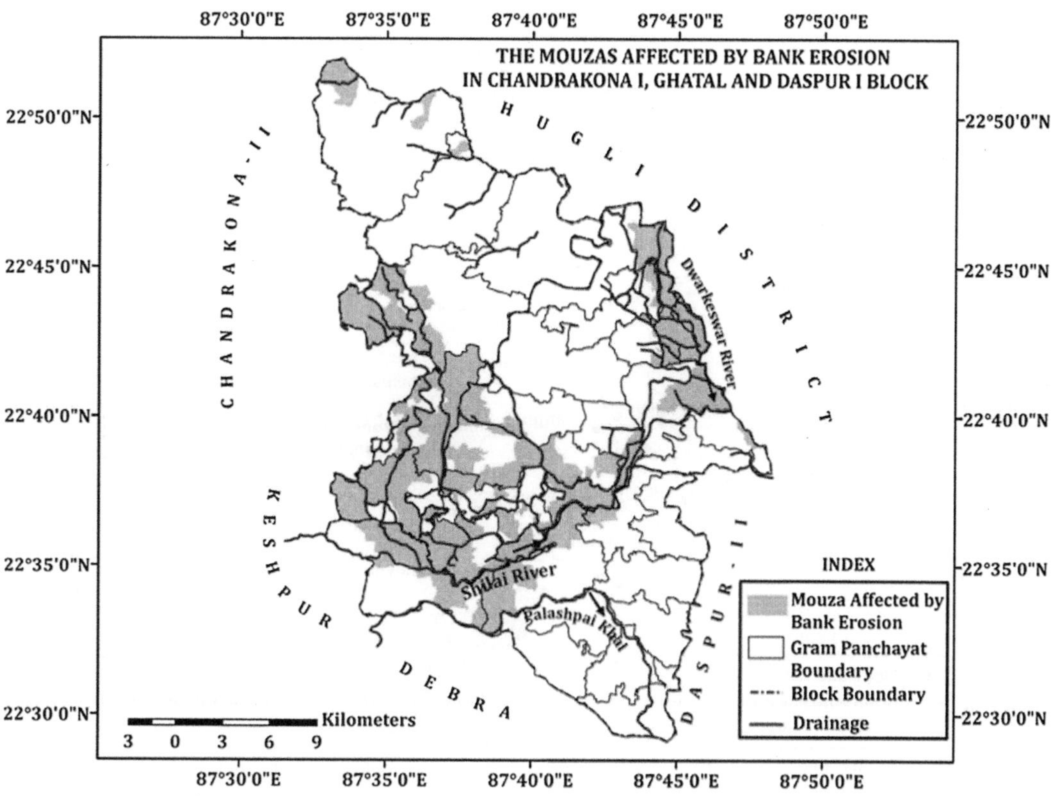

FIGURE 6.4 The mouzas affected by bank erosion in lower Shilai river reaches. *Prepared by Researcher.*

become homeless, to live in a government provided school as relief center, spending a few of the months, again they have to come back to their home. This implies that locals are actually spending a migratory life full of uncertainty due to the flooding situation. This has its impact not only economically, socially but also psychologically which sometimes is proved to be detrimental to peoples' mental health (Fig. 6.5).

9.2 Impact on human health

The severely affected sector is the health and medical infrastructure in the study area during flooded time. As a very few people in the rural households can access their home latrine during flood, majority of them have to move far from houses. The sanitization gets hampered as the drain-water and potable water pipe gets leaked and unhygienic water mix-up takes places. For these reasons people face various health diseases like Diarrhea, Cholera, Pox, Food poison and other water-borne diseases but majority of the people gets affected by snake bite during flood in the study area. Moreover, during such waterlogged period Doctors' availability both in hospitals and private nursing homes reduces remarkably, the services in hospital, primary health care centers, providing ambulance, the availability of nurses,

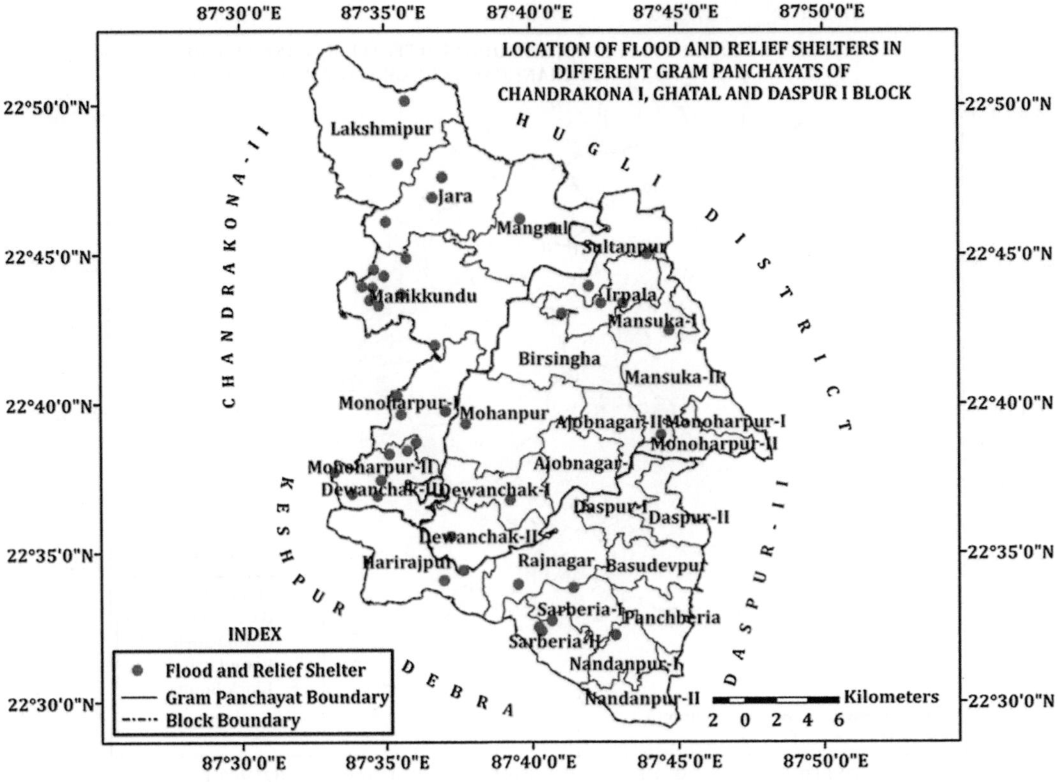

FIGURE 6.5 Location of flood and relief shelters in three blocks of lower Shilai reaches. *Prepared by Researcher.*

medicine shops also found negligible during flood time. Thus, the whole medical infrastructure tumbled down during this annual and recurring Shilai river flood (Chaudhury, 2008).

9.3 Impact on livelihood attributes

Damage to houses: Among the infrastructural impact the first place is occupied by house damage. The extension lies on riverbank household dwellers, kuchha as well as pucca house damage due to flood, even shifting from houses during flood hazard is also a common phenomenon in the study area. To overcome the situation the reconstruction of houses takes place in some cases. It is evident from the survey that in most vulnerable households are found in Daspur I block. Ghatal block exclusive of Ghatal Municipality occupies second position in terms of vulnerability of house damage then Chandrakona I block and lastly the Ghatal Municipality.

Inaccessibility to potable water: During flood the main hindrance found in accessing the potable water. In all the four administrative regions, during flood, in most of the cases, nearby sources of water like tube well, municipal tap water is inundated therefore water is only available at distance which is a major drawback for the flood prone people. Moreover, the available water is sometimes unhygienic in nature (Singh, 2006, pp. 10–12).

Damage to transport sector due to Shilai river flood: Transportation sector damage includes all the roads, power and telephone lines damage during the flooding. The main roadway is the State highway number four which connects all the three blocks with each other and during flood this road gets severely damaged. Beside these, numerous other metaled roads are also affected during inundation period. But the extreme negative impact falls upon the morrum roads and earthen roads together identified as unmetalled roads. The transport infrastructure of the locality totally breaks down during flood in the lower reaches of Shilai river. The main road that connects all the administrative localities is the State Highway 4, and due to flood, various parts of this highway get damaged and after the flooding season the government authority in transport section needs to invest a lot of money to get it reconstructed. In all the four administrative regions, the daily type transport mode becomes unavailable, or sometimes rarely available. They found the usage of boat by all, during flooded and waterlogging period in the region. People are under the threat of forcible changes of routes during floods. The most important and hazardous impact in these sectors the damage of both metaled and unmettalled roads during long period of waterlogging and flood. It becomes merely unaffordable for the local government too to reconstruct these roads every year by this annual nature of flood (Sahoo & Sivaramakrishnan, 2014).

Damage to power infrastructure: The electricity or power infrastructure during flood gets totally disrupted in the study area, i.e., in all four administrative regions the areas faces power cut during the flood sometimes even 7 consecutive days at a stretch. In that time people are compelled to use lantern, hurricane, lamp, battery driven emergency lamp, etc. Naturally the normal livelihood totally gets hampered (Singh et al., 2014).

Impact on agriculture and consequent crop damage analysis: The region is considered to be very fertile where agricultural productivity and production both are very high because of so many reasons. But annual and destructive flood occurrence in Shilai river has tremendous impact on the damage and destruction of agricultural lands, reduces its fertility and productivity and therefore hampers the locality's economic growth. It was found in 2013, a recognized flooded year, that in almost all gram panchayats of the three blocks the croplands are damaged, but maximum having more than 90% damage took place in Monoharpur I and II of Ghatal block. Then Monoharpur I and II of Chandrakona I block, the southern locating gram panchayats of Ghatal and gram panchayats located in western part of Daspur I block adjacent to Shilai river are having high rate of crop damage which ranges between 61% and 90% of the total production. The centrally located gram panchayats of Ghatal and Chandrakona I are at the next level of crop damage which is moderate in nature having 31%–60% of crop damage. Due to flood, majority of the people having less than Prepared

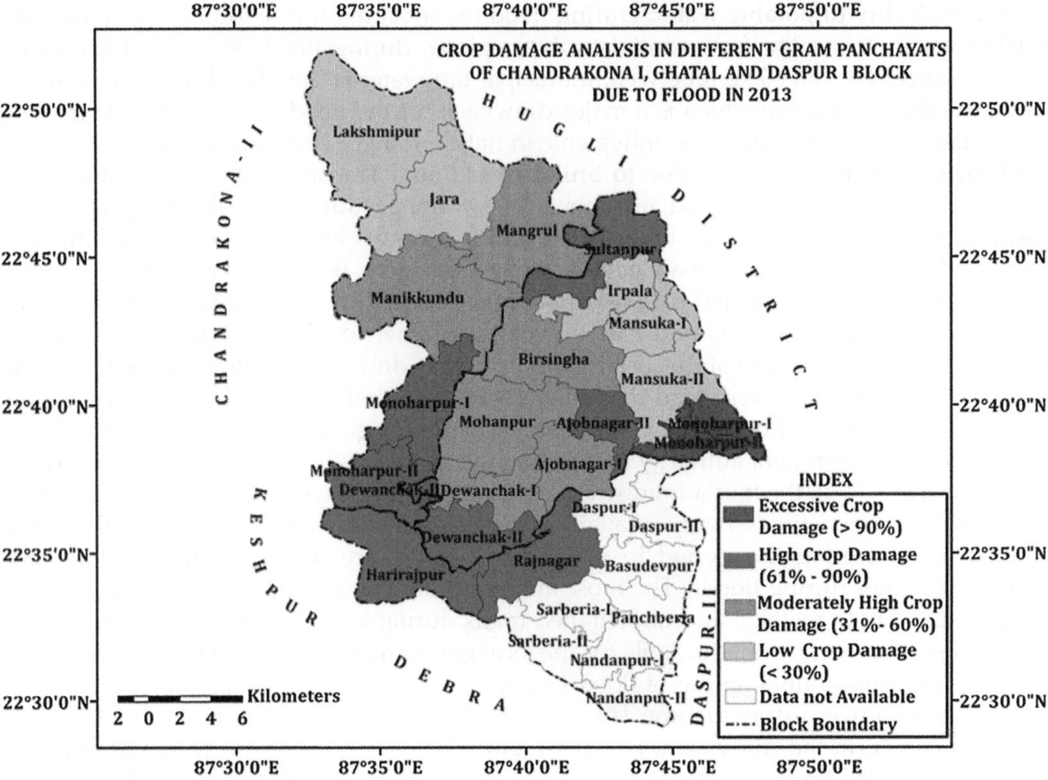

FIGURE 6.6 The mouzas affected by crop damage due to flood in lower Shilai river reaches. *Prepared by Researcher.*

by Researcher 0.0013 sq km (1 bigha) of land are affected. The surveyed households of Ghatal block are engaged in diversified agriculture, therefore it is found that 31% is engaged into single farming, about 64% in double farming and even 5% in triple farming practices, though 12% people do not have any farmland. In this block the households having less than 0.0013 sq km of land (46%) and households having less than 0.0066 sq km (5 bigha) (45%) of land are affected due to flood. Ghatal Municipality area too has some agricultural lands in its interior. With 30% households only having farmlands, about 71% of them are engaged in single farming and the rest in double cropland farming and households only posses less than 0.0013 sq km of land are majorly affected by flood. On the contrary, Chandrakona I block has 70% of farmland out of which majority of households are engaged in double cropping, only 22% in single farming and a very little about 6% are engaged in triple farming but here also people having small fragmentation of land measured nearly 0.0013 sq km of land are mostly affected due to flood (Fig. 6.6) (Table 6.1).

TABLE 6.1 Characteristics of land and flood impact on farmland.

Administrative region	Size of farmland affected					Characteristics		
Block/Town	No. farmland	Less than 1 bigha	Less than 5 bighas	Less than 10 bighas	Less than 15 bighas	Single farming	Double farming	Tripple farming
Daspur I	9%	55%	30%	6%	-	79%	21%	
Ghatal	12%	46%	45%	9%	-	31%	64%	5%
Ghatal municipality	70%	79%	14%	7%	-	71%	29%	
Chandrakona I	30%	66%	30%	2%	2%	22%	72%	6%

Prepared by Researcher. Primary Survey.

Impact on other service sectors: Beside this specifically the fishermen, the businessmen and the householders' economic loss during flood is tremendous in all the four administrative regions. The former is threatened by fish death, loss of fishes, mixing up of impurities during flood which results in huge economic loss in pisciculture sector and the Dingal village of Chandrakona I block is one of such examples. The businessmen lose in terms of his property, bought items, mode of transportation for transferring goods, which ultimately incurs huge economic loss in a few lakhs every year. Similarly the householders may loss its houses, household goods, crops, livestock, purchased owned vehicles, and therefore their income gets affected sometimes to a few lakhs too which is merely impossible for them to reimburse (Burton & Cutter, 2008).

Impact on livestock: Observing the overall livestock scenario it is evident that the rural households of the study area have more livestock in their houses compared to the town householders. In Chandrakona I block, apart from other blocks, surveyed householders have majority of more than five livestock in per households. Naturally they are prone to flood hazard by Shilai river. It is found that due to flood a huge number of livestock in all these areas are in the attack of diseases, some of them become ill due to unavailability of drinking water and food and a few also faces death due to tremendous flood hazard (Dolui & Ghosh, 2012) (Table 6.2).

Therefore, to assess the combined impact of flood on the surveyed households, Principal Component Analysis method has been adopted, and there by identifying the level of flood vulnerability in various dimension and in various places.

TABLE 6.2 Characteristics of livestock.

	Number of livestock		
Block/Town	**Absence of livestock(percent)**	**Less than 5(percent)**	**More than 5(percent)**
Daspur I	26	83	17
Ghatal	44	60	40
Ghatal municipality	59	70	30
Chandrakona I	42	34	66

Prepared by Researcher. Primary Survey.

10. The multidimensional impact of flood in the lower reaches of Shilai river

The Shilai river flood impact in different arena of the livelihood of both human beings as well as of other animals are being identified by primary surveying. Therefore, it is necessary to identify the highest affected flood vulnerable administrative region among them, by assessing the proportional flood impact which makes a region most floods vulnerable. It is done through a quantitative assessment called *Factor Analysis*. Initially from primary survey, the flood impact in each sector (in percent) is being identified. Then after standardization, the data is utilized for calculation purpose (Figs. 6.7–6.10).

Observing the loadings of each variables in three factor components, it is clear from the component matrix that **Factor 1** is combined (is most impacted in) of and are strongly explained by ***crop loss, field damage and profit reduction, sedimentation helps in cultivation, house damage and reconstruction, shifting from houses or homelessness, tube-well inundated and potable water at distance, suffering from snake bites and to move for outside latrine during flood or health deterioration***. However, in **Factor 2** component matrix, the impactful variables are **power-cut during flood time and road damage** and usage of boat during flood whereas in **Factor 3** component matrix the ***variable medical infrastructure disrupted*** got the highest importance in flood impact. Rest of the cross-loadings can be ignored. With respect to Factor 1 component, the high positive scores are found in Rajnagar of Daspur I block, and Dewanchak I, Dewanchak II, Ajobnagar II, Monohorpur I, in Ghatal block which implies that this region is mostly affected and vulnerable to Shilai river flood in the lower reaches, with the impact of the Factor 1 components. Whereas Monohorpur I of Chandrakona I block, Manikkundu, Harirajpur I, Ajobnagar I, Monohorpur II have moderate impact due to annual and recurrent Shilai river flood. Whereas lesser score of Ghatal Municipality, Monohorpur II of Chandrakona I block has implied the lesser propensity to flood which is found through the multivariate factor analysis method (Fig. 6.11).

Block	Sl No.	Location	Croploss, Field damage & Profit Reduction (%)	Sedimentation helps in cultivation (%)	Death of Livestocks (%)	House damage & Re-construction (%)	Shifting from House (%)	Tubewell inundated & Potable water available at distance(%)	Suffering from food poison and Snake Bite(%)	Unavailability of Doctors & Medical infrastructure disrupted (%)	Power cut during flooding (%)	Need to move outside for latrine (%)	Road Damage & usage of boat (%)
Daspur I	1	Harirajpur I	63	9	5	26	9	29	81	32	79	42	58
	2	Rajnagar	70	8	9	24	15	33	75	47	78	54	40
Ghatal	3	Dewanchak I	67	13	11	30	7	48	65	52	86	35	81
	4	Dewanchak II	60	15	12	25	8	50	74	50	89	26	64
	5	Ajobnagar I	66	14	10	31	6	31	67	38	85	34	68
	6	Ajobnagar II	67	8	6	28	10	47	58	64	87	29	61
	7	Monohorpur I	70	7	10	35	8	32	69	51	86	36	63
	8	Monohorpur II	71	8	12	26	9	50	72	32	89	18	68
	9	Ghatal Municipality	45	2	1	18	2	13	78	65	85	7	72
Chandrakona I	10	Monohorpur I	63	4	10	23	6	24	86	50	92	10	65
	11	Monohorpur II	57	3	7	21	6	19	71	53	90	12	58
	12	Manikkundu	58	3	11	18	5	20	87	51	91	11	45

Source: primary survey

FIGURE 6.7 Flood damage analysis of Shilai river on various parameters (primary survey). *Prepared by Researcher.*

		CropLoss	Sedimentatn	DeathLvstk	HouseDmg	Shifting	Tubewllnun	Snakebite	Dsruptd Mdcl	Powercut	OutsidLtrn	RoadDmg
Correlation	CropLoss	1	0.485	0.623	0.736	0.74	0.66	-0.41	-0.495	-0.202	0.648	-0.096
	Sedimentatn	0.485	1	0.469	0.635	0.339	0.744	-0.478	-0.379	-0.256	0.59	0.321
	DeathLvstk	0.623	0.469	1	0.352	0.271	0.539	-0.018	-0.417	0.376	0.137	-0.038
	HouseDmg	0.736	0.635	0.352	1	0.341	0.562	-0.644	-0.271	-0.232	0.61	0.354
	Shifting	0.74	0.339	0.271	0.341	1	0.517	-0.269	-0.305	-0.545	0.785	-0.497
	Tubewllnun	0.66	0.744	0.539	0.562	0.517	1	-0.587	-0.226	-0.014	0.41	0.29
	Snakebite	-0.41	-0.478	-0.018	-0.644	-0.269	-0.587	1	-0.201	0.125	-0.36	-0.382
	Dsruptd Mdcl	-0.495	-0.379	-0.417	-0.271	-0.305	-0.226	-0.201	1	0.258	-0.347	0.08
	Powercut	-0.202	-0.256	0.376	-0.232	-0.545	-0.014	0.125	0.258	1	-0.811	0.242
	OutsidLtrn	0.648	0.59	0.137	0.61	0.785	0.41	-0.36	-0.347	-0.811	1	-0.238
	RoadDmg	-0.096	0.321	-0.038	0.354	-0.497	0.29	-0.382	0.08	0.242	-0.238	1

a. Determinant = 2.21E-007 b. Kaiser-Meyer-Olkin Measure of Sampling Adequacy = .545

Source: Calculated from the values generated in Table 12.1 through primary survey

FIGURE 6.8 Correlation matrix of all parameters. *Prepared by Researcher.*

	Initial	Extraction		Initial	Extraction		Initial	Extraction
CropLoss	1	0.852	Shifting	1	0.828	Powercut	1	0.891
Sedimentatn	1	0.709	Tubewllnun	1	0.778	OutsidLtrn	1	0.952
DeathLvstk	1	0.911	Snakebite	1	0.786	RoadDmg	1	0.778
HouseDmg	1	0.771	Dsruptd Mdcl	1	0.514			

Source: Calculated from the values generated in Table 12.1 through primary survey

FIGURE 6.9 Communalities table of factor analysis. *Prepared by Researcher.*

(A)

Component	Initial Eigenvalues			Extraction Sums of Squared Loadings		
	Total	% of Variance	Cumulative %	Total	% of Variance	Cumulative %
1	4.897	44.521	44.521	4.897	44.521	44.521
2	2.227	20.242	64.763	2.227	20.242	64.763
3	1.646	14.966	79.729	1.646	14.966	79.729
4	0.985	8.953	88.682			
5	0.547	4.977	93.659			
6	0.354	3.218	96.877			
7	0.205	1.863	98.74			
8	0.078	0.708	99.448			
9	0.036	0.33	99.778			
10	0.018	0.168	99.945			
11	0.006	0.055	100			
Extraction Method: Principal Component Analysis.						

Source: Calculated from the values generated in Table 12.1 through primary survey

(B)

Variables	Component		
	1	2	3
CropLoss, Field Damage & Profit reduction	**0.885**	-0.038	0.261
Sedimnentation helps in Cultivation	**0.79**	0.285	-0.052
Death of Livestocks	0.512	0.291	**0.751**
House Damage & Reconstruction	**0.806**	0.291	-0.194
Shifting from houses	**0.743**	-0.523	0.054
Tubewell Inundated and potable water at distance	**0.79**	0.385	0.069
Suffering from Snake bites	**-0.565**	-0.413	0.544
Medical infrastructure Disrupted	-0.475	0.215	**-0.491**
Powercut during flood time	-0.421	**0.675**	0.508
To move for outside Latrine During Flood	**0.823**	-0.456	-0.257
Road Damage and usag of boat during flood	0.045	**0.82**	-0.323

(C)

Location	CropLoss	Sedimentatn	DeathLvstk	HouseDmg	Shifting	Tubewl Inun	Snakebite	Dsruptd Mdcl	Powercut	OutsidLtrn	RoadDmg	SUM	Eigen value of FC1	Factor Score= SUM/ Eigen value of FC1
Harirajpur I	55.755	7.11	2.56	20.956	6.687	22.91	45.765	15.2	33.259	34.566	2.61	247.378	4.897	50.51623443
Rajnagar	61.95	6.32	4.608	19.344	11.145	26.07	42.375	22.325	32.838	44.442	1.8	273.217	4.897	55.79273024
Dewanchak I	59.295	10.27	5.632	24.18	5.201	37.92	36.725	24.7	36.206	28.805	3.645	272.579	4.897	55.6624464
Dewanchak II	53.1	11.85	6.144	20.15	5.944	39.5	41.81	23.75	37.469	21.398	2.88	263.995	4.897	53.90953645
Ajobnagar I	58.41	11.06	5.12	24.986	4.458	24.49	37.855	18.05	35.785	27.982	3.06	251.256	4.897	51.30814785
Ajobnagar II	59.295	6.32	3.072	22.568	7.43	37.13	32.77	30.4	36.627	23.867	2.745	262.224	4.897	53.54788646
Monohorpur I	61.95	5.53	5.12	28.21	5.944	25.28	38.985	24.225	36.206	29.628	2.835	263.913	4.897	53.89279151
Monohorpur II	62.835	6.32	6.144	20.956	6.687	39.5	40.68	15.2	37.469	14.814	3.06	253.665	4.897	51.80008168
Ghatal Municipality	39.825	1.58	0.512	14.508	1.486	10.27	44.07	30.875	35.785	5.761	3.24	187.912	4.897	38.37288136
Monohorpur I	55.755	3.16	5.12	18.538	4.458	18.96	48.59	23.75	38.732	8.23	2.925	228.218	4.897	46.60363488
Monohorpur II	50.445	2.37	3.584	16.926	4.458	15.01	40.115	25.175	37.89	9.876	2.61	208.459	4.897	42.56871554
Manikkundu	51.33	2.37	5.632	14.508	3.715	15.8	49.155	24.225	38.311	9.053	2.025	216.124	4.897	44.13395957

(D)

Factor Score	Surveyed villages of selected Gram Panchayats
Less than 44	Ghatal Municipality, Monohorpur II(of Chandrakona I)
44.1-48	Monohorpur I (of Chandrakona I) , Manikkundu
48.1-52	Harirajpur I , Ajobnagar I, Monohorpur II
>52	Rajnagar, Dewanchak I, Dewanchak II, Ajobnagar II, Monohorpur I,

FIGURE 6.10 (A) Total variance explained. *Prepared by Researcher.* (B) The component matrix of factor analysis. (C) Factor load values for factor component 1 and factor score of each station. (D) Flood vulnerable zone based on factor score.

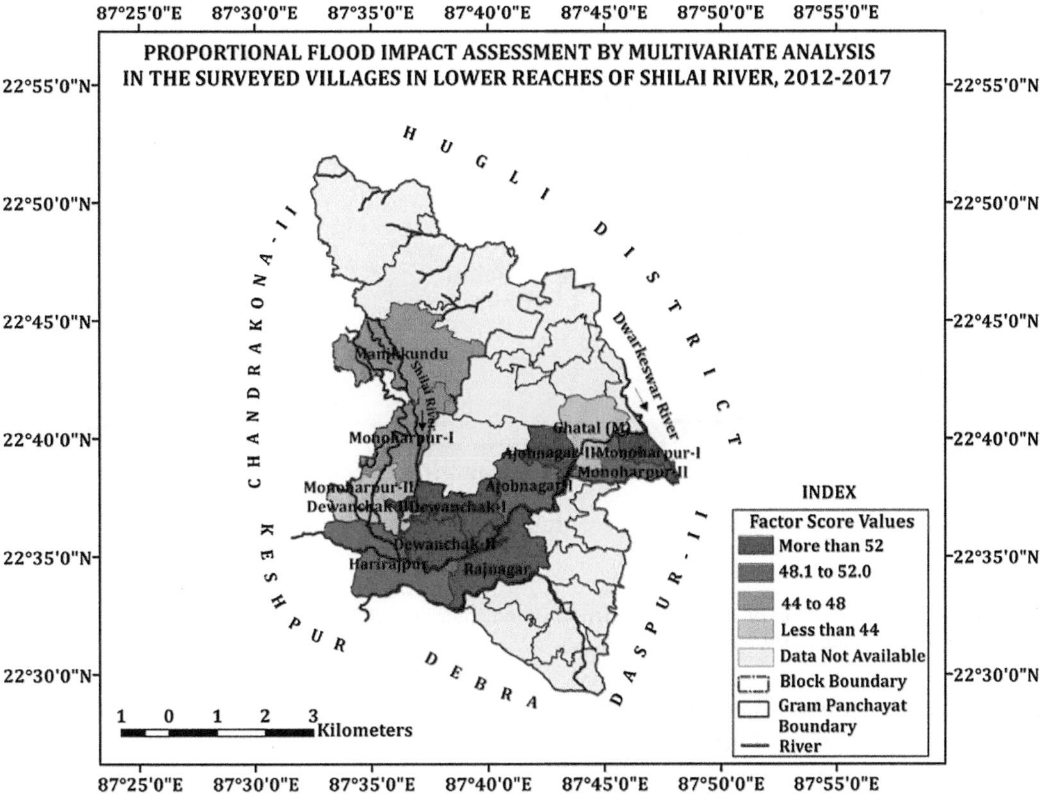

FIGURE 6.11 Proportional flood impact assessment by multivariate analysis in the surveyed villages in lower reaches of Shilai river (factor component 1 loading). *Prepared by Researcher.*

11. Livelihood status of the surveyed households

11.1 Measurement of multidimensional poverty index

Poverty has traditionally been measured in one dimension, usually income or consumption. In this analysis, a basket of goods and services considered the minimum requirement to live a nonimpoverished life is valued at the current prices. People who do not have an income sufficient to cover that basket are deemed poor. Income poverty certainly provides very useful information. Yet poor people themselves define their poverty much more broadly to include lack of education, health, housing, empowerment, employment, personal security and more. No one indicator, such as income, is uniquely able to capture the multiple aspects that contribute to poverty. The Multidimensional Poverty Index (MPI) is an index designed to measure acute poverty by measuring the proportion of people who experience multiple deprivations and the intensity of such deprivations. Therefore, the MPI combines two key pieces of information to measure acute poverty: the incidence of poverty, or the proportion of people (within a given population) who experience multiple deprivations, and the intensity of their deprivation—the average proportion of (weighted) deprivations they experience (Alkire & Foster, 2011) (Fig. 6.12).

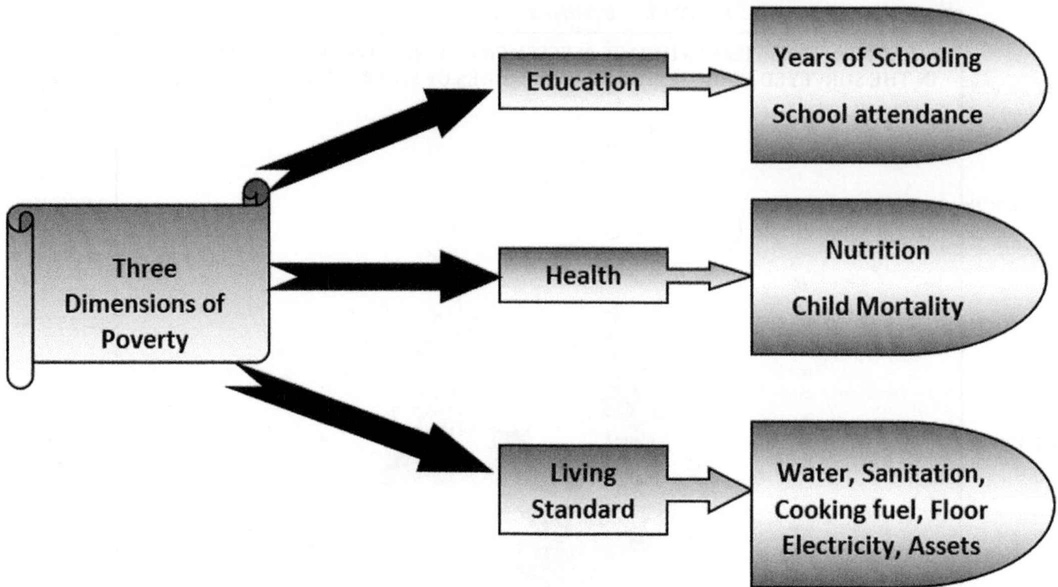

FIGURE 6.12 Composition of the MPI—dimensions and indicators. 1 indicates deprivation in the indicator; 0 indicates nondeprivation. *Prepared by Researcher.*

Vulnerability to flood hazards is much influenced by age, gender, caste, permanency of stay in flood prone areas, Literacy rate, household income and assets, the quality of housing and basic services, environmental health risks within the workplace, all round human development status and multidimensional poverty situations of the flood hazard prone people (Fig. 6.13) (Table 6.3) (Smit & Wandel, 2006). So, among the most vulnerable groups are: **Individuals/households living in poor quality homes and neighborhoods that lack adequate provision for water, sanitation, and drainage, Senior citizen, women and young children, social groups facing discrimination on the basis of class/caste and the Income-earning groups** (Patra, 2008).

11.2 Identification of human development index of the flood affected population

Appropriate indicators need to be combined to form composite index so that variation across the flood affected study area can be explained. This is done by forming Life expectancy index (LEI), Education Index (EI), and economic livelihood index or Income Index (II). The indices have been combined together to analyze the Human Development Index (HDI) of the different administrative units. However, in November 2010, The Human Development index combines three dimensions:

- A Long and healthy life expressed by Life expectancy Index: Life expectancy at birth
- Education Index: Mean years of schooling and expected years of schooling
- A decent standard of living by Income index: GNI per capita (PPPUS$)

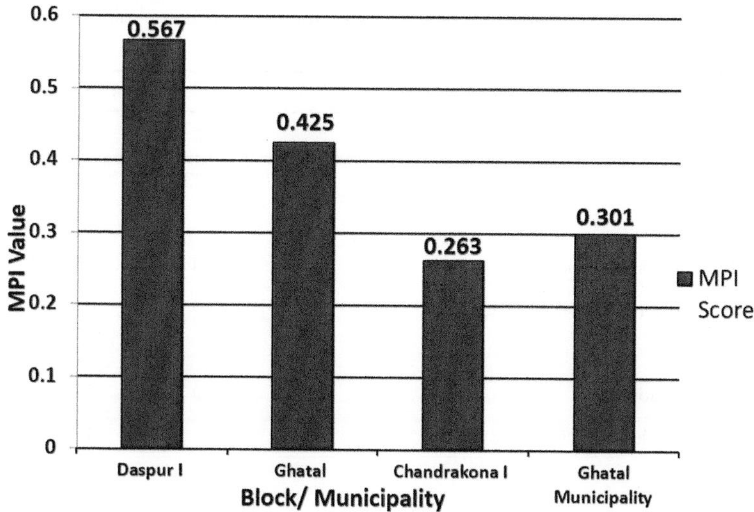

FIGURE 6.13 Multidimensional poverty index of four flood prone administrative areas of lower Shilai reaches. *Prepared by Researcher.*

In its 2010 Human Development Report, the UNDP began using a new method of calculating the HDI. The following three indices are used:

1. Life expectancy Index (LEI) = LE-20/85−20
2. Education Index (EI) = (MYSI + EYSI)/2

   Where, Mean Year of Schooling Index (MYSI) = MYS/15.
   Expected Year of Schooling Index (EYSI) = EYS/18

1. Income Index (II) = {In (GNIpc)−In (100)}/In (75,000)−In (100)

Therefore, it was stated that The Human Development Index (HDI) is the geometric mean of the previous three normalized indices.

$$\text{HDI} = \sqrt[3]{\text{LEI. EI. II}}$$

Following this methodology the all the three indices for Daspur I, Ghatal, Ghatal Municipality, and Chandrakona I blocks' surveyed households are being calculated. Later a composite index of human development has been derived for each administrative unit to develop a comparative study (Fig. 6.14) (Table 6.4).

12. Resilience strategies: Pre-flood preparedness and post-flood relief and rehabilitation

Flood resilience can be defined as the capacity that people or groups may possess to withstand or recover from flood emergencies. This resilience incorporates the active and passive preparedness measure, structural and nonstructural mitigation, and rehabilitation measure as

TABLE 6.3 Calculation of multidimensional poverty index for surveyed households of Daspur I block (sample).

Indicators	People in households					
Household size	**2**	**3**	**4**	**5**	**7**	**Weights**
Household numbers	23	52	63	15	10	
Education						**1/6**
No one has completed 5 years of schooling	0	1	1	0	1	
At least one school-age child not enrolled in school	0	1	0	1	1	
Health						**1/6**
At least one member is malnourished	0	1	1	1	1	
One or more children have died		0	1	1	0	
Living standards						**1/18**
No electricity	1	0	1	1	0	
No access to clean drinking water	0	0	1	1	1	
No access to adequate sanitation	1	0	0	1	1	
House has dirt floor	1	1	0	0	1	
Household uses "dirty" cooking fuel (dung, firewood, or charcoal)	1	0	1	1	0	
Household has no car and owns at most one bicycle, motorcycle, radio, refrigerator, telephone, or television	0	1	0	1	1	
Score *ci* (sum of each deprivation multiplied by its weight)	0.222	0.611	0.666	0.777	0.722	
Is the household poor ($c \geq 1/3 = 0.333$)**?**	No	**Yes**	**Yes**	**Yes**	**Yes**	
Censored score *ci(k)*	0	0.611	0.666	0.777	0.722	
Multidimensional head count ratio (H)	0.858					
Intensity of poverty (A)	0.661					
MPI (H x A)	**0.567**					

Prepared by Researcher.

well as the risk reduction programs. While preparedness assumes that certain groups of people or property will remain vulnerable and that preparedness is necessary to address the consequences of such flood hazards. It includes the formulation of viable emergency plans, the development of warning systems, the maintenance of inventories and the training of personnel. It may also embrace search and rescue measures as well as evacuation plans for the areas those are at risk from a recurring flood disaster. However, mitigation embraces the measures that are aimed at reducing the physical, economic and social vulnerability to threats and the underlying causes of this vulnerability (Buckle et al., 2000). Beside various structural mitigation measures, there are various nonstructural mitigation measures which

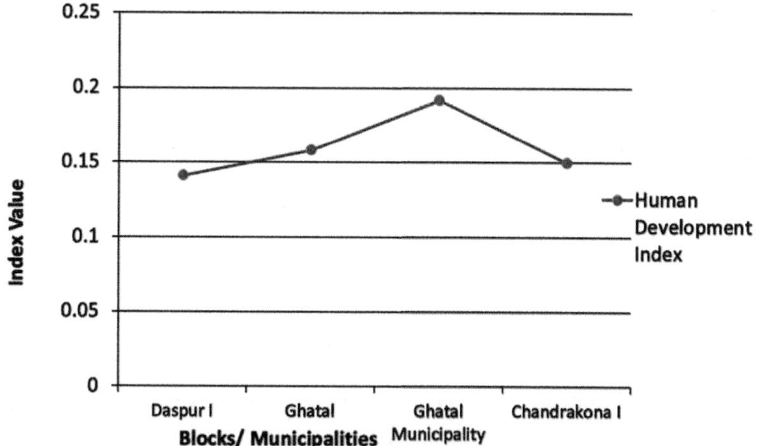

FIGURE 6.14 Human development index of the flood affected locations. *Prepared by Researcher.*

TABLE 6.4 Human development index, indicators, and their corresponding ranks in the study area.

Blocks/Municipality	LEI	Rank	MYSI	Rank	EYSI	Rank	EI	Rank	II	Rank	HDI	Rank
Daspur I	0.846	4	0.866	2	0.888	4	0.877	3	0.0038	4	0.141	4
Ghatal	0.923	2	0.800	4	0.944	2	0.872	4	0.0049	2	0.158	2
Ghatal municipality	1	1	0.933	1	0.972	1	0.953	1	0.0074	1	0.192	1
Chandrakona I	0.877	3	0.866	2	0.917	3	0.892	2	0.0043	3	0.150	3

Prepared by Researcher. Primary Survey.

act effectively like training and education of personnel, public awareness, evacuation planning, warning system etc. From the primary survey in the study area, it is found that as a resilience strategy from flood, various means which are about of 12 types are adapted by the local flood affected communities as preflood preparedness (Kar, 2008). Similarly, the post disaster relief and rehabilitation in various forms are being provided to the affected people by the local government or from the NGOs. Though in both the cases it is found that the three blocks respond at various levels in terms of preparing themselves for facing the flood hazard or getting the relief measures in post flood situation. It can be analyzed that out of 12 measures, the main emphasis given on Flood forecasting, government fund allocation, community participation, importance of dykes to protect the floodable areas in all three blocks which are adopted hugely and widely by most of the people in that region (Table 6.5) (Sampath, 2010).

Moreover, for post flood rehabilitation, a few measures have been adopted which are shown in the following table. Among those, the government help in the form of food, medicine, providing flood shelter and community participation in rehabilitation comprise the majority share as post flood rehabilitation measures (Rahman, 2010).

TABLE 6.5 Various forms of flood preparedness activities done by local people.

A	Flood forecasting messages in TV, radio, mobile phone, newspaper, etc.
B	Useful and punctual flood warning system in the study area
C	Management of riverbed to remove the obstruction of Shilai river flow
D	Government plan to stabilize materials in the upstream of riverbed
E	Presence of dykes or heightened roads to protect the floodable areas
F	Government help in river flow diversion and river course modulation
G	Government help in canal development from river to field for irrigation
H	Government housing act for riverbank dwellers
I	Community participation in maintenance of embankments
J	Sufficient fund allocation to conduct maintenance of the flood embankments
K	Preparation of suitable evacuation plan to escape flood occurrence
L	Beneficial flood evacuation system

Prepared by Researcher. Primary Survey.

In the lower Shilai watershed as local people are recurrently and frequently flood affected, they have invested their thought in various resilience strategies to withstand flood. At some places, during the flood time the community trust is being set up (with government collaboration) with the objective of providing affordable housing access specially during flood peak time away from the place of direct flood hit. Though the entire initiative has not started in full swing, still the community collectively try to sustain this process. Some people who get very badly affected every year, and do not get the community trust support or other help, try to find out the cheap rent room or migrate to the houses of their relatives at a nearby places to save their livelihood at least for the few days. Rest of them take the shelters at government relief centers in schools, etc. Moreover, the flood victim people are nowadays more keen to take and maintain the insurance policy specially the medium income group people to withstand the loss of such recurrent flood. Most of the people in flood affected locality nowadays afford the boats, which have become essential during flood time to migrate, to collect relief or to get access to safe places even along with their livestocks in times of emergency. Therefore, in every house, one or two pump boats are available which are maintained and used during flood and water stagnation period in the locality. Beside these, some of the effective flood controlling resilience strategies which can be effectively implemented in the study area are: *Flood Routing* which is the technique in hydrology to compute the effect of storage on the shape and movement of the flood wave, *Channel Improvement* by deepening, widening straightening lining and cleaning out of vegetation and debris from the river channel, *Dredging of sediment from river bed*, *River course modulation* in a meandering river where meanders becomes extremely sharp, they can be straightened by artificially cutting individual or a series of bends, *Construction of permanent embankment*, *Community participation in flood disaster management*, *Housing act for riverbank dwellers* by which people should construct their houses at a distance from river bank, Education programs through television media, *Initiation of flood insurance scheme*, etc.

13. Conclusion

The lower reaches of Shilai river are a chronically flood-affected areas. Shilai river flowing through these blocks is the main cause of such recurrent and annual flood in the said area. It causes thousands of people temporarily homeless, deteriorate their health condition, reducing their livelihood status by continuous engagement in the flood rehabilitation process for this chronical and repetitive flood, but it is also identified as a region of fertile land and high crop productivity. Therefore, people have adopted various methods to cope up this flooding situation and farmers have indigenously developed various strategies to maintain the cultivation process less affected by flood. Therefore, these strategies have some economic benefits for the study area. Though it has been found that the health deterioration and homelessness problem of flood affected people are not being totally overcome still in recent days the initiatives taken by the local community with government assistance to address such problems will definitely finds its solution to the temporary homeless people on every year specially during flood time and water stagnation period. Beside this, local government and community has also adopted several methods to withstand flood to minimize its effect. Generally, it has been observed that people's livelihood resilience and their adaptive capacities are interdependent. Therefore, it is necessary to regularly assess the surface and ground water nature considering various physicochemical parameters, for the sustainable water resource management by reducing the propensity of flood in lower reaches of Shilai river, which has its impact over future generations.

References

Alkire, S., & Foster, J. (2011). Understandings and misunderstandings of multidimensional poverty measurement. *The Journal of Economic Inequality, 9*(2), 289–314. https://doi.org/10.1007/s10888-011-9181-4

Buckle, P., Mars, G., & Smale, R. S. (2000). New approaches to assessing vulnerability and resilience. *Australian Journal of Emergency Management, 15*(2), 8–15.

Burton, C., & Cutter, S. L. (2008). Levee failures and social vulnerability in the sacramento-san joaquin delta area, California. *Natural Hazards Review, 9*(3), 136–149. https://doi.org/10.1061/(ASCE)1527-6988(2008)9:3(136)

Chaudhury, S. (2008). Floods: Impact on human health. *River Floods: A Socio- Technical Approach*, 101–112.

Chowdhury, S. (2008). Living with floods- rethinking strategy in an age of society. *River Floods: A Socio- Technical Approach*, 250–253.

Das, T. K., Lohar, D., Gupta, I., & Bhattacharya, B. (2011). *Disasters in West Bengal an interdisciplinary study*. ACB Publications.

Dolui, G., & Ghosh, S. (2012). Flood and its effects: A case study of Ghatal block, Paschim Medinipur, West Bengal. *International Journal of Scientific Research, 2*(11), 248–252. https://doi.org/10.15373/22778179/NOV2013/79

Kar, A. M. (2008). Mitigation of flood problem through participatory natural resource management. *River Floods: A Socio- Technical Approach*, 206–218.

Kolay, A. K. (2007). *Soil genesis, classification survey and evaluation*. Atlantic Publishers & Distributers (P) Ltd.

Messner, F., & Meyer, V. (2006). *Flood damage, vulnerability and risk perception – challenges for flood damage research* (Vol 67, pp. 149–167). Springer Science and Business Media LLC. https://doi.org/10.1007/978-1-4020-4598-1_13

Murty, J. V. (1998). *Watershed management in India*. Wiley Eastern Limited.

Patra, P. K. (2008). Flood management: A GIS approach. *River Floods: A Socio- Technical Approach*, 195–205.

Rahman, M. M. (2010). *Community capacity builiding on disaster preparedness in disaster mitigation: Experiences and reflections*. In P. Sahni, A. Dhameja, & U. Medury (Eds.) (pp. 107–111). New Delhi: PHI Learning Private Limited.

Singh, R. B. (2006). *Introduction to natural hazards and disasters. Conceptual framework, vulnerability, spatial pattern, impacts and mitigation strategy: Indian experience in natural hazards and disaster management*. Rawat Publications.

Sahoo, P. M., & Sivaramakrishnan, L. (2014). Vulnerability of flood prone communities in the lower reaches of Shilai river. *International Journal of Development Research, 4*, 1393–1400.

Sampath, P. (2010). Vulnerability reduction at community level: The new global paradigm. In P. Sahni, A. Dhameja, & U. Medury (Eds.), *Disaster mitigation: Experiences and reflections* (pp. 202–209). New Delhi: PHI Learning Private Limited.

Singh, S. R., Eghdami, M. R., & Singh, S. (2014). The concept of social vulnerability: A review from disasters perspectives. *International Journal of Interdisciplinary and Multidisciplinary Studies (IJIMS), 1*, 2348.

Smit, B., & Wandel, J. (2006). Adaptation, adaptive capacity and vulnerability. *Global Environmental Change, 16*(3), 282–292. https://doi.org/10.1016/j.gloenvcha.2006.03.008

Sukul, N. C. (2008). Annual floods of the silabati and the surrounding ecosystem. *River Floods: A Socio- Technical Approach*, 170–173.

CHAPTER

7

Climate-change induced out-migration and resultant women empowerment: Glimpses of resilience from Indian Sundarbans

Anindya Basu[1], *Diotima Chattoraj*[2] *and Adrija Bhattacharjee*[1]

[1]**Department of Geography, Diamond Harbour Women's University, West Bengal, India;** [2]**Wee Kim Wee School of Communication and Information, Nanyang Technological University, Singapore, Singapore**

1. Introduction

Since time immemorial, human beings have been on the move. In recent times, migration has become one of the most crucial issues because of the ubiquity, upshot, and the attendant impacts (Chattoraj, 2020). Be it in search of economic opportunities, for educational quest and family pursuit, or to escape severe conflicts, human rights violation and adverse climatic impacts, migration is about people, their aspirations and fears, triumphs, and tragedies, which makes this topic not only interesting and relevant but also compelling and fascinating. In case of rural India, the quantum of migration significantly increased after the economic liberalization and that predominantly involved males. The Indian Sundarbans, too, was not an exception. In search of sustainable livelihood, a large-scale migration was noticed; extreme climatic events and geographical isolation limited the scope of employment opportunities for the locals. With the male members of the working age group moving out, the women were left behind to take care of the family, farming, and finance and, this, in turn kind of empowered the women as they started to have "say" and played role in decision-making. While migration with its causes, factors, and impact has a long story to tell, the tale of the left-behind women is also equally fascinating to look into. A kind of great divide can be noticed between the "migrants' wives" and "non-migrant's wives" from the point of view of engagement and empowerment.

Homelessness to Hope
https://doi.org/10.1016/B978-0-443-14052-5.00007-0

The factors pondered over regarding migration were migration history, destination, educational attainment, acquired skill, kind of landholding, stage of life, family composition, spousal age-difference, nature of job, and remittance amount. As these not only explained the characteristics of the migrants but determined the situation of the nonmigrant, left behind women. Initially the women empowerment seemed to be liked with these migratory factors but with the passage of time the decision-making parameters and ability did put each woman in their own position; their individual profile along with how they handled household issues, reproductive choices, agricultural practices, business matters, financial transactions, asset creation, political exercises, social networking, so on and so forth projected themselves as strong individuals. So, in a gender-sensitive socioeconomic setup often it is believed that women empowerment is connected with the initiatives taken by the immediate surroundings, governmental support but, in this case of rural interiors of Indian Sundarbans achievement of an organic sustainable women's empowerment can be seen—where the women themselves are the primary architects.

1.1 Migrants, migration, and homelessness/unhoming: Setting up the context

(2023) defined migrant as "a person who moves away from his or her place of usual residence, whether within a country or across an international border, temporarily or permanently, and for a variety of reasons." As per Indian standards, if an individual stayed continuously for a period of 6 months or more or intends to stay for 6 months of more in a place other than his/her usual place of residence then the person is termed as migrant (GoI, 2021). The Census of India takes into account the primary causes of migration as moved after birth, Work/Employment, Business, Education, Marriage Moved with household and any other (Census of India, 2011). About 2.3% of the world's population 184 million people, contributes to the global diaspora (World Development Report, 2023). As per World Migration Report (2022), India topped the global chart with highest number of emigrant populations followed by Mexico, Russia, and China. The World Development Report, 2023 echoes similar notion as it denoted that international migrant from India scaled from 8 million to 18 million between 2000 and 2022.

Tracing back to the seminal work of Ravenstein (1885), it can be derived that physical distance between the source and destination had an inverse relationship with the volume of migration. Even today the migratory laborers are more comfortable in delving into local or national employment opportunities first rather than international ones; this has given rise to the step migration scenario. Stark (1991) pointed out that in case of "push" migration due to high unemployment or extreme uncertainty the workforce cannot be choosy. The developing nations have contributed much to the quantum of workforce migration in the postliberalized economic regime (Massey et al., 1990). Since 1991, as the market opportunities increased temporary and long-term skilled migration was on the rise resulting in substantial increase in remittances in India (Mani, 2012). Moreover, the wage differential determines the rural to urban pull in most cases of migration (Pieke & Biao, 2009; Wang et al., 2005). Following the lines of New Economics of Labor Migration of 1990s, de Haan (1999), Mosse et al. (2002) mirrored the same opinion that assurance higher income generates the voluntary labor movement from the rural belts. De Haan & Rogaly (2002) added the issue of class domination and exploitation in the rural belts as another trigger factor behind such migratory trends.

The opinions of scholars which are very closely linked with the case-study at hand, involves an "Neo-Malthusian" explanation of migratory factors like ecological stress, extreme weather events, decreasing agricultural output, flawed public policy, and debt cycle (Deshingkar, 2008; Gemenne & Brücker, 2015; Laube et al., 2012; Manandhar et al., 2011; Parkins, 2010). However, Rowthorn (2010) provided a new dimension that rural migration not only has "push" factors behind but better opportunities are often the actual cause.

1.1.1 *Unhoming or homelessness*

There is no internationally agreed definition of homelessness (Ortiz-Ospina & Roser, 2017): Different governments and organizations use different definitions. In most countries, different terms are used for different types of situations. The concept "unhoming" or "homelessness" literally means staying away or fleeing one's own home due to "war, civil conflict, political strife, gross human rights abuse" or natural disasters, or sometimes in order to improve one's lifestyle or to pursue higher education (Chattoraj, 2016). One would be considered as unhomed or homeless when he or she fails to belong or relate to the environment, culture, traditions, persons, foods, etc., in which he or she resides or once resided. Therefore, homelessness, as put by Bramley, is a "lack of a right or access to their own secure and minimally adequate housing space" (1988: 26). However, Bramley neglects the emotive aspects of homelessness as identified by Somerville (1992). Marx and Weber place homelessness as total lack of property and powerlessness (Somerville, 1992). Homelessness is also defined by fear, danger, the unknown, foreign places, traditions, unfamiliar faces, and habits. Recently, migrant homelessness has become an increasingly visible issue in many developed countries. Homelessness is a multidisciplinary concept. In the light of migration, the concept of homelessness becomes radically changed and redefined. With the passing of time, physical and cultural values prove to be inadequate in the transformations of homes. It appears inevitable that past homes develop "strange, unusual, and alien elements" in the eyes of those who have migrated abroad (Al-Ali & Koser, 2003, pp. 1–246). Home, according to Watkins and Hosier, "permeates society, and evokes such feelings as belonging, control, comfort or security whether it involves individuals or much larger groups of people" (Watkins & Hosier, 2005). While homelessness evokes certain emotions like despair, isolation, hopelessness, and grief, it also presents images as poverty, alcoholism, mental illness, and social deviance (Watkins & Hosier, 2005). Just as having and being at home equates with life stability and some measure of success, being homeless translates into transience, turmoil, and failure in life as argued by Watkins and Hosier (ibid.: 197). Home and homelessness are addressed historically across cultures, across ages, within cities, rural areas, and by gender in relation to psychological and social pathologies. Wardhaugh (1999) exclaims that spaces generally understood as home may not be experienced as such: the statement that "I do not have a home" may equally mean that "I have no house in which to live" or "the house in which I live does not feel like a home" (Dovey, 1985 cited in Wardhaugh (1999)). Furthermore, a sense of homelessness may be felt by communities and nations as well as individuals, arising from marginalization, alienation, migrations, expulsions, and the creation of national or cultural diasporas (Chattoraj, 2016).

In search of better livelihood and working environment, often large-scale exodus of working population can be noticed (Dubey & Mallah, 2015). Mobility of male population for temporary or permanent labor purposes is common across rural areas of West Bengal

(Ghosh & Mal, 2017) and Indian Sundarbans, a remote area is no exception (Rudra, 2010). Migration has always been an effective tool for the human population to adapt to environmental changes (McLeman & Hunter, 2010). Migration has always been an effective tool for the human population to adapt to environmental changes (McLeman & Hunter, 2010). Migration has always been an effective tool for the human population to adapt to environmental changes (McLeman & Hunter, 2010). In Sundarbans region too, migration is utilized as a strategy to reduce ecological and socioeconomical vulnerability in the place of origin (Banu, 2016; Mistri, 2013) by the migrants themselves, to avert the immediate danger. The main factors behind migration as enumerated by Bala (2017) were limited employment opportunities and lack of basic amenities.

1.1.2 Migration as a tool: Ecological refugees of Sundarbans

The settlement history of Sundarbans is long and since the British period large-scale clearance of mangrove has been underway to create avenues for agriculture for the expanding population, from then the continuous human intervention might have accentuated the challenges posed by climate change (Banerjee, 2013; Islam, 2014). The increased cyclonic storms, resultant salinity affected the food security of Sundarbans (Neogi et al., 2017) and led the local community to abject poverty (Dasgupta et al., 2020). Often the severe storm surges breach the embankment, salt water inundates the agricultural land putting the productivity under stress (Haldar & Debnath, 2014). The newly formed mudflats from tidal river accretion cannot be utilized to accommodate the landless marginal farmers as those are government-owned where the locals have no usage right (Raha et al., 2012). With the agricultural activities not being very rewarding, forest and aquatic resources were targeted for supplementary income (Ghimire & Vikas, 2012). While accessing the forest resources like honey, wax many become victims of tiger attack (Das, 2018) and, thus, make it a nonlucrative livelihood option for the youth. Shrimp aquaculture provided viable alternatives initially (Hoq, 2007) but extreme salinity, nonjudicious exploitation rate and indiscriminate shrimp seed collection affected the pisciculture production too (Gopal & Chauhan, 2006; Chauhan & Gopal, 2014; Hoque Mozumder et al., 2018). However, in spite of gradual decrease in fish production, Chand et al. (2012), Dutta et al. (2017), Ghosh (2017) estimated that still around 20% of the locals are associated with fishing activities. Mud crab cultivation has gained grounds in recent times due to its resilience to salinity and high return from international market (Ghosh & Roy, 2021; Nandi & Pramanik, 2017) pointed out that the younger generation of Sundarbans subjected to ecological stress with exposure to education have more ability to gather information about better employment opportunity and thus are moving out in great numbers.

1.1.3 Role of remittance vis-à-vis labor of left behind women

Often this internal or transnational labor migration is male-dominated (Massey et al., 1994) and the female members stay back to fight out the adverse situation back home, probably bit more well-equipped and less vulnerable with the remittance (de la Fuente, 2010; Samal, 2006). This often encourages folks in similar situation to take the leap of faith for the overall well-being (Andersson, 2014). Scholars like (Bhugra, 2004; Kollmair et al., 2006) have studied how remittance in terms of foreign exchange have helped in reducing poverty especially in rural areas, shaping up local livelihoods and even national progress. The role of transnational labor migration is prominent in case of India's economy as it accounts for almost 3% of the country's GDP. As per World Development Report (2022), India became the first nation to receive 100 billion dollars in the form of remittances from migrant workers abroad, with staggering annual increase of 12%, leaving countries next in the chart—Mexico, China, and the Philippines far behind. Remittances in most cases lead to inequality in income and assets between households with and without migrant laborers (Acosta et al., 2008; Barham & Boucher, 1998).

Conventionally, empowerment refers to the decision-making ability in social, economic, and political spheres. Since 1970s women empowerment started to gain ground in development theory realm. Concepts like Women in Development (WID), Gender and Development (GAD), Feminization of migration started to find a ground for women in the patriarchal society (Alasah, 2008). In the rural areas, in the initial neo-liberal phase, amidst the patriarchal framework, women were mainly confined within the household or at best took up marginal occupations with their limited skill set (Mathew, 1995) but with increased male-dominated rural—urban migration, the situation started to change gradually. Toyota et al. (2007) very rightly pointed out that discourse on migration is not only about the migrants or role of remittance but also about the left behind female counterparts who are equally affected by the phenomenon of migration and often the traditional gender roles in the rural settings gets altered. They have to bear the "desirable (by negotiating the patriarchal gender archetypes) and or undesirable (by creating undesirable work pressure)" impacts of migration. The wives of the migrant workers in addition to their regular household chores have to take part in the activities of the outer world, widening their sphere of influence (Radel et al., 2010). This expanded role of women leads to feminization of economic sector and the increased at times affects their health and wellbeing (Guerny, 1995; Lastarria-Cornhiel, 2006; Rao, 2006). Most of the available literature concentrate on the life-experiences of the left behind women facing transnational migration (Gartaula et al., 2012; Sarker & Islam, 2014; Toyota et al., 2007) through the concept "migration left-behind nexus" tried to gauge how women are psychologically impacted. Dealing with diverse situations of the outer world provide the women a level of confidence translating to empowerment (Adhikari & Hobley, 2015; Bose et al., 2017; Sinha et al., 2012). It was also noted that at times women who are in abusive relationship see male migration as a relief and partial freedom (Kasper, 2005).

The degree of stress to be handled by a left behind wife of a migrant labor depends to the nature of the family also. If it is of a nuclear structure then though the stress is more, the chance to play the de-facto household is also there; in case of joint family structure the workload gets divided but the chance to voice opinions freely is stifled (Adhikari & Hobley, 2011; Brown, 1983; de Haas & van Rooij, 2010). Initially with the trajectory of male migration, women empowerment in rural areas was mostly limited to feminization of farming practices (Taylor & Martin, 2001) but long-term migrations lead to social and political dimensions of emancipation too (Moghadam & Senftova, 2005). Still, how far this multi-tasking actually emancipates women in the long run within the prevalent social fabric remains a question (Kasper, 2005; Massey et al., 2009) and how that affects the power relations within the family framework (Castles, 1992; Rajkarnikar, 2017) highlighted the two cases of settling down or semipermanent migration and temporary migration, where in the first case the left behind women after being accustomed to changes have to get back in playing the second fiddle in the family as the migratory spouse returns and in the latter case the transition is for a longer a duration providing a scope for real empowerment. As in case of Indian Sundarbans, most of the males of the working age group were attached to agricultural and aquacultural practices, increasing salinity and resultant decreasing yield fueled the migratory stream (Hajra et al., 2018; Mistri & Das, 2015) and it most seen as the only viable coping measure by the less educated and ill-equipped locals (Ghosh, 2013). Almost 95% of the migrants are male and mostly they travel interstate seasonally to serve mostly as agricultural and construction laborers. Guha and Roy (2016), Hazra et al. (2014), Mistri (2013) noted that 34% of the residents of Sundarbans are reeling under poverty and at least one member of 75% of the households are migratory laborers. Migration-induced remittance acts as a "pathway out of poverty those who leave and also for those who stay behind" (Deepa et al., 2009). Remittance sent by the male helps in building capital base in rural areas and at times increases the degree of inequality (Haas, 2007). Though dipping agricultural productivity often induces migration (Mandal & Mandal, 2012), migration generated remittance is often used to do away with the constraints associated with farming (Brauw, 2020; Damon, 2010; Miluka et al., 2020).

2. Research question

Work participation rate for women has been traditionally low across India and, in case, of rural areas the economic isolation was more. Women of Sundarbans region, which is ecologically and economically vulnerable, are pushed into the workspace due to increasing male migration. The migration of the working age group male as semiskilled or unskilled laborers within the country or overseas gradually amplified as the study area faced cyclone strikes, tidal surge, embankment breaching. The chapter tries to inquire into the role of migration in shaping up left behind women's life. Initially the women were compelled to take part in agriculture, pisciculture, negotiate with the other players in the market and this gradually infused confidence in them to deal with the situation. Whether this participation in all spheres of life made them truly empowered or they had to face more adversities or it is only a cosmetic change in the social fabric in the study area is the moot question.

3. Study area

Sundarbans, deltaic ecosystem formed at the confluence of the Ganges, the Brahmaputra, and Meghna Rivers, spread over India and Bangladesh, is famous as the largest mangrove forest globally and is designated as UNESCO World Heritage Site for its rich biodiversity. The Indian section of Sundarbans comprises of 102 islands, out of which 54 are inhabited, housing over 4.5 million people. This, the mangrove dominated deltaic area, is guarded by mangroves and surrounded by earthen embankment to absorb storm surge impacts during extreme weather events and to regulate the ingression of saline water (Sandilyan & Kathiresan, 2015). Residents, who have settled in here are primarily dependent on agriculture and aquaculture (Raha et al., 2013). The recurrence cyclonic storm and its associated adversities like saline water intrusion, bank erosion, and embankment breaching are common at in Sundarbans (Ghosh & Roy, 2021). This repeated cyclonic disaster not only damages housing or disrupts transport communication but hinders the livelihood opportunities like agriculture and fish farming pushing the residents at the brink of abject poverty. The loss of land and gradual salinization have forced many to change their vocation from primary activity–based laborer to migrated laborer, be it national or international, long-term or seasonal. The marginalization of the ecological refugees due to the climatic vagaries, gradual change in livelihood practices have taken place in tune with that (Basu et al., 2021). To address the said issue a survey has been done in Gosaba Community Development Block, situated at the eastern side of river Bidyadhari, during 2021–22. The villages taken into account were Rangabelia, Gosaba, Satjelia, ChotoMollakhali, and Kumirmari (Fig. 7.1). Total population of entire South 24 Parganas district is 81,61,961 out of which in Gosaba Community Development blocks it is 246,598. The sex ratio of the block is 959, while the percentage of literate population and percentage of people belonging to below poverty line (BPL) are 78.98% and 38.02%, respectively (Census of India, 2011). As the block is vulnerable in terms of socioeconomic and climatic criteria, the residents have undertaken migration to ensure a better livelihood and, thus, is fit for migration-induced women empowerment study.

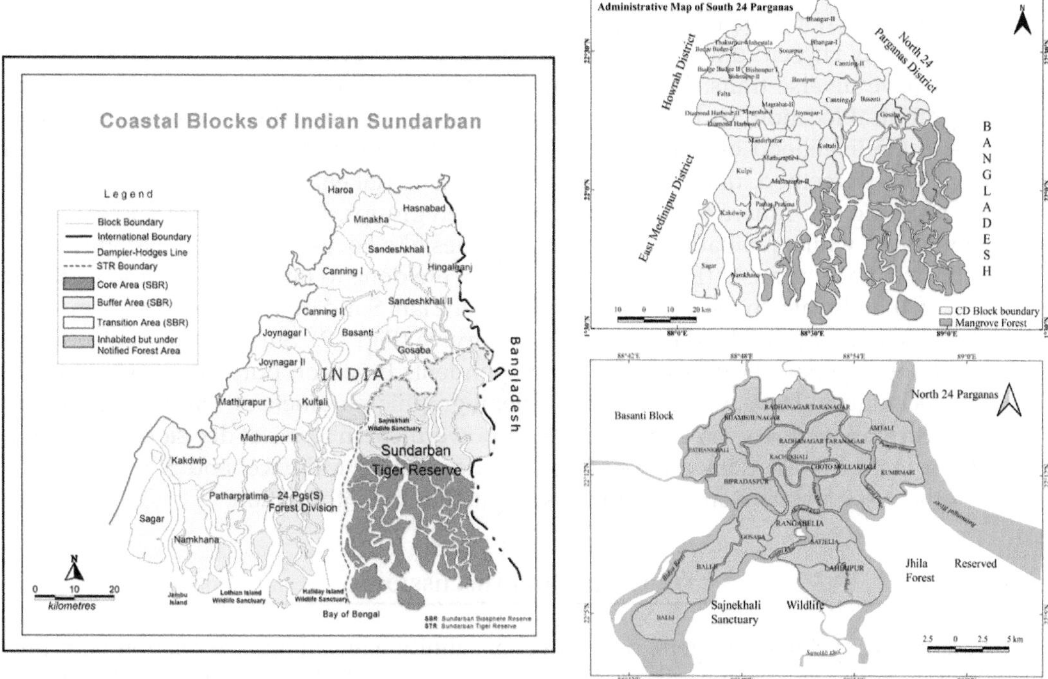

FIGURE 7.1 Location map of the study area—Gosaba community development (CD) block, south 24 Parganas district, West Bengal.

4. Objectives

The main thrust areas of the chapter are:

- To study the nature and causes behind the large-scale male out-migration in the study area
- To examine the various effects of remittance on the place of origin
- To delve into the impact of migration on the left behind wives of the migrant laborers and their resultant empowerment, if any

5. Materials and methods

The study for background analysis and contextualization of the increasing out-migratory trend involved detailed literature review. To bring out the clear picture, triangulation of data has been done from secondary sources like reports, policy documents, and primary survey.

For the primary survey, respondents were categorized into two groups, 50 male migrant laborers, staying away for around 5–7 years, who were back in their native place on a break and 150 women who were left behind wives of migrant laborers from villages of Rangabelia, Gosaba, Satjelia, ChotoMollakhali, and Kumirmari of Gosaba Community Development Block purposively chosen to lend credence to the empirical evidence as the villages though contiguous have diverse traits. The selection of GP and villages were taken place based on their geographical location, socioeconomic backwardness, vulnerability to natural calamities and risks to the livelihood of those areas. Data collection took place since the late 2021 till mid-2022. Mixed method data analysis has been undertaken; the quantitative methods involved profiling, chi-square test to compare between groups, ANOVA, Likert scale–based perception study while qualitative research techniques like Focus Group Discussion (FGDs), Participatory Rural Appraisal (PRAs), case studies, were done to understand the struggle for existence of the climate-change ridden people of Indian Sundarbans specifically, most of whom have opted for some kind of migration to survive (Fig. 7.2). Survey data was collected through in-depth personal interviews of the respondents based on purposive and snowball sampling techniques. Data was gathered on demographic parameters, remittances, nature of work participation of women and women's empowerment in terms of decision-making power. After organizing the data, analysis has been done using MS Excel and SPSS v23 while quotes of the respondents were used to pass off qualitative information.

6. Result and discussion

The increasing cyclonic disasters, rising level of salinity have an inverse relation with productivity rate and after cyclone Aila in 2009, it was noted by Debnath (2013) that the paddy production dipped by 40 quintals per 1.6 ha. Besides, the loopholes in agricultural procedures that have been highlighted were lack of judicious usage of fertilizers (Hajra et al., 2016), dearth of irrigation infrastructure, absence of permanent crop market, and issue of insufficient storage (Hazra et al., 2014). These matters lead to mono-crop culture in the area further limiting the earnings of the farmers and forcing them to move out.

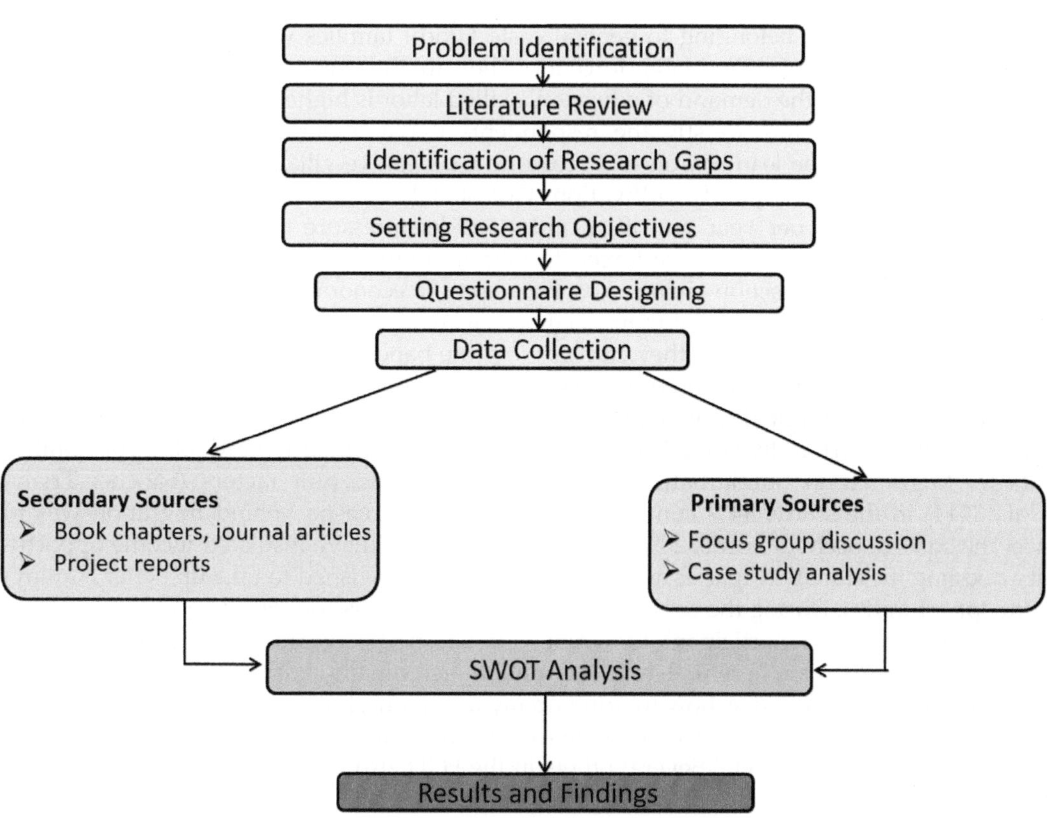

FIGURE 7.2 Methodological framework of the study.

6.1 Demography of migrant laborers

The average household size of the study area was of five to six family members, including the dependent population—the young and the old. To take care of the entire family the male members of the working age group had to undertake migration as the primary activities did not generate the required income. The interaction with the present migrant respondents ($n = 50$) reflected that all were male, of mixed religious and caste background and of varied age category. The matured ones (aged above 50 years) were lesser in number; occupational mobility was higher among the ones in early adulthood (18–35 years) and mid-life (36–49 years). Educational attainment of 80% of the respondents were up to higher secondary level (22% were even below junior high school level) and the drop out level after that was high due to accessibility issues, personal problems, and need to join the workforce to earn money for the family. The educational achievement of the residents, in general, and especially those belonging to the Muslim community was pretty low (about 68% of the Muslim respondents were of below junior high school level) and thus they are mostly engaged as semiskilled or unskilled laborers in informal sectors (Hajra & Ghosh, 2016); most of them are engaged as agricultural, plantation, or construction laborers. The highest educational attainments were postgraduation, but the percentage was very low (4%). Comparatively the respondents of early adulthood phase belonging to general caste Hindu families were more educated and few of them were part of skilled labor force too. They are the ones who are involved in transnational migration as the demand of educated, skilled labor is higher overseas and are handsomely remunerative too. Mostly the respondents (80%) undertake regular ventures as seasonal laborers in the lean season and return to their native village during the monsoons (July—August) to peruse paddy cultivation. Occasional migrants, i.e., the ones migrating for less than 90 days per year are often under sudden pressure to pay off some debt or have their savings run out and are forced to migrate immediately. Mostly they were from comparatively well-off section trying to tackle a specific economic problem. All the respondents reported that at least one of their neighbors were attracted by their improved lifestyle and asked for assistance so that they can also join the bandwagon of migration.

The ones having their own land or fishery ponds were bit reluctant to undertake migration while the landless marginal farmers, laborers, forest resource collectors (55% of the respondents) having uncertain livelihood happened to be more inclined toward job opportunities outside. Migration in Sundarbans involve both push and pull factors (Sánchez-Triana et al., 2014) in the study, 44% confided that the primary reason behind migration was to meet the basic household needs, 20% opted for migration to have insulated income opportunity dodging the climatic vagaries and seasonality and 16% wished to take up better employment opportunities. During the course of a Focus Group Discussion (FGD) conducted with 14 left behind women, a participant remarked, "It has been 3 years that my husband has migrated to the city. I am here with my three daughters. Initially, I did not want my husband to migrate as I was worried as how to bring up my kids alone, but we need money so that our kids get a good life. I started working in the local farm 2 years back and now am really confident to deal with the market." Several others in the FGD also reiterated somewhat the same thing.

In one of the interviews conducted with a left behind women who owns a small grocery shop in her neighborhood, stated, "I own this shop. I bought this shop with the money sent by my husband from Qatar. He migrated in 2018 and since he could not come to visit us. But we are okay. He has to work. We need money. Recently, our elder daughter got married and thanks to my husband we did not have any problem in arranging for her wedding. It's true, my husband could not be there, we were upset about that. My shop also fetches me decent amount of money with which I bear all the daily expenses can save whatever my husband remits for the wedding of our second daughter." In contrast, few other left behind women said, "it is not easy to stay alone as a woman and take care of your parents, and in laws and kids. But what can we do. We have chosen this life and now have to deal with it." During the in-depth interviews with the male migrant laborers, a significant number of them expressed their sorrow and pain of leaving behind their families and migrating elsewhere to earn money: "It is not that easy to leave behind your family here and go and work so far away. It pains when you cannot see them for several months and sometimes for several years. But we cannot do anything. We need to earn money to give our families a better life and it is only possible with migration. Here, situation is not good for any kind of work. We have been devastated by the cyclones so many times [...]." Another laborer added to this: "I used to own a field and work there, but the storm, 4 years ago, destroyed everything. We had nothing left. I need to feed my family of 5. So, I had no choice but to migrate to Kerala. I paid a lumpsum amount to the agent to get this job. I am happy about my decision. My family is leading a good life." One of the other male migrant laborer remarked, "I want to earn more, get my son well-educated and well placed in life so that he does not have suffer like me." Another one also followed the suit "I wish that my only daughter becomes a government service holder. I have to earn more so that I can provide her the required coaching, reading materials. The next generation should get a better life." These prove that the migrant laborers have realized the power of education and are willing to go an extra mile to help their wards to get educated and skilled.

During the COVID-19 crisis, all were back home and tried to get into meaningful job in their native places but either they failed to get any opportunity or the emoluments received were way too low to sustain the family, so all of them again tried their luck through migration. Destination of the majority of the respondents was within the country (76%), most of them moved in group under an agent/contractor or popularly known as recruiter and common destination was toward south Indian states. A small section (12%) went in and around Kolkata city as their primary destination and mostly they were on step migration mode, going through acclimatization. 72% of the migration was rural to urban where the migrants worked as construction worker, industrial laborer, mechanic or in some capacities in service sector. The rural-to-rural migration involved jobs as agricultural or plantation laborers. The size of the landholdings even for the landowners were quite small (60% of the respondents had landholdings below 0.25 ha) to sustain their acquired lifestyle that their household became accustomed to after being migratory laborers. As the income difference between the two phases (migratory and nonmigratory) are quite glaring Fig. 7.3, migration kind of becomes a unanimous choice.

To find out the relationship between occupation of migrant workers and their monthly income a one-way ANOVA has been calculated (Tables 7.1 and 7.2). The hypotheses considered

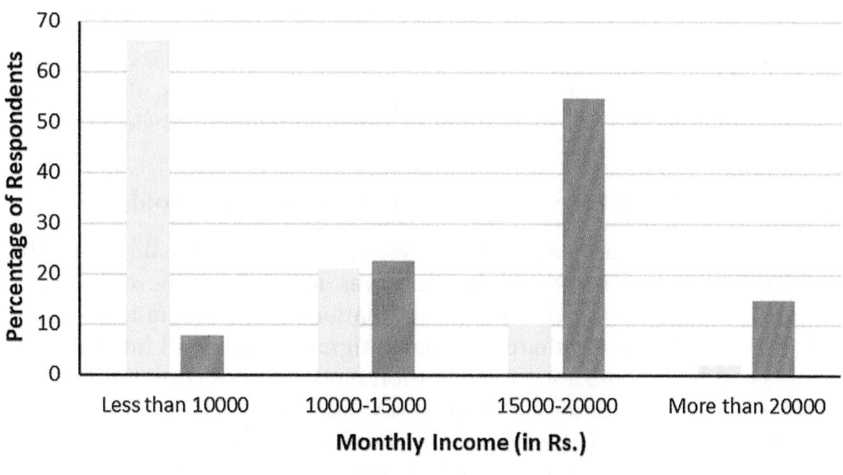

FIGURE 7.3 Change in monthly income scenario of the migratory laborers of Gosaba CD block, South 24 Parganas district, West Bengal. *Based on Primary Survey. (2022).*

TABLE 7.1 Descriptives regarding categories and monthly income of the migrant laborers of Gosaba CD block, South 24 Parganas district, West Bengal.

	N	Mean	Std. deviation	Std. error	95% confidence interval for mean		Mini mum	Maxi mum
Lower bound	**Upper bound**							
Agricultural laborer	8	9375.00	1187.74	419.93	8382.03	10,367.97	8000	11,000
Construction workers	15	17,666.67	3829.71	988.83	15,545.84	19,787.49	12,000	25,000
Other unskilled laborer	12	11,916.67	4399.55	1270.04	9121.32	14,712.01	8000	25,000
Skilled laborer	15	36,733.33	8539.54	2204.90	32,004.29	41,462.38	20,000	50,000
Total	50	20,680.00	12,284.78	1737.33	17,188.70	24,171.30	8000	50,000

TABLE 7.2 Table for ANOVA calculation.

	Sum of squares	df	Mean square	F	Sig.
Between groups	5.946	3	1.982	62.916	0.000
Within groups	1.449	46	31,501,268.116		
Total	7.395	49			

Based on Primary Survey (2022).

are: H_0: Occupation of migrant workers is not interrelated to their monthly income and H_1: Occupation of migrant workers is interrelated to their monthly income. The test results F-value of 62.916 and a *P*-value of 0.000. As the *P*-value is less than 0.05% of significance level so, there is enough evidence to reject the null hypothesis and accepting alternative hypothesis and it can be said that occupation of migrant workers is related to their monthly income.

6.2 State of affairs of left behind women of migrant households

Three decades back, international migration was perceived as a male issue. This assumption was particularly prevalent when attention was focused on the economic aspects of migration, because women's participation in international labor migration was insignificant (Ullah, 2017). Since the 1970s, the share of female migrants among all international migrants has been rising steadily. Feminization of migration has become a core dimension of the new age of international migration and globalization (Donato, 2016). The most important rationalizing factor for migration is the betterment of the families. However, the absence of husbands necessarily changes the pattern of everyday life of the wives (Lan et al., 2015). Family structures get changed as responsibilities are reassigned (Gamburd, 2000; Rigg, 2007).

Though there are estimates of total international migrants, however, the number of accompanied or dependent migrant family members are not yet known. The impact of migration on the left-behind family members has long been debated. The impact of absence of husbands on marital life is profound. When husbands cross the border to work, they leave their families behind. Wives and children do not accompany them for varied reasons. The familiar pattern is for the husbands to leave their countries, work overseas for a period of time as long as they can stay, and feel is economically feasible and return to their families (Ullah, 2017). Of the total, low- and semiskilled ones form the majority of the migrants (2023). This implies that either these migrants are not allowed to bring their family with them, or they cannot afford. This means, their wives and, of course, along with their children remain behind. In traditional societies, for the wives, this is a fresh experience of living on their own. Some studies bear out that men's migration leads to higher autonomy, independence and decision-making authority of their wives (Yabiku et al., 2010). As they are remained-behind, they start interacting with so many people, who they never talked and worked with, organizations, schools, and banks (Abadan-Unat, 1977). They take responsibilities of agricultural decisions (Boehm, 2008; Ullah, 2017).

Motivations for migration are with the primary aim of invigorating economy of their families; however, it always remains an issue that as they leave the country, they leave behind many issues that are to be dealt with by left-behind family members who may be or may not be ready for this. Some tend to say that financial flows help solve all issues they may face in absence of the male counterparts. Financial gains alone do not solve the social problems faced by left-behind ones. When household head leaves, the aftermath situations compel them to take on unfamiliar and new responsibilities. In this new role, how they do behave over time? Does this new role prepare them to become empowered or leaders? Or they become over-burdened that lead them to frustration and fatigued. Historically, in patriarchal societies, women have less personal autonomy (Bloom et al., 2001), fewer resources at their disposal and limited influence over the decision-making processes than males (Ikuomola, 2015; Ullah, 2017) which means that they are inexperienced and unprepared for taking over new roles. Physical absence of a parent, especially the male parent who has traditionally been holding the household head position and breadwinner in the family, has myriad of consequences on the decision-making processes about the family matters. However, it is important to delve into what happens to women who assume new role and more responsibilities than before as a result of their husbands' truancy.

"Home" or "inside" is equated with security, certainty, order, family, and femaleness, while "outside" or "journey" becomes synonymous with risk, strangeness, chaos, masculinity, and the public realm (Altman & Werner, 1985). An understanding of home as a "haven in a heartless world" relies on a rigid separation of inside and outside, with safety and security to be found inside, and fear and danger remaining outside (Wardhaugh, 1999). Such a definition of home can be said to contribute to the creation of homelessness, in that those who are abused and violated within are likely to feel "homeless at home," and many subsequently become homeless in an objective sense, in that they escape—or are ejected from—their violent homes (Chattoraj, 2022; Wardhaugh, 1999). Furthermore, those who are not able, or choose not to, conform to the gender, class, and sexuality deals inherent in establishing a conventional household, find themselves symbolically (and often literally) excluded from any notion or semblance of home (Smailes, 1994). The impulse to exclude and expel deviant others from conventional notions of home and neighborhood is an essential, although perpetually uncompleted function. The Other is excluded from purified space but does not cease to exist: they remain present and visible and thus continue to be dangerous. The Other has its own power, the ability to inspire fear (Sibley, 1995). Home is widely, and often unproblematically associated with femaleness (Wardhaugh, 1999): both with the women who are expected to maintain hearth and home, and with the presumed feminine principles of boundedness, physicality and nurturance. It should follow, then, that male and female relations to the home will differ radically. Object relations theory, for example, holds that females identify with their mothers (and by implication with home), and thereby establish a sense of connectedness to people and places. It is argued that males from an early age, in contrast, learn to differentiate themselves from mother and home, establish boundaries between themselves and the world, and thus establish a sense of themselves as separate and disconnected (Massey, 1992).

Many young women and men can trace the roots of their homelessness to their experience of home: family or partnership breakdowns are consistently cited as the most common causes of homelessness, with escaping violent situations and leaving institutional care and control also significant factors (Hutson & Liddiard, 1994). Different routes are taken by those women who have experienced themselves to be "homeless at home," by those gender renegades and transgressors who were unable or unwilling to conform to traditional family forms and norms, and by those whose "homes" were in effect prison cells. Just as there are many routes to becoming homeless, so there are various ways of being homeless. It is important to capture a sense of agency among such women, but they should not be portrayed as protofeminist revolutionaries. Some were more or less consciously rejecting gender norms, but others retained an attachment to conventional domesticity and traditional "family values": these women were, at most, reluctant rebels. Mainstream traditions of research hold that the homeless population is anomic, retreatist and disaffiliated (Bahr & Caplow, 1973), although later interactionist research emphasizes the social networks, sense of agency, and levels of resistance among homeless populations (Harper et al., 1994; Wagner, 2019, pp. 1–200).

Due to prevalent societal norms, the girls in this region are married off at a young age and consider child-bearing, taking care of the family to be their primary occupation. As their educational attainment is limited, the women have very inadequate exposure about the outer world and are not equipped to pursue their own careers. The respondents, all of whom were wives of migrant laborers ($n = 150$), mostly (60%) represented the young (early adulthood) age group from 18 to 35 years, followed by middle age group from 36 to 49 years (28.7%). The representation of mature adulthood belonging to more than 50 years of age was comparatively low (11.3%) as mobile, migratory males in this age bracket is harder to get. 82% of the women respondents have migrated at some point of their life—with marriage being the main driving factor (66%). Dutta (2019) stated that female migration beyond marital reasons is becoming common. Similar instance was seen in the study area too, a formidable section in pursuit of higher education (14.6%) migrated while rest either became a part of husband's migration in few occasions or women themselves migrated outside for occupational commitments (household maid, care-givers, attendants). Through Participatory Rural Appraisal conducted on the women of the migrants' household an attempt was made to find out the important factors behind the migratory trend and the ones which came up were size of the household, average age of the household, nature and extent of landownership, educational level and occupation pursued before migration.

The respondent of the middle age group shared that they are well settled in a life of a home-maker dealing with extreme poverty but after their husbands suddenly out-migrated for work, they had to fill in the shoes as the head of the family. Due to the prolonged absence of the males, the left behind wives had to take up most of the familial responsibilities. Though they mostly started as temporary, pseudo-heads with the passage of time carrying out of the duties, tasks and transactions gradually make them confident and the leadership role gets rubbed off genuinely. About 32% of the respondents reported to have a household male member as a permanent migrant and in most of those cases the wives of the migrant became the head of the household and became more conversant with the outer world. The age-sex pyramid of the area gets skewed as mostly the dependents, both old and young along with female members of the working age group are left behind by the migrant males of working age group. So, the women of migrant households have to start taking part in the agricultural activities directly, the ones who are comparatively better off employ agricultural laborers. Naturally, the work participation of the women increased over the period of time and this change in occupation pattern mostly affected the local economy positively. There is a kind of livelihood mix on the go, apart from the involvement in the primary sector related works, the women also dabbled in small scale business or acted as self-employed professionals using the surplus income generated from migratory activities. From the study it was clear that the remittance was mostly sent to wives (85.3%) by the migrant laborers. Remittances sent were mostly spent on household expenses (44.7%) and education of children (26.7%) Fig. 7.4. It is interesting to note that in present times, no distinction is made between boy and girl child in terms of providing educational opportunities. This generational shift in mindset is heartening to see which can be attributed to migratory exposure of the fathers and the gradual empowerment of the left behind mothers.

Ghosh and Roy (2021) made a very interesting observation that over a period of time (2012–19) the average number of earning members per household in Sundarbans increased from 1.91 to 2.34. This study also reflects this trend, in the migratory households the number of earning members increases steadily as the women in the house become economically active be it wife or daughter or both, members of next generation also follow the migratory path having similar or better career trajectory and adds to the family remittances. To gauge the perception of the respondents on how they viewed migration in terms of economic context,

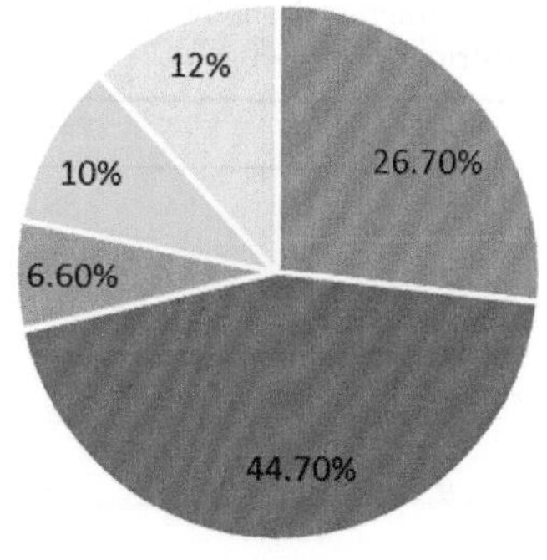

FIGURE 7.4 Usage pattern of remittance received from the migrant laborers of Gosaba CD block, South 24 Parganas district, West Bengal. *Based on Primary Survey. (2022).*

the sweeping answer was highly positive (78%). This mirrored the observation of Ghosh (2012), Hajra and Ghosh (2016) that the household which had at least one migrated member was financially much well off due to the receipt of remittances. However, in the perception study, another aspect got highlighted that a small section of respondents (8%) feared that this improved economic condition might not be viable in the long run as the demand for unskilled laborer is going to decline over time. During the Focused Group Discussion, the respondents also members of a Self Help Group (24.7%) confided that before their husbands undertook migration, they faced an uncertain future with weaker economic condition which is a widely accepted fact but besides their life was almost stifled due to the male domination in their personal space. When the women respondents were asked how their situation changed during postmigration phase, they pointed out few main facets Table 7.3, with the financial improvement aspect topping the chart.

Increased work participation and decision-making ability does not always denote true empowerment. In most cases, the decisions are limited to mundane daily activities like child education, food preference, taking part in routine economic activities. But certain indicators like ability to take decisions relating to marital issues and childbearing (25.3%), taking part in political activities (15.3%) and having completely free mobility without being answerable to anyone (2%) are more akin to empowerment i.e., being confident about one's right and in control of one's life. The façade of empowerment for many is shattered when few respondents (16%) expressed that they do not look forward to the return of their migrant husbands as they are subjected to physical or sexual violence and/or exposed to health hazards like HIV or other communicable diseases. Another chunk of respondents did not feel themselves empowered and was rather bogged down by the long working hours (28%) or took resort to greater usage of social networking sites to avoid loneliness (25%). A chi-square test of independence has been calculated to check whether there is any association between the two categorical variables i.e., duration of husbands' migration and wife's changing level of empowerment Tables 7.4 and 7.5.

TABLE 7.3 Changed situation of left behind women of migrant households of Gosaba CD block, South 24 Parganas district, West Bengal.

Post-migration noticeable changes	Frequency	Percent
Improved financial condition	58	38.7
Greater involvement in economic activities	54	36.0
Acting as decision-maker in family matters	19	12.7
Upgradation of educational attainment	7	4.7
Participation in self help groups	9	6.0
Attainment of political empowerment	3	2.0
Total	150	100.0

Based on Primary Survey (2022).

TABLE 7.4 Duration of migration by the migrant laborer husbands of Gosaba CD block, South 24 Parganas district, West Bengal.

		Financial empowerment	Social empowerment	Political empowerment	Total
Duration of migration	<1 year	17	4	2	23
	1–5 years>5 years 4324		3711	93	8938
	Total	84	52	14	150

TABLE 7.5 Calculation table for Chi-Square tests.

	Value	df	Asymp. Sig. (2-Sided)
Pearson chi-square	7.480[a]	5	0.187
Likelihood ratio	5.822	5	0.324
Linear-by-linear association	3.873	1	0.049
Number of valid cases	150		

[a]*5 cells (41.7%) have expected count less than 5. The minimum expected count is 0.46.*
Based on Primary Survey (2022).

Here the value of test statistics is 7.480 and *P*-value is 0.187 which is greater than the chosen significance level of 0.05%. So, here the null hypothesis is accepted and it can be stated that there is no association between duration of husbands' migration and women's changing level of empowerment. The intention and capacity of women are more important in changing the empowerment level. Male out-migration leads to un-homing of the male members as more time they spent outside their motherland, the connections and interactions become more remittance-centric rather than that of physical and psychological reliance while the left out female members become more dynamic entities in the home front.

7. Conclusion

The economy of Sundarbans is driven by the triad "Productivity-Poverty-Outmigration"; with decreasing yield rate, indigence of the inhabitants increased and to counter that outmigration has been used as a weapon. The findings of this study very well match to that of (Hajra et al., 2018) where they pointed out associations between decreasing agricultural productivity and increasing out-migration and highlight how remittances can do away with the household poverty. But migration has both positive and negative impact on local economy of the study area. Migration definitely helped to cope with economic distress and improved the financial condition. With sharply increasing monthly income, growing household assets, mounting surplus income and savings not only the household conditions improved, the situation of the entire deltaic region improved. Land values increased, modernization of agriculture and aquaculture became possible and even service-sector like small businesses, construction work flourished creating few job opportunities as the purchasing power amplified with increasing remittance and is having a multiplier effect including women empowerment. Emphasis on child's education ensured a more educated, skilled next generation ready to take on the world. Due to the out-migration of their husbands, the left behind wives took up most of the familial responsibilities and with the passage of time they gained confidence and became the head of the household and became more conversant with the outer world. Additionally, the migratory households saw a steady increase in the number of earning members as the female members in the house became economically active with the out-migration of their male counterparts. In Sundarbans, due to lack of exposure to the outside world, female members were mostly responsible for home-making and child-bearing and were not equipped to pursue their own careers. And most of the women migrated due to marriage. Our data shows that, in recent days, the region is experiencing a large number of female out-migration in pursuit of higher education and occupational commitments. But everything is not rosy about migration. Out-migration created a shortage of labor availability and the wage rates increased manifold. As agriculture on saline soil involves lot of hard work and comparatively low return, the new generation is not much interested to stay back in rural interiors and toil for farming and is attracted by the quick buck earning through migratory pursuits in urban areas. Besides, there are only small businesses or household industries coming up which fail to generate adequate job opportunities. Another serious concern is the health issue, often the migratory laborers contract HIV contamination or other communicable diseases and put their wives back at home at risk.

More than economic, migration in Sundarbans has an ecological connotation. The socioeconomic vulnerability is magnified by the ecosystem vulnerability. Migration can lead to having more self-reliant, confident, and empowered womenfolk; remittances can solve pressing economic issues, help in educational progress; governmental initiatives can improve connectivity, accessibility but the ecological distress—cyclone, storm surges, sea level rise, embankment breaching, saline water intrusion, flooding, coastal erosion does not have a quick fix. Banerjee et al. (2023) tried to construct an impact chain outline highlighting the socioecological susceptibilities, vulnerabilities of the inhabitants of Sundarbans keeping faith on short term coping capacities and long-term adaptive capacities to overcome those. If the climate induced out-migration is on the rise, then the issue will not only be limited to migration but will become ecological refugee. Presently, few well-off migrants are even planning to complete relocation along with their family in sub-urban Kolkata so to avoid the climatic wrath and livelihood uncertainties. It is, thus, paving the way for rural–urban migration, referred as the movement of people from rural to urban areas in search of a better life. Though one cannot really predict whether through judicious livelihood mix, proper governmental interventions and positive intention of the residents Sundarbans will be able to overcome the ecological distress and avert the largescale permanent depopulation which seems to be on the card but in the due course of fighting for existence, the largescale out-migration have definitely made the womenfolk independent and poised infusing a sense of empowerment into them.

References

Abadan-Unat, N. (1977). Implications of migration on emancipation and pseudo-emancipation of Turkish women. *International Migration Review, 11*(1), 31–58. https://doi.org/10.1177/019791837701100102

Acosta, P., Calderón, C., Fajnzylber, P., & Lopez, H. (2008). What is the impact of international remittances on poverty and inequality in Latin America? *World Development, 36*(1), 89–114. https://doi.org/10.1016/j.worlddev.2007.02.016

Adhikari, J., & Hobley, M. (2011). Everyone is leaving. Who will sow our fields?" the livelihood effects on women of male migration from Khotang districts to the Gulf countries and Malaysia. *Swiss Agency for Development and Cooperation (SDC).*

Adhikari, J., & Hobley, M. (2015). Everyone is leaving. Who will sow our fields?" the livelihood effects on women of male migration from Khotang and Udaypur districts, Nepal, to the Gulf countries and Malaysia. *Himalaya, 35*(1).

Al-Ali, N., & Koser, K. (2003). *New approaches to migration?: Transnational communities and the transformation of home.* United Kingdom: Routledge Taylor & Francis Group. https://doi.org/10.4324/9780203167144

Alasah, A. A. (2008). *Women's empowerment and community development in cameroon. A case study of NGOs and women's organisations in the Northwest Province.* Doctoral thesis).

Altman, I., & Werner, C. M. (1985). *Home environments.* Springer.

Andersson, L. (2014). *Migration, remittances and household welfare in Ethiopia.*

Bahr, H. M., & Caplow, T. (1973). *Old men drunk and sober.* New York University Press.

Bala, A. (2017). Migration in India: Causes and consequences. *International Journal of Advanced Educational Research, 2*(4), 54–56.

Banerjee, K. (2013). Decadal change in the surface water salinity profile of Indian Sundarbans: A potential indicator of climate change. *Journal of Marine Science: Research and Development, 01*(S11). https://doi.org/10.4172/2155-9910.S11-002

Banerjee, S., Chanda, A., Ghosh, T., Cremin, E., & Renaud, F. G. (2023). A qualitative assessment of natural and anthropogenic drivers of risk to sustainable livelihoods in the Indian sundarban. *Sustainability, 15*(7), 6146. https://doi.org/10.3390/su15076146

Banu, N. (2016). Pattern of rural- urban migration and socio-economic transformation in West Bengal. In *Proceedings of 8 Research World International Conference.*

Barham, B., & Boucher, S. (1998). Migration, remittances, and inequality: Estimating the net effects of migration on income distribution. *Journal of Development Economics, 55*(2), 307–331. https://doi.org/10.1016/S0304-3878(98)90038-4

Basu, A., Kar, N. S., & Nandy, G. (2021). Evaluating the viability of shrimp aquaculture to impede climate change in the Sundarbans: Experiences from Bangladesh. In *South Asia and climate change: Unravelling the conundrum* (pp. 156–188). India: Taylor and Francis. https://www.routledge.com/South-Asia-and-Climate-Change-Unravelling-the-Conundrum/Kar-Mukhopadhyay-Sarkar/p/book/9780367478124.

Bhugra, D. (2004). Migration and mental health. *Acta Psychiatrica Scandinavica, 109*(4), 243–258. https://doi.org/10.1046/j.0001-690X.2003.00246.x

Bloom, S. S., Wypij, D., & Das Gupta, M. (2001). Dimensions of women's autonomy and the influence on maternal health care utilization in a North Indian city. *Demography, 38*(1), 67–78. https://doi.org/10.2307/3088289

Boehm, D. A. (2008). \Now I am a man and a woman!\: Gendered moves and migrations in a transnational Mexican community. *Latin American Perspectives, 35*(1), 16–30. https://doi.org/10.1177/0094582X07310843

Bose, Saha, S., & Goswami, R. (2017). *Male out-migration and empowerment of women left behind in Indian sunderbans.* Refugeewatchonline. Retrieved on. 8.

Bramley, G. (1988). The extent and character of London's growing homelessness problem. In *Homelessness and the London Housing Market.* University of Bristol, School for Advanced Urban Studies.

Brauw, A. D. (2020). Seasonal migration and agricultural production in Vietnam. *Migration, transfers and economic decision making among agricultural households* (pp. 114–139). United States: Taylor and Francis. https://www.routledge.com/products/9780415495134.

Brown, B. B. (1983). The impact of male labour migration on women in Botswana. *African Affairs, 82*(328), 367–388. https://doi.org/10.1093/oxfordjournals.afraf.a097536

Castles. (1992). The 'new' migration and Australian immigration policy. In *Asians in Australia: The dynamics of migration and settlement* (pp. 45–72).

Census of India. (2011). *District census handbook, north and south twenty four parganas, West Bengal, series-20 Part XII-A & B, directorate of census operations, West Bengal. Registrar general of India, ministry of home affairs*. Government of India.

Chand, B. K., Trivedi, R. K., Biswas, A., Dubey, S. K., & Beg, M. M. (2012). Study on impact of saline water inundation on freshwater aquaculture in Sundarban using risk analysis tools. *Exploratory Animal and Medical Research, 2*, 170–178.

Chattoraj, D. (2016). *Ambivalent attachments: Shifting notions of home among displaced Sri Lankan Tamils* (Doctoral dissertation).

Chattoraj, D. (2020). AKM ahsan Ullah and Md. Shahidul haque (2020). The migration myth in policy and practice: Dreams, development and despair. *Migration Letters, 17*(3), 473–475. https://doi.org/10.33182/ml.v17i3.936

Chattoraj, D. (2022). Underlying theoretical aspects. *Asia in Transition, 11*, 55–85. https://doi.org/10.1007/978-981-16-8132-5_3

Chauhan, M., & Gopal, B. (2014). Sundarban mangroves: Impact of water management in the Ganga river basin. In *Our National River Ganga* (pp. 143–166). Cham: Springer. https://doi.org/10.1007/978-3-319-00530-0_5

Damon, A. L. (2010). Agricultural land use and asset accumulation in migrant households: The case of El salvador. *Journal of Development Studies, 46*(1), 162–189. https://doi.org/10.1080/00220380903197994

Das, C. S. (2018). Pattern and characterisation of human casualties in sundarban by tiger attacks, India. *Sustainable Forestry, 1*(4). https://doi.org/10.24294/sf.v1i2.873

Dasgupta, S., Wheeler, D., Sobhan, M. I., Bandyopadhyay, S., Nishat, A., & Paul, T. (2020). *Coping with the vulnerability of the sundarban in a changing climate: Lessons from multidisciplinary studies.*

De Haan, A., & Rogaly, B. (2002). Introduction: Migrant workers and their role in rural change. *Journal of Development Studies, 38*(5), 1–14. https://doi.org/10.1080/00220380412331322481

de Haan, A. (1999). Livelihoods and poverty: The role of migration – A critical review of the migration literature. *Journal of Development Studies, 36*(2), 1–47. https://doi.org/10.1080/00220389908422619

de Haas, H., & van Rooij, A. (2010). Migration as emancipation? The impact of internal and international migration on the position of women left behind in rural Morocco. *Oxford Development Studies, 38*(1), 43–62. https://doi.org/10.1080/13600810903551603

Debnath, A. (2013). Condition of agricultural productivity of gosaba CD block, south24 parganas, West Bengal, India after severe cyclone Aila. *International Journal of Scientific and Research Publications, 3*(7), 1–4.

Deepa, N., Lant, P., & Soumya, K. (2009). *Moving out of poverty: Volume 2. Success from the bottom up*. World Bank Publications. https://doi.org/10.1596/978-0-8213-7215-9

Deshingkar. (2008). Migration and development within and across broader: Research and policy perspectives on internal and international migration. *International Organization of Migration*, 161–188.

Donato, K. M. (2016). *The global feminization of migration: Past, present, and future*. Migration Policy Institute.

Dubey, S., & Mallah, V. (2015). Migration: Causes and effects. *Business and Management Review, 5*(4), 228–232.

Dutta, S., Chakraborty, K., & Hazra, S. (2017). Ecosystem structure and trophic dynamics of an exploited ecosystem of Bay of Bengal, Sundarban Estuary, India. *Fisheries Science, 83*(2), 145–159. https://doi.org/10.1007/s12562-016-1060-2

Dutta, U. (2019). Migration and its impact on socio-economic and demographic structure of Sundarban region of south 24 parganas. *Indian Journal of Spatial Science, 10*, 55–59.

de la Fuente, A. (2010). Remittances and vulnerability to poverty in rural Mexico. *World Development, 38*(6), 828–839. https://doi.org/10.1016/j.worlddev.2010.02.002

Gamburd, M. (2000). Nurture for sale: Sri Lankan housemaids and the work of mothering. *Home and hegemony: Domestic service and identity politics in South and Southeast Asia* (pp. 179–205). University of Michigan Press.

Gartaula, H. N., Visser, L., & Niehof, A. (2012). Socio-cultural dispositions and wellbeing of the women left behind: A case of migrant households in Nepal. *Social Indicators Research, 108*(3), 401–420. https://doi.org/10.1007/s11205-011-9883-9

Gemenne, F., & Brücker, P. (2015). From the guiding principles on internal displacement to the nansen initiative: What the governance of environmental migration can learn from the governance of internal displacement. *International Journal of Refugee Law, 27*(2), 245–263. https://doi.org/10.1093/ijrl/eev021

Ghimire, M., & Vikas, M. (2012). Climate change—impact on the Sundarbans, a case study. *International Scientific Journal: Environmental Science, 2*(1), 7–15.

Ghosh, A. (2012). *Living with changing climate—impact, vulnerability and adaptation challenges in Indian Sundarbans.*

Ghosh, R., & Mal, S. (2017). Impacts of rural labour migration of south bengal: A case study of Bankura and Purulia districts of West Bengal, India. *Global Journal of Human-Social Science:Interdisciplinary, 17*(7), 24–32.

Ghosh, S. (2017). *Coping with a natural disaster: Sundarbans after cyclone Aila* (pp. 116–136). SAGE Publications. https://doi.org/10.4135/9789353280284.n11

Ghosh, S., & Roy, S. (2021). *Climate change, ecological stress and livelihood choices in Indian Sundarban* (pp. 399–413). Springer Science and Business Media LLC. https://doi.org/10.1007/978-981-16-0680-9_26

Ghosh, S. (2013). Extreme event, anthropogenic stress and ecological sustainability in Sundarban Islands. *Vidyasagar University Journal of Economics, XVII*, 132–148.

GoI. (2021). *Periodic labour force survey (PLFS), ministry of statistics and programme implementation.*

Gopal, B., & Chauhan, M. (2006). Biodiversity and its conservation in the Sundarban mangrove ecosystem. *Aquatic Sciences, 68*(3), 338–354. https://doi.org/10.1007/s00027-006-0868-8

Guerny. (1995). *Gender migration, farming systems and land tenure.*

Guha, I., & Roy, C. (2016). Climate change, migration and food security: Evidence from Indian Sundarbans. *International Journal of Theoretical and Applied Sciences, 8*(2), 45–49.

Haas, H. D. (2007). *Remittances, migration and social development: A conceptual review of the literature.*

Hajra, R., & Ghosh, T. (2016). Migration pattern of Ghoramara Island of Indian Sundarban-identification of push and pull factors. *Asian Academic Research Journal of Social Sciences and Humanities, 3*(6), 2278–2859.

Hajra, R., Ghosh, A., & Ghosh, T. (2016). *Comparative assessment of morphological and landuse/landcover change pattern of Sagar, Ghoramara, and Mousani island of Indian sundarban delta through remote sensing* (pp. 153–172). Springer Science and Business Media LLC. https://doi.org/10.1007/978-3-319-46010-9_11

Hajra, R., Ghosh, T., Chadwick, O., & Renaud, F. (2018). Agricultural productivity, household poverty and migration in the Indian Sundarban Delta. *Elementa: Science of the Anthropocene, 6*. https://doi.org/10.1525/elementa.196

Haldar, A., & Debnath, A. (2014). Assessment of climate induced soil salinity conditions of Gosaba Island, West Bengal and its influence on local livelihood. In *Climate Change and Biodiversity* (pp. 27–44). Springer Science and Business Media LLC. https://doi.org/10.1007/978-4-431-54838-6_3

Harper, D., Snow, D. A., & Anderson, L. (1994). Down on their luck: A study of homeless street people. *Contemporary Sociology, 23*(1), 42. https://doi.org/10.2307/2074854

Hazra, S., Das, I., Samanta, K., & Bhadra, T. (2014). Impact of climate change in Sundarban area West Bengal, India. School of oceanographic studies. *Earth Science and Climate Book, 9326*.

Hoq, M. E. (2007). An analysis of fisheries exploitation and management practices in Sundarbans mangrove ecosystem, Bangladesh. *Ocean and Coastal Management, 50*(5–6), 411–427. https://doi.org/10.1016/j.ocecoaman.2006.11.001

Hoque Mozumder, M. M., Shamsuzzaman, M. M., Rashed-Un-Nabi, M., & Karim, E. (2018). Social-ecological dynamics of the small scale fisheries in Sundarban Mangrove forest, Bangladesh. *Aquaculture and Fisheries, 3*(1), 38–49. https://doi.org/10.1016/j.aaf.2017.12.002

Hutson, S., & Liddiard, M. (1994). *Youth homelessness: The construction of a social issue.* Bloomsbury Publishing.

Ikuomola, A. D. (2015). An exploration of life experiences of left-behind wives in Edo State. *Nigeria. Journal of Comparative Research in Anthropology and Sociology, 6*(1), 289–307.

Islam, S. N. (2014). An analysis of the damages of Chakoria Sundarban mangrove wetlands and consequences on community livelihoods in south east coast of Bangladesh. *International Journal of Environment and Sustainable Development, 13*(2), 153–171. https://doi.org/10.1504/IJESD.2014.060196

Kasper, H. (2005). *'I Am the household head now!': Gender aspects of out-migration for labour in Nepal.* Kathmandu: Nepal Institute of Development Studies. NIDS.

Kollmair, M., Manandhar, S., Subedi, B., & Thieme, S. (2006). New figures for old stories: migration and remittances in Nepal. *Migration Letters, 3*(2), 151–160. https://doi.org/10.33182/ml.v3i2.66

Lan, A. H., Theodora, L., Brenda, S. A. Y., & Elspeth, G. (2015). Transnational migration, changing care arrangements and left-behind children's responses in South-east Asia. *Children's Geographies, 13*(3), 263–277.

Lastarria-Cornhiel, S. (2006). Feminization of agriculture: Trends and driving forces. In *Background paper for world development report.*

Laube, W., Schraven, B., & Awo, M. (2012). Smallholder adaptation to climate change: Dynamics and limits in Northern Ghana. *Climatic Change, 111*(3), 753–774. https://doi.org/10.1007/s10584-011-0199-1

Manandhar, S., Vogt, D. S., Perret, S. R., & Kazama, F. (2011). Adapting cropping systems to climate change in Nepal: A cross-regional study of farmers' perception and practices. *Regional Environmental Change, 11*(2), 335–348. https://doi.org/10.1007/s10113-010-0137-1

Mandal, M., & Mandal, S. (2012). Impact of natural resource on socio-economic status of Sagar Island. *International Journal of Engineering Science, 2*(1), 76–80.

Mani, S. (2012). High skilled migration and remittances. In *Growth, development, and diversity: India's rcord since liberalization* (pp. 181–210). India: Oxford University Press. https://doi.org/10.1093/acprof:oso/9780198077992.003.0008

Massey, D. (1992). A place called home. *New Formations, 17*(3), 3–15.

Massey, D., Rafique, A., Rogaly, B., & Seeley, J. (2009). Staying behind when husbands move: Women's experiences in India and Bangladesh. In *Development research centre on migration, globalisation and poverty briefing* (p. 18).

Massey, D. S., Alarcon, R., Durand, J., & Gonzalez, H. (1990). *Return to Aztlan: The social process of international migration from Western Mexico.* University of California Press.

Massey, D. S., Arango, J., Hugo, G., Kouaouci, A., Pellegrino, A., & Taylor, J. E. (1994). An evaluation of international migration theory: The North American case. *Population and Development Review, 20*(4), 699–751. https://doi.org/10.2307/2137660

Mathew, G. (1995). New economic policy, social development and sociology. *Sociological Bulletin, 44*(1), 63–77. https://doi.org/10.1177/0038022919950104

McLeman, R. A., & Hunter, L. M. (2010). Migration in the context of vulnerability and adaptation to climate change: Insights from analogues. *Wiley Interdisciplinary Reviews: Climate Change, 1*(3), 450–461. https://doi.org/10.1002/wcc.51

Miluka, J., Carletto, G., Davis, B., & Zezza, A. (2020). The vanishing farms? The impact of international migration on albanian family farming. In *Migration, Transfers and Economic Decision Making among Agricultural Households* (pp. 140–161). Albania: Taylor and Francis. https://www.routledge.com/products/9780415495134.

Mistri, A., & Das, B. (2015). Environmental legislations and livelihood conflicts of fishermen in Sundarban. *India Asian Profile, 43*, 389–400.

Mistri, A. (2013). Migration and sustainable livelihoods: A study from Sundarban biosphere reserve. *Asia Pacific Journal of Social Sciences, 5*(2), 76–102.

Moghadam, V. M., & Senftova, L. (2005). Measuring women's empowerment: Participation and rights in civil, political, social, economic, and cultural domains. *International Social Science Journal*, (184), 389–412. https://doi.org/10.1111/j.1468-2451.2005.00557.x

Mosse, D., Gupta, S., Mehta, M., Shah, V., Rees, J.f., & Team, KRIBP Project (2002). Brokered livelihoods: Debt, labour migration and development in tribal western India. *Journal of Development Studies, 38*(5), 59–88. https://doi.org/10.1080/00220380412331322511

Nandi, N. C., & Pramanik, S. K. (2017). Livelihood on mud crab catchment: A case study of sundarban coast, West Bengal, India. *Journal of Fisheries and Aquaculture Development, 1*(1). https://doi.org/10.29011/2577-1493.100003

Neogi, S. B., Dey, M., Kabir, S. M. L., Masum, S. J. H., Kopprio, G., Yamasaki, S., & Lara, R. (2017). Sundarban mangroves: Diversity, ecosystem services and climate change impacts. *Asian Journal of Medical and Biological Research, 2*(4), 488–507. https://doi.org/10.3329/ajmbr.v2i4.30988

Ortiz-Ospina, E., & Roser, M. (2017). Homelessness. In *Our world in data.*

Parkins, N. C. (2010). Push and pull factors of migration. *American Review of Political Economy, 8*(2). https://doi.org/10.38024/arpe.119

Pieke, F. N., & Biao, X. (2009). Legality and labour: Chinese migration, neoliberalism and the state in the UK and China. Geopolitics. *History and International Relations, 1*, 11–45.

Radel, C., Schmook, B., & McCandless, S. (2010). Environment, transnational labor migration, and gender: Case studies from southern Yucatán, Mexico and Vermont, USA. *Population and Environment, 32*(2–3), 177–197. https://doi.org/10.1007/s11111-010-0124-y

Raha, A., Das, S., Banerjee, K., & Mitra, A. (2012). Climate change impacts on Indian Sunderbans: A time series analysis (1924-2008). *Biodiversity and Conservation, 21*(5), 1289–1307. https://doi.org/10.1007/s10531-012-0260-z

Raha, A. K., Zaman, S., Sengupta, K., Bhattacharyya, S. B., Raha, S., Banerjee, K., & Mitra, A. (2013). Climate change and sustainable livelihood programme: A case study from Indian sundarban. *Journal of Ecology, 107*(335348).

Rajkarnikar, P. J. (2017). *The impacts of foreign labor migration of men on women's empowerment in Nepal.*

Rao, N. (2006). Land rights, gender equality and household food security: Exploring the conceptual links in the case of India. *Food Policy, 31*(2), 180–193. https://doi.org/10.1016/j.foodpol.2005.10.006

Ravenstein, E. G. (1885). The laws of migration. *Journal of the Statistical Society of London, 48*(2), 167–227. https://doi.org/10.2307/2979181

Rigg, J. (2007). Moving lives: Migration and livelihoods in the Lao PDR. *Population, Space and Place, 13*(3), 163–178. https://doi.org/10.1002/psp.438

Rowthorn, R. (2010). Combined and uneven development: Reflections on the north south divide. *Spatial Economic Analysis, 5*(4), 363–388. https://doi.org/10.1080/17421772.2010.516445

Rudra, K. (2010). The proposal of strengthening embankment in Sundarbans: Myth and reality. *A South Asian Journal on Forced Migration*, 86–93.

Samal, C. K. (2006). Remittances and sustainable livelihoods in semi-arid areas. *Asia-Pacific Development Journal, 13*(2).

Sandilyan, S., & Kathiresan, K. (2015). Mangroves as bioshield: An undisputable fact. *Ocean and Coastal Management, 103*, 94–96. https://doi.org/10.1016/j.ocecoaman.2014.11.011

Sarker, M., & Islam, S. (2014). Husbands' international labour migration and the change of wives' position among the left-behind in rural Bangladesh. *Research on Humanities and Social Science, 4*(16), 57–62.

Sánchez-Triana, E., Paul, T., & Leonard, O. (2014). *Building resilience for sustainable development of the Sundarbans. The international bank for reconstruction and development*. The World Bank.

Sibley, D. (1995). *Geographies of exclusion: Society and difference in the west*. Psychology Press.

Sinha, B., Jha, S., & Negi, N. S. (2012). Migration and empowerment: The experience of women in households in India where migration of a husband has occurred. *Journal of Gender Studies, 21*(1), 61–76. https://doi.org/10.1080/09589236.2012.639551

Smailes, J. (1994). The struggle has never been simply about bricks and mortar. In *Housing women*.

Somerville, P. (1992). Homelessness and the meaning of home: Rooflessness or rootlessness? *International Journal of Urban and Regional Research, 16*(4), 529–539. https://doi.org/10.1111/j.1468-2427.1992.tb00194.x

Stark, O. (1991). *The migration of labor*. Basil Blackwell (undefined).

Taylor, J. E., & Martin, P. L. (2001). Chapter 9 Human capital: Migration and rural population change. *Handbook of Agricultural Economics, 1*, 457–511. https://doi.org/10.1016/S1574-0072(01)10012-5

Toyota, M., Yeoh, B. S. A., & Nguyen, L. (2007). Bringing the 'left behind' back into view in Asia: A framework for understanding the 'migration—left behind nexus'. *Population, Space and Place, 13*(3), 157–161. https://doi.org/10.1002/psp.433

Ullah, A. K. M. A. (2017). Male migration and 'left—behind' women: Bane or boon? *Environment and Urbanization ASIA, 8*(1), 59–73. https://doi.org/10.1177/0975425316683862

Wagner, D. (2019). *Checkerboard square: Culture and resistance in a homeless community*. Taylor and Francis. https://doi.org/10.4324/9780429034640

Wang, T., Maruayama, A., & Kikuchi, M. (2005). Rural-urban migration and labour market in China: A case study in a northeastern province. *The Developing Economies, 37*(1), 80–104.

Wardhaugh, J. (1999). The unaccommodated woman: Home, homelessness and identity. *The Sociological Review, 47*(1), 91–109. https://doi.org/10.1111/1467-954x.00164

Watkins, J. F., & Hosier, A. F. (2005). Conceptualizing home and homelessness: A life course perspective. In *Home and identity in late life: International perspectives* (pp. 197–216).

World Development Report. (2022). *Finance for an equitable recovery*. Washington: The World Bank. https://www.worldbank.org/en/publication/wdr2022#downloads.

World Development Report. (2023). *Migrants, refugees and societies*. Washington: The World Bank. https://www.worldbank.org/en/publication/wdr2023.

World Migration Report. (2022). *International organization for migration*. Geneva https://publications.iom.int/books/world-migration-report-2022.

Yabiku, S. T., Agadjanian, V., & Sevoyan, A. (2010). Husbands' labour migration and wives' autonomy, Mozambique 2000-2006. *Population Studies, 64*(3), 293–306. https://doi.org/10.1080/00324728.2010.510200

CHAPTER

8

Comparing rural and urban homelessness in the context of environmental vulnerability in India's hill states

Nipun Behl[1] *and Uttam Roy*[2]

[1]IIT Roorkee, Department of Architecture & Planning, Roorkee, Uttarakhand, India; [2]Department of Architecture and Planning, Centre for Transportation Systems (CTRANS), Indian Institute of Technology Roorkee, Roorkee, Uttarakhand, India

1. Introduction

In 2023, India surpassed China as the world's most populous country, with an estimated population of approximately 14.2 billion (UN-Habitat, 2020). The consistent population rise has caused significant stresses in developmental attributes like housing and land, infrastructure, and environment. Such pressures have been more vulnerable for hill areas due to the limited land availability and consequent rise in housing shortage (Lakshmana, 2013). The rise in property prices in hilly regions, driven by economic growth and the popularity of these areas as tourist destinations, has made housing unaffordable for many (Kumar & Mohanta, 2003). Consequently, homelessness has become a pressing issue, particularly among low-income individuals and families struggling to find affordable housing. Geographical constraints, such as difficult terrain and limited land availability, compound the challenges of urbanization and homelessness in hilly regions. These factors hinder urban development and exacerbate the housing crisis, necessitating targeted interventions and comprehensive urban planning (Wright, 1990). Homelessness is a complex and multidimensional social issue that refers to the lack of access to a permanent, safe, and secure place to live. According to the United Nations, approximately 150 million people worldwide are estimated to be homeless or living in inadequate housing, with the number of homeless individuals and families rising in many countries. Homelessness is a growing problem in developing nations, where

Homelessness to Hope
https://doi.org/10.1016/B978-0-443-14052-5.00008-2

the lack of resources and support services exacerbates the challenges of homeless individuals and families.

Homelessness encompasses more than the absence of shelter. It involves the loss of a physical home, legal rights, and social connections, causing isolation and vulnerability. Poverty and housing supply failures are primary causes, along with factors like domestic violence, support system erosion, socio-political instability, and health issues (Batterham, 2018). The homeless population is diverse in age, gender, livelihoods, and reasons for street living. Definitions include those lacking adequate nighttime shelter, temporary accommodations, or substandard housing (Springer, 2000). Quantifying homelessness is challenging due to invisibility and limited survey methods. The definition of homelessness varies internationally but typically emphasizes the absence of a permanent and adequate place to live, along with safety and security. For example, the US Department of Housing and Urban Development (HUD) defines homelessness as "an individual without a fixed, regular and adequate nighttime residence" (Delhi, 2013). The National Alliance to End Homelessness defines it as "the condition of people without a permanent dwelling" (Dikmen & Elias-Ozkan, 2016). The United Nations and the World Health Organization similarly define it as a situation in which people lack a place to live that is safe and secure and protects against environmental elements (UN-Habitat, 2020).

This chapter explores the interrelationship between homelessness in rural and urban settings within the hill states of India. The primary aim of the chapter is to gain a comprehensive understanding of the multifaceted challenges faced by homeless individuals and families in these regions, as well as to analyze the underlying factors and wide-ranging ramifications of homelessness. By examining these aspects, the objective is to illuminate the complex dynamics and implications associated with homelessness in both rural and urban contexts within the hilly landscapes of India. The significance of exploring homelessness in these settings is underscored by the paucity of existing literature, which hampers a thorough comprehension of the unique adversities experienced by homeless individuals. This shortage of literature also impedes the identification of effective strategies and interventions tailored to the distinctive context of hill states. This chapter emphasizes the critical importance of generating and disseminating knowledge to inform evidence-based policy formulation, raise awareness, advocate for evidence-driven practices and facilitate meaningful interventions to combat homelessness in both rural and urban settings of hill states in India. This study focuses on the hill states within the Great Himalayan region, which extend from the northern to the country's northeastern border. The study aims to shed light on the specific challenges homeless individuals and families face in these regions, including the lack of access to resources, support services, and economic opportunities. By understanding these challenges, the study aims to inform policymakers and stakeholders about the best approaches to address homelessness and improve the overall quality of life for those affected.

1.1 Homelessness in hilly regions

In India, the problem of homelessness is critical in the country's environmentally fragile hill states, where the limited availability of resources and support services are common. Individuals and families in these hill states face numerous challenges related to their basic needs and well-being, including limited access to food, shelter, and healthcare, limited employment and educational opportunities, and increased risk of exploitation and abuse

(Shim et al., 2018). The lack of affordable housing in these regions limited access to basic necessities, and limited access to basic necessities can result in widespread poverty and social marginalization (Williams, 2005). The difficulties homeless individuals and families face in environmentally fragile hill states of the developing nation underline the significance of a comprehensive and concerted approach from policymakers and relevant stakeholders. This approach should encompass investments in affordable housing, providing support services, and creating economic opportunities. Addressing the underlying causes of homelessness and enhancing the standard of living for those impacted is crucial for effectively addressing this intricate issue in these regions.

1.2 Overview Himalayan hill states of India

India is home to a diverse range of hilly regions spread across different states, known for their captivating landscapes, rich biodiversity, and distinct cultural heritage. These regions include the Eastern Ghats, Western Ghats, Cardamom Hills, and the renowned Himalayan Ranges. The Great Himalayan Range spans approximately 2400 km (1500 miles) across multiple states in northern and northeastern India, encompassing Jammu and Kashmir, Himachal Pradesh, Uttarakhand, Meghalaya, Nagaland, Arunachal Pradesh, Manipur, Tripura, and Sikkim (Long et al., 2011). These hill states boast unique indigenous communities and cultural heritage, presenting their own challenges regarding economic development and resource accessibility. Approximately 34 million people inhabit this expansive region. Addressing homelessness in these areas requires tailored and comprehensive approaches considering the region's specific geographical, social, and economic characteristics (Wilson, 2017). The Himalayan Hill region is renowned for its majestic mountain ranges, including famous peaks like Mount Everest, its breathtaking landscapes, diverse ecosystems, and vibrant cultural heritage. This region offers a blend of natural beauty, adventure opportunities, and spiritual destinations, making it an integral and cherished part of India's hilly landscape (Kumar, 2016).

However, the hill states, characterized by steep slopes, high altitudes, and fragile ecosystems, are prone to environmental degradation. These regions face challenges related to limited resources, delicate ecosystems, and susceptibility to natural disasters and environmental hazards. Understanding these challenges is crucial for developing effective solutions to address environmental degradation and its associated issues. Urban development in the Himalayas presents unique opportunities and challenges. With an influx of urbanization, hill towns, and cities have experienced population growth and infrastructure development, becoming important economic, administrative, and tourism hubs. These urban centers provide essential services such as employment opportunities, educational institutions, and healthcare facilities to both residents and visitors (Mitterer et al., 2012). The rugged terrain and ecological sensitivity of the Himalayas pose specific challenges to urban development. The steep slopes, limited flat land, and geological vulnerabilities necessitate careful planning and innovative engineering solutions to ensure sustainable and resilient urban infrastructure. Building roads, structures, and amenities in the hilly terrain requires expertise and adherence to environmental regulations to minimize the impact on the fragile ecosystem (Berhane & Walraevens, 2013).

Preserving the natural and cultural heritage of the Himalayas is essential in urban development. Balancing economic growth with conservation is crucial for protecting the region's unique biodiversity, landscapes, and cultural traditions. Sustainable tourism practices, heritage conservation initiatives, and responsible urban planning play a significant role in ensuring the long-term viability of urban centers and their surrounding ecosystems (Dikmen & Elias-Ozkan, 2016). The Hilly regions face specific challenges related to water management, disaster resilience, and climate change adaptation. Melting glaciers, changing precipitation patterns, and increased vulnerability to natural disasters necessitate the incorporation of climate-resilient strategies into infrastructure development and urban design. Managing water resources, waste, and energy efficiency are vital considerations for sustainable urban development in these regions (Maina, 2013). Considering the intricate interplay of various factors, homelessness assumes an exceedingly intricate nature in hilly areas. The dynamics of urban development, sociocultural attributes, and a range of environmental, climatic, and geographical constraints influence the issue. The unique topographical features, ecological sensitivity, and extreme weather conditions of hilly regions contribute significantly to the complexity of the problem.

In hilly areas, the challenges associated with homelessness are further compounded by the scarcity of suitable land for housing, limited infrastructure, and restricted access to basic services. Steep slopes and rugged terrain present substantial obstacles to constructing stable and sustainable housing solutions. Moreover, the delicate ecosystems of these regions are vulnerable to degradation and disruption, exacerbating the difficulties homeless individuals and families face (Anand et al., 2014). The influence of climate and weather patterns on homelessness in hilly areas cannot be understated. These regions are prone to intense precipitation, including heavy rains, snowfall, and avalanches, leading to increased landslide risks and flash floods (Bisht et al., 2018). Such natural disasters can displace populations and destroy existing housing structures, leaving many people without a place to live. Additionally, extreme temperatures and harsh climatic conditions pose significant challenges for individuals without shelter, jeopardizing their health and overall well-being (Mitterer et al., 2012). The geographical constraints of hilly areas add to the complexities of addressing homelessness. Limited accessibility and connectivity make it challenging to provide adequate support services, resources, and opportunities for economic development. The remote locations of many hilly regions create logistical difficulties in delivering essential supplies and services to homeless populations, further exacerbating their vulnerability.

2. Overview of homelessness in rural and urban areas

In India, the definition of homelessness is unclear and varies by source. The Ministry of Housing and Urban Affairs (MoHUA) defines it as the lack of access to safe and adequate housing, which aligns with the UN definition. However, the lack of a clear definition has made estimating the extent of homelessness in India challenging. It is believed to affect millions of people, particularly in urban areas. Urban homelessness is linked to poverty, limited affordable housing, job shortages, limited access to services, and social stigma and discrimination (Meert & Bourgeois, 2005). Addressing it requires a comprehensive approach that addresses poverty and housing, and provides support and services to the homeless. Rural homelessness, on the other hand, is caused by rural poverty, limited economic opportunities, environmental hazards, scarce housing, and limited access to services. Addressing it requires a specialized approach that takes into account the unique challenges and interdependence of rural and urban areas (Fitchen, 1992).

To effectively tackle homelessness, it is crucial to understand the challenges faced by the homeless in both rural and urban areas and develop targeted solutions to address the root causes of homelessness in each setting. According to the 2011 Census, the houseless population in India was 1.82 million. The trend in homelessness has shown variations over time. From 1961 to 1981, there was a consistent rise in the homeless population, with a notable spike between 1961 and 1971. However, from 1991 to 2011, there was a decline in the overall houseless population, although the decrease was less pronounced in 2001 and 2011 (Nambiar, 2020). In rural areas, the homeless population followed an upward trend until 1981 but has since decreased. Conversely, urban areas witnessed a continuous increase in the homeless population from 1961 to 2011. The difference in the proportion of the houseless population between rural and urban areas has been diminishing over the years. However, when considering the share of the homeless population in relation to the total population, urban areas consistently exhibited a higher prevalence of homelessness in both 2001 and 2011. Between 2001 and 2011, see Figs. 8.1 and 8.2, the share of the urban houseless population increased, while there was a slight decline in rural areas (Kumuda, 2012). Overall, homelessness has declined in rural areas compared to urban areas, where it remains a growing concern.

The Himalayan hill states of India demonstrate a parallel trend to the rest of the country, characterized by an overall decline in the total rate of homelessness. However, a closer examination of each state reveals distinct dynamics at play. Notably, there has been a significant upswing in homelessness within urban areas, accompanied by a corresponding decrease in the rate of homelessness in rural areas. This indicates a notable spatial redistribution of homelessness within the region. To gain a comprehensive understanding of this phenomenon, conducting an in-depth analysis of the specific factors and dynamics influencing homelessness in each of these hill states becomes crucial. By delving into the unique characteristics and

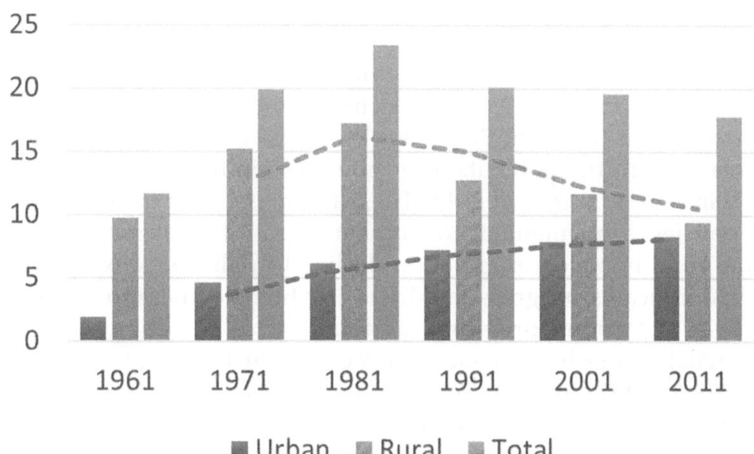

FIGURE 8.1 Urban and rural houseless population in India. *Source: Census 2011.*

FIGURE 8.2 Urban and rural houseless population trends in India. *Source: Census 2011.*

Year	Rural Houseless Population	Urban Houseless Population
1961	Rising trend	Rising trend
1971	Sharp increase	Rising trend
1981	Rising trend	Rising trend
1991	Decline	Decline
2001	Decline (lower)	Rising trend
2011	Decline	Rising trend

complexities of the local contexts, policymakers and stakeholders can better address the emerging challenges associated with homelessness in these regions.

Figure Urban and Rural Houseless Population and Trends in Himalayan Hill in addition to the overall analysis of homelessness in the Himalayan hill states, further examination reveals interesting trends regarding the distribution of homeless populations in urban and rural areas.

Himachal Pradesh and Sikkim stand out as states that experienced a steep increase in urban homelessness. In Himachal Pradesh, the percentage of houseless individuals in urban areas rose from 0.13% in 2001 to 0.37% in 2011, reflecting a significant upward trend. Similarly, Sikkim witnessed a notable increase in urban homelessness, with the percentage of houseless individuals in urban areas growing from 0.05% in 2001 to 0.1% in 2011. These findings highlight the urgent need for targeted interventions in urban settings to address the growing challenge of homelessness in these states. Conversely, Nagaland, Manipur, and Uttarakhand witnessed a steep decline in the urban homeless population. Nagaland saw a decrease in the percentage of houseless individuals in urban areas from 0.22% in 2001 to 0.13% in 2011, indicating significant progress in addressing urban homelessness. Similarly, Manipur experienced a decline in the percentage of houseless individuals in urban areas from 0.14% in 2001 to 0.1% in 2011. Uttarakhand also displayed a significant decrease, with the percentage of houseless individuals in urban areas dropping from 0.22% in 2001 to 0.06% in 2011. These trends suggest the effectiveness of interventions and policies implemented to combat urban homelessness in these states. Interestingly, the rural homeless population declined in all states except Arunachal Pradesh. This decline indicates progress in addressing homelessness in rural areas, emphasizing the success of various programs and initiatives targeting rural communities. However, Arunachal Pradesh witnessed an increase in the rural homeless population from 0.18% in 2001 to 0.18% in 2011, highlighting the need for targeted efforts to address the specific challenges faced in this region.

Overall, the data showcases the complex dynamics of homelessness in the Himalayan hill states. While Himachal Pradesh and Sikkim grapple with increasing urban homelessness,

Nagaland, Manipur, and Uttarakhand demonstrate significant progress in reducing homelessness in urban areas. The declining rural homeless population in most states reflects positive outcomes of interventions, while the situation in Arunachal Pradesh warrants further attention. These findings emphasize the importance of tailored strategies, focusing on both urban and rural contexts, to effectively combat homelessness in the Himalayan hill states. Arunachal Pradesh saw an increase in homelessness from 442 individuals in 2001 to 1522 individuals in 2011. The percentage of houseless individuals to the total population increased from 0.22% to 0.3%. The distribution of houseless individuals was relatively balanced between rural and urban areas, highlighting the need for comprehensive strategies addressing homelessness in both contexts (Kumar, 2016). These technical analyses of the Himalayan hill states' homelessness situation provide insights into the trends and challenges faced in these regions. The findings underscore the importance of targeted interventions, particularly in rural areas, to address these states' unique dynamics of homelessness.

3. Factors contributing to homelessness in hill states

The Himalayan hill states in India face a range of challenges that contribute to the issue of homelessness. These factors play a significant role in exacerbating vulnerability and hindering the overall well-being of individuals and communities in these regions. Understanding these contributing factors is crucial for developing effective strategies to address homelessness. The major factors influencing homelessness in the Himalayan hill states include environmental degradation, infrastructure deficiencies, political instability, and the proximity to international borders (Morrison, 2009). Environmental degradation is a pressing concern in states like Meghalaya, Arunachal Pradesh, and Sikkim. Deforestation and overexploitation of natural resources not only harm the environment but also disrupt the livelihoods of local communities (Nygren, 2000). These ecological challenges lead to increased poverty and unemployment, creating a conducive environment for homelessness.

Inadequate infrastructure and limited access to basic amenities further compound the problem of homelessness in these regions (Sewell et al., 2019). States such as Uttarakhand and Tripura struggle with deficient infrastructure, including poor connectivity and limited access to essential services. Insufficient housing and basic amenities worsen the homelessness condition and create barriers to social and economic development. Political instability and frequent outbreaks of violence and unrest pose significant challenges (Jong-A-Pin, 2009) in Jammu and Kashmir, Nagaland, and Manipur. The resulting instability hampers overall development, contributes to poverty and unemployment, and increases the risk of homelessness in these states. The proximity of the Himalayan hill states to international borders significantly influences the homelessness situation. The presence of borders brings unique challenges, particularly related to security concerns and restrictions on land usage. Heightened security measures and limitations on land availability for housing and development projects impact the housing situation in these areas. Additionally, restricted infrastructure development, including limited access to essential services, further exacerbates socioeconomic disparities and increases the risk of homelessness (Delhi, 2013).

International borders also impact the economic activities and livelihood opportunities in these regions (Sohn, 2014). Fluctuations and restrictions in trade and commerce due to geopolitical factors affect the local economy, leading to reduced job opportunities, income disparities, and economic instability. The movement of people across borders, including migrants, refugees, and displaced populations, adds to the complexities of addressing homelessness, as it places additional strain on already limited resources and infrastructure. The table Table 8.1 presented above offers a holistic overview of challenges such as Environmental Degradation, Infrastructure, Political Instability, and Development, along with the underlying factors that contribute to these challenges across different states. Addressing homelessness in the Himalayan hill states requires a comprehensive understanding of the factors shaped by environmental degradation, infrastructure deficiencies, political instability, and the impact of international borders (Delhi, 2013). Effective strategies should prioritize sustainable environmental practices, infrastructure development, access to basic amenities, political stability, poverty alleviation, job creation, and affordable housing initiatives. Collaborative approaches involving government agencies, local communities, and international stakeholders are essential to mitigate security concerns, promote sustainable development, and create an enabling environment that supports reducing and preventing homelessness in the Himalayan hill states.

4. Homelessness trends and patterns in the Himalayan hill states

Homelessness in the hill states of the Himalayas refers to the absence of stable and permanent housing for individuals and families residing in hilly regions. It encompasses those living on the streets, in shelters, transitional housing, or other unstable living arrangements. The causes of homelessness in these states are diverse and multifaceted, ranging from poverty and unemployment to displacement due to development and a lack of affordable housing.

4.1 Urban homelessness

4.1.1 Increase in urban homelessness

Urban homelessness has witnessed a significant rise in Himachal Pradesh and Sikkim, primarily attributed to major urban development projects undertaken in these states. These projects have attracted a growing number of people to migrate toward urban centers seeking opportunities and improved living conditions. In Himachal Pradesh, the percentage of houseless individuals in urban areas increased substantially from 0.13% in 2001 to 0.37% in 2011. Similarly, Sikkim experienced an upward trend in urban homelessness, with the percentage of houseless individuals in urban areas growing from 0.05% in 2001 to 0.1% in 2011 as per census data. The data highlights the urgent need for targeted interventions and focused efforts in addressing the growing challenge of homelessness in urban settings within these states. The lack of adequate infrastructure and limited access to basic amenities further exacerbates the problem of homelessness in the Himalayan hill states. Deficient infrastructure, including poor connectivity and limited access to essential services, hampers social and economic development, ultimately contributing to the prevalence of homelessness. This situation is particularly evident in states such as Uttarakhand and Tripura, where insufficient infrastructure and limited access to essential services compound the homelessness situation.

TABLE 8.1 Drivers and challenges of homelessness in the Himalayan hill states of India.

Hill state	Environmental degradation	Infrastructure	Political instability	Development	Drivers
Himachal Pradesh	N/A	N/A	N/A	N/A	Poverty, unemployment, health and mental health issues, lack of affordable housing, displacement due to development
Uttarakhand	Significant challenge	Poor access to basic amenities	N/A	N/A	Poverty, unemployment, lack of job opportunities
Jammu and Kashmir	N/A	Poor access to basic amenities	Significant challenge, frequent outbreaks of violence and unrest	Lack of development	Poverty, unemployment
Meghalaya	Significant challenge, deforestation, overexploitation of natural resources	Poor access to basic amenities	N/A	N/A	Poverty, unemployment
Nagaland	N/A	Poor access to basic amenities	Significant challenge, frequent outbreaks of violence and unrest	Lack of development	Poverty, unemployment
Arunachal Pradesh	Significant challenge, deforestation, overexploitation of natural resources	Poor access to basic amenities	N/A	N/A	Poverty, unemployment
Manipur	N/A	N/A	Significant challenge, frequent outbreaks of violence and unrest	N/A	N/A
Tripura	N/A	Poor access to basic amenities	Significant challenge, frequent outbreaks of violence and unrest	Lack of development	Poverty, unemployment
Sikkim	Significant challenge, deforestation, overexploitation of natural resources	Poor connectivity, limited access to basic amenities	N/A	Lack of development	Poverty, unemployment

4.1.2 Decline in urban homelessness

In contrast to the rising trend of urban homelessness in some Himalayan hill states, significant progress has been made in addressing this issue in Nagaland, Manipur, and Uttarakhand. These states have witnessed a notable decline in the urban homeless population, reflecting the effectiveness of interventions and policies implemented to combat homelessness. Nagaland, for instance, experienced a considerable decrease in the percentage of houseless individuals in urban areas, dropping from 0.22% in 2001 to 0.13% in 2011. This decline highlights the positive outcomes of initiatives aimed at addressing urban homelessness and promoting social stability. Similarly, Manipur witnessed a decline from 0.14% in 2001 to 0.1% in 2011, while Uttarakhand displayed a significant decrease with the percentage dropping from 0.22% in 2001 to 0.06% in 2011 . The declining trend in urban homelessness in these states can be attributed to various factors, including targeted interventions, policy measures, and urban development initiatives. Efforts focused on upgrading urban infrastructure and creating a conducive environment for equitable development have played a crucial role in reducing homelessness and improving the overall well-being of individuals and communities. It is worth noting that while these states have implemented measures to combat urban homelessness, they have also prioritized ensuring equitable development for all residents. This approach emphasizes the importance of addressing socio-economic disparities and promoting inclusive growth as essential components of comprehensive strategies to tackle homelessness in urban areas.

4.2 Rural homelessness

4.2.1 Decline in rural homelessness

With the **exception of Arunachal Pradesh**, there has been a general decline in the rural homeless population across the mentioned states, although specific figures are not provided. This overall trend signifies progress in addressing homelessness in rural areas and highlights the effectiveness of various programs and initiatives specifically targeting rural communities. In addition to the efforts made to combat rural homelessness, several factors have contributed to this decline. Political instability in Jammu and Kashmir, Nagaland, and Manipur has influenced out-migration, which has played a role in reducing homelessness in these states. Furthermore, other states have implemented critical measures such as rural development programs, expanding urban boundaries, and addressing migration issues, all of which have significantly declined rural homelessness. Acknowledging the pressing concerns regarding environmental degradation in states like Meghalaya, Arunachal Pradesh, and Sikkim is crucial. These regions experience the adverse effects of deforestation and the overexploitation of natural resources, leading to disruptions in the livelihoods of local communities. The consequences of environmental degradation include increased poverty and unemployment, ultimately contributing to the prevalence of homelessness in these areas.

4.2.2 Increase in rural homelessness

In contrast to the general trend of declining rural homelessness in the mentioned states, Arunachal Pradesh experienced an increase in the rural homeless population from 0.18%

in 2001 to 0.18% in 2011. The proximity of the Himalayan hill states to international borders plays a significant role in shaping the homelessness situation in this region. The presence of international borders brings about security concerns and leads to restrictions on land usage, which directly impact the availability of housing options. These limitations further exacerbate the challenges faced by the population, creating socioeconomic disparities and increasing the risk of homelessness. Additionally, the lack of adequate infrastructure development and restricted access to essential services contribute to the complex dynamics of homelessness in Arunachal Pradesh.

4.3 Comparative analysis of rural and urban homelessness

The population and economy of hill states in India are divided between rural and urban areas, with a majority of the population residing in rural areas. According to data from the Census of India, in 2011, the rural population of Himachal Pradesh was over 70%, while the urban population was just over 28%. In Uttarakhand, the rural population was over 80%, and the urban population was just over 19%. This section analyzes the differences and similarities between homelessness in rural and urban areas in hilly regions. The causes of homelessness in these two areas may vary, with poverty, unemployment, and lack of affordable housing being the main drivers in rural areas, and a combination of poverty, health issues, and lack of support services driving homelessness in urban areas. The total homeless population of the Hill states is 83,766, with the urban homeless population being 41,816 and the rural homeless population being 41,950. Comparing this data with the national average, we see that the total homeless population of the Hill states is around 4.7% of the national average, with the urban homeless population being 4.4% and the rural homeless population being 5.0%. The data shows that the hill states have a lower homeless population compared to the national average, but there are still disparities between urban and rural regions, with a higher proportion of the rural homeless population compared to urban (Fig. 8.3).

The table provides information about the population and homelessness data for nine hill states in India. The total population of these states is approximately 33.8 million, with Jammu & Kashmir having the largest population of 12.5 million and Sikkim having the least population of 0.6 million. The data also shows the number of homeless individuals in each state, with Jammu & Kashmir having the highest number of 19,047 individuals experiencing homelessness, followed by Uttarakhand with 11,824 individuals. The data also compares the homelessness trends and patterns in rural and urban areas. The results show that there is a disparity between the urban and rural homeless population in each state, with the majority of the homeless population located in urban areas in some states and in rural areas in others. For example, in Arunachal Pradesh, 77% of the homeless population is located in rural areas, while in Tripura, 73% is in urban areas (Fig. 8.4).

The disparities between urban and rural areas are not solely a result of population size, but also of poverty, unemployment, lack of access to basic amenities, and environmental degradation, among other factors. The area of each state also plays a role in determining the population density of the region. Hill states like Arunachal Pradesh and Meghalaya have large areas but comparatively lower populations, indicating a lower population density. On the other hand, states like Uttarakhand and Himachal Pradesh have smaller areas but higher

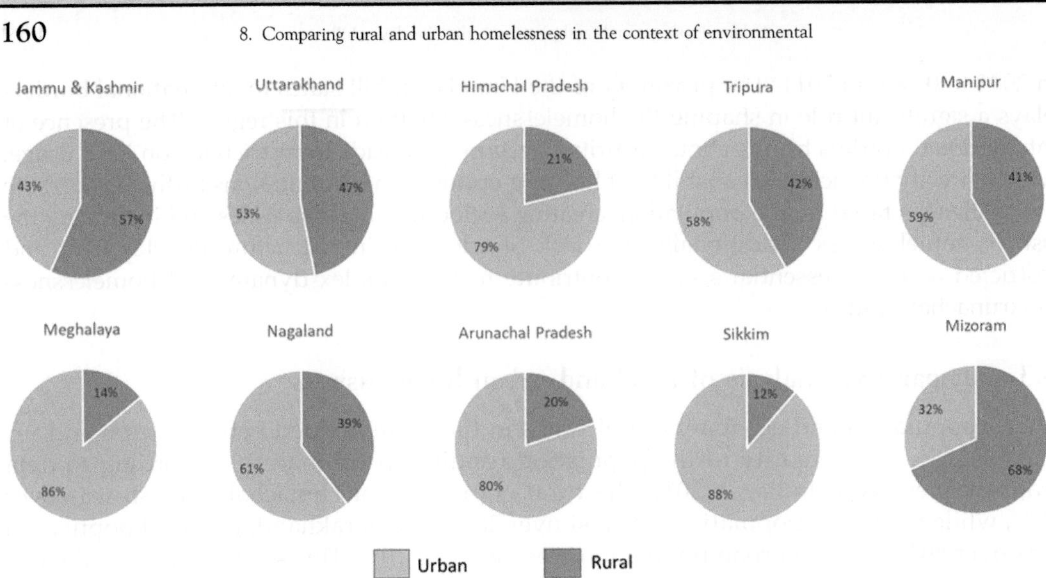

FIGURE 8.3 Percentage distribution of urban and rural homeless population in Himalayan hill states of India.

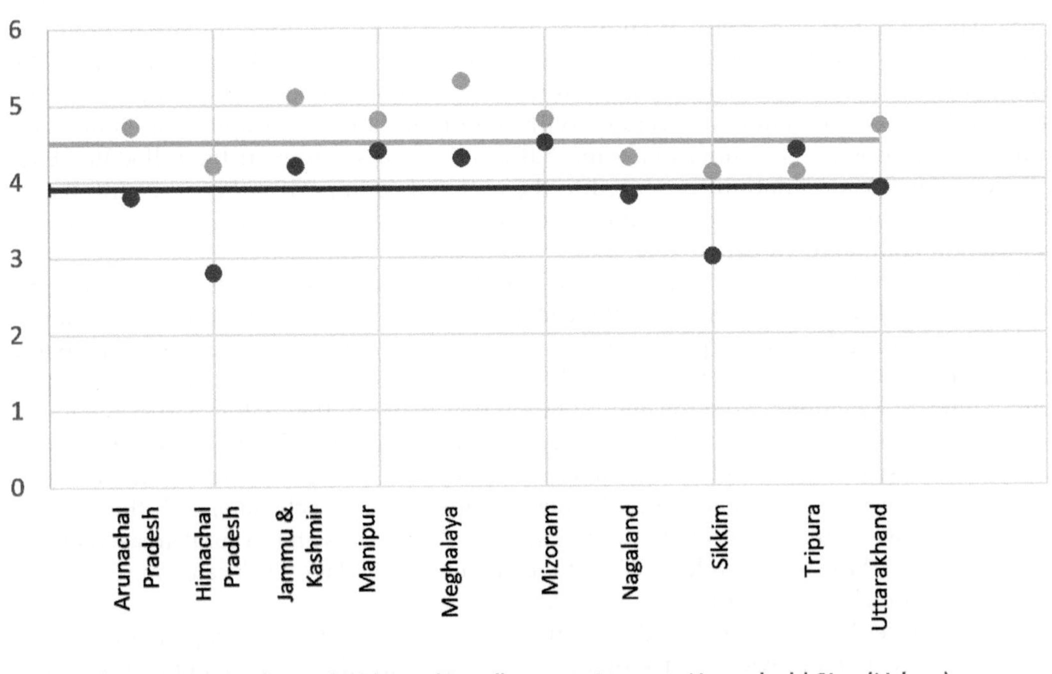

FIGURE 8.4 Average household size in urban and rural areas of hill states of India.

populations, indicating higher population densities. In terms of the economy, rural areas in hill states are primarily dependent on agriculture and forestry, while urban areas have a more diversified economy, with industries such as tourism, manufacturing, and services. However, the economic development in both rural and urban areas of hill states is impacted by factors such as lack of infrastructure, limited access to markets and financial services, and challenging geographic conditions. Migrating people from rural to urban areas in search of better economic opportunities and improved quality of life can lead to urbanization and strain on limited resources in the cities. This can contribute to the concentration of poverty and unemployment in urban areas, affecting the overall economic development of the hill states. The division between urban and rural areas can also impact access to education and healthcare, with urban areas often having better access to educational institutions and healthcare facilities. Rural areas may struggle with limited resources and a lack of infrastructure, resulting in lower enrollment rates in schools and limited access to medical care. This disparity can further perpetuate poverty and inequality in rural areas, impacting the overall health and well-being of the population. In addition, the study describes the trends and patterns of homelessness in these environmentally fragile hill states, including the number and characteristics of homeless individuals and families. The causes of homelessness can be diverse, from poverty and unemployment to displacement due to development and lack of affordable housing.

Factors, Effects, and Support Services of Homelessness

The factors driving homelessness in hill states can differ based on the particular state and region, including poverty, unemployment, health issues, lack of affordable housing, displacement due to development, limited access to support services, lack of infrastructure and services, remote location, and environmental hazards. Addressing these factors will be crucial in promoting stable communities and improving the quality of life of those experiencing homelessness (Fig. 8.5).

4.4 Factors contributing to homelessness in environmentally fragile hill states

The hill states of India face unique and complex challenges that contribute to the problem of homelessness. These challenges are largely influenced by the region's topography, climate,

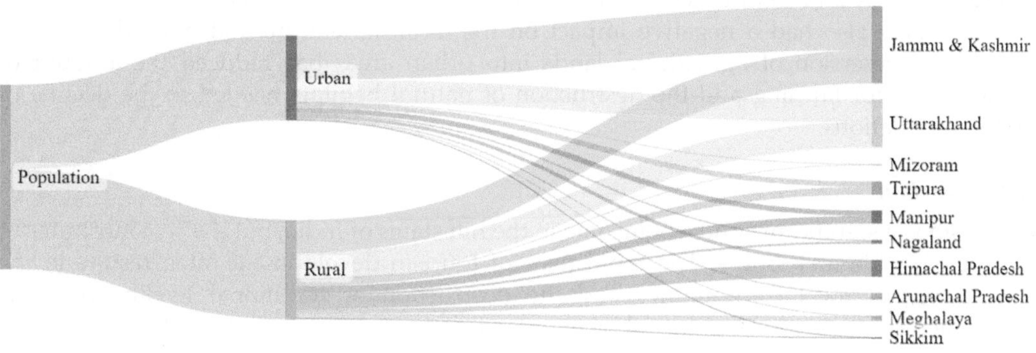

FIGURE 8.5 Homeless population flow diagram for urban and rural areas in Himalayan hill states of India.

cultural and economic diversity. The combination of environmental hazards, economic downturns, displacement due to development, and rural–urban migration has resulted in an increase in the number of homeless people in the region.

4.4.1 *Rural–urban migration*

The hill states of India have seen a steady increase in rural-urban migration in recent years. This is due to various reasons, including the lack of job opportunities in rural areas, limited access to basic amenities, and better educational and healthcare facilities in urban areas. This migration puts pressure on the already limited urban infrastructure and housing, leading to an increase in homelessness.

4.4.2 *Displacement due to development*

The rapid pace of development in the hill states has led to the displacement of large numbers of people, who are often unable to find alternative housing. This can be due to the construction of large hydroelectric projects, tourism developments, and other infrastructure projects. The displacement often results in the loss of traditional livelihoods and sources of income, which contributes to homelessness.

4.4.3 *Economic downturn*

The hill states of India are often susceptible to economic downturns, which can result in job losses, reduced income, and a decline in living standards. The combination of high poverty levels and a lack of employment opportunities leads to an increase in homelessness. In addition to the other challenges faced by the hill states, political unrest can also contribute to homelessness. Political instability can result in the disruption of economic activity, loss of jobs, and reduced income levels.

4.4.4 *Environmental degradation*

Development and urbanization have had a significant impact on the environment in the hill states. Agricultural lands have been transformed into urban areas, leading to the destruction of natural habitats and the decline of agricultural productivity. This degradation has resulted in the loss of biodiversity and the decline of natural resources, which in turn affects the local economy and contributes to homelessness. The development and urbanization in the hill states have also had a negative impact on the local agriculture and natural flora and fauna. The conversion of agricultural lands into urban areas has reduced the amount of land available for farming and the destruction of natural habitats has led to the decline of wildlife populations.

4.4.5 *Tourism*

Tourism is a significant source of income in the hill states of India, but it also contributes to the problem of homelessness. The rapid pace of tourism development often results in the displacement of local communities and the destruction of traditional livelihoods. This displacement puts pressure on local housing, leading to an increase in homelessness.

4.4.6 Environmental hazards

The hill states of India are vulnerable to a number of environmental hazards, including cloudbursts, landslides, and flash floods. These hazards can result in the destruction of homes and displacement of populations, which in turn contributes to homelessness.

4.5 Impact of challenges faced by homeless individuals and families

The impact of these challenges faced by homeless individuals and families in hill states of India can be significant and far-reaching. Here are a few examples:

Lack of access to necessities: Homeless individuals and families often struggle to access basic necessities such as food, shelter, and healthcare. This can lead to hunger, malnutrition, and a range of health problems.

Insecurity and vulnerability: Homelessness can make individuals and families more vulnerable to physical and psychological harm, such as crime, abuse, and exploitation.

Reduced economic opportunities: The lack of stable housing and access to resources can limit economic opportunities for homeless individuals and families, leading to poverty and unemployment.

Disruption of social networks and support systems: Homelessness can also disrupt social networks and support systems, leading to feelings of isolation and hopelessness.

Negative impact on children: Homelessness can have particularly damaging effects on children, affecting their education, mental health, and future opportunities.

Increased burden on emergency and healthcare services: Homeless individuals and families often need emergency and healthcare services, placing additional strain on these already-overburdened systems.

Environmental degradation: Unplanned and rapid urbanization can lead to environmental degradation, affecting the natural flora and fauna of the area and making it vulnerable to environmental hazards such as cloudbursts.

Seismic vulnerability: The high seismic vulnerability of hill states makes them prone to earthquakes and other natural disasters, further exacerbating the challenges homeless individuals and families face.

4.6 Importance of support services in addressing homelessness

Support services play a critical role in addressing homelessness, especially in hilly regions where the challenges are unique and complex. Homeless individuals and families in the hill states are often faced with a range of obstacles, including displacement due to development, economic downturns, environmental hazards, and rural-urban migration. Support services provide critical assistance in overcoming these obstacles and addressing the underlying causes of homelessness. Support services' first and foremost role is to provide immediate assistance to homeless individuals and families. This can include providing temporary shelter, food, medical care, and referrals to other social services such as employment training, financial assistance, and mental health support. By addressing the immediate needs of the homeless population, support services provide a foundation for addressing the underlying causes of homelessness. In hilly regions, support services also play a crucial role in addressing the challenges of displacement due to development and environmental hazards. These challenges can result in the loss of traditional livelihoods and sources of income, which can contribute to homelessness. Support services can provide job training, financial assistance, and other support programs to help individuals and families overcome these challenges and secure stable housing. Another important role of support services is to provide educational and vocational training to the homeless population. In hilly regions, limited employment opportunities and low levels of education often contribute to homelessness. Support services can provide training programs that equip individuals with the skills and knowledge needed to secure employment and improve their standard of living. Finally, support services play a critical role in raising awareness and advocacy on the issue of homelessness in hilly regions. Through public education and outreach campaigns, support services can raise awareness of the challenges the homeless population faces and encourage greater support and investment in addressing the issue. By working in partnership with local organizations and governments, support services can help to create a more supportive and inclusive environment for homeless individuals and families. India is a vast country with diverse geographical features that significantly influence the sources of drinking water available to households. The four main categories are urban hill areas, rural hill areas, urban plain areas, and rural plain areas. Each of these areas has unique challenges in terms of providing access to clean drinking water.

5. Discussion

Comprehensive strategies are crucial to combat urban homelessness in states like Himachal Pradesh, Sikkim, Uttarakhand, Tripura, and others. These strategies should focus on infrastructure development, improved connectivity, and enhanced access to essential services. Lessons can be learned from successful cases in Nagaland, Manipur, and Uttarakhand, where interventions and policies have led to a decline in urban homelessness. Addressing rural homelessness requires comprehensive approaches that consider environmental degradation, socio-economic factors, and sustainable solutions. Tailored strategies should go beyond housing solutions and encompass sustainable practices, economic development, and social support systems. The specific challenges in regions like Arunachal Pradesh need to be addressed through targeted interventions, including land usage policies, security concerns, infrastructure development, and access to essential services. These findings underscore the importance of tailored strategies, focusing on both urban and rural contexts, to effectively combat homelessness in the Himalayan hill states. By acknowledging the data and trends, policymakers and stakeholders can develop targeted interventions and comprehensive approaches to address the unique challenges of urban and rural homelessness.

5.1 Comparison of housing adequacy parameters

Housing conditions vary across different geographical locations in India. Table 8.2 A comparison of housing adequacy (Table 8.2) in urban hill areas, rural hill areas, urban plain areas, and rural plain areas reveals significant variations in access to basic amenities, housing size, and tenancy status.

TABLE 8.2 Comparison of basic adequacy parameters in urban and rural areas of hill regions and plain areas in India.

Basic adequacy parameters	Urban hill areas	Rural hill areas	Urban plain areas	Rural plain areas
Availability of water supply	Usually available through piped water supply from municipal corporations	Usually available from natural springs, streams, and wells, but often contaminated	Usually available through piped water supply from municipal corporations, supplemented by groundwater sources like wells and borewells	Mainly rely on groundwater sources like wells, borewells, and hand pumps, which are often unreliable and contaminated
Availability of electricity	Generally available in power surplus states, with limited power cuts can be frequent	Generally available in power surplus states, although power cuts can be frequent	Generally available, although power cuts can be frequent	Erratic or limited availability
Sanitation facilities	Generally available, although may be limited in some areas	Often lack basic sanitation facilities like toilets and bathing facilities	Generally available, although may be limited in some areas	Often lack basic sanitation facilities like toilets and bathing facilities
Living space	Generally smaller due to high population density and limited space	More spacious due to lower population density	Generally smaller due to high population density and limited space	More spacious due to lower population density
Housing condition	Can be cramped and poorly maintained in some areas	Housing quality can vary widely, with some homes being poorly constructed or in disrepair	Can be cramped and poorly maintained in some areas	Housing quality can vary widely, with some homes being poorly constructed or in disrepair
Housing tenure	Generally rented, with limited ownership	Mainly owned or built by the households themselves	Generally rented, with limited ownership	Mainly owned or built by the households themselves

Access to basic amenities such as water supply, sanitation, and electricity is much better in urban areas compared to rural areas. In rural hill areas, for instance, only a small percentage of households have access to tap water and electricity, while almost all households in urban plain areas have access to these amenities. Housing size tends to be smaller in urban areas compared to rural areas. In urban hill areas, for example, the average household size is around two to three members, and the average house size is less than 50 square meters. In contrast, households in rural areas tend to have larger houses, with an average house size of around 70–80 square meters. Ownership status of housing is higher in rural areas compared to urban areas. The majority of households in rural areas own their houses, while a significant proportion of households in urban areas live in rented accommodation.

5.2 Homelessness in urban, rural, and hill areas: Differences and similarities

Homelessness is a critical issue affecting individuals and families across the world. While homelessness is often associated with urban areas, it is also prevalent in rural and hill areas. However, there are differences in how homelessness manifests in urban, rural, and hill areas due to factors such as population density, access to services, environmental factors, and community responses Table 8.3. In this section, we aim to explore the differences and similarities

TABLE 8.3 Comparison of homelessness and related factors in urban areas and hill areas.

Aspect	Urban areas	Hill areas
Homelessness	Homelessness is often more visible due to the density of the population and the prevalence of street homelessness. Homeless individuals may have greater access to emergency shelter services and support programs.	Homelessness may be less visible due to the remote and isolated nature of hill areas. Homeless individuals may have limited access to emergency shelter services and support programs.
	Homelessness may be less visible due to the spread-out nature of rural communities. Homeless individuals may have limited access to emergency shelter services and support programs.	Homelessness may be more visible in rural areas due to the lack of affordable housing and limited employment opportunities. Homeless individuals may have limited access to emergency shelter services and support programs.
Gender and family	Homelessness may disproportionately affect women and children in urban areas. Family shelters may be available to provide temporary housing and support services.	Homelessness may disproportionately affect men and families in hill areas. Family shelters may be limited or nonexistent, and men may have to fend for themselves.
Access to healthcare	Homeless individuals in urban areas may have greater access to healthcare facilities and services.	Homeless individuals in hill areas may have limited access to healthcare facilities and services due to the remote and isolated nature of these regions.
Mental health and addiction	Homeless individuals in urban areas may have access to mental health and addiction support services.	Homeless individuals in hill areas may have limited access to mental health and addiction support services.
Environmental factors	Overcrowding, high pollution levels, lack of green space.	Lack of basic amenities like water and sanitation, exposure to natural disasters.
Access to services	Greater access to healthcare, social services, and emergency shelters.	Limited access to healthcare, social services, and emergency shelters, lack of transportation options.
Causes of homelessness	Higher cost of living, economic opportunities, domestic violence, mental health issues, substance abuse.	Poverty, displacement due to natural disasters or development projects, lack of affordable housing, seasonal employment.
Community response	Greater availability of support services, community engagement, and outreach programs.	Limited community resources and support services, lack of community engagement.
Impacts on quality of life	Increased risk of physical and mental health problems, exposure to crime and violence.	Limited access to education and job opportunities, increased risk of physical and mental health problems, exposure to crime and violence.

in homelessness across urban, rural, and hill areas and shed light on the challenges homeless individuals face in accessing basic necessities and support services.

Homelessness is a complex issue that affects individuals in urban, rural, and hill areas. Homeless individuals face significant challenges in accessing emergency shelter services, healthcare, and support programs, and are often stigmatized and discriminated against. While the causes and consequences of homelessness may vary across different regions, it is essential to understand the differences and similarities in order to develop effective policies and support services to combat this issue. Addressing homelessness requires a coordinated effort from governments, nonprofit organizations, and communities to provide adequate housing, social services, and job opportunities to individuals facing homelessness, regardless of their location. Based on the data presented in the Table, it can be observed that there are significant differences in various categories between plain areas and hill areas. Hill areas generally have better environmental conditions and rural hill areas have better infrastructure compared to plain areas. However, plain areas have higher incomes and better healthcare facilities, especially in urban areas. In terms of education, there are mixed results, with high education levels in urban plain areas and rural hill areas, while medium education levels are observed in rural plain areas and low education levels in urban hill areas. Overall, it can be concluded that there are trade-offs between different categories when comparing plain areas and hill areas, and a comprehensive approach is needed for balanced development in both types of regions. Hill areas generally have better environmental conditions and rural hill areas have better infrastructure compared to plain areas. However, plain areas have higher income and better healthcare facilities, especially in urban areas. In terms of education, there are mixed results, with high education levels in urban plain areas and rural hill areas, while medium education levels are observed in rural plain areas and low education levels in urban hill areas.

6. Conclusion

The research aims to scrutinize the multifaceted interrelations between homelessness and various contributing factors in ecologically vulnerable hill states. Key findings elucidate that a myriad of interconnected variables substantially impacts the well-being of homeless individuals and families in these regions. Among these variables are rural-to-urban migration patterns, which are often instigated by a lack of local economic opportunities and access to essential services. Further exacerbating the situation is involuntary displacement arising from large-scale development projects, often undertaken without adequate social impact assessments. Economic downturns, occurring in cycles, add another layer of complexity by further marginalizing these already vulnerable communities. Additionally, environmental degradation—often accelerated by unsustainable tourism practices—negatively impacts traditional livelihoods dependent on natural resources. The study also finds that these regions are disproportionately exposed to environmental hazards and possess a high level of seismic vulnerability, thereby amplifying the challenges faced by homeless populations.

The investigation reveals that the implications of homelessness manifest differently when contrasting hill states with plain areas. The unique geographical and socioeconomic characteristics of hill states necessitate a bespoke policy approach. Hence, the adoption of a "one-size-fits-all" strategy is not only inefficacious but risks exacerbating existing socioeconomic imbalances. To foster equitable growth and sustainable development across these divergent landscapes, an integrated policy approach that accommodates the specific challenges of each region is imperative.

A significant portion of the analysis underscores the necessity to address the foundational determinants of homelessness in hill states. These core issues encompass limited access to essential amenities, including water sanitation, educational institutions, healthcare services, affordable housing, and employment opportunities. Neglecting these elementary needs perpetuates the cyclical nature of homelessness and undermines the efficacy of any targeted interventions.

A pivotal finding of the study emphasizes the indispensability of grassroots-level community engagement for the formulation of effective and culturally sensitive policies. Such local involvement is instrumental in securing community buy-in, thereby ensuring that policies and interventions are not solely top-down impositions but also possess local relevance and acceptance.

Furthermore, the research posits that empirical data collection and evidence-based analyses serve as cornerstones for the conceptualization and implementation of homelessness alleviation strategies. Robust, scientifically gathered data facilitates the nuanced understanding necessary for resource allocation and intervention efficacy, thereby informing policy decisions and bolstering the effectiveness of homelessness support services.

These findings offer manifold implications for a diverse array of stakeholders, ranging from policymakers to nonprofit organizations and community leaders. The study strongly advocates for policy strategies that prioritize affordable housing solutions, such as subsidized housing initiatives, and comprehensive support services, including mental health programs and vocational training. Community-centric policy development, underpinned by robust empirical evidence, emerges as a key recommendation for achieving long-term, sustainable solutions to the complex issue of homelessness in ecologically sensitive hill states. This collective approach, synergizing efforts across governmental bodies, nonprofit organizations, and local communities, presents a viable pathway for the creation of more stable, sustainable, and inclusive social ecosystems.

References

Anand, S., Bhan, G., Idicheria, C., Jana, A., & Koduganti, J. (2014). Locating the debate: Poverty and vulnerability in urban India. *Indian Institute for Human Settlements*. https://doi.org/10.24943/iihsrfpps4.2014

Batterham, D. (2018). Homelessness as capability deprivation: A conceptual model. *Housing, Theory and Society, 36*(3), 274–297. https://doi.org/10.1080/14036096.2018.1481142

Berhane, G., & Walraevens, K. (2013). Geological and geotechnical constraints for urban planning and natural environment protection: A case study from mekelle city, northern Ethiopia. *Environmental Earth Sciences, 69*(3), 783–798. https://doi.org/10.1007/s12665-012-1963-x

Bisht, S., Chaudhry, S., Sharma, S., & Soni, S. (2018). Assessment of flash flood vulnerability zonation through geospatial technique in high altitude Himalayan watershed, Himachal Pradesh India. *Remote Sensing Applications: Society and Environment, 12*, 35–47. https://doi.org/10.1016/j.rsase.2018.09.001

Dikmen, N., & Elias-Ozkan, S. T. (2016). Housing after disaster: A post occupancy evaluation of a reconstruction project. *International Journal of Disaster Risk Reduction, 19*, 167–178. https://doi.org/10.1016/j.ijdrr.2016.08.020

Delhi, N. (2013). *The committee to study developmentdevelopment in hill states arising from management of forest lands with special focus on creation of infrastructure, livelihood and human development.*

Kumuda, D. (2012). Homeless population in India: A study. *Global Journal for Research Analysis, 3*(8), 54–55. https://doi.org/10.15373/22778160/August2014/16

Fitchen, J. M. (1992). On the edge of homelessness: Rural poverty and housing insecurity. *Rural Sociology, 57*(2), 173–193. https://doi.org/10.1111/j.1549-0831.1992.tb00462.x

Jong-A-Pin, R. (2009). On the measurement of political instability and its impact on economic growth. *European Journal of Political Economy, 25*(1), 15–29. https://doi.org/10.1016/j.ejpoleco.2008.09.010

Kumar, A. (2016). Impact of building regulations on Indian hill towns. *HBRC Journal, 12*(3), 316–326. https://doi.org/10.1016/j.hbrcj.2015.02.002

Kumar, A. K., & Mohanta, D. L. (2003). Population, environment and development in India. *Journal of Human Ecology, 14*, 383–392.

Lakshmana, C. M. (2013). Population, development, and environment in India. *Chinese Journal of Population Resources and Environment, 11*(4), 367–374. https://doi.org/10.1080/10042857.2013.874517

Long, S., McQuarrie, N., Tobgay, T., & Grujic, D. (2011). Geometry and crustal shortening of the Himalayan fold-thrust belt, eastern and central Bhutan. *Geological Society of America Bulletin, 123*(7–8), 1427–1447. https://doi.org/10.1130/B30203.1

Maina, J. J. (2013). Uncomfortable prototypes: Rethinking socio-cultural factors for the design of public housing in Billiri, north east Nigeria. *Frontiers of Architectural Research, 2*(3), 310–321. https://doi.org/10.1016/j.foar.2013.04.004

Meert, H., & Bourgeois, M. (2005). Between rural and urban slums: A geography of pathways through homelessness. *Housing Studies, 20*(1), 107–125. https://doi.org/10.1080/0267303042000308750

Mitterer, C., Künzel, H. M., Herkel, S., & Holm, A. (2012). Optimizing energy efficiency and occupant comfort with climate specific design of the building. *Frontiers of Architectural Research, 1*(3), 229–235. https://doi.org/10.1016/j.foar.2012.06.002

Morrison, D. S. (2009). Homelessness as an independent risk factor for mortality: Results from a retrospective cohort study. *International Journal of Epidemiology, 38*(3), 877–883. https://doi.org/10.1093/ije/dyp160

Nambiar, D. (2020). Social determinants of health among urban homeless in Delhi, India. *The European Journal of Public Health, 30*(Suppl. 5). https://doi.org/10.1093/eurpub/ckaa166.329

Nygren, A. (2000). Development discourses and peasant-forest relations: Natural resource utilization as social process. *Development and Change, 31*(1), 11–34. https://doi.org/10.1111/1467-7660.00145

Sewell, S. J., Desai, S. A., Mutsaa, E., & Lottering, R. T. (2019). A comparative study of community perceptions regarding the role of roads as a poverty alleviation strategy in rural areas. *Journal of Rural Studies, 71*, 73–84. https://doi.org/10.1016/j.jrurstud.2019.09.001

Shim, J. E., Kim, S. J., Kim, K., & Hwang, J. Y. (2018). Spatial disparity in food environment and household economic resources related to food insecurity in rural Korean households with older adults. *Nutrients, 10*(10). https://doi.org/10.3390/nu10101514

Sohn, C. (2014). The border as a resource in the global urban space: A contribution to the cross-border metropolis hypothesis. *International Journal of Urban and Regional Research, 38*(5), 1697–1711. https://doi.org/10.1111/1468-2427.12071

Springer, S. (2000). Homelessness: A proposal for a global definition and classification. *Habitat International, 24*(4), 475–484. https://doi.org/10.1016/S0197-3975(00)00010-2

UN-Habitat. (2020). World cities Report 2020 the Value of sustainable urbanization key findings and messages. *Sereal Untuk, 51*, 1.

Williams, J. C. (2005). The politics of homelessness: Shelter now and political protest. *Political Research Quarterly, 58*(3), 497. https://doi.org/10.2307/3595618

Wilson, H. F. (2017). On geography and encounter. *Progress in Human Geography, 41*(4), 451–471. https://doi.org/10.1177/0309132516645958

Wright, J. D. (1990). Poor people, poor health: The health status of the homeless. *Journal of Social Issues, 46*(4), 49–64. https://doi.org/10.1111/j.1540-4560.1990.tb01798.x

CHAPTER

9

The price of a roof: How rental stress is fueling hidden homelessness in Khulna City, Bangladesh

Kazi Saiful Islam

Urban and Rural Planning Discipline, Khulna University, Khulna, Bangladesh

1. Introduction

With 16.52 million people, Bangladesh is one of the most densely populated countries in the world. Roughly 31.66% of the total population lives in urban areas of Bangladesh (BBS, 2022). Nevertheless, their contribution to the national GDP is more than 65% (UNDP, 2019). Urban population continues to rise at an extraordinary rate of 3.4% each year (GED, 2020). This has a significant impact on the current housing stock. In Bangladeshi cities, the disparity between demand and supply of appropriate housing units has led to the proliferation of slums and terrible housing stress (Rahman, 1985; Rahman & Hill, 2019). Approximately 40% of urban poor households lack access to appropriate housing (Ahmad, 2015). Slums are their only remaining option. In 2021, the urban housing shortfall was predicted to be 8.5 million (GED, 2020). After the liberation, Bangladesh adopted socialist approach to nation-building for a short period of time. But very soon adopted neoliberal policies. It had a detrimental impact on housing affordability (Farzana, 2022). Although Bangladesh enacted the Premises Rent Control Act in 1991, due to the lack of its enforcement and conflict with the housing policy, the grassroots people are not benefiting from it. Home ownership is also promoted through the housing finance market, which is also aggravating the Bangladeshi housing crisis. People from various socioeconomic backgrounds reside in various types of dwellings. House rent also differs based on housing types, construction materials, location, esthetics, etc. Because of financial constraints, the vast majority of people in the middle- and lower-income brackets do not own their own houses. They are down to renting as their final option. Rental housing offers flexibility in housing choice (HUD, 2010). It also provides access to housing services because it is much less expensive than purchasing a property. Tenants are not required to pay for unforeseen but necessary home repairs. However, the

Homelessness to Hope
https://doi.org/10.1016/B978-0-443-14052-5.00009-4

accessibility of the lower- and lower-middle income groups to the existing rental housing stock is hindered by the unprecedented growth of house rent in recent years (Chowdhury, 2013).

Rapid urbanization has led to a dramatic uptick in the need for new houses, yet construction has not kept pace (Mansur & Alam, 2023). A definitive answer to the question of how many dwellings are needed for various socioeconomic classes has not been reached as of yet (Ahmad, 2015). People are having trouble finding affordable housing that meets their needs because builders rarely take into account the varying requirements of different socioeconomic groups (Mansur & Alam, 2023). When laws aren't strictly enforced (as in the case of rental control mechanisms), tenants are charged exorbitant sums of money for dwelling units that often fail to satisfy their needs (Ahmad, 2015). Mismatch among the affordability of the households, house rent, requirements of the households and characteristics of the housing units has given birth to a situation where people are forced to live in substandard housing that does not satisfy their needs, thereby causing hidden homelessness (Fiedler et al., 2006; Levinson, 2004; Murdie, 2011; Savage et al., 1992). Like many other developing countries, homelessness-related scientific studies are almost nonexistent in Bangladesh. Instead of analyzing macroeconomic and political issues that are responsible for worsening the housing situation in Bangladesh, this study's main goal is to discover the relationship and gap between the characteristics of different socioeconomic groups and the characteristics of the houses in which they live.

2. (Re)conceptualizing hidden homelessness for Bangladesh

The concepts of "Home" and "Homelessness" are commonly perceived as diametrically opposed to one another in public discourse (McCarthy, 2018). The complexity of distinguishing between these two entities has been further exacerbated by the incorporation of additional dimensions (Tipple & Speak, 2005). The concept of home is inherently subjective, individualized, and significantly shaped by one's life satisfaction (Clapham et al., 2018; O'Grady et al., 2020; Somerville, 1992; Walter et al., 2016) identified seven key signifiers of home- Shelter, Hearth, Heart, Privacy, Roots, Abode and Paradise. Shelter is just one of the signifiers of home. The European Typology of Homelessness (ETHOS) study conducted by Edgar and Meert (2005) identified seven broad types of homelessness based on three domains (physical, legal, and social). Despite being heavily criticized, this is one of the most significant studies in advancing the debate on the concept of homelessness. In general, homelessness refers to more than just a lack of a place to live; it also refers to psychological stress, insecurity, a lack of privacy, feeling vulnerable and uncomfortable, and so on. Attempting to encapsulate the multifaceted nature of homelessness in a singular description proves to be a challenging endeavor. There is, in fact, no universal definition of homelessness. It is also neither necessary (Haile et al., 2020), nor within the purview of this chapter.

Hidden homelessness is deeply ingrained in the definition of "Home". At the same time, "Home relies on homelessness to construct and define itself" (McCarthy, 2018). The depth and dimensions of hidden homelessness yet to be fathomed properly (Deleu et al., 2023). Hidden homeless people are typically not roofless (Levinson, 2004). Housing (Homeless Persons)

Act (1977) of the United Kingdom defined someone "homeless" if s/he has no accommodation that they could reasonably be expected to occupy. Thus, different people perceive homelessness differently depending on their socio-economic circumstances. Hidden homelessness is often considered the first step toward homelessness (Demaerschalk et al., 2019). The term "hidden homeless" refers to individuals who meet the criteria for homelessness, although their circumstances are not readily apparent in public spaces or captured in official data (Gray et al., 2022). Rodrigue (2016) defined the hidden homeless as people who had nowhere else to live and had to temporarily live with family, friends, in their car, or anywhere else because they had nowhere else to live. Overcrowding and unsuitable housing were also mentioned as characteristics of hidden homelessness by Barnardos et al. (2018). These people usually do not have a fixed address or home (Hans et al., 2004). Poverty and unaffordable rental housing play the most important role beneath all of this (Levinson, 2004; Murdie, 2011). Rental housing affordability analysis is not only about income-to-rent ratio, but it also involves subjective and contextual judgments (Chowdhury, 2013). The income-to-rent ratio exhibits significant variation across different geographical locations. Furthermore, it should be noted that it does not serve as a comprehensive measure of housing stress (Begum & Rahaman, 2023). The concept of "Sustainable housing affordability" emerged out of these criticisms (Mulliner et al., 2013). This concept proposed to include economic, social, and environmental sustainability indicators of housing affordability. Lack of sustainable housing affordability is also responsible for hidden homelessness. Then again, people living in affordable houses may also feel homeless because of so many variables (Clapham et al., 2018). Unfortunately, not many studies on hidden homelessness can be found in developing countries. As, socioeconomic characteristics and housing characteristics are interrelated (Ahmad, 2015); any household who deserves to live in a house but are living in a substandard one is defined in this chapter as hidden homeless.

3. Methodology

At the absence of any control of the government, house rents are determined by demand and supply in the rental market. On the supply side, landlords seek to maximize the capital return from their housing assets as a part of their investment (Henderson & Ioannides, 1983). Renters, on the other hand, strive to maximize the utilities for the money they can afford to pay as rent. Segmenting the housing units is easier in theory than in practice. There are so many variables at play here. Researchers used a variety of indicators depending on the nature of the study and its setting. Price and characteristics of housing units (Goodman & Thibodeau, 1998), dwelling size and transaction price (Goodman & Thibodeau, 2007), geographic household properties (Watkins, 2001), housing attributes factor (Dale-Johnson, 1982), racial discrimination (Follain & Malpezzi, 1981), distance from different land uses (i.e., train station, shopping mall, etc.) (Islam & Asami, 2011) are some of the dominants used by different researchers. It is beyond the scope of this chapter to review the factors and processes that contribute to the segmentation of the rental housing market. Islam and Asami (2009) comprehensively reviewed literature on housing market segmentation. Avid readers are encouraged to go through this article. One's socioeconomic status can be thought of as a measure of how well off they are in terms of material possessions, political influence, financial stability, social support systems, health, leisure, and educational opportunities. The level of education within a household (particularly that of the head of the household), income, employment pattern, housing, household assets, etc., are all crucial indicators of a family's socioeconomic standing. The present study utilizes two distinct datasets, namely the properties of the rental housing unit and the household characteristics, which were both obtained from the same housing units. Both sets of data are partitioned into clusters according to their degree of similarity. Subsequently, an analysis is conducted to discover the associations between clusters of residential units and households, with the aim of identifying instances of hidden homelessness. Please refer to Fig. 9.1 for detailed, sequential procedure adopted in this study.

There is a plethora of potential variables that can be incorporated into the analysis of both the housing units and the households' datasets. However, the ability of models to explain data is diminished when too many variables are included (often regarded as overfitting). Factor analysis was utilized to identify the primary determinants of house rents. Hedonic regression (linear regression) was then applied to these variables to better understand their predictive power. Two-step clustering method was used in this investigation to group dwellings and households together. The wealthy typically reside in high-rise apartments, while those of a middle income prefer to live close to their places of employment, and the poor are typically located in slums. However, many people have no choice but to settle for substandard housing because of financial constraints and other causes. Correspondence analysis is used in this chapter to make sense of the situation. The measure of correspondence is an indication of the similarity, affinity, confusion, association, or interaction between two sets of variables (IBM, 2022) (in this case, attributes of housing units and the characteristics of the households living in these units).

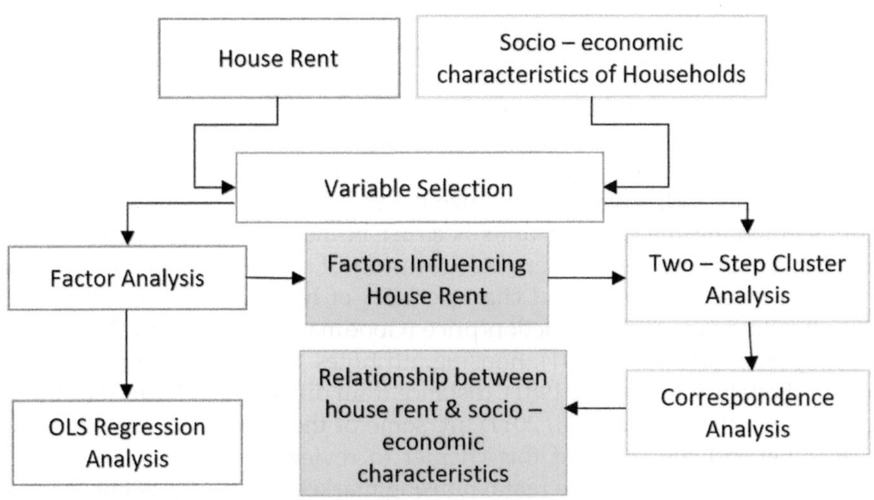

FIGURE 9.1 Methodology.

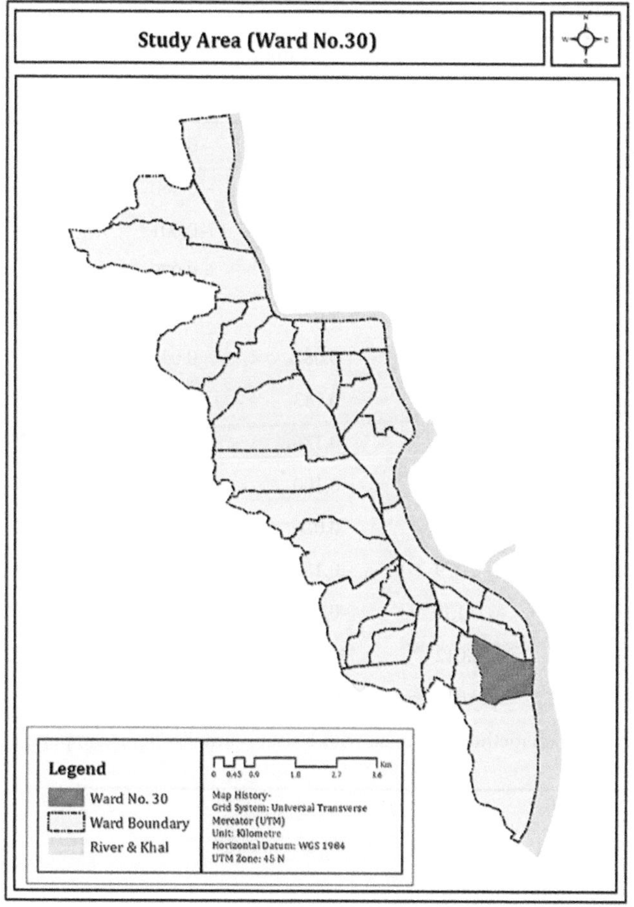

FIGURE 9.2 Map of the study area (ward no 30 of Khulna city, Bangladesh).

3.1 Study area and data collection

Khulna City, the third largest metropolitan city in Bangladesh, is the focus of this research. Specifically, ward no 30 (Fig. 9.2) is being analyzed. It is an unplanned region with basic services that reflects the natural rate of migration and urbanization. Residents of this region are expected to represent all socioeconomic categories in Khulna, and all types of dwelling structures and units seen in Khulna are also present. It is an ideal unplanned residential area because of its distance from CBD, diversity, population density, and number of households.

A total of 197 houses were chosen for the questionnaire survey out of a total of 5899. The systematic random sampling method was utilized for sample selection. It began with a random selection of one house and progressed to every 30th house after that. A structured questionnaire was used to collect two sets of variables (characteristics of housing units and households residing in these units).

TABLE 9.1 Component score coefficient matrix.

Variables	Component 1	2	3	4
Total income of the family	−0.139	−0.002	−0.018	−0.173
Size of the housing unit	−0.058	0.144	−0.060	−0.091
Number of bedrooms	0.102	**0.444**	−0.001	0.252
Number of persons in the family	0.147	**0.501**	−0.078	0.222
Material used for floor construction	**0.274**	0.028	0.017	−0.176
Material used for wall	**0.307**	0.082	0.046	−0.077
Material used for roof	**0.285**	0.143	0.021	0.099
Age of house	−0.142	0.176	−0.026	**0.627**
Availability of veranda	−0.130	−0.100	−0.393	−0.112
Availability of open space	0.085	0.026	0.077	**0.563**
Location of the housing unit (floor)	0.079	−0.133	**0.602**	0.160
Distance between residence and working place in km	0.131	−0.006	−0.015	0.157
Current value of land per katha (=720 sq.ft.)	−0.012	0.153	−0.121	−0.213
Type of structures	−0.013	−0.088	0.361	0.045

Extraction method: Principal component analysis. Rotation method: Varimax with Kaiser normalization. Component scores.

4. Results and discussion

4.1 Factor analysis

KMO (Kaiser–Meyer–Olkin) Measure of Sampling Adequacy for the data set is 0.808, meaning the sample size of the dataset is large enough to conduct factor analysis. Bartlett's Test of sphericity is also significant. Meaning there is sufficient correlation in the database. Based on the component score coefficient (Table 9.1), 08 (eight) variables were selected as the determinants of the house rent model (eigenvalue cut point was 0.75).

4.2 Ordinary least squares (OLS) regression

Variables chosen in the preceding phase were used as independent variables in an ordinary least-squares (OLS) regression. The variables predicted dwelling rent relatively well (R = 0.755, R-square = 0.570, and adjusted R-square = 0.551). Table 9.2 displays the outcome of the OLS regression. Clearly, the most important variables responsible for determining

TABLE 9.2 Result of the OLS regression.

Model	Unstandardized coefficients		Standardized coefficients			Collinearity statistics	
	B	Std. error	Beta	T	Sig.	Tolerance	VIF
(Constant)	0.363	0.408		0.889	0.375		
Material used for floor construction	0.223	0.196	0.104	1.140	0.256	0.276	3.625
Material used for wall	−0.084	0.066	−0.148	−1.263	0.208	0.168	5.962
Material used for roof	0.046	0.047	0.088	0.975	0.331	0.284	3.520
Number of bedrooms	0.336	0.076	0.385	4.449	0.000	0.306	3.267
Number of persons in the family	0.094	0.050	0.152	1.868	0.063	0.348	2.875
Location of the housing unit (floor)	0.174	0.040	0.233	4.380	0.000	0.807	1.239
Age of house	−0.010	0.005	−0.128	−2.222	0.027	0.689	1.451
Availability of open space	−0.329	0.089	−0.227	−3.682	0.000	0.601	1.664

house rent are the number of bedrooms, the floor of the dwelling units, and the availability of open space. The age of the property and the number of family members residing in the dwelling unit are also significant factors.

4.3 Two-step cluster analysis

The Two-Step Cluster Analysis procedure is a methodological approach employed for exploratory purposes, with the aim of discerning inherent groups, commonly referred to as clusters, within a given dataset. It is utilized to analyze the monthly house rent, using the independent variables obtained via component analysis. Based on six variables, housing units were grouped into three clusters. Thirty two percent data (63) were assigned to the first cluster. This figure is 18.3% (36) and 49.7% (98), respectively, for the second and third cluster. From socioeconomic variables, four clusters were extracted. The first cluster contains 54.8% of the records (108). This figure is 22.8% (45), 11.7% (23) and 10.7% (21) for the second, third, and fourth clusters, respectively. The log-likelihood distance metric was employed to determine the distance between cases due to the presence of both continuous and categorical input variables for housing rent. The Euclidean distance measure has been employed to assess socioeconomic attributes, as these are continuous in nature. As a clustering criterion, Schwarz's Bayesian Criterion (BIC) was applied. This is a relative measure of goodness-of-fit that is used to compare solutions with varying numbers of segments. "Relative" signifies that these criteria are not scaled from 0 to 1 but can generally take any value. The cluster quality graph (silhouette measure of cohesion and separation) shows that the overall model quality is "Fair" for both house rent and socioeconomic characteristics of the households living in these units, indicating that the output produced by this model is acceptable. The silhouette measure

TABLE 9.3 Cluster profile of housing units.

	Two-step cluster number		
	1	2	3
Expenditure on house rent	5708.1633	3507.9365	1073.6111
Material used for floor construction	2.0000	2.0000	2.5000
Material used for wall	2.0000	1.8730	4.4167
Material used for roof	2.0000	1.8730	5.4444
Number of bedrooms	2.5102	2.0476	1.5833
Number of persons in the family	3.7959	3.0159	2.8611
Location of the housing unit (floor)	2.2551	1.1905	1.0000
Age of house	9.2959	21.1746	12.1667
Availability of open space	1.2143	1.6984	2.0000

of cohesiveness and separation is based on the average distances between the objects and can range from −1 to +1. A silhouette measure of less than 0.20 indicates a bad solution quality, a measure between 0.20 and 0.50 indicates a fair solution, and values greater than 0.50 indicate a good solution.

4.3.1 *Cluster profiles of house rent*

The cluster profile (Table 9.3) illustrates the significance of various predictors for each cluster. Based on mean house rent, clusters 1, 2, and 3 have been designated as high, medium, and low rent clusters, respectively. From the mean values of the predictors, characteristics of each cluster can be inferred. High-rent housing units are typically constructed of concrete, are primarily located on the first floor, have three bedrooms and sufficient open space. In contrast, low-rent housing clusters are typically constructed of mud, bamboo, tin, and sometimes concrete. Typically, these residences are occupied by households consisting of three individuals and are frequently located on the lower levels without access to open space.

Housing units with a medium house rent have characteristics of both high- and low-rent housing. These structures are constructed of brick and concrete. In these units with 02 (two) bedrooms, households of 3 (three) people live on average. In these units, there is very little or no open space.

4.3.2 *Cluster profiles of the socioeconomic characteristics*

The mean values of the predictors were used to separate the clusters (Table 9.4). Educational status of the household head is clearly the most important predictor in defining a socioeconomic cluster. Monthly income and the occupation of the household head are also important predictors, with current land value contributing the least. Based on these variables, the four socioeconomic clusters have been labeled as high, higher-middle, middle-, and low class.

TABLE 9.4 Profile of the socioeconomic clusters.

	Two-step cluster number			
	1	2	3	4
Educational status of household head	5.4259	4.3778	2.1739	1.2857
Monthly income household head	5.8519	4.7333	6.2174	2.4286
Occupation of household head	2.787	1.9778	1.1304	5.3333
Duration of living in the current location	3.3519	9.9667	3.2609	3.5714
Current value of land per katha (720 sq. ft.)	1,050,833	1,000,000	1,245,217	998,571

Remarkably, it is the education level and the stability of job that ultimately determine a socioeconomic class, rather than the income. Household heads of the high-class are typically employed in government or semigovernment organizations. Their monthly income ranges from BDT 20,000 to BDT 30,000. Although the household heads of middle socioeconomic cluster earn the same amount of money, they do so through business. They are only literate (can read and write). Heads of the upper-middle class earn between BDT 15,001 and BDT 20,000 per month. Although these heads are generally government employees, they are less educated than the upper class. Heads of low-income households are illiterate, with monthly earnings ranging from BDT 4001 to 10,000. They are typically day workers.

4.4 Relationship between house rent and the characteristics of different socioeconomic groups

4.4.1 *Correspondence analysis*

Correspondence analysis was used to determine the relationship between clusters of housing units and various socioeconomic groups. It visualizes the relationship between two nominal variables in a multidimensional space. A correspondence table is a two-way table in which each cell contains some measure of correspondence between the rows and columns. According to the correspondence table (Table 9.5), cluster 1 (high rent) of house rent is most closely associated with cluster 1 (high class) of socioeconomic characteristics. Evidently, people from the upper crust are assumed to pay high rent. However, people belonging to the middle class (cluster 3 of socioeconomic characteristics) also pay a high rent.

It can be presumed that a particular socioeconomic group would exhibit a strong association with only one cluster of dwelling units. In contrast, the correspondence table reveals that clusters 2 (upper-middle class) and 4 (poor class) within the socioeconomic category have equal associations with all three clusters of house units. Upper-middle- and low-class people apparently face hidden homelessness. Evidently, income is not the only determinant of affordability. The affordability of upper-middle class households exhibits great variation depending on factors such as the number of family members and consumption patterns. Consequently, some of these households are able to avail themselves of the facilities

TABLE 9.5 Correspondence table.

	Clusters of socioeconomic characteristics				
Clusters of house rent	**1 (High class)**	**2 (Higher-middle class)**	**3 (Middle class)**	**4 (Low class)**	**Active margin**
1 (High rent)	62	13	15	8	98
2 (Medium rent)	34	17	6	6	63
3 (Low rent)	12	15	2	7	36
Active margin	108	45	23	21	197

associated with high-rent dwellings, while others are compelled to reside in low-rent accommodations. Educational qualification, employment status, social background, etc., may have raised the expectation of the households and the society (peer pressure) that forced these people to occupy high-rent houses. Households belonging to the lower socioeconomic class (specifically, cluster 4) residing in high-rent dwellings often find themselves in a situation where they are either providing services to homes of higher socioeconomic class, particularly those in the high-class or higher-middle class or experiencing significant levels of stress. Clearly low-class groups are residing in substandard dwelling units paying higher rent (Ahmad, 2015; Bashar, 2022; Sultana & Nazem, 2020).

Each of these families consists of multiple members who contribute to the household income. Hence, income status of households is not solely determined by the income of the household head. This scenario necessitates additional inquiry. Quantitative analysis can only partially unpack the answer. Residual income approach (Stone et al., 2011) would be the best method for explaining this situation. Nevertheless, the complexities of housing preferences in relation to life expectations, societal pressures (such as the potential for social stigma associated with one's choice of housing), the influence of social stereotypes (such as those pertaining to individuals identifying as a third gender, scavengers, or prostitutes), as well as the impact of ethnicity and minority status (including the reluctance of homeowners to rent to these individuals), present significant challenges when attempting to uncover these phenomena through quantitative research methods. Because different housing units provide varying levels of utilities, there is a significant difference between the actual house rent and the rent that should be paid due to a lack of enforcement of the Premises Rent Control Act (1991). This difference varies greatly from place to place, forcing some people to live in substandard and unsuitable housing units (with fewer than required services and facilities). This exacerbates the situation of hidden homelessness. In our case, people from the higher-middle and lower classes are experiencing housing stress.

4.4.2 Row and column profiles

The weighted average of the column scores is used to calculate the row scores. The weighted sum of squared distances to the centroid is standardized at 1. Likewise weighted average of the row scores is used to calculate column scores. It is also standardized, like row score, to have a weighted sum of squared distances to the centroid of 1. These two profiles (Tables 9.6 and 9.7) demonstrate how row and column variables differ from one another.

TABLE 9.6 Row profiles.

Clusters of house rent	Clusters of socioeconomic characteristics				
	1 (High class)	2 (Higher-middle class)	3 (Middle class)	4 (Low class)	Active margin
1 (High rent)	0.633	0.133	0.153	0.082	1.000
2 (Medium rent)	0.540	0.270	0.095	0.095	1.000
3 (Low rent)	0.333	0.417	0.056	0.194	1.000
Mass	0.548	0.228	0.117	0.107	

TABLE 9.7 Column profiles.

Clusters of house rent	Clusters of socioeconomic characteristics				
	1 (High class)	2 (Higher-middle class)	3 (Middle class)	4 (Low class)	Mass
1 (High rent)	0.574	0.289	0.652	0.381	0.497
2 (Medium rent)	0.315	0.378	0.261	0.286	0.320
3 (Low rent)	0.111	0.333	0.087	0.333	0.183
Active margin	1.000	1.000	1.000	1.000	

The row and column scores are plotted in Fig. 9.3. Categories that are similar to each other appear close to each other in the plot. Here, clusters of house rent are considered as row profile and clusters of socioeconomic groups are considered as column profile. The plot show that cluster 1 of house rent is close to cluster 1 and 3 of socioeconomic characteristics. Other clusters are well separated.

5. Discussion

The definition of homelessness adopted by any country is deeply rooted in its sociocultural, economic and political ideology (Tipple & Speak, 2009). Sadly, the Government of Bangladesh, like many other countries do not acknowledge the complicated character of homelessness (Tipple & Speak, 2005). In the developing countries, like Bangladesh, homelessness is very poorly conceptualized (Chowdhury, 2013; Ghafur, 2002; Tipple & Speak, 2005) and highly politicized (Speak, 2019). Thus, the data related to homelessness are often not reliable. This is one of the main reasons behind the lack of sufficient number of research on this topic. In fact, the concept and categorization of homelessness in developing countries is much more complex compared to developed world. Ghafur (2002) broadly classified homelessness from Bangladeshi perspective into three types, e.g., extreme homelessness, passive

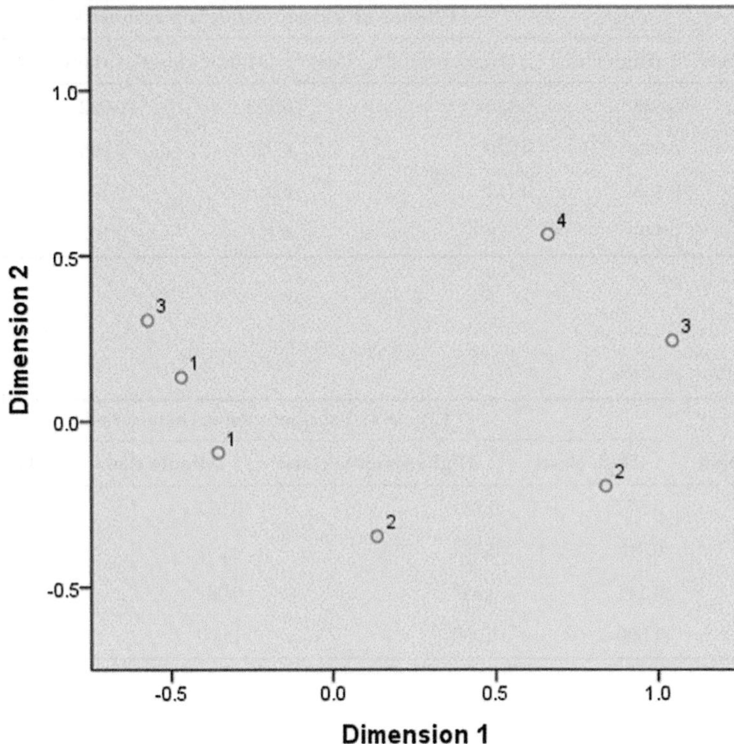

FIGURE 9.3 Rent clusters versus socioeconomic clusters.

homelessness, and potential homelessness. However, hidden homelessness remains an unexplored phenomenon. The etiology of hidden homelessness is a highly intricate matter, characterized by the interplay of various sociostructural factors (such as the economy, housing markets, and social security) and individual life circumstances that heighten the probability of entering a path that ultimately results in homelessness (Fiedler et al., 2006; Fitzpatrick, 2000; Gray et al., 2022). So many complicated factors, including ethnicity (Bramley et al., 2022), social exclusion and poverty (Watson et al., 2016), tenure insecurity (Ali, 2018; Demaerschalk et al., 2019), inadequate safety and security measures (Reeve, 2011), insufficient social infrastructure (Dowhaniuk, 2020; Forchuk et al., 2010; Karabanow et al., 2014), and a scarcity of available housing stock (Forchuk et al., 2010) influence the hidden homelessness situation. One can easily correlate the causes of hidden homelessness with homelessness. Poverty and affordability (economic perspective) are often cited as the reasons behind homelessness. Weaknesses in the national policy framework, poor institutional and financial management, weak governance are also responsible for further aggravating the homelessness and hidden homelessness situation. Unfortunately, these issues do not receive expected attention in the research landscape.

6. Conclusion

Tipple and Speak (2009) presented evidence concerning the gravity of hidden homelessness in Bangladesh. Underlying factors contributing to the severity of the issue remain elusive. Without understanding social norms and values, it is almost impossible to understand hidden homelessness. The influence of social perception and stereotyping (such as the bad perception toward scavengers), peer pressure, social marginalization of groups like the third gender, political oppression, criminality, and social unrest, are firmly ingrained in Bangladeshi society and contribute to the prevalence of hidden homelessness. Separating the influence of these variables from politico-economic variables poses a significant challenge. The extent and characteristics of hidden homelessness are also contingent upon factors such as geographical location, socioeconomic status, and political landscape. Like many developing countries, Bangladesh also has a very strong legal and policy framework. Because of weak democracy, poor governance, and institutional arrangements; developing countries often fail to address homelessness (let alone hidden homelessness). Homelessness is a complex, multifarious issue. It should be dealt with holistically.

References

Ahmad, S. (2015). Housing demand and housing policy in urban Bangladesh. *Urban Studies*. ISSN: 1360063X, *52*(4), 738–755. https://doi.org/10.1177/0042098014528547

Ali, Nadia (2018). *Understanding hidden homelessness*. https://www.homelesshub.ca/blog/understanding-hidden-homelessness. (Accessed 10 October 2023).

Barnardos, Focus Ireland, Simon Communities, Society of St Vincent de Paul. (2018). *Hidden homelessness: Key recommendations*.

Bashar, Toriqul (2022). Residential stability of the urban poor in Bangladesh: The roles of social capital. *Cities*. ISSN: 02642751, *126*, 103695. https://doi.org/10.1016/j.cities.2022.103695

BBS. (2022). *Population and housing census 2022 (preliminary report)*. Dhaka, Bangladesh: Bangladesh Bureau of Statistics.

Begum, Halima, & Rahaman, Md Mustafizur (2023). Approaches of measuring housing affordability: Retrofitting affordability approach from replicability to reality. *Jahangirnagar University Planning Review, 21*, 81–94.

Bramley, Glen, Fitzpatrick, Suzanne, McIntyre, Jill, & Johnsen, Sarah (2022). *Homelessness amongst black and minoritised ethnic communities in the UK: A statistical report on the state of the nation* (pp. 1–35). Edinburgh: Heriot-Watt University.

Chowdhury, M. Z. S. (2013). *The housing affordability problems of the middle-income groups in Dhaka: A policy environment analysis the University of Hong Kong*.

Clapham, David, Foye, Chris, & Christian, Julie (2018). The concept of subjective well-being in housing research. *Housing, Theory and Society, 35*(3), 261–280. https://doi.org/10.1080/14036096.2017.1348391

Dale-Johnson, D. (1982). An alternative approach to housing market segmentation using hedonic price data. *Journal of Urban Economics, 11*(3), 311–332. https://doi.org/10.1016/0094-1190(82)90078-X

Deleu, Harm, Schrooten, Mieke, & Hermans, Koen (2023). Hidden homelessness: A scoping review and avenues for further inquiry. *Social Policy and Society, 22*(2), 282–298. https://doi.org/10.1017/S1474746421000476

Demaerschalk, Evelien, Hermans, Koen, Steenssens, Katrien, & Regenmortel, Tine Van (2019). Homelessness merely an urban phenomenon? Exploring hidden homelessness in rural Belgium. *European Journal of Homelessness, 13*(1), 99–118.

Dowhaniuk, Nadia L. (2020). *Social infrastructure: A way to see hidden homelessness in rural and Northern Ontario Towns*. Toronto, Ontario: Ryerson University. https://rshare.library.torontomu.ca/articles/thesis/Social_infrastructure_a_way_to_see_hidden_homelessness_in_rural_and_Northern_Ontario_towns/19067540. (Accessed 18 October 2023).

Edgar, B., & Meert, H. (2005). *Fourth review of statistics on homelessness in Europe: The ETHOS definition of homelessness*. Brussels, Belgium: Feantsa.

Farzana, F. (2022). Neoliberalism and housing affordability crisis in Dhaka where market-enabling efforts failed. In *Accessible housing for South Asia* (pp. 85–102). Springer.

Fiedler, Rob, Schuurman, Nadine, & Hyndman, Jennifer (2006). Hidden homelessness: An indicator-based approach for examining the geographies of recent immigrants at-risk of homelessness in Greater Vancouver. *The World Urban Forum III – Vancouver, 23*(3), 205–216. https://doi.org/10.1016/j.cities.2006.03.004

Fitzpatrick, S. (2000). *Young homeless people*. London: Palgrave Macmillan. https://link.springer.com/book/10.1057/9780230509931.

Follain, J. R., & Malpezzi, S. (1981). Another look at racial differences in housing prices. *Urban Studies, 18*(2), 195–203. https://doi.org/10.1080/00420988120080351

Forchuk, Cheryl, Montgomery, Phyllis, Berman, Helene, Ward-Griffin, Catherine, Csiernik, Rick, Gorlick, Carolyne, Jensen, Elsabeth, & Riesterer, Patrick (2010). Gaining ground, losing ground: The paradoxes of rural homelessness. *The Canadian Journal of Nursing Research = Revue canadienne de recherche en sciences infirmieres, 42*(2), 138–152.

GED. (2020). *Eighth five year plan (July 2020–June 2025)*. Dhaka, Bangladesh: General Economics Division (GED), Bangladesh Planning Commission, Government of the People's Republic of Bangladesh.

Ghafur, S. (2002). *Informe final del proyecto R7905 del Economic and Social Research Programme (ESCOR)*. Department for International Development (DFID).

Goodman, A. C., & Thibodeau, T. G. (1998). Housing market segmentation. *Journal of Housing Economics, 7*(2), 121–143. https://doi.org/10.1006/jhec.1998.0229

Goodman, A. C., & Thibodeau, T. G. (2007). The spatial proximity of metropolitan area housing submarkets. *Real Estate Economics, 35*(2), 209–232. https://doi.org/10.1111/j.1540-6229.2007.00188.x

Gray, A.-M., Hamilton, J., Bell, J., Faulkner-Byrne, L., & McCready, P. (2022). *'Hidden' homelessness in Northern Ireland* (pp. 1–83). Belfast, Northern Ireland: Simon Community Northern Ireland. https://simoncommunity.org/homelessness/knowledge-hub/hidden-homelessness. (Accessed 15 September 2023).

Haile, K., Umer, H., Fanta, T., Birhanu, A., Fejo, E., Tilahun, Y., Derajew, H., Tadesse, A., Zienawi, G., Chaka, A., Damene, W., & Withers, M. H. (2020). Pathways through homelessness among women in Addis Ababa, Ethiopia: A qualitative study. *PLoS One, 15*(9), e0238571. https://doi.org/10.1371/journal.pone.0238571. http://www.ncbi.nlm.nih.gov/pmc/articles/pmc7467327/

Hans, Baer, Singer, M., & Susser, I. (2004). *Medical anthropology and the world system: A critical perspective*. Bergin & Garvey.

Henderson, J. V., & Ioannides, Y. M. (1983). A model of housing tenure choice. *The American Economic Review, 73*(1), 98–113.

HUD. (2010). *Study of rents and rent flexibility*. U.S. Department of Housing and Urban Development. https://www.huduser.gov/publications/pdf/Rent%20Study_Final%20Report_05-26-10.pdf. (Accessed 15 October 2023).

IBM. (2022). Correspondence analysis. In *IBM documentation*. International Business Machines. https://www.ibm.com/docs/en/spss-statistics/29.0.0?topic=application-correspondence-analysis. (Accessed 5 April 2023).

Islam, K. S., & Asami, Y. (2011). Addressing structural instability in housing market segmentation of the used houses of Tokyo, Japan. *Procedia - Social and Behavioral Sciences, 21*, 33–42. https://doi.org/10.1016/j.sbspro.2011.07.021

Islam, K. S., & Asami, Y. (2009). Housing market segmentation: A review. *Review of Urban & Regional Development Studies, 21*(2–3), 93–109. https://doi.org/10.1111/j.1467-940X.2009.00161.x

Karabanow, Jeff, Caila, Aube, & Naylor, Ted D. (2014). From place to space: Exploring youth migration and homelessness in rural Nova Scotia. *Journal of Rural and Community Development, 9*(2). https://journals.brandonu.ca/jrcd/article/view/905.

Levinson, D. (2004). *Encyclopedia of homelessness*. SAGE Publications, Inc.

Mansur, A. H., & Alam, H. (2023). *Rapid urbanisation and growing demand for affordable housing in Bangladesh*. Policy Insight. https://policyinsightsonline.com/2023/01/rapid-urbanisation-and-growing-demand-for-affordable-housing-in-bangladesh/. (Accessed 25 July 2023).

McCarthy, Lindsey (2018). (Re)conceptualising the boundaries between home and homelessness: The unheimlich. *Housing Studies, 33*(6), 960–985. https://doi.org/10.1080/02673037.2017.1408780

Mulliner, Emma, Smallbone, Kieran, & Maliene, Vida (2013). An assessment of sustainable housing affordability using a multiple criteria decision making method. *Management Science and Environmental Issues, 41*(2), 270–279. https://doi.org/10.1016/j.omega.2012.05.002

Murdie, R. A. (2011). *Precarious housing & hidden homelessness among refugees, asylum seekers, and immigrants: Bibliography and review of Canadian literature from 2005 to 2010*. CERIS.

O'Grady, Bill, Kidd, Sean, & Gaetz, Stephen (2020). Youth homelessness and self identity: A view from Canada. *Journal of Youth Studies, 23*(4), 499–510. https://doi.org/10.1080/13676261.2019.1621997

Rahman, Habibur (1985). Urbanisation and the problem of slums in Bangladesh. *Community Development Journal, 20*(1), 52–57. https://doi.org/10.1093/cdj/20.1.52

Rahman, H. Z., & Hill, R. (2019). Poverty in urban Bangladesh: Trends, profiles and spatial differences. *Bangladesh Development Studies, XLII*(2 & 3), 131–171.

Reeve, K. (2011). *The hidden truth about homelessness: Experiences of single homelessness in England*. London: Centre for Regional Economic and Social Research.

Rodrigue, S. (2016). *Hidden homelessness in Canada*. Statistics Canada. https://www150.statcan.gc.ca/n1/pub/75-006-x/2016001/article/14678-eng.htm. (Accessed 23 July 2023).

Savage, Mike, Watt, Paul, & Arber, Sara (1992). Social class, consumption divisions and housing mobility. In Roger Burrows, & Catherine Marsh (Eds.), *Consumption and class - Divisions and change*. London: Palgrave Macmillan London. Explorations in Sociology https://link.springer.com/book/10.1007/978-1-349-21725-0#about-this-book.

Somerville, Peter (1992). Homelessness and the meaning of home: Rooflessness or rootlessness? *International Journal of Urban and Regional Research, 16*(4), 529–539. https://doi.org/10.1111/j.1468-2427.1992.tb00194.x. (Accessed 19 October 2023)

Speak, S. (2019). *The state of homelessness in developing countries expert group meeting on "Affordable housing and social protection systems for all to address homelessness*. Nairobi, Kenya: United Nations Office at Nairobi. (Accessed 10 July 2023).

Stone, Michael E., Burke, Terry, & Ralston, Liss (2011). The residual income approach to housing affordability: The theory and the practice. In *AHURI Positioning Paper*. Australian Housing and Urban Research Institute. https://scholarworks.umb.edu/communitystudies_faculty_pubs/4/.

Sultana, S., & Nazem, N. I. (2020). Housing affordability of ready-made garment workers in the Dhaka metropolitan area. *Environment and Urbanization ASIA, 11*(2), 313–325. https://doi.org/10.1177/0975425320938581

Tipple, Graham, & Speak, S. (2005). Definitions of homelessness in developing countries. *Habitat International, 29*(2), 337–352. https://doi.org/10.1016/j.habitatint.2003.11.002

Tipple, G., & Speak, S. (2009). *The hidden millions: Homelessness in developing countries (Housing and Society Series)*. Routledge.

UNDP. (2019). *Why a national urban policy should be our top priority*. UNDP Bangladesh. https://www.undp.org/bangladesh/blog/why-national-urban-policy-should-be-our-top-priority. (Accessed 12 May 2023).

Walter, Zoe C., Jetten, Jolanda, Dingle, Genevieve A., Parsell, Cameron, & Johnstone, Melissa (2016). Two pathways through adversity: Predicting well-being and housing outcomes among homeless service users. *British Journal of Social Psychology, 55*(2), 357–374. https://doi.org/10.1111/bjso.12127. . (Accessed 19 October 2023)

Watkins, Craig A. (2001). The definition and identification of housing submarkets. *Environment and Planning A: Economy and Space, 33*(12), 2235–2253. https://doi.org/10.1068/a34162

Watson, J., Crawley, J., & Kane, D. (2016). Social exclusion, health and hidden homelessness. *Public Health, 139*, 96–102. https://doi.org/10.1016/j.puhe.2016.05.017

CHAPTER

10

Urbanization, homelessness, and climate change: Urban resilience challenges in Indonesia

Abdillah Abdillah[1,3], *Ida Widianingsih*[2,3], *Rd Ahmad Buchari*[2,3] *and Heru Nurasa*[2]

[1]Universitas Padjadjaran, Gradute Program in Administrative Science, Bandung, Indonesia; [2]Universitas Padjadjaran, Public Administration, Bandung, Indonesia; [3]Center for Decentralization & Participatory Development Research, Faculty of Social and Political Sciences, Universitas Padjadjaran, Bandung, Indonesia

1. Introduction

Urbanization displays a massive influx of urban residents and causes an increase in population, crime, congestion, and demand for land, demand for drinking water, and urban services (Ooi, 2009; Pee & Pan, 2022). The disturbing reality is that the city's rapid growth rate has created unprecedented poverty and the pace of development has slowed or stopped dramatically, compounded by weak management (Ooi, 2009; Pee & Pan, 2022; Yeh & Chen, 2020). The rise of development in big cities in Indonesia can spur economic growth. However, another impact is that these cities will become magnets for residents looking for work and residence, often called urbanization (Yeh & Chen, 2020; Zubaidah et al., 2023). However, the impact of urbanization causes various kinds of problems because there is no control over it. These problems include slums and squatter settlements, because planning for services lags far behind the pace of urbanization, resulting in densely populated cities riddled with high levels of poverty, weak infrastructure, and inadequate access to public services (Njoh, 2003; Ooi, 2009; Pee & Pan, 2022; Yeh & Chen, 2020). With a largely poor population, the rapid growth of cities has had a negative impact on providing housing for low-income groups, homeless, and protecting the urban environment (Adetokunbo & Emeka, 2015; Njoh, 2003). In addition, urban areas often experience the unique impacts of climate change, such as rampant poverty, social vulnerability, and economic inefficiency (Adetokunbo & Emeka, 2015; Kidd et al., 2023; Satterthwaite, 2009).

Homelessness to Hope
https://doi.org/10.1016/B978-0-443-14052-5.00010-0

Urbanization is triggered by differences in growth or inequality of development facilities, especially between rural and urban areas. As a result, urban areas become attractive magnets for urbanites to find work (Adetokunbo & Emeka, 2015; Kidd et al., 2023; Satterthwaite, 2009). Thus, true urbanization is a natural process of change in an effort to improve the welfare of the population or society (Kidd et al., 2023; Satterthwaite, 2009). The development of urbanization in Indonesia itself needs serious attention (Fahmi et al., 2014; Jones, 2015, pp. 271–276; Lewis, 2014; Mardiansjah et al., 2021; Prastiyo et al., 2020).

Many studies show that the level of population concentration in big cities in Indonesia has grown rapidly. The study conducted by Jones (2015, pp. 271–276) shows that high population movement in Indonesia has the impact of high vulnerability. In addition, research conducted by Lewis (2014) shows that urbanization also positively impacts economic growth in Indonesia. Research conducted by Mardiansjah et al. (2021) shows that the urbanization process has increased the number of cities in Indonesia from 50 to 94 and expanded large urban areas including DKI Jakarta and Bandung City. This condition means that vulnerability to big cities tends to be uncontrolled. There is a phenomenon that big cities will always grow and develop, then form cities called metropolitan cities. One of the cities that has experienced this is DKI Jakarta and Bandung City, Indonesia. It started as a big city and then developed into a metropolitan city and is currently leading to become a megapolitan city (Harahap, 2013; Mardiansjah et al., 2021; Prastiyo et al., 2020; Widiawaty, 2019, pp. 1–10).

Various studies show that in Indonesia it is known that the factors that greatly influence the occurrence of urbanization in large cities in Indonesia are economic factors (Abdillah, Widianingsih, Buhari, et al., 2023; Zubaidah et al., 2023). Like DKI Jakarta because the city's income is high, namely DKI Jakarta of IDR. 4,416,186,548 in 2021 (Anggraeni, 2022). Bandung City is caused by several factors such as economic problems, lack of jobs in the village, minimal facilities and infrastructure in the village, minimal educational and development facilities, and the pull factor, namely the influence of people who have already urbanized to the city they are aiming for. The frills of living in the city are much better and it is also easy to get the desired job or it is also easier to open a business, the wages in the city are much higher, there is more entertainment, tourist attractions or buildings are more and nice to visit and individual freedom is more flexible (Anggraeni, 2022; Harahap, 2013; Mardiansjah et al., 2021; Sony, 2023; Widiawaty, 2019, pp. 1–10).

Urban conditions that are increasingly out of control due to excessive urbanization have given rise to various new problems such as increased crime due to poverty, large-scale actions, increased slum settlements, and so on (Abdillah, Widianingsih, Buhari, et al., 2023; Malik et al., 2021; Prianto & Abdillah, 2023; Zubaidah et al., 2023). Therefore, urbanization will be seen as a scene-making factor for a city to develop both physically and socially. In this way, the form or meaning of urbanization can be seen more clearly as well as the impact it has on life in the city.

The phenomenon of urbanization and climate change is a challenge for urban areas in Indonesia (Harahap, 2013; Mardiansjah et al., 2021; Prastiyo et al., 2020). As a result, the occurrence of insecurity in urban areas in Indonesia causes urbanization to become chaotic and many natural disasters occur due to climate change without the support of capacity which can give rise to the phenomenon of homelessness in urban areas as expressed by this problem (Harahap, 2013; Mardiansjah et al., 2021; Prastiyo et al., 2020; Zhu & Simarmata, 2015). The purpose of this chapter is to examine the vulnerability due to urbanization, climate change, and the homeless crisis as challenges to realizing urban resilience in the context of big cities (DKI Jakarta and Bandung City) in Indonesia. The contribution of this chapter is to obtain potential lessons that can be drawn from urban experiences in Indonesia to overcome insecurity and face threats from the impacts of urbanization, the homeless crisis, and climate change so as to create resilience. This chapter conducted an analytical study in DKI Jakarta and Bandung City, Indonesia.

2. Materials and methods

This chapter method that the author uses is explorative-qualitative (Creswell & Poth, 2016) as well as evaluation using various research data such as field notes, government documents, government websites, online news, literature, and documentation. This is where this data does not come directly from the author, but which is obtained or has been collected from various reliable sources and can be said from certain parties with validated sources. The exploratory-qualitative method itself is a research method for exploring cases and getting to describe and showing facts about phenomena that are clear or what they are, then the data is evaluated to get new conclusions and findings. The process is carried out through the conceptual analysis approach "urban " which focuses on efforts to face and survive all kinds of threats and disturbances in the future (Abdillah, Buchari, et al., 2023; Abdillah, Widianingsih, Buchari, et al., 2023). This evaluation method is used to find out what triggers or factors influence the increase in urbanization, especially DKI Jakarta and Bandung City in 2021–23.

Data collection techniques in this study used field observations, documentation studies, literature reviews, policy surveys, and online media. Meanwhile, the analytical technique used in this research is to collect reliable and validated data or literature such as from journal articles, government documents, government websites, online news, books, and several documents that are appropriate to the topic of research. Then the data was analyzed using interactive qualitative data analysis techniques from Miles et al. (2018). And guided by using the qualitative data analysis tool Nvivo 12 Pro (Woolf & Silver, 2017, pp. 1–208) to get the best findings and conclusions (see Fig. 10.1). The aim of this analysis technique is to obtain in detail the updates or emergence of new ideas that are useful and provide some new contribution in providing information for further research.

3. Results and discussion

3.1 The existing condition of cities in Indonesia: An overview from DKI Jakarta and Bandung City, Indonesia

In Indonesia, the problem of urbanization has started with the rolling out. First, there is a macroeconomic policy (from 1967 to 2023) in Indonesia, in which cities are the center of the economy (Kominfo, 2022). Second, the combination of import substitution policies and foreign investment in the manufacturing sector, in fact, triggered a polarization of development centered on metropolitan Jakarta and not much different from the city of Bandung

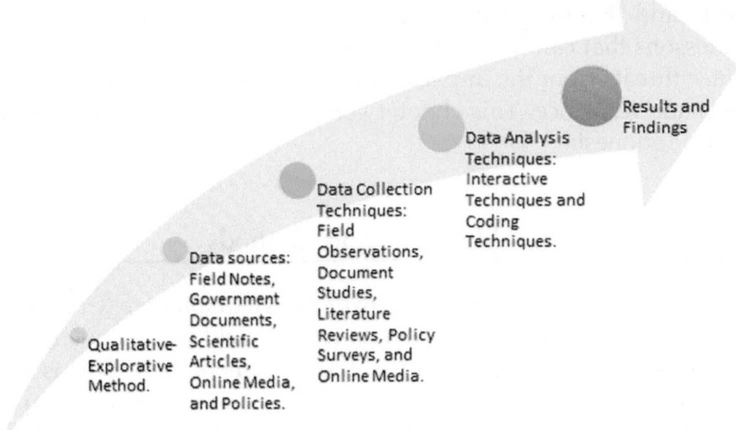

FIGURE 10.1 Steps of the research method. *Processed by the author, 2023.*

(Pemerintah DKI Jakarta, 2022; Pemerintah Kota Bandung, 2018). Third, poverty and facilities cause many young people and graduates to go to the city to look for work and improve their welfare (Pemerintah DKI Jakarta, 2022; Pemerintah Kota Bandung, 2018).

This uncontrolled flow of urbanization is a significant challenge for urban development planning strategies and absorbs urban facilities beyond the control of the city government. The impact of urbanization in DKI Jakarta and Bandung City, Indonesia, increased crime, decreasing levels of welfare, inequality in per capita income, unemployment, air and noise pollution, growth of urban slum areas in DKI Jakarta and Bandung City, Indonesia (Evasentia, 2023; Pemerintah DKI Jakarta, 2022; Pemerintah Kota Bandung, 2018).

The urban conditions which are increasingly out of control due to excessive urbanization have given rise to various new problems such as increased crime due to poverty, massive unemployment, an increase in slum settlements, and development challenges (see Fig. 10.2). Therefore, urbanization will be seen as a determining factor in how a city can develop both physically and socially. That way, the form or understanding of urbanization can be seen more clearly as well as the consequences/impacts it has on life in the city (Evasentia, 2023; Pemerintah DKI Jakarta, 2022; Pemerintah Kota Bandung, 2018).

Population growth and also an increase in population density over the past 10 years in DKI Jakarta and Bandung City have experienced a rapid increase, this is considered a challenge and obstacle to sustainable development in Indonesia (Abdillah, Buchari, et al., 2023; Abdillah, Widianingsih, Buchari, et al., 2023; Widianingsih et al., 2023). As in the case, increasing population density in DKI Jakarta and Bandung City is very important in efforts to maintain the quality and capacity of the environment while meeting the needs of the city's residents (Kusuma et al., 2022; Provinsi, 2019; Sony, 2023). In terms of educational indicators, the number of educational facilities, which reflects the ratio of the number of educational facilities to the total student age population, has decreased in value. This indicator has a significant influence on urban sustainability (29.46%) (Kusuma et al., 2022). This decline not only reflects the low number of educational facilities in DKI Jakarta but also disparities in access to

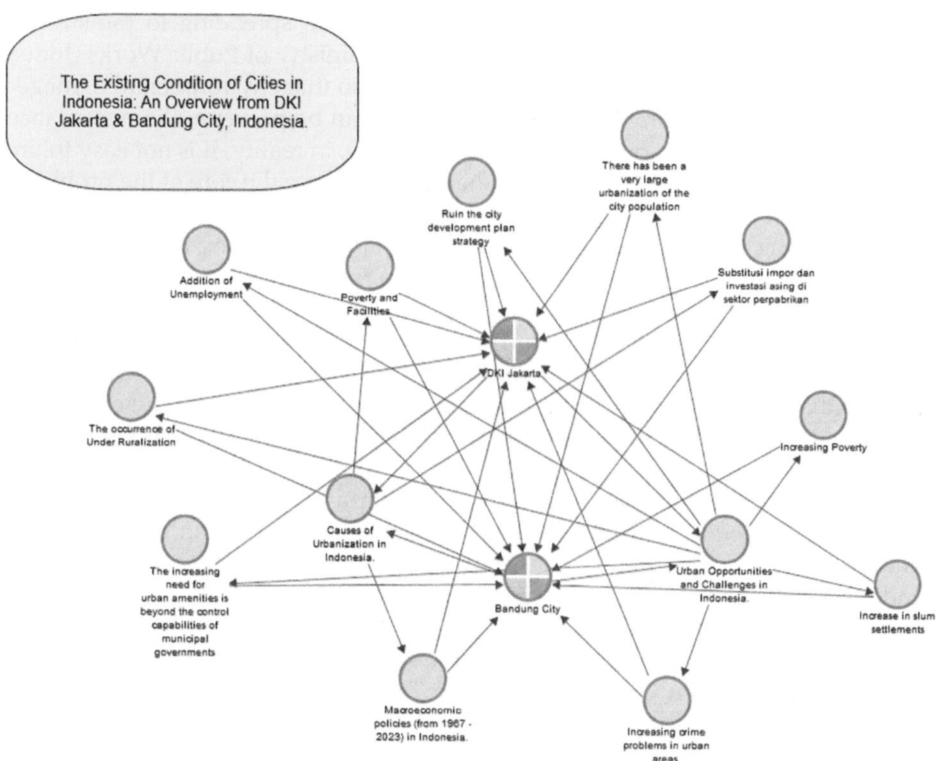

FIGURE 10.2 Existing urban conditions and impacts of urbanization, in DKI Jakarta and Bandung City, Indonesia. *Processed from Nvivo 12 Plus (2023).*

public education. Reviewing and prioritizing access to education (availability, quality, regulations and policies, etc.) must be the government's task so that all children can receive an education (Kusuma et al., 2022; Provinsi, 2019; Sony, 2023).

DKI Jakarta as the nation's capital is currently an urban area with urbanization originating from the surrounding area. DKI Jakarta is currently a metropolitan city with an area of 653.73 km^2 with an urbanization rate of 100%, and a population growth rate of 0.86% (Badan Pusat Statistik Provinsi DKI Jakarta, 2019). DKI Jakarta for 10 years tends to improve economic aspects compared to environmental and social aspects. This shows that the sustainability aspects of the city are experiencing an imbalance. As a result, it can be said that DKI Jakarta Province has not been responsive to the urbanization phenomenon in its development from 2010 to 2019, resulting in a lack of city sustainability. This is caused by the uneven development of these three characteristics from year to year, and the city tries to improve the quality of these three aspects together to achieve city sustainability (Provinsi, 2019).

In the last 20 years, the city of Bandung has faced high urban governance challenges. This happened because the development of the city of Bandung was out of control and violated spatial planning (Sony, 2023). The emergence of new settlements along with the mushrooming of commercial centers in this city causes serious traffic jams, especially during the holiday season (Sony, 2023). To prevent environmental clutter from spreading to the surrounding areas, the central government of Indonesia through the Ministry of Public Works (Indonesian Central Government) is helping to find the best solution so that environmental damage in the Bandung City Area and its surroundings in the future can be prevented in accordance with national spatial conditions. Responding to these problems, in reality, it is not easy to improve environmental conditions that are already in shambles. Because the core of the problem lies in the high will and awareness of all parties involved. Another factor that also has a big influence, namely funding and priority scale, is very crucial in encouraging resilience in Bandung City facing all kinds of threats and disturbances in the future. What's more, the condition of Bandung City and its surroundings is a disaster-prone location. This means that intensive use of space as built-up areas and cultivation areas will cause urban physical conditions to become increasingly vulnerable (Batubara et al., 2023; Dai et al., 2023; Evasentia, 2023; Sony, 2023; Widiawaty, 2019, pp. 1–10). And if it is not anticipated from now on, it is possible that in 20 years it will disrupt the comfort of life in Bandung City (Batubara et al., 2023; Dai et al., 2023; Evasentia, 2023; Sony, 2023).

Utilization of good spatial planning in DKI Jakarta and Bandung City as an effort to encourage urban resilience by establishing many city parks, allocation of SMEs, and slum area planning programs. This step was taken to deal with the impact of increasing urbanization in Indonesia which is manifested in city park developers, especially in facilitating beautiful, comfortable, and productive residences, such as in DKI Jakarta which promotes green cities as sustainable urban concept, and also the City of Bandung which encourages the concept of developing Green Open Space (*Ruang Terbuka Hijau*, RTU) with city parks for a beautiful and disaster-resilient city (Batubara et al., 2023; Dai et al., 2023; Evasentia, 2023; Provinsi, 2019; Sony, 2023).

3.2 Urbanization, homelessness, and climate change: Impacts in DKI Jakarta and Bandung City, Indonesia

As a study of urbanization, urbanization is defined as a phenomenon of increasing population in urban areas in line with the level of welfare and economic development of the population in a country (Cohen, 2006; Ventriglio et al., 2021; Xian et al., 2023). Urbanization is simply defined as the movement of people from rural areas to urban areas. Urbanization is heavily influenced by natural processes which include. birth and death rates, natural disasters, environmental changes and socioeconomic problems including individual income, education, health, basic facilities, industrialization, and government policies (Cohen, 2006; Ventriglio et al., 2021; Xian et al., 2023). There are factors that cause urbanization in Indonesia which include pull and push factors. The pull factor is a condition that causes a person to be interested in moving to an urban area because there is an attraction to offer. Driving factors consist of adequate health facilities, high standards of living, high standards of education, recreational facilities, employment opportunities, better security of life and property, and a better social environment. Meanwhile, push factors are factors that cause someone to move to an urban area because rural conditions are no longer supportive. Driving factors include poverty, low living standards, low life security, minimal transportation, and communication facilities, lack of employment opportunities, minimal health facilities, low-quality education (Batubara et al., 2023; Dai et al., 2023; Evasentia, 2023; Provinsi, 2019; Sony, 2023; Suhartini & Jones, 2023).

The characteristics of urbanization that occur in DKI are almost the same as those of large cities in the world, namely that it is characterized by an increase in the number of city residents that occurs every year (Cahyani, 2022; World Bank, 2010). This was then continued with the concentration of all community activities in one area, thus radically changing the spatial structure of the city. These changes can be seen in the pattern of changes in land use as indicated by the intensity of built-up land, the distribution of urban facilities, the transportation network system and movement patterns to the city center, as well as the development of land use, the development of the level of urbanization and migration of city residents, and subsequently the development of city economic activity. Metropolitan Jakarta has a very high and complex city development rate (Cahyani, 2022; World Bank, 2010). These symptoms began to be felt in the late 60s until now. To date, urbanization in Jakarta has swelled to more than 10 million people with relatively high population growth. As a result, there have been traffic jams, crime, environmental pollution, pollution, flooding and uncontrolled land use. Conditions like this have become a daily phenomenon for the growth of DKI Jakarta (Cahyani, 2022; Harahap, 2013; Idris & Triani, 2023). The city of Bandung is an urban area that is developing rapidly and is the center of the Greater Bandung region. Regional expansion and the emergence of peri-urban zones indicate that urbanization in this city has crossed formal administrative boundaries (BI Institute, 2020; Idris & Triani, 2023; Ismail et al., 2020). Urbanization is not only meant as a movement of rural residents to cities, but as a series of processes of development of urban areas which are always accompanied by a process of population growth both internally and externally (BI Institute, 2020; Idris & Triani, 2023; Ismail et al., 2020). Urbanization in the city of Bandung has led to an increase in population due to the migration process that has been going on since the colonial era to the present. As a result, the existence of built-up land is increasing and the vegetated area is decreasing. The lack of available land also causes density in Bandung City to continue to increase and creates slum areas around the "central business " (CBD) (Idris & Triani, 2023; Ismail et al., 2020). Thus, urbanization in the city of Bandung not only causes complex physical environmental problems but also social and economic problems that impact population density, homelessness and street children in Bandung City, Indonesia (BI Institute, 2020; Idris & Triani, 2023; Ismail et al., 2020). The impact of high urbanization in Indonesia creates environmental (pollution and pollution), social (poverty, harassment, and homelessness), and urban community life (health and office facilities) problems in DKI Jakarta and Bandung City (see Table 10.1).

The impact of urbanization on urban development in DKI Jakarta and Bandung City is: (1) Physically: (a) Built-up land versus green/open land. It is certain that almost all urban land has been built for residential buildings, trade and service areas, industry, offices, and other buildings. The intensity of built-up land in DKI Jakarta continues to increase, making it difficult to find green/open land that functions as a public space, although in Bandung City there are several green open spaces such as city parks, but Bandung City is also one of the most populous cities in Indonesia which is facing many impacts. negative for high urbanization; (b) distribution of urban facilities, provincial government centers, trade centers, service activity centers, DKI Jakarta and Bandung City as well as entry and exit points for international and national transportation with fairly high mobility. Due to its nature, various potentials and problems arise in trade areas, recreational areas, and the economy; (c) transportation networks and urban movement patterns trigger adjustments, improvements and additions of

TABLE 10.1 Impact of urbanization, homelessness, and climate change in DKI Jakarta & Bandung City, Indonesia.

DKI Jakarta	Bandung City
Land problems for buildings and green open spaces.	Slums.
Inadequate distribution of urban facilities.	Unemployment, homelessness, street children, and high poverty.
Natural disasters such as earthquakes and floods.	Natural disasters such as earthquakes, landslides, earthquakes, and floods.
Crimes.	Crimes.
Environmental pollution and pollution (air pollution).	Increase in HIV disease.
Street children.	Clean water problems.
Homelessness.	Urban temperature rise.

Processed from various sources, 2023.

roads and new modes of transportation; (d) development of land use, intensive and extensive construction and development of settlements or housing carried out by the government and the private sector which has an impact on changes in the urban spatial structure in DKI Jakarta and Bandung City; (e) environmental problems, decreasing carrying capacity and increasing pollution as a result of unplanned development, transportation mobility, and population density as well as chaotic management of city facilities and infrastructure giving rise to increasingly serious environmental, social, and urban governance problems such as flooding, land landslides, air, soil, water, and air pollution; and (f) in slum settlements, more and more city residents live crowded together in various residential centers and the number of settlers continues to increase; and (2) Socially: (a) Unemployment, poverty, and homelessness are increasingly becoming problems, resulting in an increase in urban poverty rates. Limited education, abilities, and skills also become an obstacle for job seekers to get a job; and (b) crime and street children caused by pressure to survive, for example, will encourage people to take any action, including committing criminal acts. This is also the reason why the number of crime, homelessness, and street children in DKI Jakarta and the city of Bandung is increasing day by day (BI Institute, 2020; Cahyani, 2022; Harahap, 2013; Idris & Triani, 2023; Ismail et al., 2020; Pemerintah DKI Jakarta, 2022; Pemerintah Kota Bandung, 2018; World Bank, 2010).

This excessive urbanization has caused various problems in Indonesia. Not only does it cause problems in the target city but it also causes problems in the abandoned village. Problems that occur in cities include increasing poverty rates so that slum settlements also increase, an increase in urban crime, and many other problems. There will also be problems in the village, including a reduction in human resources because the population has gone to the city, so the village does not experience real development. Urban conditions that are increasingly out of control due to excessive urbanization have given rise to various new problems such as increased crime due to poverty, massive unemployment, an increase in slum settlements, and so on. Therefore, urbanization will be seen as a determining factor in how a city can develop both physically and socially. That way, the form or understanding of urbanization can be seen more clearly as well as the consequences/impacts it has on life in the city (Batubara et al., 2023; Dai et al., 2023; Evasentia, 2023; Hidayati, 2021; Idris & Triani, 2023; Ismail et al., 2020; Sony, 2023; Widiawaty, 2019, pp. 1–10).

3.3 Social vulnerability, homelessness, and the right to housing in Indonesia: An evaluation

In the evaluation of this study, there was concern that urbanization would be highlighted from the negative side (Batubara et al., 2023; Cahyani, 2022; Cheerli, 2019; Dai et al., 2023; Evasentia, 2023; Harahap, 2013; Idris & Triani, 2023; Ismail et al., 2020; Pemerintah DKI Jakarta, 2022; Pemerintah Kota Bandung, 2018; Prianto & Abdillah, 2023; Provinsi, 2019; Widianingsih et al., 2023; World Bank, 2010). People who come to urban areas are considered to increase the number of unemployed, destroy the environment, add to the chaotic atmosphere, encourage increased crime, and other things that disrupt development activities. On the other hand, they are also considered to be an obstacle to development activities in rural areas, the cause of a shortage of labor in the agricultural sector, carriers of juvenile delinquency in villages, and so on. The most important thing that the DKI Jakarta government and the Bandung City government, Indonesia, need to pay attention to in dealing with urban problems is by encouraging urban resilience in good urban governance in the economic, social, and ecological dimensions. Urban governance pursued in DKI Jakarta and Bandung City pays attention to resilience values such as system and individual capabilities as seen from resilience values: Robustness, redundancy, flexibility, and capacity against threats and disasters in urban areas such as poverty, crime, environmental pollution, pollution, health, and natural disasters (Batubara et al., 2023; Cahyani, 2022; Cheerli, 2019; Dai et al., 2023; Evasentia, 2023; Harahap, 2013; Idris & Triani, 2023; Ismail et al., 2020; Pemerintah DKI Jakarta, 2022; Pemerintah Kota Bandung, 2018; Prianto & Abdillah, 2023; Provinsi, 2019; Widianingsih et al., 2023; World Bank, 2010). Even though urban resilience efforts in the city of Bandung still have weaknesses in terms of flexibility and urban redundancy, they have attempted robustness and capacity through various policies and development programs as shown in Fig. 10.3.

Currently, these efforts to minimize disasters are often not implemented optimally in DKI Jakarta and Bandung City, Indonesia. These efforts are prepared in the regional development master plan (Regional Medium Term Development Plan, RPJMD) in the form of development programs and strategies, although their implementation still needs to be evaluated. Several efforts were made in DKI Jakarta and Bandung City to deal with the impact of urbanization and climate change which resulted in many homeless and street children in the city (Batubara et al., 2023; Dai et al., 2023; Evasentia, 2023; Hidayati, 2021; Idris & Triani, 2023; Ismail et al., 2020; Sony, 2023). Like, there are two steps that will be pursued namely, technical and socialization. Technical steps include setting zoning regulations, completing facilities and infrastructure, mapping disaster areas, limiting occupancy, and preparing shelter locations (Batubara et al., 2023; Dai et al., 2023; Pemerintah DKI Jakarta, 2022; Pemerintah Kota Bandung, 2018; Sony, 2023). The socialization steps include encouraging village tourism to

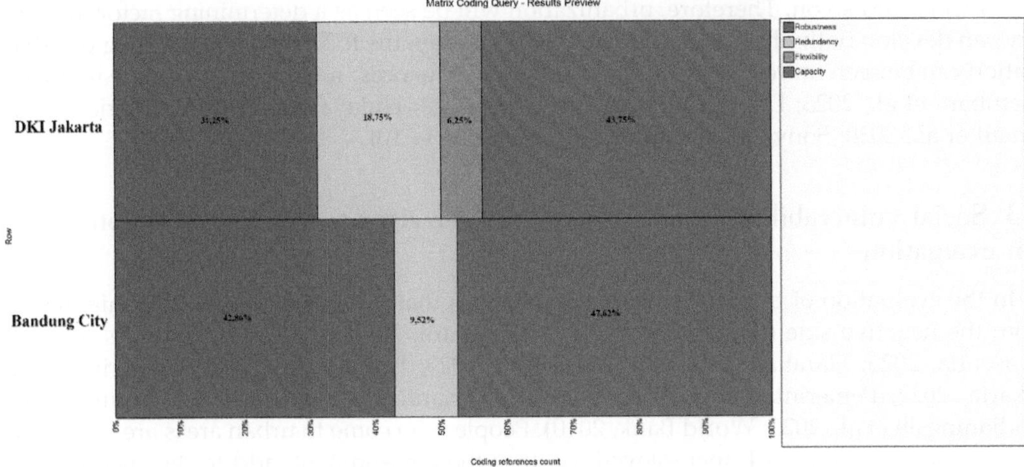

FIGURE 10.3 Urban resilience to urbanization, homelessness, and climate change in DKI Jakarta Bandung City, Indonesia. *Reworked via Nvivo 12 Plus, 2023.*

improve the economic sector, educating schools, preparing databases/information, limiting occupancy in disaster-prone areas, and improving environmental quality including community participation (Batubara et al., 2023; Dai et al., 2023; Pemerintah DKI Jakarta, 2022; Pemerintah Kota Bandung, 2018; Sony, 2023). In addition, the increasing number of homeless people poses challenges for DKI Jakarta and Bandung City, Indonesia. This makes it necessary to encourage management collaboration from various parties such as the private sector, civil society, NGOs, central and regional governments and academics in various important sectors by paying attention to resilience in the economic, social, and ecological dimensions to realize sustainable development to equal social prosperity (Abdillah, Buchari, et al., 2023; Abdillah, Widianingsih, Buchari, et al., 2023; Batubara et al., 2023; Cheerli, 2019; Dai et al., 2023; Kemen PUPR RI, 2021).

The increase in homelessness in DKI Jakarta and Bandung City, Indonesia, is due to the majority of people living on the streets, whether living under bridges, under the shade of large trees, or in the shade of buildings, more or less having the same characteristics as those who live in lying organization (Rahardjo, 2006; Tipple & Speak, 2005). The residences for most homeless people in these two cities are mostly less permanent, they have almost the same informal jobs as scavengers, street vendors, pedicab drivers, construction workers, and other unskilled jobs, although some street homeless earn a living by begging (Rahardjo, 2006; Speak & Tipple, 2006; Tipple & Speak, 2005).

Both squatters and homeless people are often the targets of raids by Social Services in DKI Jakarta to Bandung City, Indonesia. More violent evictions usually occur for squatters because they usually involve sensitive issues regarding land ownership. The residents usually refuse to be moved on the grounds that they have lived there long before the land had any commercial value. More peaceful raids on homeless people are carried out because they are considered to cause a nuisance or—a term often used by city officials and planners—"disrupt the attractiveness of the city" (Rahardjo, 2006; Speak & Tipple, 2006; Tipple & Speak, 2005). Homeless people, unlike squatters, usually do not fight back. They view their move as only temporary (such as when an official visits or a national day is celebrated). After things return to "normal," they are usually allowed to return to their original place, but sometimes there is conflict because the place previously used has been used by another user.

In Indonesia, Law (UU) no. 4 of 1992 concerning housing and settlements which was changed to Law (UU) No. 1 of 2011 concerning housing and settlement areas; Government Regulation (PP) no. 88 of 2014 concerning development, implementation of housing and residential areas; to Government Regulation no. 44 of 2016 concerning the implementation of housing and settlement areas, recognizes the right of every citizen to live and/or use and/or own a house that is adequate and located in a healthy, safe, harmonious and orderly environment.

Furthermore, the UU and PP define adequate housing as a house structure that at least meets the requirements for building safety, minimum floor area, and health. This definition of housing adequacy, when referring to various previous literature, shows that the Indonesian government still views housing as only a basic need, not a strategic need, and not yet a human right for every person. Housing is considered a strategic basic need, because it is seen as an entry point for fulfilling other basic needs (Rahardjo, 2006; Speak & Tipple, 2006; Tipple & Speak, 2005). This is why the problem of homelessness in DKI Jakarta and Bandung City is still not being handled properly. We emphasize that the governance of homelessness in Indonesia has not become a serious concern for local and national governments, both in the policy-making paradigm and in a just governance system. In fact, good handling of homeless people in Indonesia is one of the keys to overcoming social vulnerabilities that occur and can encourage urban resilience in Indonesia.

The strategies for overcoming the social problem of homelessness in Indonesia that we can offer in this chapter can be carried out by the Indonesian government, such as the strategies offered by Evans et al. (2019) who say that homelessness can be reduced and prevented by doing (1) emergency financial assistance and more comprehensive interventions that provide a range of financial assistance, counseling, and legal support can prevent homelessness among families at risk of losing their home, but more research is needed on how to prevent it. Programs can be delivered well and targeted to those who need them most; (2) legal representation for renters facing eviction promises to improve court-related outcomes for renters and reduce evictions, although more research is needed on which types of legal tactics and programs are effective; (3) permanent supportive housing increases housing stability for individuals with serious mental illness and for veterans experiencing homelessness. There is little robust evidence on the impact of permanent supportive housing for other groups in society; (4) although rapid housing redevelopment has the potential to be a cost-effective solution for providing immediate access to housing, there is little robust evidence regarding the impact of rapid housing redevelopment on long-term housing stability; (5) subsidized long-term housing assistance in the form of Housing Choice Vouchers helps low-income families avoid homelessness and remain stable; (6) additional research is needed regarding the effectiveness of other strategies to reduce homelessness. This review identifies gaps in the literature and raises several new questions that need to be considered when evaluating homelessness prevention or assistance programs. These steps are more effective if they are integrated into local policies/regulations or included in priority development programs in Indonesia.

4. Conclusions

In Indonesia, the problem of urbanization has begun with the introduction of the first, macroeconomic policies (from 1967 to 2023), where cities are the center of the economy. Second, the combination of import substitution policies and foreign investment in the manufacturing sector, in fact, triggered a polarization of development centered on metropolitan Jakarta and not much different from the city of Bandung. Third, poverty and facilities cause many young people and graduates to go to the city to look for work and improve their welfare. Urbanization in DKI Jakarta and Bandung City, Indonesia, is essentially a reflection of differences in growth and inequality of development facilities between one region and another in Indonesia. So there is a great need for reform in terms of village development to encourage equitable growth in various regions in Indonesia. The updates can be in any way, one of which is the improvement of transportation and communication infrastructure, encouraging empowerment programs for the welfare of farmers, to assistance for small businesses in rural areas. The problem of urban resilience arises due to the high impact of urbanization in Indonesia due to a lack of attention to resilience values in development programs and policy formulation in DKI Jakarta and Bandung City, Indonesia. Therefore, encouraging urban resilience is in line with the goals of sustainable development programs and solutions to face the impact of high urbanization in Indonesia which cannot be stopped. Efforts to minimize the negative impacts due to urbanization and natural disasters which often befall DKI Jakarta and Bandung City are classified into two steps, namely, technical steps and outreach steps. Technical steps include setting up zoning regulations, mapping disaster areas, limiting occupancy, and preparing shelter locations. The socialization steps include educating schools, preparing databases/information, limiting occupancy in disaster-prone areas, and improving environmental quality including community participation. Adequate housing which is not seen as a strategic need and as a human right is the main cause of increasing social vulnerability (on the issue of homelessness) in Indonesia. This is what causes the homeless problem in DKI Jakarta and Bandung City to still not be handled properly. We emphasize that the governance of homelessness in Indonesia has not yet become a serious concern for local and national governments, both in the policy-making paradigm and in a just governance system. In fact, good handling of homelessness in Indonesia is one of the keys to overcoming social vulnerabilities that occur and can encourage urban resilience in Indonesia.

The limitation of this research is that it only studies two big cities in Indonesia, it does not rule out the possibility that other big cities in Indonesia are equally important to study so that they can be taken into consideration for policy formulation to encourage resilience and sustainable development in Indonesia. So the authors suggest that future researchers be able to study urbanization issues, the increase in homelessness and street children, poverty, and the impact of climate change in the Cities of Surabaya, Semarang, Makassar, and Yogyakarta, Indonesia.

Acknowledgments

Thanks to Center for Decentralization and Participatory Development Research, Faculty of Social and Political Sciences, Universitas Padjadjara, and Universitas Padjadjaran, Indonesia, for supporting this research, so that this research can be completed and published.

AI Disclosure

During the preparation of this work the author(s) used grammarly in order to grammarly helps you write mistake-free in chapter. After using this tool/service, the author(s) reviewed and edited the content as needed and take(s) full responsibility for the content of the publication.

References

Abdillah, A., Buchari, R. A., Widianingsih, I., & Nurasa, H. (2023). Climate change governance for urban resilience for Indonesia: A systematic literature review. *Cogent Social Sciences, 9*(1). https://doi.org/10.1080/23311886.2023.2235170

Abdillah, A., Widianingsih, I., Buchari, R. A., & Nurasa, H. (2023). Implications of urban farming on urban resilience in Indonesia: Systematic literature review and research identification. *Cogent Food and Agriculture, 9*(1). https://doi.org/10.1080/23311932.2023.2216484

Abdillah, A., Widianingsih, I., Buhari, R. A., & Rusliadi, R. (2023). Collaborative communication models in non-cash food assistance (bantuan pangan non-tunai, BPNT) program: Toward community resilience. In *Lecture notes on data engineering and communications technologies* (Vol 171, pp. 75–91). Indonesia: Springer Science and Business Media Deutschland GmbH. https://doi.org/10.1007/978-981-99-1767-9_6

Adetokunbo, I., & Emeka, M. (2015). Urbanization, housing, homelessness and climate change. *Journal of Design and Built Environment, 15*(2), 15–28. https://doi.org/10.22452/jdbe.vol15no2.3

Anggraeni, F. A. (2022). Analisis faktor yang mempengaruhi peningkatan urbanisasi di kota jakarta dan surabaya pada tahun 2020-2021. *Jurnal Ekonomi Bisnis dan Akuntansi, 2*(2), 41–53. https://doi.org/10.55606/jebaku.v2i2.115

Batubara, B., Kooy, M., Leynseele, Y. V., Zwarteveen, M., & Ujianto, A. (2023). Urbanization in (post-) new order Indonesia: Connecting unevenness in the city with that in the countryside. *Journal of Peasant Studies, 50*(3), 1207–1226. https://doi.org/10.1080/03066150.2021.2000399

BI Institute. (2020). *Geliat Kota Bandung: Dari Kota tradisional menuju modern.* https://www.bi.go.id/id/bi-institute/publikasi/Documents/Buku_Sejarah_KPwBI_BANDUNG.pdf. (Accessed 15 September 2023).

Cahyani, A. P. (2022). *Urbanisasi dan struktur ekonomi.* https://www.kompasiana.com/aisyah77883/63ab1e4508a8b5112f479974/urbanisasi-dan-struktur-sosial-ekonomi?page=2&page_images=2. (Accessed 10 September 2023).

Cheerli, Cheerli (2019). *Hadapi tantangan urbanisasi, Indonesia, pererat kolaborasi.* https://www.nawasis.org/portal/berita/read/hadapi-tantangan-urbanisasi-indonesia-pererat-kolaborasi/51260. (Accessed 10 September 2023).

Cohen, B. (2006). Urbanization in developing countries: Current trends, future projections, and key challenges for sustainability. *Technology in Society, 28*(1–2), 63–80. https://doi.org/10.1016/j.techsoc.2005.10.005

Creswell, J. W., & Poth, C. N. (2016). *Qualitative inquiry and research design: Choosing among five approaches.* Sage publications.

Dai, X., Chen, J. G., & Xue, C. (2023). Spatiotemporal patterns and driving factors of the ecological environmental quality along the jakarta–bandung high-speed railway in Indonesia. *Sustainability, 15*(16), 12426. https://doi.org/10.3390/su151612426

Evans, W. N., Philips, D. C., & Ruffini, K. J. (2019). *Reducing and preventing homelessness: A review of the evidence and charting a research agenda.* SSRN. https://www.ssrn.com/index.cfm/en/.

Evasentia, Y. (2023). *Dampak urbanisasi bagi perkembangan kota di Indonesia.* https://www.floresnews.id/opini-cerita/pr-4996499772/dampak-urbanisasi-bagi-perkembangan-kota-di-indonesia. (Accessed 10 September 2023).

Fahmi, F. Z., Hudalah, D., Rahayu, P., & Woltjer, J. (2014). Extended urbanization in small and medium-sized cities: The case of cirebon, Indonesia. *Habitat International, 42*, 1–10. https://doi.org/10.1016/j.habitatint.2013.10.003

Harahap, F. R. (2013). Dampak urbanisasi bagi perkembangan kota di Indonesia. *Society, 1*(1), 35–45. https://doi.org/10.33019/society.v1i1.40

Hidayati, I. (2021). Urbanisasi dan dampak sosial di Kota besar: Sebuah tinjauan. *Jurnal Ilmiah Ilmu Sosial, 7*(2), 212. https://doi.org/10.23887/jiis.v7i2.40517

Idris, M., & Triani, M. E. (2023). Kausalitas sumber daya manusia, urbanisasi dan modal manusia di Indonesia. *Jurnal Kajian Ekonomi dan Pembangunan, 5*(1), 45. https://doi.org/10.24036/jkep.v5i1.14420

Ismail, A., Dede, M., & Widiawaty, M. A. (2020). Urbanisasi dan HIV di kota bandung (perspektif geografi kesehatan). *Buletin Penelitian Kesehatan, 48*(2), 139–146. https://doi.org/10.22435/bpk.v48i2.2921

Jones, G. W. (2015). *Migration and urbanization in China, India and Indonesia: An overview*. Springer Science and Business Media LLC. https://doi.org/10.1007/978-3-319-24783-0_17

Kemen PUPR RI. (2021). *Hadapi tantangan perubahan iklim, Indonesia perkuat kerjasama dalam manajemen air di Asia*. https://www.pu.go.id/berita/hadapi-tantangan-perubahan-iklim-indonesia-perkuat-kerjasama-dalam-manajemen-air-di-asia. (Accessed 1 September 2023).

Kidd, S. A., Bezgrebelna, M., Hajat, S., Keevers, L., Ravindran, A., Stergiopoulos, V., Wells, S., Yamamoto, S., Galvao, L. A., Hale, M., Njengah, S., Settembrino, M., Vickery, J., & McKenzie, K. (2023). A response framework for addressing the risks of climate change for homeless populations. *Climate Policy, 23*(5), 623–636. https://doi.org/10.1080/14693062.2023.2194280

Kominfo. (2022). *Kebijakan ekonomi makro tahun 2023 akan dorong pemulihan dari sumber non-APBN*. https://www.kominfo.go.id/content/detail/40038/kebijakan-ekonomi-makro-tahun-2023-akan-dorong-pemulihan-dari-sumber-non-apbn/0/berita. (Accessed 2 September 2023).

Kusuma, M. E., Situmorang, R., & Ramadhani, A. (2022). Faktor yang berpengaruh dalam indeks keberlanjutan kota di Provinsi DKI Jakarta. *TATALOKA, 24*(4), 312–320. https://doi.org/10.14710/tataloka.24.4.312-320

Lewis, B. D. (2014). Urbanization and economic growth in Indonesia: Good news, bad news and (possible) local government mitigation. *Regional Studies, 48*(1), 192–207. https://doi.org/10.1080/00343404.2012.748980

Malik, I., Abdillah, Rusnaedy, Z., Khaerah, N., Harakan, A., & Jermsittiparsert, K. (2021). Coastal women's resilience strategy against climate change vulnerability in Makassar, Indonesia. In *E3S web of conferences* (Vol 277, p. 01003). EDP Sciences. https://doi.org/10.1051/e3sconf/202127701003

Mardiansjah, F. H., Rahayu, P., & Rukmana, D. (2021). New patterns of urbanization in Indonesia: Emergence of non-statutory towns and new extended urban regions. *Environment and Urbanization ASIA, 12*(1), 11–26. https://doi.org/10.1177/0975425321990384

Miles, M. B., Huberman, A. M., & Saldaña, J. (2018). *Qualitative data analysis: A methods sourcebook*. Sage Publications.

Njoh, A. J. (2003). Urbanization and development in sub-saharan Africa. *Cities, 20*(3), 167–174. https://doi.org/10.1016/S0264-2751(03)00010-6

Ooi, G. L. (2009). Challenges of sustainability for Asian urbanisation. *Current Opinion in Environmental Sustainability, 1*(2), 187–191. https://doi.org/10.1016/j.cosust.2009.09.001

Pee, L. G., & Pan, S. L. (2022). Climate-intelligent cities and resilient urbanisation: Challenges and opportunities for information research. *International Journal of Information Management, 63*, 102446. https://doi.org/10.1016/j.ijinfomgt.2021.102446

Pemerintah DKI Jakarta. (2022). *Rencana pembangunan daerah ((RPJMD)) tahun 2023-2026*. https://drive.google.com/file/d/1U7Bb3SLeMHcPHffIDkx6qOswdW1WQdA8/view.

Pemerintah Kota Bandung. (2018). *Rencana pembangunan jangka menengah daerah (rpjmd) tahun 2018-2023*. http://file:///C:/Users/HP/Downloads/AUTENTIFIKASI%20PERDA%203%20TAHUN%202019%20RPJMD%202018%20-%202023%20[EDIT%206%20MEI%202019]%20CETAK%20(4)%20(2).pdf. (Accessed 2 September 2023).

Prastiyo, S. E., Irham, Hardyastuti, S., & Jamhari, F. (2020). How agriculture, manufacture, and urbanization induced carbon emission? the case of Indonesia. *Environmental Science and Pollution Research, 27*(33), 42092–42103. https://doi.org/10.1007/s11356-020-10148-w

Prianto, A. L., & Abdillah, A. (2023). Vulnerable countries, resilient communities: Climate change governance in the coastal communities in Indonesia. In *Climate change, community response and resilience: Insight for socio-ecological sustainability* (pp. 135–152). Elsevier. https://doi.org/10.1016/B978-0-443-18707-0.00007-2

Provinsi, DKI Jakarta. (2019). *Statistik daerah provinsi DKI Jakarta*. Jakarta: BPS Provinsi DKI Jakarta. https://drive.google.com/file/d/1U7Bb3SLeMHcPHffIDkx6qOswdW1WQdA8/view. (Accessed 23 September 2023).

Rahardjo, T. (2006). Forced eviction, homelessness and the right to housing in Indonesia. In *Conference on Homelessness: A Global Perspective* (pp. 9–13).

Satterthwaite, D. (2009). The implications of population growth and urbanization for climate change. *Environment and Urbanization, 21*(2), 545–567. https://doi.org/10.1177/0956247809344361

Sony. (2023). *Kondisi tara ruang Kota Bandung semakin amburadul*. https://pu.go.id/berita/kondisi-tara-ruang-kota-bandung-semakin-amburadul. (Accessed 3 September 2023).

Speak, S., & Tipple, G. (2006). Perceptions, persecution and pity: The limitations of interventions for homelessness in developing countries. *International Journal of Urban and Regional Research, 30*(1), 172–188. https://doi.org/10.1111/j.1468-2427.2006.00641.x

Suhartini, N., & Jones, P. (2023). Urbanization and the development of the kampung in Indonesia. In *Urban book series* (pp. 57–77). Indonesia: Springer Science and Business Media Deutschland GmbH. https://doi.org/10.1007/978-3-031-22239-9_4

Tipple, G., & Speak, S. (2005). Definitions of homelessness in developing countries. *Habitat International, 29*(2), 337–352. https://doi.org/10.1016/j.habitatint.2003.11.002

Ventriglio, A., Torales, J., Castaldelli-Maia, J. M., De Berardis, D., & Bhugra, D. (2021). Urbanization and emerging mental health issues. *CNS Spectrums, 26*(1), 43–50. https://doi.org/10.1017/S1092852920001236

Widianingsih, I., Abdillah, A., Herawati, E., Dewi, A. U., Miftah, A. Z., Adikancana, Q. M., Pratama, M. N., & Sasmono, S. (2023). Sport tourism, regional development, and urban resilience: A focus on regional economic development in Lake Toba district, North Sumatra, Indonesia. *Sustainability, 15*(7). https://doi.org/10.3390/su15075960

Widiawaty, M. A. (2019). *Faktor-faktor urbanisasi di Indonesia.* Pendidikan Geografi UPI. https://doi.org/10.31227/osf.io/vzpsw

Woolf, N. H., & Silver, C. (2017). *Qualitative analysis using MAXQDA: The five-level QDA method.* United States: Taylor and Francis. https://doi.org/10.4324/9781315268569

World Bank. (2010). *Jakarta tantangan perkotaan seiring perubahan iklim.* https://documents1.worldbank.org/curated/en/566051468267612011/pdf/650180Replacem00Change0BHS0LR0rev3c.pdf. (Accessed 3 September 2023).

Xian, C., Gong, C., Lu, F., Wu, H., & Ouyang, Z. (2023). The evaluation of greenhouse gas emissions from sewage treatment with urbanization: Understanding the opportunities and challenges for climate change mitigation in China's low-carbon pilot city, Shenzhen. *Science of The Total Environment, 855*, 158629. https://doi.org/10.1016/j.scitotenv.2022.158629

Yeh, A. G. O., & Chen, Z. (2020). From cities to super mega city regions in China in a new wave of urbanisation and economic transition: Issues and challenges. *Urban Studies, 57*(3), 636–654. https://doi.org/10.1177/0042098019879566

Zhu, J., & Simarmata, H. A. (2015). Formal land rights versus informal land rights: Governance for sustainable urbanization in the Jakarta metropolitan region, Indonesia. *Land Use Policy, 43*, 63–73. https://doi.org/10.1016/j.landusepol.2014.10.016

Zubaidah, S., Widianingsih, I., Rusli, B., & Saefullah, A. D. (2023). Policy network on the kotaku program in the global south: Findings from Palembang, Indonesia. *Sustainability, 15*(6), 4784. https://doi.org/10.3390/su15064784

CHAPTER

11

Contextualizing the drivers of homelessness among women in Kolkata megacity: An exploratory study

Margubur Rahaman[1], *Kailash Chandra Das*[1] *and Md. Juel Rana*[2]

[1]Department of Migration & Urban Studies, International Institute for Population Sciences, Mumbai, Maharashtra, India; [2]Govind Ballabh Pant Social Science Institute, Prayagraj, Uttar Pradesh, India

1. Introduction

Homelessness is an urgent issue in our modern world, stemming from social injustice, human rights violations, economic and spatial inequalities, and inadequate government policies and programs (Klodawsky, 2009). It is indeed a tragic and profoundly concerning reality that a significant portion of the population in modern and democratic nations continues to grapple with homelessness. This issue underscores the complex interplay between modernity and homelessness in contemporary societies. The process of modernization, characterized by urbanization and significant economic shifts, is often identified as a contributing factor to the problem of homelessness (Batterham et al., 2022). This is primarily because modernization tends to drive up the cost of living, particularly in urban areas (Cebula & Alexander, 2020). While democratic governments often speak of their commitment to safeguarding fundamental human rights and providing social safety nets such as homeless shelters and initiatives to ensure housing for all (Toro, 2007), the stark truth remains that millions of individuals still find themselves without a stable place to call home (Internal Displacement Monitoring Centre, 2023). This persistent problem persists due to deep-rooted socioeconomic disparities and the inadequacy of these social safety nets to fully address the complex web of

Homelessness to Hope
https://doi.org/10.1016/B978-0-443-14052-5.00011-2

challenges that lead to homelessness (Barile et al., 2018). This ongoing debate is a stark reminder that homelessness is not a matter of choice in the contemporary world. It is a multifaceted issue intricately woven into the fabric of modern society, driven by a range of factors that extend far beyond the individual's control (Kidd & Davidson, 2007). Important drivers of homelessness include natural disasters and hazards driven displacement, socioeconomic deprivation, discrimination, interpersonal violence, and chronic mental illness (Bezgrebelna et al., 2021; Internal Displacement Monitoring Centre, 2023; Singh et al., 2018). Manufactured disasters, such as communal or racial violence, geopolitical unrest, and accidents, also contribute to homelessness (Allen et al., 2017). The latest figure shows that 122.3 million individuals worldwide are homeless, with an exceptionally high concentration in African and South Asian countries (World Population Review, 2023). Nigeria tops the list with an alarming figure of 24.4 million homeless people, followed by Pakistan with 20 million and Egypt with 12 million. India is ranked 16th on the list, with 1.8 million homeless individuals. The report concludes that homelessness is a critical global issue in nearly all countries (World Population Review, 2023). However, defining homelessness, understanding its levels, patterns, and growth rates, as well as identifying its underlying drivers, is complex and varies depending on the specific spatial, demographic, and temporal contexts (Cowan et al., 1988; Internal Displacement Monitoring Centre, 2023). For instance, since 1900, 65 million individuals in South Asia have experienced homelessness due to natural disasters and hazards, with India alone accounting for 35 million (Ritchie et al., 2022). Further, homelessness hotspots have been expanding in Central African nations due to civil wars and the adverse effects of climate change (Internal Displacement Monitoring Centre, 2023). Similarly, the south-central Asian and Mediterranean Muslim world is characterized by frequent civil wars, territorial aggression, terrorism activities, and political unrest, resulting in severe homelessness. Therefore, comprehending the issue of homelessness requires a comprehensive study that considers spatial, environmental, demographic, and temporal dynamics, as it is unclear and incomplete without such an approach.

The issue of homelessness in India exhibits spatial variations, with more pronounced visibility in urban areas compared to rural regions (Sahoo & Jeermison, 2018; Singh et al., 2018). Each day, countless individuals migrate to urban centers in search of diverse job opportunities, and a significant portion of them need a home in the fiercely competitive urban environment (Goel et al., 2017). The soaring living costs driven by urban competitiveness and modernization compel many migrant workers in the informal sector, impoverished urban residents, and other marginalized socioeconomic groups to reside in slums, on pavements, or as homeless individuals (Brown et al., 2020; Goel et al., 2017). Despite the implementation of the Pradhan Mantri Awas Yojana (Urban) Mission, which aims to ensure "Housing for all," challenges such as the cyclical nature of homelessness, pseudo-homelessness, relative homelessness, and lack of proper citizenship documentation act as barriers to eradicating homelessness (Government of India Ministry of Housing & Urban Poverty Alleviation, 2013). Additionally, the Indian government has introduced the National Urban Livelihoods Mission (NULM) to provide permanent shelters with essential services for the urban homeless in a phased manner through the Scheme of Shelter for Urban Homeless (SUH). However, the inadequate number of shelters, subpar quality of services at existing shelters, the absence of homeless-specific shelters in hotspot areas, and the lack of gender-specific provisions for homeless individuals still leave millions without a roof over their heads (Bhattacharjee, 2023; Singh et al., 2018).

Living without shelters in urban settings exposes individuals to many challenges, including environmental pollution, social-cultural stigma, discrimination, violence, and exploitation (Goel & Chowdhary, 2017). Gender-sensitive studies have shed light on the heightened vulnerability of homeless women compared to men in urban areas, primarily due to the additional risks of gender-based sexual violence and exploitation (Bhattacharya, 2022). Moreover, homeless women experience more complex, multifaceted, and dynamic factors contributing to their homelessness (Bhattacharya, 2022). The gender dimension of homelessness emphasizes the significance of factors such as marriage, marital dissolution, sex-selective violence, trafficking, honey traps, and exploitation by unorganized employment agents as predictors of homelessness among women in urban areas (Haile et al., 2020). While urban homelessness is generally male-dominated, the proportion of homeless women in metropolitan India ranges from 10% to 55% (Office of the Registrar General & Census Commissioner, 2011). Hence, conducting gender-specific studies is crucial in Indian urban settings to comprehensively understand the issue of homelessness and develop gender-sensitive policies and programs that promote equity among all genders.

In India, the total homeless population stands at 1.8 million, with West Bengal contributing 8% to this number, placing it the seventh-largest state with a homeless population (Office of the Registrar General & Census Commissioner, 2011). Within West Bengal, Kolkata takes the lead among districts, with 69,798 homeless individuals. Likewise, among the Indian megacities, Kolkata ranks first in terms of the size of its homeless women population. Despite Kolkata being the largest city in terms of homeless women (Office of the Registrar General & Census Commissioner, 2011), few studies have explicitly focused on the pathways to homelessness among women in Kolkata (Arindam Roy & Siddique, 2018; Bhattacharya, 2022; Roy & Bailey, 2021; Roy et al., 2023). The realm of women-centric literature focusing on homeless women in Kolkata has been noticeably limited (Bagchi, 2016; Bhattacharya, 2022; Chakravarti, 2014, pp. 117–137; Haile et al., 2020; Klodawsky, 2009; Krishnadas et al., 2021; Narasimhan et al., 2020; Speak, 2005), often restricted to specific subjects like the livelihood challenges faced by these marginalized individuals. In this context, Chakravarti's work in 2014 stands out as a significant contribution (Chakravarti, 2014, pp. 117–137). Chakravarti's research shed light on the emergence of the urbanism movement, a social movement that drew attention to critical issues confronting the urban poor, including the hardships associated with livelihood struggles, shelter insecurity, exploitation, and the distressing specter of police brutality. Notably, the study underscored the heightened vulnerability of homeless women in these urban settings (Chakravarti, 2014, pp. 117–137). Similarly, Bagchi (2016) conducted a comprehensive study within the vibrant backdrop of Kolkata city, specifically delving into the lives of women waste pickers. Their findings offered a stark and unsettling glimpse into the harsh reality of these women waste pickers, portraying dire living conditions and alarmingly low incomes (Bagchi, 2016). Despite their efforts in the city, many of these hardworking women found themselves grappling with the formidable challenges of securing even a rented room or accessing government homeless shelters. Adding another layer to the multifaceted issue, Bhattacharya's research in 2022 spotlighted the distressing prevalence of sexual violence targeting homeless women. This violence posed a grave physical threat and inflicted significant harm on the mental well-being of these vulnerable individuals (Bhattacharya, 2022).

Nevertheless, a notable research gap exists in understanding the intricate pathways that lead women into homelessness within the dynamic and diverse landscape of Kolkata. It is within this void that the present chapter endeavors to make a meaningful contribution. By contextualizing the pathways to homelessness among women in Kolkata, this chapter aspires to examine the complex factors and circumstances resulting in women becoming homeless in Kolkata megacity. Such insights hold the potential to inform and shape future interventions and policies, ultimately working toward a more equitable and compassionate society that provides support and dignity to all its residents, regardless of their circumstances.

2. Methodology

2.1 Study area

The present study focuses on understanding the homelessness pathways among women in Kolkata. In this study, the term "homeless women" refers to women who do not have a structured house, as defined by the Indian census (Office of the Registrar General & Census Commissioner, 2011). These women live in various open spaces, such as roadsides, pavements, religious places, and unstructured tents. The study specifically selected Kolkata Municipal Corporation (KMC) for several reasons: Firstly, the KMC is India's second most homeless populous urban center, where nearly 50% are females. Secondly, it has the highest concentration of homeless women among urban centers in India. Finally, the homeless population in KMC has shown a positive growth trend in recent census data (Office of the Registrar General & Census Commissioner, 2011). The visual representation of the study area and its boundaries is presented in Fig. 11.1.

2.1.1 Study setting, sampling, and participants

This study employed a qualitative approach, adopting key informant interviews (KIIs) and In-depth Interviews (IDIs). The interviews were conducted in 26 KMC wards to ensure diverse and varied results (Fig. 11.1). The wards were selected based on a pilot field survey and observation, focusing on areas with a high concentration of homeless population—a total of 6 KIIs conducted to understand the overview of the study subjects. A purposive sampling technique was used to conduct 52 IDIs, with 2 IDIs from each selected ward (Fig. 11.1). Face-to-face IDIs were conducted to gather data from participants. The data analysis focused on the respondents' life experiences, emotions, perceptions, understanding, and expectations. The interpretation of each experience provided by the participants was also considered during the data analysis. For the IDIs, the inclusion criteria required women who had experienced at least 1 month of homelessness, were above 18 years of age, and were proficient in either Bengali or Hindi language. The KIIs involved individuals at least 25 years old and engaged in activities related to the welfare of homeless women, such as NGO workers and ward counselors.

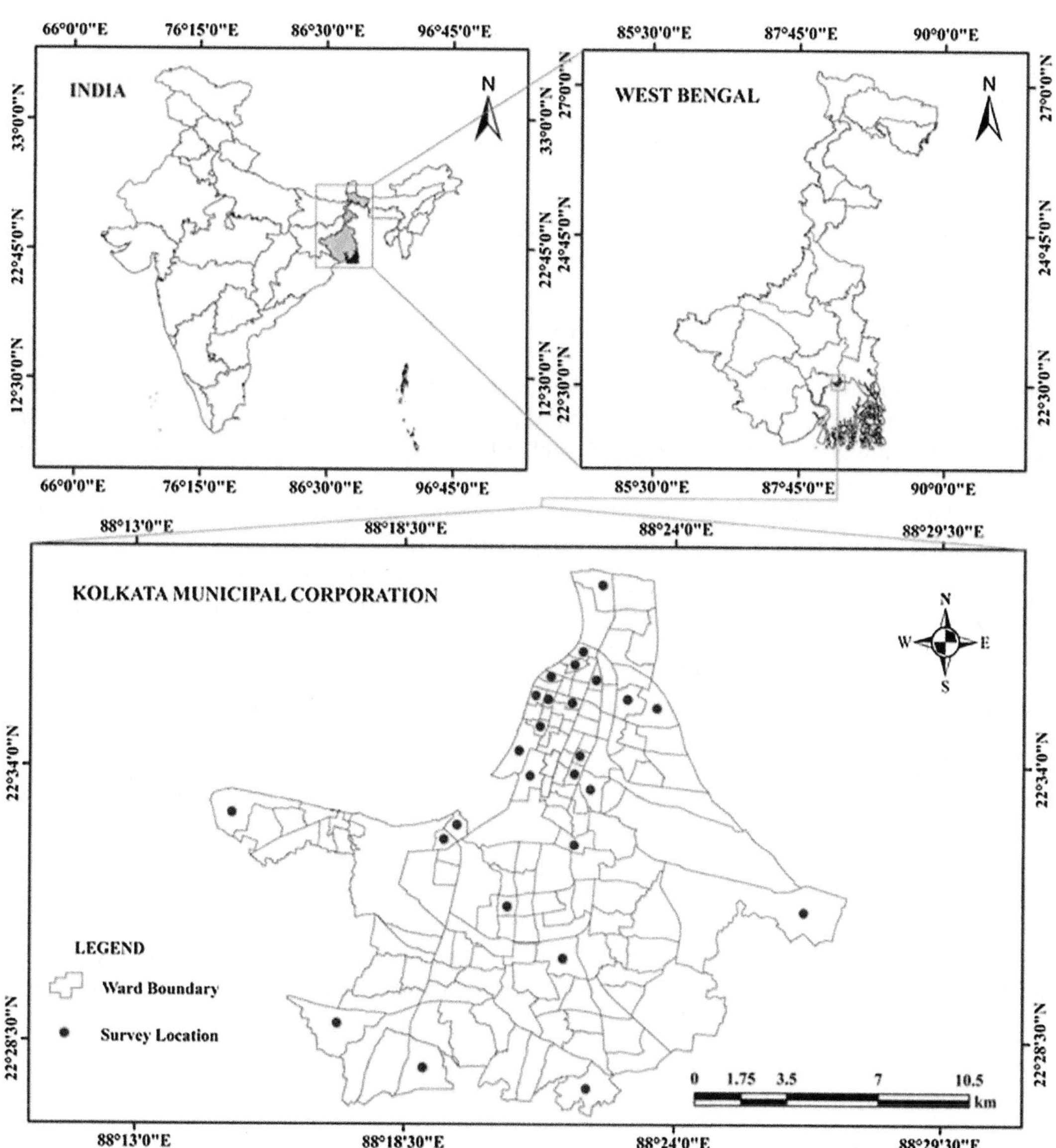

FIGURE 11.1 Location map of the study area.

2.2 Data collection procedure

Data collection involved conducting IDIs with participants until theoretical saturation was reached (Flick, 2022). Two individuals conducted the interviews: the first author and a female survey assistant. One of the interviewers possessed expertise in homeless surveys, while the other was a senior research fellow. To maintain consistency, an open-ended interview guide was used during the interviews (Flick, 2022). On average, the interviews lasted approximately 35–45 min. Digital audio recorders on smartphones were utilized to capture and preserve the interviews as audio recordings.

2.3 Data analysis

The qualitative data analysis software QDA Miner 5.0.30 was employed to analyze the collected data. Initially, the recorded audio was transcribed into text, written in the local languages of Bengali and Hindi. The text was carefully read line by line and then translated into English. Each respondent's data was compiled into individual Microsoft Word documents and subsequently imported into the software as separate "cases." Specific predetermined topics were included in the interview guide questions to explore particular areas of interest, while additional issues emerged from the data. Thematic analysis was adopted as the approach for data analysis (Flick, 2022). A meticulous line-by-line examination of each document was conducted to identify the main topics within the data. These topics were then assigned labels and entered into the software's "variables" section. The responses were coded and organized under the relevant topics or variables for further analysis. During the data analysis, themes were identified and grouped under the appropriate topic or "label." Similar themes were merged and emerged as new patterns. Thereafter, new insights were discovered. It is worth noting that while some themes may have overlapped for specific participants, each factor was separately counted to ensure a comprehensive list of possibilities (Flick, 2022). The flow chart of thematic analysis presented in Fig. 11.2.

2.3.1 Ethical considerations

The study received ethical approval from the institutional review board (Approval Number: IIPS/AC/MR/IO-889/2022) at the International Institute for Population Sciences. The researchers made several revisions to the questionnaire and research objectives based on rigorous screening. The study obtained an ethical certificate from the institute, reflecting the researchers' commitment to maintaining ethical standards (Flick, 2022). Informed consent was provided and read by the researcher before interviewing participants to uphold ethical guidelines during the survey. For illiterate participants, the documents were read aloud in the

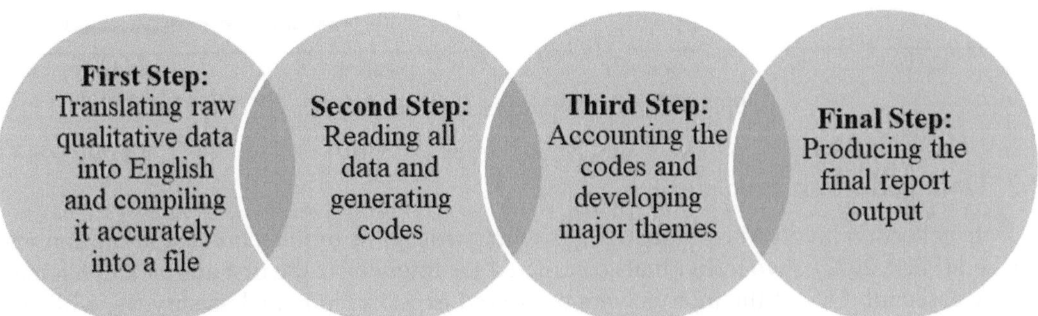

FIGURE 11.2 Flow chart of the thematic analysis used in the study.

presence of a witness. Interviews were conducted at the participants' convenience, and they could discontinue the interview at any time. To protect anonymity, participating homeless women were assured that their responses would remain confidential, and pseudonyms were used to maintain their privacy. The researchers created a comfortable and safe environment throughout the study to ensure the well-being of the participants.

3. Results

3.1 Background characteristics of the study participants

Most women were married and belonged to the age group of 25–35 years, the Muslim religion, and the backward classes (Table 11.1). They have severe socioeconomic deprivation, including high illiteracy rates and poor income levels. Most of them relied on low status occupations such as rag picking, construction labor, begging, and working as domestic maids to sustain themselves. Furthermore, chronic homelessness was a prevalent issue among them.

3.2 Predictors of homelessness

The predictors of homelessness are categorized into four main themes: interpersonal, environmental, economic, and structural factors.

3.2.1 Interpersonal factors

3.2.1.1 Marriage

The present study has identified marriage as a pivotal factor that can act as a doorway, leading individuals into or out of homelessness. Marriage is one of the most frequently reported predisposing factors. Some participants expressed that their household members arranged marriages with unskilled migrant laborers without considering their living conditions at the intended destination. The prevailing extreme poverty at the household level compelled family members to alleviate their burdens by arranging marriages for their daughters without thoroughly assessing the suitability of the prospective grooms. Consequently, after a short period following their marriages, these individuals were relocated to cities with their spouses, ultimately resulting in homelessness due to their partners' existing homeless status.

> At just 16, my dreams of a bright future collided with a harsh reality. As I entered married life, the promise of a spousal home quickly crumbled, replaced by the temporary shelter of a polythene tent near the railway tracks at Park Circus. Since that day, the grip of homelessness has never loosened. ***Based on the response of the IDI respondent, 24 years old.***

3.2.1.2 Family dissolution

The participants emphasized how the dissolution of their families played a crucial role in their journey toward homelessness. They shared deeply personal stories of marital discord, separation, or divorce, which ultimately led to their experience of homelessness. Their narratives revealed a profound emotional and financial upheaval caused by the breakdown of their family units. Feelings of abandonment and the loss of crucial support systems emerged as

TABLE 11.1 Background characteristics in-depth interviews (IDI) participants in the study, Kolkata municipal corporation, field survey, 2022–23.

Background characteristics (n = 52)	Mean	%
Age (mean)	25.7	
Currently married		69.7%
Years of schooling (mean)	3.2	
Religion		
Hindu		50.0
Muslim		42.3
Others		7.7
Social group		
No caste		23.1
Scheduled castes		26.9
Tribes		7.7
Backward classes		42.3
Average monthly income (mean)	₹8300 ($100.53)	
Occupation		
Rag picker		32.7
Home maid		25.0
Beggar		21.2
Construction labor		11.5
Others		9.6
Migration status		
Migrants		65.4
Nonmigrants		23.1

recurring themes, leaving these women vulnerable and without a stable foundation. The participants described how, following their marital dissolution, they were often perceived as unlucky and ominous by their maternal and married households, resulting in the loss of financial and livelihood support. As a result, homelessness became their inevitable destination, and they resorted to activities such as begging to survive.

> My husband's accidental death shattered my world. Overwhelmed by grief, I sank into deep despair. With him gone, the sole breadwinner, survival became a daunting challenge. Desperate, I journeyed to the city, seeking work. Yet, caring for my two children while earning a living proved impossible. Pushed to the brink, I

reluctantly resorted to begging. Stripped of pride, I joined people experiencing homelessness, finding solace among fellow beggars. ***Reported by the IDI respondent, 22 years old.***

3.2.1.3 Experience with spousal and domestic violence

Among the respondents, an essential factor contributing to their homelessness was the experience of severe spousal and domestic violence. These individuals shared traumatic stories of abuse and mistreatment within their households, ultimately forcing them to flee for safety. After leaving their homes, they embarked on a challenging journey, wandering through various places until they reached Kolkata. Due to financial constraints and a lack of support from others, they engaged in multiple types of work to make ends meet. These women connected with other homeless individuals who had undergone similar experiences during their struggles. Bonds were forged, and a community of homeless women emerged, providing support and companionship during their shared journey of homelessness. Together, they navigated life's difficulties on the streets, relying on each other for strength and solidarity.

> Trapped in a tormenting cycle, my husband's daily drunkenness brought forth relentless physical and emotional abuse. I endured his daily pushes and kicks, pleading for help from my marital and maternal families. Their response was a deafening silence, urging me to bear it all and adjust. But the pain became unbearable, and with no other choice, I fled, finding myself thrust into the world of homelessness. ***Based on the IDI respondent, 19 years old.***

3.2.1.4 Mental and physical abnormality and abuse

The interviews revealed a distressing pattern among respondents who were living alone. Many shared histories of mental or physical illness and recounted experiences of abuse and mistreatment within their households. These traumatic circumstances eventually drove them to flee their homes for safety. Once they left their homes, these individuals embarked on a challenging and nomadic journey, wandering through various places. Disconnected from their families and lacking stable support networks, they lived alone, often on footpaths or other makeshift locations. Engaging in begging became a standard means of survival, and they were frequently found seeking refuge in the premises of religious places like mosques, temples, and others. The KII also revealed that mental and physical illness is highly prevalent among homeless women. Additionally, many of these women had limited or no knowledge about their background history.

> I was mentally and physically weak during my teenage years, unlike a normal child, and I faced everyday abuse within my household, from neighbors, and even from friends. As a result, I decided to flee my house, and since then, I have been homeless. I entered into a marriage with someone, but unfortunately, he left me when I became pregnant. ***Responsed by IDI respondent, 16 years old.***

3.2.2 Environmental risk factors of homelessness

3.2.2.1 Climatic catastrophes

The narratives uncovered the stark reality of the complex interplay between natural disasters, climate change, and poverty, pushing these individuals to the brink of homelessness. The unrelenting impacts of climate change notably heightened the frequency of extreme climatic events, prominently floods, droughts, and storms, as recounted by our respondents. These extreme weather phenomena amplified the occurrence of crop failures, leading to a decline in the agricultural workforce at their places of origin and adversely affecting their means of livelihood. Consequently, small-scale farm owners and agricultural laborers embarked on journeys to urban areas in search of hope. Due to their lack of skills and limited educational attainment, the majority found themselves in the realm of informal, low-paying employment. Ultimately, the persistent state of poverty and meager income in urban settings became the decisive factor driving them into homelessness.

> Our lives were anchored in the tranquility of the Sundarban delta until the merciless Cyclonic Storm Amphan violently ripped through our world. Our once cherished home was mercilessly destroyed, leaving us displaced and shattered. In search of survival, we were forced to seek refuge in the sprawling city of Kolkata. But our story is not an isolated one. Countless families, like ours, find themselves uprooted and homeless in this metropolis, victims of the relentless onslaught of damaging floods, droughts, and storms. ***Based on the IDI respondent, 17 years old.***

3.2.2.2 Unregulated mining and land collapse

The study revealed a prominent linkage between land collapses and the loss of housing, leading to homelessness, particularly among Hindi-speaking respondents. These individuals primarily migrated from neighboring states such as Odisha and Jharkhand. According to their accounts, illegal mining activities, specifically coal, rock, and other minerals mining, were identified as significant factors contributing to land collapses. As a consequence of these collapses, their homes were engulfed, leaving them without a place to live and resulting in homelessness.

> Illegal mining tore through our peaceful tribal village, leaving destruction in its wake. Once sturdy and secure, our homes crumbled under the weight of greed. We were left displaced, stripped of our roots, and wandering as homeless souls. In search of survival, we decided to uproot ourselves and migrate to this unfamiliar city. ***From the IDI respondent, 25 years old.***

3.2.2.3 River bank erosion

The IDIs vividly portray the harsh reality those uprooted by riverbank erosion face—losing their homes and becoming homeless. This poignant narrative mainly unfolds in the districts of Malda and Murshidabad, which stand out as the epicenter of this devastating crisis. Respondents reveal that houses along the riverbanks were destroyed, wreaking havoc and disrupting their livelihoods. Faced with the desperate need for survival, homeless families migrated to Kolkata, seeking any means of sustenance through various unorganized jobs. However, their low incomes make it impossible to afford rent, leaving them with no choice but to endure life as people experiencing homelessness.

> Swallowed by the Ganga River's erosion, our village vanished, taking our homes and land with it. Seeking shelter and work, our family migrated to Kolkata. We are among many victims of this relentless force, forever changed by the unforgiving power of nature. ***Reproduced from the IDI respondent's story, 31 years old.***

3.2.3 *Economic aspects*

3.2.3.1 Occupational and economic deprivation

Many respondents were engaged in informal occupations without guaranteed daily work participation, resulting in nonpermanent job and a low income. As a result, the idea of residing in a rented flat remained a distant dream rather than a reality. However, some reported that they occasionally secured cheap rented rooms in slum areas, particularly during the rainy seasons, childbirth, and postnatal periods. In most cases, they preferred to reside alongside other homeless individuals under flyovers, on footpaths, or in market sheds. Clusters of homelessness predominantly emerged close to workplaces, such as nearby wholesale markets, railway tracks, platforms, dumping grounds, and religious sites. These locations offered some semblance of convenience and proximity to potential sources of income, albeit under highly challenging living conditions.

> Trapped in low wages and a scarcity of work, paying rent became impossible. Day after day, our meager earnings couldn't keep up with the soaring costs, and homelessness engulfed us like a relentless tide. ***Based on the IDI respondent, 23 years old.***

3.2.3.2 Minimizing living costs and maximizing remittances

Some respondents expressed that they had become accustomed to living as homeless to save on living costs and send more remittances to their native places. This unique factor can contribute to their experience of homelessness. To prioritize financial support for their left-behind families, they consciously decided to forgo standard arrangements for a transient lifestyle to send maximum remittances. It is essential to recognize that a complex web of economic and social circumstances shapes this factor of homelessness.

> As migrant workers, we sacrifice our shelter to save on living costs and send more remittances to our families left behind. It's a miserable existence, living as homeless souls, but the hope for a better life for our loved ones drives our selflessness. ***From the IDI respondent, 25 years old.***

3.2.4 *Structural elements*

3.2.4.1 Intergenerational homelessness

Some respondents responded that they were homeless because they were born in a lost community structure. Their families had been trapped in a cycle of homelessness that spanned multiple generations. These individuals described the profound impact of being born and raised in unstable living conditions with no permanent place to call home. The participants revealed the hardships and challenges they faced from a young age, growing up without the stability and security that a home provides. They described the constant struggle to find shelter, often relying on temporary accommodations, ceilings, or even living on the streets.

> Since birth, the burden of homelessness has been my constant companion in this unforgiving city. It is a burden that has haunted my family across generations, an intergenerational cycle perpetuating our destitution state. It is no mystery why I find myself trapped in the same plight, for this is the only life I have ever known. ***Modified from IDI respondent's opinion, 31 years old.***

3.2.4.2 Inadequate capacity and poor functionality of the homeless shelters

The current study revealed that some respondents had sought refuge in homeless shelters provided by government and nongovernmental organizations. However, their experiences shed light on the need for improvements in the functionality and operation of these shelters. Concerns were raised regarding the lack of privacy, overcrowding, and safety risks, including the potential for violence, which discouraged them from utilizing the shelter services. Additionally, respondents highlighted the limited capacity within the homeless shelters, making securing a spot for residence highly competitive. The narratives also pointed to corruption in the allocation of shelter seats. Furthermore, the considerable distance between the working sites and the shelters was a significant barrier to accessing shelter facilities. Consequently, despite experiencing homelessness, these individuals chose not to stay in shelters and continued to live without a stable place to call home.

> In my search for shelter, I turned to government night stay facilities, but I found a lack of privacy, overcrowding, and safety risks that deterred me from utilizing their services. The fear of violence loomed overhead, casting a shadow of unease. Moreover, the distance between the shelters and my workplace made the daily journey impractical. ***From the IDI respondent, 31 years old.***

3.2.4.3 Gender-exclusive work site shelters

A significant presence of construction laborers was observed in the study. They raised a pressing concern: while there were designated sheds for male workers at construction sites, no separate shelters were provided for the female laborers. This oversight by the construction employment authorities left female laborers needing appropriate accommodations. The women face safety risks when they reside alone without the company of relatives or family members in male-dominated sheds. Consequently, most female laborers are compelled to embrace homelessness, seeking solace in female homeless clusters or within family-oriented communities.

> In my workplace, the absence of separate resting and accommodation facilities for female labor compelled me to avoid staying among the male labor crowd. The lack of safe and gender-segregated spaces left me with no choice but to live as a homeless individual. ***Based on the IDI respondent, 18 years old.***

4. Discussions

This chapter illuminates the underlying factors contributing to the plight of homeless women in Kolkata using primary qualitative survey data and research methods. Kolkata, one of India's oldest and most renowned urban spaces, presents a stark duality. On one side, it is famous as the "City of Joy." On the other hand, it harbors one of the highest populations of homeless women (Office of the Registrar General & Census Commissioner, 2011). This contrasting reality highlights the complex nature of the city, where vibrancy and hardship coexist. Our findings uncover various factors, including individual circumstances, socioeconomic conditions, environmental influences, and structural challenges. It becomes abundantly clear that the pathway to women's homelessness is far from a simple, one-way road; instead, it emerges as a complex, multifaceted roadmap that calls for a systematic and scientific discussion. Most previous studies in India and elsewhere also suggested that the reasons for homelessness are multifaceted and vary with individuals, clusters, space, and culture (Bhattacharjee, 2023; Clapham, 2003; Goel & Chowdhary, 2017; Goel et al., 2017; Klodawsky, 2009; Singh et al., 2018).

Similar to previous studies conducted in India and other regions (Cross et al., 2010; Haile et al., 2020; Krishnadas et al., 2021), the present research delves into the intricate relationship between marriage, family dissolution, and domestic violence as contributory factors to homelessness. Through IDIs and insights from critical informants, our findings illuminate the crucial role that marriage plays in the lives of homeless individuals. While some find solace in transitioning from homelessness to a married home, others tragically face eviction from their maternal homes, thrusting them into the streets. The study in developing nations and Ethiopia underscores similar findings on family dynamics and homelessness (Haile et al., 2020; Speak, 2005). Similar to previous studies (Haile et al., 2020; Roy et al., 2023), the fragmentation of families and the pervasive occurrence of domestic violence contribute to a cycle of uncertainty and homelessness among women. This issue is further exacerbated by physical and mental illnesses, particularly among women (Gopikumar et al., 2015; Narasimhan et al., 2020). It was found that women with mental or physical diseases were more susceptible to experiencing gender-based violence and mistreatment. A previous study also portrayed similar findings, which found a strong nexus between mental illness, discrimination, and homelessness (Narasimhan et al., 2019). Consequently, women are plunged into multidimensional poverty, compelled to flee their homes to escape spousal or domestic violence (Narasimhan et al., 2019). Lamentably, women with limited resources and lower education levels are left with no alternative but to confront the harsh realities of homelessness. Although they may escape household violence, their lives on the streets expose them to many adversities, including physical and sexual violation, social discrimination, and profound isolation (Bhattacharya, 2022). The finding is a stern reminder of the low empowerment and limited rights afforded women within societies entrenched in patriarchal norms. Numerous research studies have consistently underscored how patriarchal norms constrict women's rights and choices (Bretherton, 2020). These norms create an environment where discrimination against women is pervasive, particularly after the dissolution of a marriage. Even within the confines of a marital union, the risk of spousal or domestic violence is alarmingly high, leading to severe consequences, including suicidal thoughts and actions and women fleeing their homes for safety and freedom (Haile et al., 2020). For many of these women who flee, the harsh reality of homelessness becomes their daily struggle. They are burdened not only by the absence of a stable place to live but also by the enduring mental trauma resulting from the abusive relationships they left behind (Krishnadas et al., 2021). It is a painful testament to the urgent need for societal transformation, where women are granted their rightful empowerment and rights and where more equitable and compassionate values replace patriarchal norms. Only through such changes can we uproot one crucial aspect of the cycle of homelessness, suffering, and despair among women.

In India, the country's climatic diversity and the periodic nature of the monsoon climate increase the risk of climatic catastrophes. Similar to previous studies (Rayhan & Grote, 2007), the current study documented homelessness resulting from displacement caused by river bank erosion, devastating floods, severe droughts, and cyclones. In particular, storms have been identified as significant risk factors that drive people into homelessness, which aligns with findings from a previous study (Mukhopadhyay, 2009, pp. 18–20). The narratives of individuals affected by these environmental forces paint a stark picture of their challenges. The destructive power of these extreme climatic events, intensified by the effects of climate change, is a recurring theme in the accounts provided by the respondents. The disruption of livelihoods caused by crop failures further compounds the hardships individuals face, especially those who rely on agriculture found in the study (Senapati & Das, 2022). Similarly, many previous studies also highlighted the issue of climate change and its impact on primary economic activities, resulting in migration for survival and becoming homeless, displaced, and rootless (Bezgrebelna et al., 2021). The climate change-driven extreme event increases the risk of displacement and homelessness, also displayed in the latest report published by the Our World in Data Organization (Ritchie et al., 2022). Consequently, many migrate to urban areas for survival (Bezgrebelna et al., 2021). However, chronic poverty and limited income often result in homelessness rather than offering a solution. The study also highlights the issue of unregulated mining activities, which contribute to environmental risks such as land collapses and subsequently lead to the loss of housing and homelessness. Illegal mining practices have been found to cause extensive damage, engulfing homes and leaving individuals without a place to live, pushing them toward homelessness. Similar to prior studies (Das & Samanta, 2022; Singha et al., 2020), river bank erosion is another weighty risk factor identified in the study. The destruction of homes along riverbanks disrupts lives and livelihoods, forcing families to immigrate to urban centers. Kolkata is a significant economic hub in eastern India, well connected to several climate-vulnerable areas like the cyclone-prone Sundarbans, drought-prone Bankura, and the flood-prone middle Ganga plain (Senapati & Das, 2022). Consequently, individuals migrate to Kolkata for survival after losing their livelihood due to natural extreme events. However, the low incomes of these individuals make it challenging for them to afford housing, leaving them with no choice but to endure life on the streets (Singha et al., 2020). As a result, this influx of climate refugees has significantly contributed to the growing homeless population in the city. The interplay of climate change, environmental factors, and economic disparities underscores the urgent need for comprehensive strategies to address homelessness and the broader challenges posed by climate-induced migration. It is also crucial to focus on climate resilience strategies for livelihood stability (Singha et al., 2020). Additionally, establishing support systems and providing affordable housing options are vital in assisting those affected by these environmental hazards to rebuild their lives and regain stability. Policymakers can effectively address the pressing issue of shelter provision for households affected by disasters by adopting a comprehensive approach that considers the site, situation, and standards outlined in the UNHCR emergency shelter handbook (United Nations High Commissioner for Refugees, 2020). By implementing this approach and ensuring the provision of emergency shelters and other essential basic needs, the flow of displaced individuals from disaster-affected areas to urban centers can be curtailed, thereby alleviating the scourge of homelessness.

The study reveals that occupational and economic deprivation, minimizing living costs and maximizing remittances, intergenerational homelessness, inadequate capacity and functionality of homeless shelters, and gender-exclusive work site shelters are underlined factors contributing to homelessness. Socioeconomic marginalization poses significant obstacles for individuals to secure permanent housing (Goel et al., 2017; Sharam & Hulse, 2014). Some individuals are compelled to live as homeless to save money and provide for their families. Intergenerational homelessness perpetuates the cycle of poverty, with individuals growing up without a stable home. There are a limited number of homeless shelters available in the city. While some respondents sought refuge in these shelters, their experiences shed light on several shortcomings. Similar to a previous study (Goel et al., 2017), the present study highlights privacy issues, safety risks, and limited capacity in the homeless shelters are common obstructions for not accessing the services. Female laborers face safety risks due to the absence of separate shelters, leading them to embrace homelessness. Addressing these factors requires interventions to improve employment opportunities, enhance shelter functionality, and provide gender-inclusive housing options, aiming to break the cycle of homelessness and improve the lives of marginalized communities.

Interestingly, the chapter's findings indicate that homelessness among women in the Indian urban context is not expressively attributed to substance abuse. This contrasts with the prevalent trend observed in Western societies, highlighting regional variations in the factors influencing women's homelessness (Cross et al., 2010). In the Indian context, women's homelessness in urban areas is primarily driven by rural–urban migration for marriage or employment alongside family members, mental illness, and socioeconomic and environmental distress. To address the issue of homelessness in urban settings, it is crucial to focus on creating gender-inclusive labor shelters within the workplace. Additionally, providing affordable housing options for migrant workers and their families is essential. Access to mental health treatment and support services for homeless individuals is also necessary. Special housing and employment schemes should be implemented in environmental hazards and disaster-prone to solve displacement, migration, and homelessness. Increasing awareness regarding mental health and promoting gender equality will also contribute to addressing the issues of mental health stigma, gender-based violence, human rights violations, and, indirectly, women's homelessness. By adopting a comprehensive and holistic approach that addresses the underlying causes and provides targeted interventions, significant progress can be made in tackling homelessness among women in urban areas. Finally, there is a need to follow the operational guidelines for the Scheme of SUH under the NULM to ensure the provision of basic amenities for people experiencing homelessness and guarantee their well-being (Government of India Ministry of Housing & Urban Poverty Alleviation, 2013). By adhering to these operational guidelines, policymakers can effectively address the plight of homeless individuals in urban areas.

5. Strengths and limitations

The chapter provides rich insights into the pathways of homelessness among women in urban settings and can explore many new perspectives on homelessness. However, the study has several limitations (Flick, 2022). Firstly, the findings derived from IDIs and KIIs may not represent Kolkata's entire homeless women population. Purposive sampling was used in the study, which could introduce selection bias and limit the generalizability of the results. Second, negative experiences due to fear of judgment, stigma, or retribution could affect the research findings. This can lead to underreporting certain factors contributing to homelessness or a skewed representation of their experiences. Third, participants may have difficulty recalling certain events or details accurately, mainly if they relate to traumatic experiences or if the interviews are conducted months or years after. This can affect the reliability and validity of the data collected. Finally, the study provides rich insights but may need more statistical significance. Based solely on these interview methods, it can be challenging to quantify and measure the prevalence and magnitude of factors contributing to homelessness among women in Kolkata. To mitigate these limitations, there is a need for a mixed-method research approach to provide a more comprehensive understanding of the factors contributing to homelessness among women.

6. Conclusion

In conclusion, this chapter highlights the complex factors contributing to women's homelessness in Kolkata. While marriage often serves as an initial trigger, it is not the sole cause. Factors such as marital dissolution, violations of women's rights, discrimination, and domestic violence also significantly push women into homelessness. The chapter also illuminates the impact of unemployment, economic challenges, intergenerational homelessness, limited shelter options, and workplaces that do not accommodate women, all of which compound their vulnerability and make it more difficult for them to escape homelessness. To address these multifaceted challenges, comprehensive solutions are imperative. Firstly, there is a need for increased public awareness regarding legal avenues for protecting women's rights and addressing gender-based discrimination. Additionally, the government must focus on creating opportunities for economic empowerment through skill development and facilitating access to financial resources like microcredit. Ensuring the availability of gender-sensitive shelter options and workplaces is crucial for preserving women's dignity and security. Ultimately, the overarching goal should be to build a society free from violence, discrimination, and environmental risks. Achieving this necessitates a combination of legal reforms, economic empowerment initiatives, and robust support systems that target the root causes of homelessness. The Government of India has already implemented initiatives such as the Jawaharlal Nehru National Urban Renewal Mission and Pradhan Mantri Awas Yojana (Urban), which include provisions for providing affordable housing for all. Therefore, there is a need to prioritize and enhance these ongoing programs to create urban environments free from homelessness. Furthermore, it is crucial to acknowledge that women are particularly vulnerable in homeless situations. Hence, there is a pressing need for gender-inclusive shelter facilities that address the unique challenges homeless women face. By focusing on these policies and interventions, Kolkata has the potential to achieve a homelessness-free environment for women, allowing them to live with dignity and security.

References

Allen, W., Anderson, B., Van Hear, N., Sumption, M., Düvell, F., Hough, J., Rose, L., Humphris, R., & Walker, S. (2017). Who counts in crises? the new geopolitics of international migration and refugee governance. *Geopolitics, 23*(1), 217–243. https://doi.org/10.1080/14650045.2017.1327740

Bagchi, D. (2016). Street dwelling and city space: Women waste pickers in Kolkata. *Economic and Political Weekly, 51*(26–27), 63–68. http://www.epw.in/system/files/pdf/2016_51/26-27/Street_Dwelling_and_City_Space_0.pdf.

Barile, J. P., Pruitt, A. S., & Parker, J. L. (2018). A latent class analysis of self-identified reasons for experiencing homelessness: Opportunities for prevention. *Journal of Community and Applied Social Psychology, 28*(2), 94–107. https://doi.org/10.1002/casp.2343

Batterham, D., Cigdem-Bayram, M., Parkinson, S., Reynolds, M., & Wood, G. (2022). The spatial dynamics of homelessness in Australia: Urbanisation, intra-city dynamics and affordable housing. *Applied Spatial Analysis and Policy, 15*(4), 1021–1043. https://doi.org/10.1007/s12061-022-09435-5

Bezgrebelna, M., McKenzie, K., Wells, S., Ravindran, A., Kral, M., Christensen, J., Stergiopoulos, V., Gaetz, S., & Kidd, S. A. (2021). Climate change, weather, housing precarity, and homelessness: A systematic review of reviews. *International Journal of Environmental Research and Public Health, 18*(11). https://doi.org/10.3390/ijerph18115812

Bhattacharjee, S. (2023). The homeless of urban India under COVID-19 lockdown: Rethinking their rights and the role of public policy. *Asian Social Work and Policy Review, 17*(3). https://doi.org/10.1111/aswp.12285

Bhattacharya, P. (2022). "Nowhere to sleep safe": Impact of sexual violence on homeless women in India. *Journal of Psychosexual Health, 4*(4), 223–226. https://doi.org/10.1177/26318318221108521

Bretherton, J. (2020). Women's experiences of homelessness: A longitudinal study. *Social Policy and Society, 19*(2), 255–270. https://doi.org/10.1017/S1474746419000423

Brown, M., Mihelicova, M., Collins, K., Cummings, C., & Ponce, A. (2020). Housing options for individuals experiencing chronic homelessness: Subsidized and non-subsidized housing outcomes among pathways to independence participants. *Journal of Social Service Research, 46*(5), 642–657. https://doi.org/10.1080/01488376.2019.1612822

Cebula, R. J., & Alexander, G. M. (2020). Economic and noneconomic factors influencing geographic differentials in homelessness: An exploratory state-level analysis. *The American Journal of Economics and Sociology, 79*(2), 511–540. https://doi.org/10.1111/ajes.12320

Chakravarti, P. (2014). *Living on the edge: Mapping homeless women's mobilization in Kolkata, India.* Springer Science and Business Media LLC. https://doi.org/10.1057/9781137390578_8

Clapham, D. (2003). Pathways approaches to homelessness research. *Journal of Community and Applied Social Psychology, 13*(2), 119–127. https://doi.org/10.1002/casp.717

Cowan, C. D., Breakey, W. R., & Fischer, P. J. (1988). The methodology of counting the homeless. In *Homelessness, health, and human needs* (pp. 12–20). http://www.asasrms.org/Proceedings/papers/1986_029.pdf.

Cross, C., Seager, J., Erasmus, J., Ward, C., & O'Donovan, M. (2010). Skeletons at the feast: A review of street homelessness in South Africa and other world regions. *Development Southern Africa, 27*(1), 5–20. https://doi.org/10.1080/03768350903519291

Das, R., & Samanta, G. (2022). Impact of floods and river-bank erosion on the riverine people in Manikchak block of Malda district, West Bengal. *Environment, Development and Sustainability, 25*. https://doi.org/10.1007/s10668-022-02648-1

Flick, U. (2022). *An introduction to qualitative research.* sage.

Goel, G., Ghosh, P., Ojha, M. K., & Shukla, A. (2017). Urban homeless shelters in India: Miseries untold and promises unmet. *Cities, 71*, 88–96. https://doi.org/10.1016/j.cities.2017.07.006

Goel, K., & Chowdhary, R. (2017). Living homeless in urban India: State and societal responses. In *Faces of homelessness in the Asia pacific* (pp. 47–63). Australia: Taylor and Francis. https://doi.org/10.4324/9781315475257

Gopikumar, V., Narasimhan, L., Easwaran, K., Bunders, J., & Parasuraman, S. (2015). Persistent, complex and unresolved issues: Indian discourse on mental ill health and homelessness. *Economic and Political Weekly, 50*(11), 42–51. http://www.epw.in/system/files/pdf/2015_50/11/Persistent_Complex_and_Unresolved_Issues.pdf.

Government of India Ministry of Housing and Urban Poverty Alleviation. (2013). *National urban livelihoods mission and scheme of shelters for urban homeless (operational guidelines).* New Delhi: Ministry of Housing & Urban Poverty Alleviation. Government of India. Unpublished content https://mohua.gov.in/upload/uploadfiles/files/7NULM-SUH-Guidelines.pdf.

Haile, K., Umer, H., Fanta, T., Birhanu, A., Fejo, E., Tilahun, Y., Derajew, H., Tadesse, A., Zienawi, G., Chaka, A., Damene, W., & Withers, M. H. (2020). Pathways through homelessness among women in addis ababa, Ethiopia: A qualitative study. *PLoS One, 15*(9), e0238571. https://doi.org/10.1371/journal.pone.0238571

Internal Displacement Monitoring Centre. (2023). *Global internal displacement database.* Geneva: Internal Displacement Monitoring Centre (IDMC), Humanitarian Hub Office. https://www.internal-displacement.org/database/displacement-data. (Accessed 3 April 2023).

Kidd, S. A., & Davidson, L. (2007). You have to adapt because you have no other choice: The stories of strength and resilience of 200 homeless youth in New York city and Toronto. *Journal of Community Psychology, 35*(2), 219–238. https://doi.org/10.1002/jcop.20144

Klodawsky, F. (2009). Home spaces and rights to the city: Thinking social justice for chronically homeless women. *Urban Geography, 30*(6), 591–610. https://doi.org/10.2747/0272-3638.30.6.591

Krishnadas, P., Narasimhan, L., Joseph, T., Bunders, J., & Regeer, B. (2021). Factors associated with homelessness among women: A cross-sectional survey of outpatient mental health service users at the banyan, India. *Journal of Public Health, 43*(Suppl. 2), ii17–ii25. https://doi.org/10.1093/pubmed/fdab219

Mukhopadhyay, A. (2009). *Cyclone Aila and the Sundarbans: An enquiry into the disaster and politics of aid and relief.*

Narasimhan, L., Gopikumar, V., Jayakumar, V., Bunders, J., & Regeer, B. (2019). Responsive mental health systems to address the poverty, homelessness and mental illness nexus: The Banyan experience from India. *International Journal of Mental Health Systems, 13*(1). https://doi.org/10.1186/s13033-019-0313-8

Narasimhan, L., Kishore Kumar, K. V., Regeer, B., & Gopikumar, V. (2020). Homelessness and women living with mental health issues: Lessons from the banyan's experience in Chennai, Tamil Nadu. In *Gender and mental health: Combining theory and practice* (pp. 173–191). India: Springer Singapore. https://doi.org/10.1007/978-981-15-5393-6_12

Office of the Registrar General and Census Commissioner. (2011). *Primary census abstract data for houseless, India & States/UTs - district Level – 2011.* New Delhi: PCA HS. https://censusindia.gov.in/nada/index.php/catalog/5047.

Rayhan, I., & Grote, U. (2007). Coping with floods: Does rural-urban migration play any role for survival in rural Bangladesh. *Journal of Identity and Migration Studies, 1*(2), 82–98.

Ritchie, Rosado, P., & Roser, M. (2022). *Natural disasters.* http://'https://ourworldindata.org/natural-disasters. (Accessed 15 May 2023).

Roy, A., Mandal, M. H., Sahoo, K. P., & Siddique, G. (2023). Investigating the relationship between begging and homelessness: Experiences from the street beggars of Kolkata, West Bengal, India. In *Human Arenas.* India: Springer Science and Business Media B.V.. https://doi.org/10.1007/s42087-023-00330-0

Roy, A., & Siddique, G. (2018). Pavement dwellers of Kolkata: A geographical appraisal. *Researchers World - Journal of Arts Science and Commerce, 9*(1), 36. https://doi.org/10.18843/rwjasc/v9i1/05

Roy, S., & Bailey, A. (2021). Safe in the city? Negotiating safety, public space and the male gaze in Kolkata, India. *Cities, 117*, 103321. https://doi.org/10.1016/j.cities.2021.103321

Sahoo, H., & Jeermison, R. (2018). Houseless population in India: Trends, patterns and characteristics. *Population Geography, 40*(1&2), 43–52.

Senapati, U., & Das, T. K. (2022). Geospatial assessment of agricultural drought vulnerability using integrated three-dimensional model in the upper Dwarakeshwar river basin in West Bengal, India. In *Environmental Science and Pollution Research.* India: Springer Science and Business Media Deutschland GmbH. https://doi.org/10.1007/s11356-022-23663-9

Sharam, A., & Hulse, K. (2014). Understanding the nexus between poverty and homelessness: Relational poverty analysis of families experiencing homelessness in Australia. *Housing, Theory and Society, 31*(3), 294–309. https://doi.org/10.1080/14036096.2014.882405

Singh, N., Koiri, P., & Shukla, S. K. (2018). Signposting invisibles. *Chinese Sociological Dialogue, 3*(3), 179–196. https://doi.org/10.1177/2397200918763087

Singha, P., Das, P., Talukdar, S., & Pal, S. (2020). Modeling livelihood vulnerability in erosion and flooding induced river island in Ganges riparian corridor, India. *Ecological Indicators, 119*, 106825. https://doi.org/10.1016/j.ecolind.2020.106825

Speak, S. (2005). Relationship between children's homelessness in developing countries and the failure of women's rights legislation. *Housing, Theory and Society, 22*(3), 129–146. https://doi.org/10.1080/14036090510034581

Toro, P. A. (2007). Toward an international understanding of homelessness. *Journal of Social Issues, 63*(3), 461–481. https://doi.org/10.1111/j.1540-4560.2007.00519.x

United Nations High Commissioner for Refugees. (2020). *Emergency handbook of shelter solutions.* Switzerland: United Nations, Geneva. https://emergency.unhcr.org/emergency-assistance/shelter-camp-and-settlement/shelter/shelter-solutions.

World Population Review. (2023). *Homelessness by country 2023.* United States: Walnut. https://worldpopulationreview.com/country-rankings/homelessness-by-country. (Accessed 9 March 2023).

CHAPTER

12

Distress in ecological drivers and its impact on homelessness: Case study of Sagar Island

Avijit Mondal[1,2], *Trisha Barui*[1,3] *and Susmita Paul*[1,4]

[1]Council of Architecture (COA), Institute of Town Planners, India (ITPI), Kolkata, West Bengal, India; [2]Indian Institute of Technology Kharagpur, Government of India, Kolkata, West Bengal, India; [3]Indian Institute of Engineering Science and Technology, Shibpur, Government of India, Kolkata, West Bengal, India; [4]Architecture and Planning Department, Indian Institute of Technology Roorkee, Government of India, Kolkata, West Bengal, India

1. Introduction

The Sundarban Delta, a distinctive ecological system characterized by its extensive mangrove forests, is subject to various crucial ecological drivers. These drivers encompass the tidal cycle, precipitation patterns, sediment deposition and salinity levels. These factors play a pivotal role in delineating the distribution of different mangrove species and the richness and diversity of associated flora and fauna, as well as influencing the well-being of the local inhabitants. The region has historically experienced challenges, such as tropical cyclones, sea level rise, and fluctuations in rainfall. However, in recent decades, these challenges have been exacerbated by human-induced factors such as deforestation, pollution, and the impacts of climate change. Consequently, the degradation of mangroves, land loss due to erosion, homelessness, and subsequent outmigration have become prevalent phenomena in the area. This chapter seeks to examine the ecological drivers prevalent in the Sundarban Delta, with a specific focus on Sagar Island as the study area. It aims to investigate the impact of these drivers on three distinct forms of homelessness: transitional, episodic, and chronic, all stemming from the adverse consequences of tropical cyclones, rising sea levels, and soil erosion. The research methodology will employ a combination of diverse sources, including secondary data analysis, primary surveys, and focus group discussions with the locals to achieve comprehensive insights into the subject matter.

Homelessness to Hope
https://doi.org/10.1016/B978-0-443-14052-5.00012-4

Throughout history, homelessness has been perceived as a state affecting specific marginalized groups within society. These groups are often associated with deviant behaviors, distinct lifestyle preferences, and subcultural adaptations, which contribute to a persistent sense of disaffiliation. In an effort to comprehend the evolving and diverse aspects of homelessness, a model has been formulated to categorize case profiles into chronic, episodic, and transitional patterns (Kuhn & Culhane, 1998). The transitionally homeless population refers to individuals who typically enter the shelter system for a single episode and for a relatively short duration. They are more likely recent additions to the precariously housed population who have become homeless due to unforeseen catastrophic events such as unemployment, separation, the death of the primary householder, utility disconnection, or fire. It is plausible that they have already exhausted the option of seeking temporary shelter with friends or relatives (Weitzman et al., 1990). On the other hand, the episodically homeless population comprises individuals who frequently oscillate between periods of homelessness and other forms of housing (Farr et al., 1986). The chronically homeless population embodies individuals who closely align with the stereotypical profile of long-term homelessness. Members of this subgroup are extensively engaged with the shelter system, and for them, shelters may resemble more of a long-term housing arrangement rather than a temporary emergency solution (Cohen & Sokolovsky, 1989; Piliavin et al., 1993; Rossi, 1989). The Sundarban Delta is well-known for its diverse biodiversity and unique environment. However, the area has recently seen severe ecological suffering as a result of climate change and industrial activity. The lives of those who live in the Sundarban Delta have been significantly impacted by this change in ecological factors, particularly in terms of the increased rate of homelessness. Numerous ecological factors have an impact on the Sundarban Delta, a distinctive mangrove habitat. The ecological services provided by the Sundarbans are significantly impacted by these forces, both natural and man-made.

One of the major biological forces affected by the Sundarban Delta's distress is mangrove forests. These trees act as a barrier against coastal erosion and serve as a haven for a wide variety of species and flora. On the other side, the region's mangroves have suffered serious harm as a result of pollution, deforestation, and rising sea levels, which has led to an increase in coastal flooding and the loss of wildlife habitat and also contributes to homelessness. Cyclone is the main reason behind the loss of Mangrove cover. Over last few decades, higher intensity cyclone occurred 7–12 years interval, affects 20% or more the Mangroves in Sundarban. Recently Cyclone Amphan had a detected area around 550.1 sq.km and caused damage of 11.5% of existing mangrove vegetation (Sharma et al., 2022). The deterioration of ecosystem services brought on by changes in the biological processes is the main factor causing homelessness in the Sundarban Delta. A reduction in agricultural production caused by soil salinization, riverbank erosion and the destruction of natural ecosystems has forced many islands to give up their traditional ways of life. This problem has been made worse by the cyclonic storm surges' introduction of salt water into agricultural fields (Hajra & Rajarshi, 2023). A single cyclonic storm's surge can cause at least 2–3 years of total crop failure and two more years of low yielding due to the high salt; in the meantime, it can destroy the inland fisheries' infrastructure (Mistri & Das, n.d.). Homelessness has increased as a result of a lack of other sources of income and the poorer segments of the population's incapacity to adjust to these changes (Hajra & Rajarshi, 2023). If we see last 15 years, a total of 13 tropical cyclone emerged from Bay of Bengal and out of which 3 has landfall over Indian part of Sundarban delta. In due course, total number of casualties happed due to the cyclone was 4876 and total amount of damage was created around 31,598.3 million USD damage due to Cyclone Sitrang was not considered (Ghosh & Mistri, 2023).

The aforementioned causes have caused unfavorable changes in the majority of the Sundarbans ecosystem services over the past 20 years. Sea levels have risen, rainfall patterns have changed, and the frequency and severity of extreme weather events have all increased as a result of climate change. The Sundarbans have been greatly impacted by these changes, which have resulted in soil salinization, riverbank erosion and the destruction of natural ecosystems (Islam & Atiqur, 2017; Hajra & Rajarshi, 2023). The very existence of Sundarban is in danger because of rising water level. India has already lost 60% of its Sundarban delta. Indian portion of Sundarban comprises around 1330 sq.km of tiger reserve. A total of approx 3000 sq.km of remains other part of land out of which 2000 sq.km is habitable where lives 4.4 million population as per 2011 census. In past 20 years, four Island have sunk, including Lohachara Island, the first populated island in the world which was drowned due to rising water level. Ghoramara is another Island which is on the brink of the same fate as those four Island mentioned before (Mondal et al., 2022). Indian Sundarban accommodates around 4.4 million people (Banerjee et al., 2023). A large section of this population is associated with primary activity. Agriculture cultivation and horticulture plantations are the main drivers of growth in the Sundarban. Aquaculture and inland and offshore fishing both also contribute to their growth along with tourism and pilgrimage-related activities (Hajra & Rajarshi, 2023; Islam & Atiqur, 2017). In terms of income for fisherman, 70%–75% of the income comes from estuarine fishing. people associated with tourism activity around 60% of their income comes from tourism activities. In terms of agriculture, 60% of the farmers are dependent upon other social welfare schemes like the Krishak Bandhu, PM Kisan Yojana etc. Other Non-Timber Forest Products (NTFPs) like honey wax collection related activity are seasonal and happens for certain period of months, April and May mainly (Mondal et al., 2022). Homelessness in the Sundarban Delta is greatly and variably influenced by ecological distress. Many people have lost their houses and means of support due to environmental factors, which has increased the number of homeless people. The most vulnerable members of society, the poorer groups, have been hardest hit. Food insecurity has increased as a result of changes in work patterns and the deterioration of ecological services. The sectors of the population with lower incomes are the most vulnerable. Many islanders have been forced to relocate due to the damage cyclonic storm surges have caused, which resulted in salty water incursion into agricultural areas. The failure to adapt to these changes and the unavailability of other means of support have led to an upsurge in homelessness and eventually it becomes out migration. While describing migration pattern, it is observed that 60% of total sample didn't want to migrate at all, 20% migrated in Kolkata, 15% outside West Bengal and rest 5% migrated within the Sundarbans (Jamal et al., 2022).

2. Identification of study area

Sagar Island serves as an ideal study area for investigating the impact of ecological drivers, such as cyclones, soil erosion and rising water levels. These drivers are abundantly present in the vicinity of the island, making it a compelling location for research. Focusing on the tropical cyclones in the Bay of Bengal between 2005 and 2022, there have been a total of seven cyclones that made landfall along the Indian coast. Notably, three of these cyclones, namely Aila, Bulbul, and Amphan, specifically impacted Sagar Island and its surrounding areas (Ghosh & Mistri, 2023). One significant concern for Sagar Island is its susceptibility to soil erosion. This issue has been studied over the years and more recently, various ecological factors have exacerbated the erosion problem manifold. Originally, Sagar Island was a reclaimed land, with reclamation efforts initiated during the colonial government era in 1770 (Pargiter, 1934). The reclamation process continued until the 1980s (Ascoli, n.d.) in the western South 24 Parganas district. Currently, the island boasts an approximate perimeter of 71 km, primarily protected by revetment type concrete and earthen embankments. The only beach on Sagar Island is Gangasagar. Mangroves are found in the north-western and south-eastern regions of the island. Notably, significant shoreline changes have been observed along the West Bengal coast, particularly in specific areas of Sagar Island. The eastern and southern stretches, especially around Sibpur (Boatkhali), have experienced predominant erosion, while accretion has been observed at the mouth of Sikarpur creek (Shoreline Management Plan for the Coastal Areas of Sagar Island, 2019). Also, the detrimental impact of cyclonic storm surges, leading to saline water intrusion into agricultural land, thereby compelling a substantial number of islanders to initiate relocation measures.

2.1 Location

The Sagar Island is located at the south-west boundary of Gangetic Delta, extending latitudinally from 21 degrees 37′40″ N to 21 degrees 55′ 20″ N and longitudinally from 88 degrees 02′45″ E to 88 degrees 10′30″ E. The area comes under Sagar Community Development Block (CD Block), situated in Kakdwip Subdivision, South 24 Parganas district, West-Bengal, India. The Island has Muriganga River on its East, River Hooghly on West and Bay of Bengal on the South. Sagardwip (Sagar Island) has eight Gram Panchayats (GPs) which are divided into 42 Mouzas (Fig. 12.1). Total population of Sagar Island is 206,865, the total number of households is 47,020 and it has 277.09 sq.km. of land area.

2.2 Baseline details of Sagar Island

Sagar Island is situated in the southernmost part of the South 24 Pargana District of West Bengal, India. Brief details about the Island are given in Table 12.1 below.

2.3 Connectivity

In terms of connectivity from the mainland, Sagardwip can be accessible from Namkhana or Kakdwip side (East) (Fig. 12.2). This route happens to be the most popular one. We can reach Namkhana or Kakdwip through bus services which are available from Esplanade, Kolkata. Train service is also there from Sealdah to Namkhana. From the west, the Island can be accessed from Rasulpur, East Midnapur. It also has ferry service from Haldia. In terms of connectivity, the island is dependent upon ferry service.

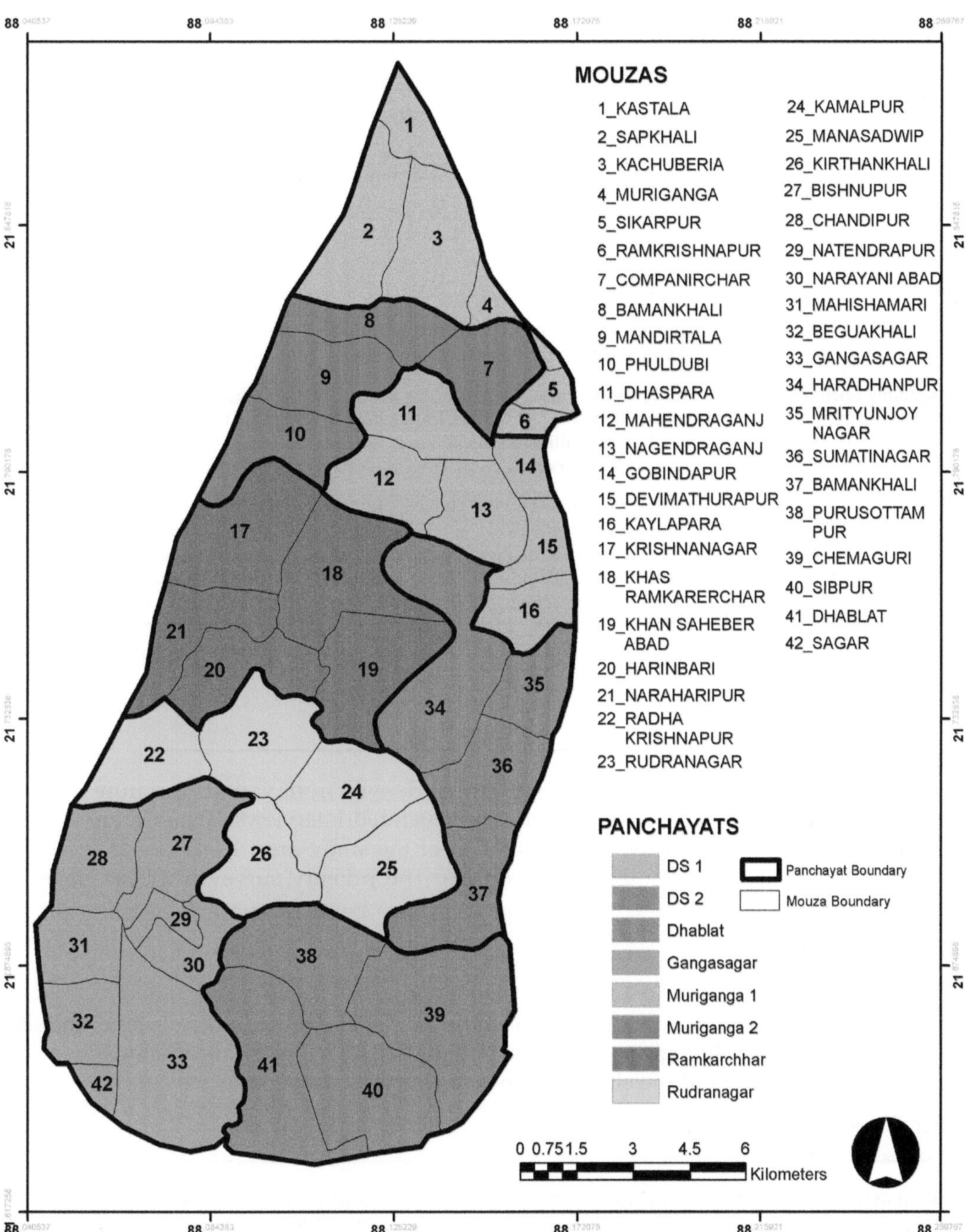

FIGURE 12.1 Map of Sagar Island with name of all villages and Panchayats.

TABLE 12.1 Basic details about Sagar Island.

Component	Brief
Location	Sagar Island, Sagar Block, Kakdwip Sub-division, South 24 Parganas district, Sundarbans, West Bengal, India.
Geography	Muriganga River on its East, River Hooghly on West and Bay of Bengal on South.
Distance from mainland	100 km south of Kolkata. 3.5 km from embarking points, Namkhana and Harwood.
Mode of travel	Ferry service.
Area	277.09 sq.km.
Topography	Highland areas, medium and lowland areas. Some lowland areas are 3 meters below sea level. Many creeks and channels; clayey silt soil with much vegetation.
Demography (including social demography)	207,904 number of population, 47,020 number of households, overall literacy—84.21%+ and women literacy—77.39%. Total worker is 40.03% of the total population out of which 24.46% are cultivators, 43.72% are cultivators, 4.18% are household workers, and 27.63% are other workers.
Livelihoods	Agriculture—Paddy, betel vine, chillies, sunflower, vegetables are major crops. Fisheries—Marine and freshwater—50% of households participate in pisciculture; labor—construction, agriculture, wage labor in mangroves and casuarinas. Tourism-based and other livelihoods—transport, rickshaw vans, beedi rolling, sea-shell handicraft, coaching centers, small trades, migration for construction works.

Author.

3. Methodology

Initial information about the study area has been derived from various literature review. Later, primary data were collected through semistructured interviews. The primary survey gives first-hand awareness of the actual situation of the study area. It also helps to verify data which are collected from secondary sources. The primary survey has been done in two segments.

1. Reconnaissance survey
2. Interviews

Reconnaissance survey is a kind of visual study for identifying various features of the site and helps to formulate strategies for field survey. It gives an overview of the study area. To get detail information about the site, two types of interviews are conducted:

1. Semistructured interview
2. Focused interview

Semistructured interviews contain a mix of open-ended and closed-ended questions and most of those questions are quantities in nature. Focused interviews are done to get an expert

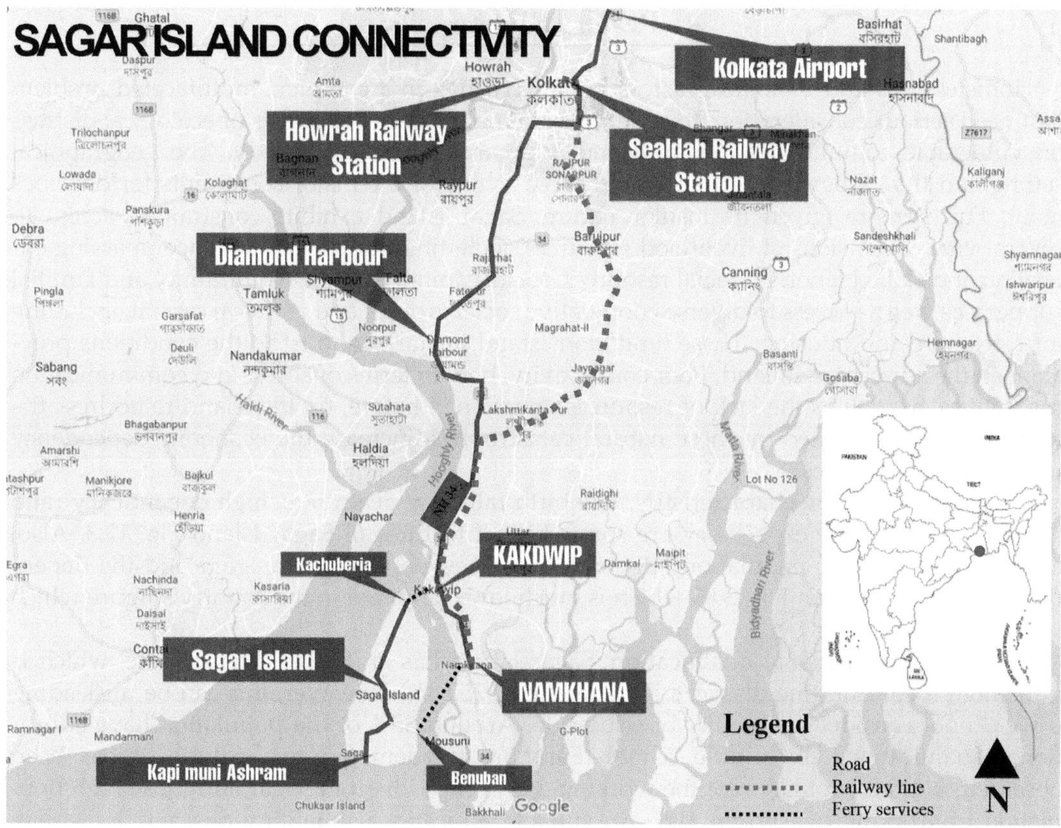

FIGURE 12.2 Connectivity map of Sagar Island connectivity through Road, Rail, and ferry has been shown on the map.

opinion about the current situation of physical infrastructure services and their thoughts of how the situation can be improved. No separate questionnaires are prepared for this kind of interviews, and the questions are all open-ended. There are 8 Panchayats and 42 Villages in Sagar Island and primary survey has been conducted in all eight of those Panchayats. The name of those Panchayats are as follows:

1. Daspara Sumatinagar – 1
2. Daspara Sumatinagar – 2
3. Dhablat
4. Gangasagar
5. Muriganga – 1
6. Muriganga – 2
7. Ramkarchar
8. Rudranagar

4. Result and discussion

Homelessness and ecological factors are intertwined in a complex, multifaceted problem that need serious consideration. In order to examine the negative effects of ecological distress on vulnerable communities, Sagar Island is an instructive case study. The geographical setting and the topography of the region serve as one of the major constraints for development. Furthermore, given its insular nature, Sagar Island exhibits constrained access to several vital dimensions of livelihood resources, including human capital, encompassing education and skilled labor; financial resources; social capital, entailing community and familial support, as well as access to diverse cooperative social groups; and physical capital, including infrastructure and housing. These limitations stand in stark contrast to the conditions prevalent on the adjacent mainland. Poor connectivity, both in terms of access and communication capabilities, as well as the lack of resources, mentioned before, on the island to address the harsh conditions created by these natural catastrophes, increase the economic dependency upon natural resources.

If we see population characteristics, low birth rate, low death rate, high dependency ratio along with longer life expectancy are some distinct features of Sagar Island Fig. 12.3. Also, around 45.72% of the population falls under working population category and the dependency ratio for the Island is 33.83 which is much low compared to the country's figure which is 59.4.

In terms of educational qualification, Sagar Island has an 84.21% literacy rate, which is quite high if consider the district's literacy rate of 75.68%. However, this can be misleading if we do not consider the quality of education. Around 85% of the population have educational qualifications till middle school. Early in the demography analysis, it has been observed that the population of the working-age group is high in Sagar and those high population of working age group lacks in suitable educational qualifications and is heavily dependent upon primary resource of the Island.

Sagar Island's local economy predominantly relies on primary activities, with agriculture being the primary occupation for its residents. Tourism also plays a role in contributing to the economy, involving various sectors such as hospitality, transportation, and cottage industry.

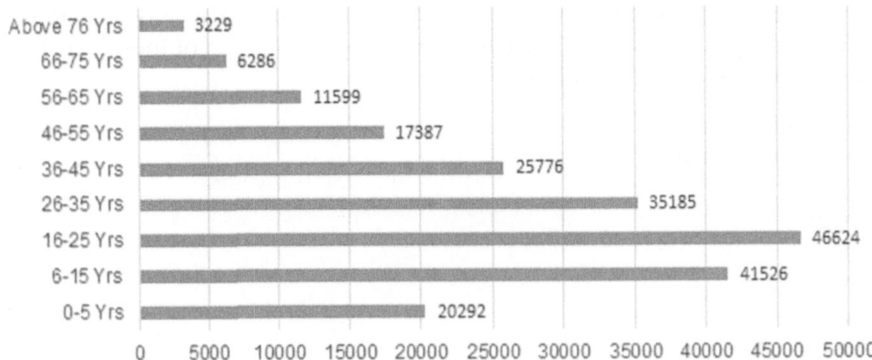

FIGURE 12.3 Distribution of population by age. The figure shows the population based on their age group. It shows dependency ratio on the island is low compared to the country.

With proper infrastructure development, tourism has the potential to create alternative job opportunities for islanders, albeit in a limited and location-specific manner. Due to the dependence on primary activities and natural resources, coupled with limited educational qualifications, most islanders prefer to remain on the island rather than seeking employment elsewhere. This preference for staying on the island exposes them to an increased risk of experiencing episodic and transitional homelessness. Through interactions with the local population, it has been observed that the majority of islanders view the island as their preferred workplace, further reinforcing their vulnerability to homelessness patterns characterized by intermittent and transitional episodes.

Fishing and honey collection are other primary livelihoods in the Sundarban area, with honey harvesting occurring from March to June. During this period, honey collectors experience episodic homelessness, relying on small boats (Fig. 12.4) as their sole shelter and means of transport for weeks. Tragically, cyclones often coincide with the harvesting season, prompting honey collectors to take greater risks by venturing deeper into the forest or treacherous waters, leading to potential loss of life. Fishing remains another primary economic activity for these communities, necessitating frequent excursions into the vast riverine ecosystem of the Sundarbans. Consequently, they too are exposed to the dangers of cyclones while navigating the hazardous waters. As with honey collectors, fishermen often take refuge

FIGURE 12.4 Boats for honey harvesting Wooden motorized boats are used by locals for honey collection.

FIGURE 12.5 Fishing boats Wooden motorized boats are used by local fishermen.

in their fishing boats (Fig. 12.5) during these excursions, leaving them susceptible to episodic homelessness during their fishing trips. Additionally, the shrinking fish sources undermine local communities' customary livelihoods, driving them into poverty and squalor. They are compelled to move to surrounding cities or towns in search of work due to a lack of alternative economic options, where they frequently wind up in slums or makeshift communities and endure even more difficulties.

Sagar Island, like other areas in the Sundarbans region, is a reclaimed land facing significant shoreline erosion. Coastal villages such as Shibpur (Botkhali), Dhablat, and Beguakhali are severely impacted by this erosion, posing a serious threat to the homes and livelihoods of the inhabitants. During monsoons and storm surges, residents living along the shoreline constantly fear losing their homes. The continuous loss of acres of land to the sea has rendered many people landless and homeless. These are chronical homelessness where people lost their home overnight. Homelessness has serious social and cultural repercussions for the communities it affects, going beyond only the loss of housing. Villages' tight-knit social networks are shattered by displacement, which causes the decline of cherished cultural norms and values. The emotional misery that the displaced people experience is further exacerbated by the loss of a feeling of identity and belonging. Survey results indicate that approximately 71% of households on the island are linked to the primary sector, mainly with agriculture.

When these individuals lose access to their land, they forfeit not only their means of livelihood but also their way of life, social safety net and in essence, the very identity that has been crafted over numerous generations. This pervasive uncertainty compels the island's inhabitants to make the difficult decision of relocating to more secure regions. In fact, 65% of the surveyed individuals expressed their readiness to move to the mainland.

Often, government and relevant agencies have responded slowly and insufficiently to address the situation. Disaster relief and rehabilitation programs are examples of initiatives that are frequently reactive and fall short of offering long-term remedies to the problem's underlying causes. The Island has several cyclone shelters (Fig. 12.6), but their number falls short of the demand. As a result, school buildings are utilized as makeshift shelters, leading to extended class suspensions. In most cases, the government's response is primarily reactive. The government endeavors to assist affected individuals by implementing financial support programs. Unfortunately, the efficacy of these efforts is hindered by a lack of a proper monitoring system, leading to difficulties in ensuring that the aid reaches those who truly need it. As a consequence, some vulnerable communities continue to experience inadequate support during times of crisis. One such initiative was "Relief at your Door Step." In the course of the survey, some respondents mentioned that such initiatives are frequently influenced by political factors when it comes to identifying victims. Here, what's needed are a set of comprehensive strategies that consider the ecological vulnerability of the region and its impact on the lives of the island's residents. Also, there is a pressing need for the development and implementation of robust monitoring mechanisms.

FIGURE 12.6 Cyclone shelters at Sagar Island: Existing cyclone shelters are not sufficient in number.

Additionally, the lack of concrete-reinforced embankments has practical, real-world repercussions (Fig. 12.7). Saline water intrusion into agricultural lands is a real, harsh reality rather than just a theoretical problem (Fig. 12.8). Significant long-term shoreline alterations have been prominently documented in Dhablat, Gangasagar, and Muriganga-1 Panchayat (Shoreline Management Plan for the Coastal Areas of Sagar Island, 2019). The impact is most pronounced on the inhabitants of the southern part of the Island. Those living along the shoreline in Shibpur village, Dhablat Panchayat, live in a constant state of apprehension throughout the

FIGURE 12.7 Wooden embankment along the shoreline: Lack of concrete embankment in Sumatinagar and Shibpur village.

FIGURE 12.8 Saline water intruding into agricultural land. Houses and agricultural lands are encroached by sea water.

year. Even during high tides, they fear the encroachment of the sea into their homes while they sleep. Large areas of farmland are rendered unproductive by this sea encroachment, forcing the heartbroken local residents to decide to leave their homes in quest of a more secure future. In the past, Sagar Island served as a refuge, a sort of safe haven during ecological crises in the region. It was a place where people from neighboring islands sought shelter during cyclones. However, the existence of Sagar Island itself is now in doubt due to global warming and rising sea levels. Some initial steps have been taken, such as the repair of damaged embankments at Bamkimnagar and Dhaspara Sumantinagar-2 panchayat, but these makeshift measures are inadequate. Sagar Island urgently requires a comprehensive action plan to safeguard its coastline from the relentless advance of the sea.

In the context of ecological distress, individuals residing in the remote interior areas of the island frequently find themselves facing situational homelessness. Those with kuchha houses (made of temporary or nonpermanent materials) often seek refuge in nearby public or commercial buildings, such as Dharamshala or hotels, during cyclonic events. Cyclone shelters have been established to provide a safe haven for people during such challenging situations. However, the duration of their stay in these facilities often extends when their homes are damaged by the cyclone.

Sagar Island is among the areas most adversely impacted by anthropogenic activity, natural disasters, and climate change. The island's delicate ecosystem is being affected primarily by rising sea levels, coastal erosion, and regular cyclones. The ecological suffering on the island is further exacerbated by human-caused issues like deforestation, overfishing, and unsustainable development methods. The loss of arable land, lowered access to natural resources, and a worsening environment are caused by the interaction of these ecological forces, posing a serious threat to the island's population's ability to support itself. The islanders struggle mightily to uphold their ancient ways of life because there are few viable options for generating revenue and a sustainable way of existence. The population figures over the past few decades have displayed a declining trend. The growth rate of the Island between 2001 and 2011 was 12%, which is lower than the state's growth rate of 13.8% and the country's growth rate of 17.7%. It is evident that people are not inclined to remain on the island.

People in erosion-affected areas, particularly in the southern and southeast regions of the island, live in constant fear of becoming homeless. Population growth in those areas is also on the lower side. They tend to gravitate toward safer areas within the island, such as the northern part where the elevation is on the higher side, or consider relocating to the mainland. It is clear that these at-risk areas exhibit slower population growth.

5. Conclusion

After discussion on ecological distress within the Sundarban region, particularly with respect to the formidable issues of homelessness and displacement using Sagar Island as study area, it can be concluded that the vulnerability of locals, the marginalized communities is exacerbated by the interplay of climate change, ecological distress, unsustainable development, and inadequate policy responses. And that leads people into a vicious cycle of poverty, forced eviction, homelessness, and migration. In response, it becomes paramount for governmental bodies and stakeholders to prioritize comprehensive environmental conservation initiatives. These efforts should be complemented by the establishment of resilient infrastructural systems and diversified livelihood opportunities tailored to the unique needs of the affected populations. However, it is crucial to recognize that the effects of climate change and ecological distress transcend geographical boundaries. Hence, while localized interventions are crucial, they might not yield immediate and effective outcomes for the island residents. A holistic strategy necessitates the formulation of nuanced mitigation policies at a governmental level to safeguard the interests of the Sundarban inhabitants. Given the persistent socioeconomic and livelihood paradigms of the region, a long-term policy-focused approach becomes imperative. This approach aims to mitigate the prevailing overreliance on finite natural resources among the local communities. Exploring alternative livelihood options such as saline-tolerant pisciculture, agro-based industries and cottage enterprises could provide viable solutions. Additionally, it is essential to identify hazard-prone areas and develop meticulous rehabilitation plans under the guidance of governmental agencies. Launching awareness campaigns to inform the local population about diverse opportunities can mitigate the risk of involuntary displacement in the pursuit of alternative livelihoods. This integrated effort leads to a mutually beneficial outcome where the region's sustainability is enhanced by alleviating the strain on natural resources, while concurrently implementing a well-structured mitigation plan to combat homelessness.

References

2019. Shoreline Management plan for the coastal areas of Sagar Island. Kolkata.

Ascoli, F. Kolkata. West Bengal District Gazetteers, Govt. of West Bengal Revenue history of Sundarbans from 1870 to 1920.

Banerjee, S., Chanda, A., Ghosh, T., Cremin, E., & Renaud, F. (2023). A qualitative assessment of natural and anthropogenic drivers of risk to sustainable livelihoods in the Indian Sundarban. *Sustainability, 15*(7), 6146. https://doi.org/10.3390/su15076146

Cohen, C., & Sokolovsky, J. (1989). *Old men of the Bowery: Strategies for survival among the homeless.*

Farr, R., Koegel, P., & Burnam, A. (1986). *A study of homelessness and mental illness in the sid roe area of Los Angeles.*

Ghosh, S., & Mistri, B. (2023). Cyclone-induced coastal vulnerability, livelihood challenges and mitigation measures of Matla—Bidya inter-estuarine area, Indian Sundarban. *Natural Hazards*, 3859—3860.

Hajra, R., & Rajarshi. (2023). Sustainability assessment of Indian Sundarban delta islands using DPSIR framework in the context of natural hazards. *National Hazards Research, 3*(1), 76—88.

Islam, M. M., & Atiqur. (2017). Drivers of mangrove ecosystem service change in the Sundarbans of Bangladesh. *Singapore Journal of Tropical Geography*, 22.

Jamal, S., Ghosh, A. G., Hazarika, R., & Sen, A. (2022). Livelihood, conflict and tourism: An assessment of livelihood impact in Sundarbans, West Bengal. *International Journal of Geoheritage and Parks, 10*(3), 383—399.

Kuhn, R., & Culhane, D. (1998). Applying cluster analysis to test a typology of homelessness by pattern of shelter utilization: Results from the analysis of administrative data. *American Journal of Community Psychology, 26*(2), 207—232.

Mistri, A., Das, B. Sundarban biosphere reserve, India, environmental change, livelihood issues and migration 978-981-13-8735-7.

Mondal, M., Biswas, A., Halder, S., Mandal, S., Mandal, P., Bhattacharya, S., & Paul, S. (2022). Climate change, multi-hazards and society: An empirical study on the coastal community of Indian Sundarban. *Natural Hazards Research,* 2(2), 84—96.

Pargiter, F. (1934). *Revenue history of the Sundarbans from 1765 to 1870.* Kolkata: West Bengal District Gazetteers, Government of West Bengal.

Piliavin, I., Sosin, M., Westerfelt, A., & Matsueda, R. (1993). The duration of homeless careers: An exploratory study. *Social Service Review, 67*(4), 576—598.

Rossi, P. (1989). *Down and out in America: The origins of homelessness.* Chicago: University of Chicago Press.

Sharma, S., Suwa, R., Ray, R., & Mandal, M. (2022). Successive cyclones attacked the World's largest mangrove forest located in the Bay of Bengal under pandemic. *Sustainability, 14*(9), 5130.

Weitzman, B., Knickman, J., & Shinn, M. (1990). Pathways to homelessness among New Yourk city families. *Journal of Social Issues, 46*(4), 125—140.

SECTION III

Homelessness and social stress

CHAPTER

13

Surviving the streets: An unequal battle for homeless women

Rahul Bhushan

Centre For Geographical Studies, Aryabhatta Knowledge University, Patna, Bihar, India

1. Introduction

While 1.6 billion people across the world live under inadequate housing condition, 150 million people have been estimated as the homeless. Homelessness can be defined as the inability of the people to enjoy permanent accommodation. Homelessness is both cause and outcome of human rights violation. Certainly, homelessness is a potent and evocative social issue that has become symbolic of social inequality and injustice in otherwise affluent societies (Barker, 2012). The definition of the homelessness must not be limited to the absence of physical house/space as home is a social phenomenon and lived space (Wardhaugh, 1999). Homelessness as the social phenomena can be inferred from the social attitude and stigmatization of the homeless population. In comparison to the population in general, the homeless population witness profound social exclusion and biases based on their housing pattern. The social exclusion stems from the structural deprivation and it reinforces and intensifies the systematic exclusion of the homeless people. It also has adverse effect on the entitlement of the homeless people. In the wake of significantly curtailed entitlement survival in the streets become difficult. The entitlement gets further reduced for the homeless women. The intricate interplay of the biases and stigmatization along with the existing gender-based discrimination results into formidable challenge for the homeless women. These challenges exhibit in various forms like insignificant access to basic services, occupational segregation, and gender-based violence. Further, available research also suggests that prospect of homeless women getting employment is far less than the homeless men (Zugazaga, 2004). Also, homeless women are also subjected to employment segregation and gender-based wage gap (Roll et al., 1999). Studies also substantiates that homeless people especially the homeless women, minor, children, and elderly have greater rate of exposure toward violence and significantly inadequate social or governmental support (Kennedy, 2007). Drastically lesser social support for the homeless women also translates into social exclusion and policy exclusion. Globally,

Homelessness to Hope
https://doi.org/10.1016/B978-0-443-14052-5.00013-6

the homeless women get access to lesser number of public services than the homeless men due to varying reasons. Certainly, gender is an important determinant of the social inequality among the homeless population, the gendered perspective of homelessness is yet to be explored and investigated in detail (Stoner, 1983). There is general consensus that women have been underrepresented in the studies concerning the issue of homelessness. Also, most of the studies which are available and highlight gendered perspective of homelessness have been conducted in the context of developed economies (de Vet et al., 2019). On the other hand, studies conducted in the developing countries like India on the homelessness are primarily centered around the gender-based violence and forceful evictions and displacement.

In this backdrop, this chapter aims to interpret the gendered experience of homelessness. Scope of the chapter includes comparative analysis of vulnerability of homeless women and men to develop a comprehensive gendered perspective on challenges faced by the homeless women in context of accessing basic services like food, shelter, etc., and level of risk of violence. This chapter also aims to identify and analyze the factors and process responsible for gender inequality among homeless people. Further, intersectional approach has been applied to interpret the relative inequality among different subgroups of homeless women. Though homeless women in general face several forms of deprivation, violence, and abuse, several subgroups withing homeless women face greater vulnerability. These includes single homeless women or single homeless mother, pregnant or lactating women, minor girls and women with disability, HIV, or mental health issues. This chapter also attempts to suggest ways forwards like shifting away from the existing linear model and creating gender responsive instead of gender-neutral policy to reduce dispossession and vulnerability of homeless women.

2. Methods and materials

The present study is based on the secondary data available in the domain of knowledge. The secondary data for the study were collected from the existing literature, a decennial report by the Census of India, a National Sample Survey, a Periodic Labor Force Survey, various reports and survey by civil societies like HLRN, IGSS, etc. While the decennial census report helps in understanding the temporal pattern of change in demographic composition of homelessness, the periodic labor force survey enables interpretation of gender-based and sectoral occupational structure. Total homeless population in India came down from 19.43 lakh in 2001 to 17.73 lakh in 2011. However, the number of houseless family in cities registered 37% growth between the two censuses (2001 and 2011). Further since the objective of the study includes both policy review of schemes like Pradhan Mantri Awas Yojana, National Food Security targeted toward homeless and urban poverty and human aspects, i.e., experience of vulnerability and fear of displacement and eviction, both quantitative and qualitative data were collected from secondary sources of data. In this chapter, Lucid-chart was used to explain the intersectional identity and gender inequality and Microsoft office has been used for data visualization.

2.1 Limitations

This chapter is primarily a secondary data–based study based on the earlier study conducted in the context of the vulnerability of homeless women. The major limitation for this study is related to the availability and suitability of the data according to the objective of the study. The availability of gender-disaggregated data (GDD) is a prerequisite for interpreting the systematic gender discrimination and biases against the homeless women. However, in both developed and developing countries, homeless population are considered as uniform unit thereby there is significant gap in availability of gender-disaggregated data. Secondly, majority of the homeless women consists of migrant workers coming from marginalized communities whose voices often remain unheard or invisible. Due to lack of permanent address, significant number of the homeless women does not have identity documents which increases their invisibility. Further, the gender alienation of the homeless women is an outcome of the intricate interplay of the social relations in Indian society and their manifestation in the economic spaces. Due to the limitation of the data in this context, a generalization or conclusive statement might present an erroneous picture of the complex reality. Another critical limitation associated with the study is the lack of uniform definition of the homelessness. In the absence of uniform definition of homelessness across the globe, the comparative analysis among the homeless population in different geographies is not viable.

3. Results and discussion

3.1 Gendering homelessness

Homelessness is a global phenomena, however it varies across the geographies in terms of determinants of homelessness and challenges associated with it. However, homelessness does not only vary on regional level but homelessness also has gendered experience as it affects men and women differently. Though women are worst affected subgroup of the homeless population due to existing gender inequality, patriarchal structure, and caste- and religion-based intersectionality and discrimination against the women in general, orthodoxical perspective of homelessness does not reflect the gendered experience of homeless population because it considers homeless as homogenous set of people. The gendered experience of homelessness manifests in various form and the elevated risk of violence against the homeless women in comparison to the male population and women population in general is most pertinent among them (Wright et al., 2003). Similarly, the general trend of gender inequality in earning as men in India capture 82% of labor income, while women earn just 18%, according to the World Inequality Report 2022, can also be observed among the homeless population. In comparison to the homeless male, homeless women find it difficult to get work and even if they get work there is significant wage disparity. Only 0.9% homeless women in Delhi are engaged in skilled workforce in comparison to 6.9% male homeless. Construction work provides maximum work opportunity to urban homeless in India, but in comparison to 30% male, only 21% female homeless get construction work (Enabling Inclusive City for the Homeless, Indo-Global Social Service Society (IGSSS), & Organisation Functioning for Eytham's (OFFER) New Delhi, 2019). The gendered experience of homeless women also reflects in several other ways like in terms of higher food insecurity among homeless women, stigmatization and victimization of homeless women. Also, homeless women also find it difficult to access social security schemes like food security and schemes related to health in general like PM Ayushman Yojana or reproductive health like Pradhan Mantri Matru Vandan Yojana due to administrative apathy and lack of agency among homeless women.

The gendered experience of homelessness in India can be attributed to the deeply rooted patriarchal values and other intersectional identity like caste, religion, and literacy. Patriarchy creates a complex structure of gender-based power relation in favor of male which results into systematic discrimination against the women. Patriarchy affects women's position in both private and public sphere. Among the women homeless domestic violence and abuse is a dominant reason behind homelessness since majority of the women population in India does not have property rights. Further, in patriarchal family system abandonment by husband, desertion or forced eviction after death of husband further increases vulnerability of women for homelessness. Apart from increasing women's vulnerability toward homelessness the patriarchal system shapes gendered experience too as it limits homeless women's access to resources. According to baseline survey by IGSS, 22% homeless women in Delhi do not have any valid identity document. Also, in Delhi, there is single shelter home for around 10,000 homeless women. Another social determinant of gendered experience in the street is the caste-based identity. Although more than 85% homeless population hails from socially disadvantaged group, among the homeless women this percentage is even higher. The caste-based prejudices result into generational deprivation and social exclusion in which homeless women find it difficult to get a work or even survive without being exploited. Further, gendering homelessness requires gender-disaggregated data of homeless population, however women are underrepresented in homelessness statistics and published findings are predominantly focused on homelessness among men (Pleace, 2017) (Fig. 13.1).

However, this underrepresentation or census silence is not unique to the homeless women. In general too the homeless population is largely underrepresented. According to Census of India, 2011, Delhi, national capital of India, hosts around 46,000 houseless population, however, according to the headcount conducted by the commissioners appointed by Supreme Court of India, the number was more than 2,50,000, i.e., 5 times of the official figure issued by the Census of India (Fig. 13.2).

Apart from the methodological barrier in enumerating homeless population, the social attitude and stigmatization toward homeless women is another important factor behind their underrepresentation. Even among the homeless women, single women have been particularly alienated. In patriarchal society the identity of a woman is so strongly attached to their family that proving identity as a single woman out of the traditional patriarchal system becomes a daunting task. In most of such cases the single women get systematically driven out of the enumeration or registration process despite 39% increase in the number of single women in India (Salve, 2015). Further, the basic difference between concept of homelessness for men and women also keep women underrepresented. While men in the streets use their numbers to assert their strength their rights, women choose invisibility and shadow to reduce their visibility and vulnerability. This fundamental difference in the coping strategy in surviving the streets among men and women stems from their social position (Wardhaugh, 1999)

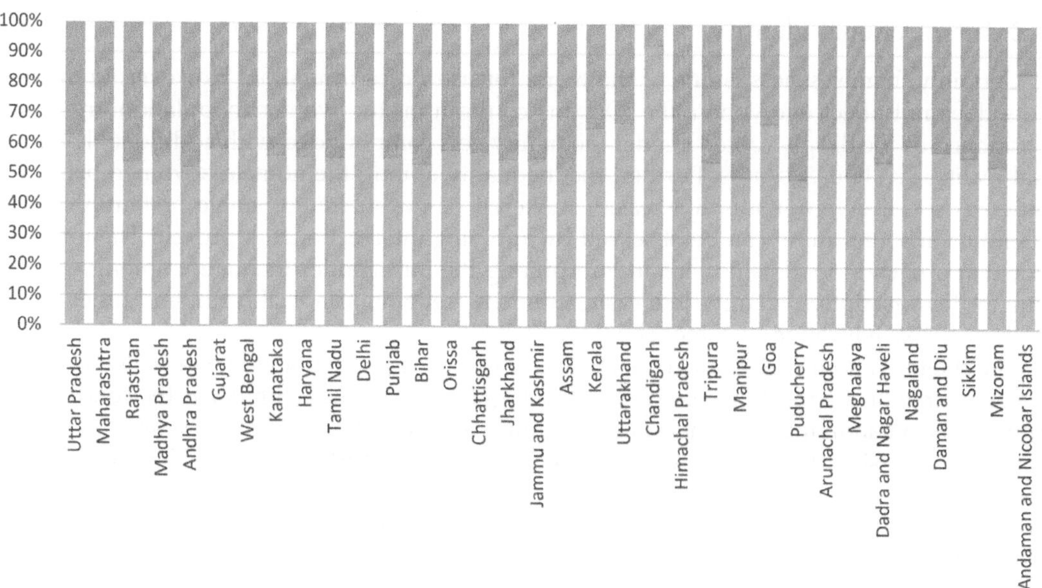

FIGURE 13.1 Male–female ratio of homeless population: Women constitute around 40% of total homeless population in India. *From Census of India, 2011.*

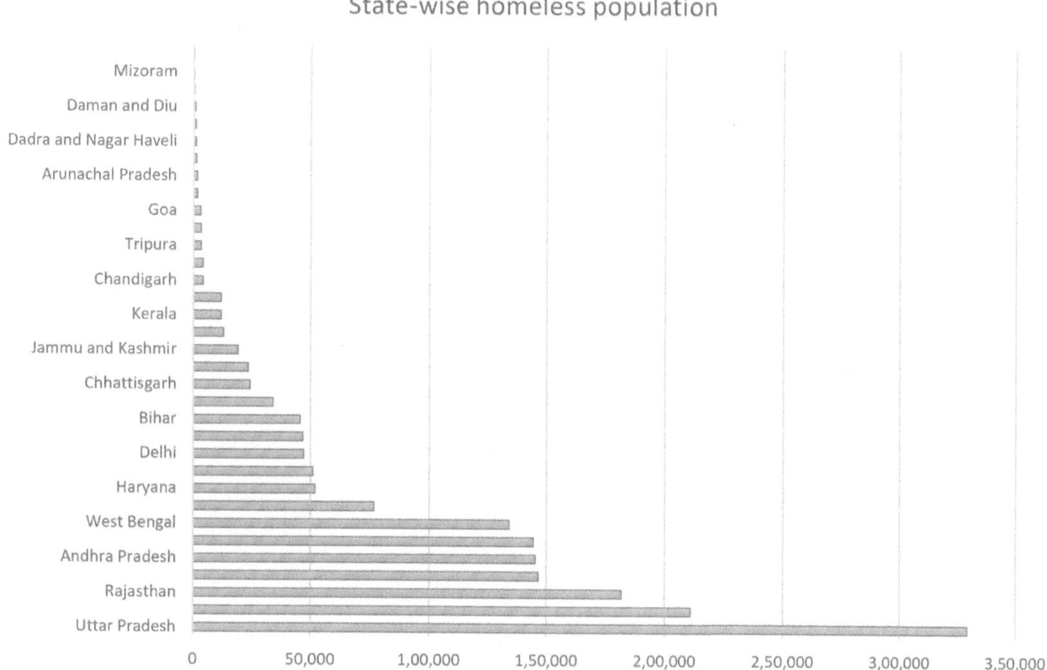

FIGURE 13.2 State-wise homeless population in India: Uttar Pradesh hosts maximum number of homeless population among different states in India. *From Census of India, 2011.*

3.2 Intersectional identity

Certainly, homelessness is one of the worst forms of discrimination. However, homelessness does not affect everyone alike. On this continuum, the women population are worst positioned as the deeply rooted gender inequality gets compounded with additional layer social reality, i.e., homelessness. Further, the inequality among the homeless women based on different social determinants like, caste, sheltered or nonsheltered, single or family-based, sound or unsound mind can be observed by applying intersectional approach. The intersectional approach in feminism helps in interpreting the intragender inequality among women due to their ascriptive identity like caste, ethnicity, religion, place of birth, and other social indicators like education, income, etc. (Fig. 13.3).

The sheltered homeless women are less likely to face violence than the nonsheltered homeless women who are forced to spend sleepless nights on the streets. Similarly, the condition of homeless woman living with their family is much secure in terms of food security, wage security, or accessing the welfare schemes run by government (de Vet et al., 2019). One of the worst affected subgroups among homeless women is the single women. Single women do not only find it difficult to secure job for them but are also at elevated risk of sexual abuse and violence. In female-headed homeless families, often the women head has to play the role of both primary care provider and sole bread earner. Further, the minor homeless girl or women with children are subjected to greater risk of violence and victimization. Similarly, surviving the streets for a homeless woman with children is formidable challenge as they face multidimensional and structural discrimination. The social inequality and stigmatization

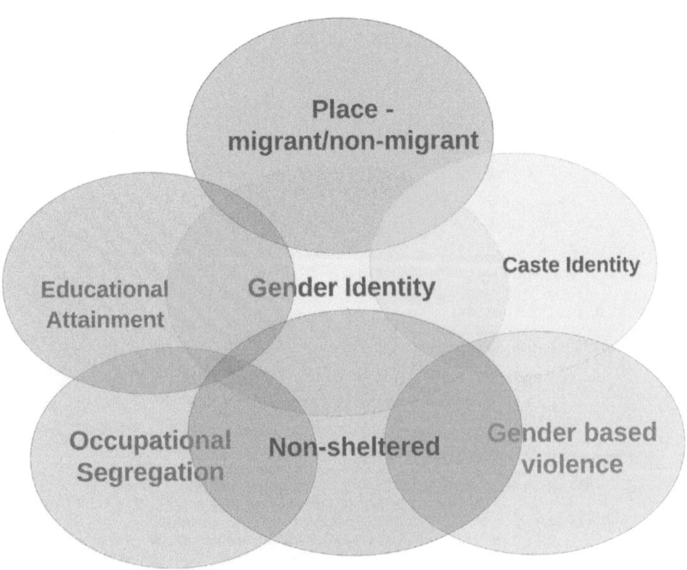

FIGURE 13.3 Intersectional identity and gender inequality: intersectional approach helps in interpreting compounding effect of gender inequality based on different identity. *Developed by Researcher.*

of the single mother can be considered as the social reality which reflects more prominently among the homeless group of demography. Marginalized groups such as low-income single mothers, typically experience multiple stigmas and sources of oppression (Sparks et al., 2005). Manifestation of this inequality can be observed in terms of occupational exclusion of single mothers, greater violence susceptibility, and skewed economic opportunities, etc. The care work along with the livelihood makes it extremely challenging for the single homeless mother to ensure the survival of her own and her children.

3.3 Surviving the streets

With extreme and multidimensional deprivation homelessness manifests worst form of human rights violation. The state of homelessness deprives the population of the most fundamental necessities of life like home, food, access to health care, etc. Vulnerability toward violence, constant fear of eviction, and significantly inadequate opportunity of social and economic mobility among this set of population makes the survival even more difficult. Surviving in the streets is comparatively more challenging for the women due to the gender-based inequalities. Chronic hunger, occupational segregation, inequality in accessing basic services, and violence have magnifying effect on the destitution of homeless women (Fig. 13.4).

3.3.1 *Accessing public services*

Homeless women usually make less use of homelessness services and postpone entering the service system until sources of informal support are depleted (Mayock et al., 2017). Due to this last-resort selection, homeless women could be a particularly vulnerable group compared to homeless men. This tendency of homeless women can be described in the context of their coping mechanism against the victimization and violence. Unlike homeless men, homeless women tend to minimize their visibility and attempts to live in shadow in

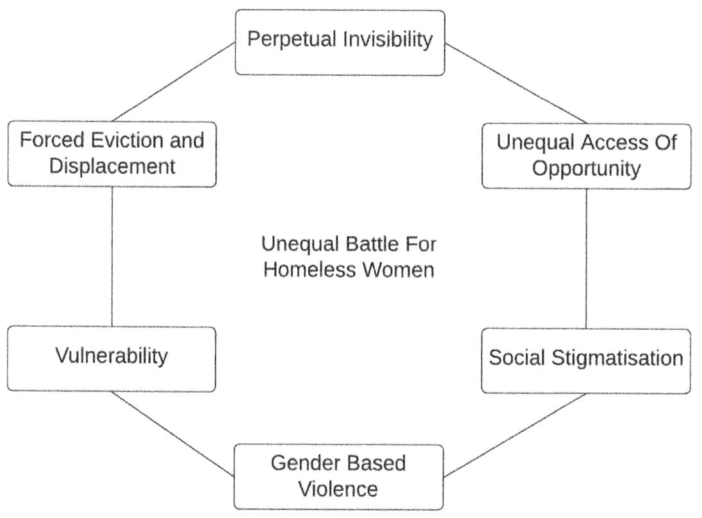

FIGURE 13.4 Multidimensional deprivation of homeless women. The state of homelessness results into multidimensional deprivation and discrimination for homeless women. *Developed by Researcher.*

order to reduce their vulnerability toward the violence (Wardhaugh, 1999). Gender intersects with the social, economic, geographical, political, and cultural factors and together these results into structural and institutionalized gender inequality. Gender inequality in accessing public services manifests in the form of prominent gender gap in education, healthcare, employment, welfare schemes, etc. As per Enabling Inclusive City For Homeless report by IGSS, only 47% homeless population is literate while more than 50% is not illiterate. Among the illiterate homeless population, homeless women with 58% is major constituent. This gender gap in literacy rate has adverse effect on the prospects of employment and public service accessibility for the women. In terms of accessing the health care facilities male population is better positioned than the female population. Around 40% homeless population does not have access to healthcare and majority of it consist of the homeless women. The distance to the healthcare facility also restricts accessibility of homeless women to the healthcare. Around 55% of homeless population does not have access to the healthcare facility within 1 km of radius. Further, the greater rate of malnutrition and disease burden among the women homeless population also substantiates the gender inequality in terms of accessing public services. However, these trend of inequality is not unique to the homeless population. It is very much evident in the Indian societies in general too and thereby it can be concluded that gender inequality among homeless population is reflection of the gender inequality in the general. However, gender-based discrimination among homeless population is unique from the population at large in two ways. Firstly, the impact of this inequality. Among homeless population which itself is one of the most underprivileged section, gender discrimination has crippling effect on the women as it makes their survival further problematic due to compounding and aggravating effect.

Secondly, the factors responsible behind the gender discrimination. Patriarchal values and deeply ingrained social attitude toward the gender inequality is the primary driver of thriving gender inequality in developing and even to an extent in the developed societies. The patriarchal norms also shape the social relations in which power equation is disproportionately balanced in the favor of male. Lack of property rights and skill development apart from the unrecognized and unpaid labor creates a structure of perpetual economic and social vulnerability for women. Homeless women living with the family, spend significant amount of time in care work and even when they are on work children often accompany them. Constant time poverty along with social and economic deprivation puts homeless women on greater risk of mental health–related issues apart from health issues in general. But among the homeless women, apart from these common factors of gender inequality, other factors like lack of permanent address, identity proof, living in the shadow as the coping mechanism, administrative apathy, and disengagement plays prominent role.

3.3.2 *The invisibles*

Isn't it ironic that homeless population who are compelled to live in open without any privacy and most of their life is open to public visibility are largely invisible in the policy process across the world? UN report on homelessness estimates 150 million people as homeless across the world. The number has been projected to witness exponential growth in the backdrop of urbanization, conflict, climate change, and pandemic. Despite, this mounting challenge, the world yet to have any uniform definition of homelessness. Similarly, in India too, in policy process the homeless population are considered as the urban poor or slum dwellers thereby doesn't have any specific policy for the homeless population. However, homeless population must not be confused with the urban poverty or slum population as the set of challenges faced by the homeless population are not in congruence with set of challenges faced by the urban poor. For example, urban poor might or might not have permanent address unlike the homeless population. As majority of the public service schemes in India are based on the linear model which requires permanent address of the beneficiary. For example, food security schemes, making Aadhaar Card, Voter ID, ration card, etc. According to baseline survey by IGSS, 2019, around 39% of the homeless population does not have valid identity proof for like Aadhaar or Voter ID for accessing the public services. Further, around 22% homeless women does not have any valid identity document which perpetuates their exclusion from the welfare schemes. The homeless people without permanent address and any agency, find it difficult to get through the administrative complexity to acquire the identity proof like Voter Identity Card and Aadhaar Card, etc. The process is even more cumbersome for the homeless women, especially the single women since administrative process is even more hostile and discouraging for acquiring identity out of the traditional patriarchal family system (Kalia, 2021). This ironic invisibility of the homeless population acts as prominent obstacle in accessing public services. Baseline survey by IGSS, 2019 suggests that only 18% of homeless population procure food from the public distribution system, while 82% population are compelled to procure food on nonsubsidized rate.

3.3.3 *Accessing food: Battling hunger on the streets*

The condition of homelessness has worsening effect on the acute food insecurity and hunger. Homeless population consist of the poorest set of demography and thereby food and nutritional security is a chronic issue. In National Capital of India, 43% homeless population faces severe food insecurity while around 42% of the population faces moderate food insecurity. The same study also suggests that among the women homeless around 50% population faces severe food insecurity in comparison to 39% of homeless male population (Prashad, 2020) (Table 13.1).

This gender inequality in terms of food insecurity can be interpreted in the context of structural discrimination against women which disallows them from social and economic mobility.

Further the procurement pattern also explains the deeply rooted gender inequality among the homeless population. As per the Baseline Survey on homelessness by IGSS, 2019 only 33% homeless female are covered under the public distribution system, i.e., food security schemes. In other words, around 67% percentage of homeless women are forced to buy food grains on normal rate rather than subsidized rate which further compounded their food insecurity.

TABLE 13.1 Food insecurity among homeless population in Delhi.

Gender	Mild	Moderate	Severe
Male	16	44	39
Female	9	41	50

From Prashad (2020)

Also, female homeless respondents in national capital of India, Delhi are nearly twice dependent on other sources of food, such as begging/gifted food/leftover food than male homeless people (Prashad, 2020). Apart from the food procurement pattern, there is also an observable distinction between the food basket of male homeless population and women homeless. Food basket can be defined as the type of foods that constitute the daily dietary intake. Food basket of 63% of the homeless female in Delhi does not includes milk. Similarly, only 30% female claimed to eat fruit on daily, weekly, or occasional. Further, 47% of them does not even have seasonal vegetable in their dietary intake (Prashad, 2020). While the food basket of the male population has greater diversity ensuring the better nutritional status, dietary intake for the female population largely constitute of the staple diet. It also reflects in the higher rate of malnutrition and disease burden among the homeless women. A quarter of women of reproductive age in India are undernourished, with a body mass index (BMI) of less than 18.5 kg/m as per the National Family Health Survey 04. The condition is even worse among the homeless women due to their marginalized position and significantly compromised bargaining capacity.

3.3.4 Accessing shelter home: The gender barrier

The Census of India, 2011, estimates that nearly 17.7 people of India are homeless, however this number is believed to highly skewed as it fails to capture the actual number of homeless surviving in streets of India. According to a third party survey conducted by Ministry of Housing and Urban Affairs about the urban homelessness in India, India's urban space alone host around 23 lakh homeless which is more than the number estimated by Census, 2011, for both rural and urban India. Similarly, the commissioners appointed by the Supreme Court the number of urban homeless in India is around 37 lakh which is about 1% of India's total urban population. However, total capacity of the shelter homes under Urban Homeless Shelter Scheme is around 1.2 lakh against the total requirement of 23 lakh. Further, in shelter homes, there is acute gender gap. In most of the Indian metro cities which host maximum number of urban homeless, there is huge gap between the shelter homes available for the male homeless and female homeless (Table 13.2). This gap results into further deprivation and victimization of the homeless women as they are forced spend sleepless nights in the streets. Possibility of violence against the homeless women who does not have access to the shelter houses is much more than those who stays in the shelter homes.

TABLE 13.2 Different types of shelters and their capacity.

Type of shelter	No. of shelter	Capacity of shelter
Men	323	16,623
Female	191	6126
Family	35	2933
General	1808	94,462
Special	220	12,237

From Deendayal Antyodaya Yojana-National Urban Livelihoods Mission (DAY-NULM), Ministry of Housing and Urban Affairs – September 2022.

Another concern related to the shelter homes in India is that most of the shelter homes are not conducive for the homeless families. While a significant number of homeless population consists of the migrant worker families. In this backdrop, the migrant homeless families often find themselves with no option other than to survive on the streets in the makeshift houses. Most of these makeshift houses consist of the wastes and plastics are not suitable for the climate extremes like cold wave, heat wave, rainfall, and water logging. Distance and accessibility of public toilet is another concern for the unsheltered homeless families especially for the female population who become further vulnerable toward sexual predation and health issues like urinary infection, etc. Apart from these structural challenges faced by nonsheltered homeless population, constant fear of eviction and displacement aggravated their plights and vulnerability. In the wake of rising wave of exclusionary urbanization, the homeless population is getting further marginalized as the contemporary urban planning consider homeless as the problem instead of homelessness as the problem. With this approach, in global south the urban infrastructure is becoming largely exclusive of the urban poor especially the urban homeless who does not have access to the shelter houses. Even in the exclusionary urbanization trend, the homeless women population are more vulnerable due to their socioeconomic position in this continuum.

3.4 Accessing employment: Occupational segregation

Indian labor market is largely fragmented on the line of social hierarchy as the people belonging to the disadvantaged section gets fewer opportunity of economic mobility and significantly concentrated in the informal, menial, and low paying jobs. This social and economic reality of Indian labor market also reflect among the homeless population since majority of the homeless population in India consist of the socially disadvantaged groups like SCs, STs, OBCs, and minority. This underlying discrimination further gets compounded with the gender identity as the homeless women undergo double whammy of occupation segregation based on the caste and gender identity. The Occupational segregation can be defined as the process of denial or allocation of a particular individual or a subset of the population in certain types of occupations. Occupational segregation occurs when workers are not allocated to occupations in proportion to their shares in the population (Chakravarty & Silber, 2007). Unrecognized labor, constant time poverty, inadequate housing facility, sexual and physical violence, neo-bondage, slavery, etc., put women workers in the most disadvantaged position in this entire spectrum. Female worker participation rate for the homeless women has been declined by almost 7% in comparison to change of almost 1 percentage point male worker participation rate (Table 13.3). The table also suggests that in comparison to less than 40% homeless male, more than 60% homeless female does not get employment or not get counted as the worker. There can be two explanations for this gender gap. Firstly, the homeless women may not be working outside and the care work provided by them does not come under the definition of workforce participation rate. Secondly, the homeless women worker may not be considered as the independent worker and thereby their work remain unrecognized.

This perpetual economic discrimination against homeless women causes aggravating effect on the social and political marginalization of these women by depriving them from their

TABLE 13.3 Temporal change in worker participation rate.

Gender	2001	2011
Male (total population)	51.7	53.3
Male (homeless population)	62.9	61.3
Female (total population)	25.6	25.5
Female (homeless population)	45.3	38.2

Data from Census of India 2001 and 2011.

bargaining capacity and increasing their vulnerability in the streets. The gender gap in terms of economic opportunity and wage results into the feminization of the poverty. Worldwide, women earn on average slightly more than 50% of what men earn. As the women living in the homelessness are often denied of their access to the resources and their work goes unrecognized their vulnerability toward getting trapped in illegal activities like compulsive prostitution, drug abuse, slavery, and neo-bondage. The lack of social support for the homeless women in comparison to their male counterpart has compelling effect. Greene et al. (1999), found that several homeless women get trapped into compulsive prostitution or survival sex. The survival sex can be defined as the form of prostitution which results out of the extreme compulsion or need. Among the homeless women survival sex not necessarily stem from the economic compulsion but sometimes due to social vulnerability and victimization they get trapped into it.

3.5 The violence and homeless women

The gender-based violence is a common social evil of Indian not just in comparison of their male counterpart, homeless women face elevated risk of violence than women in the general too. This elevated risk of gender-based violence against the homeless women can be described in the context of the socioeconomic position of the homeless women. Absence of shelter, ladies washroom, economic security, and administrative apathy together make the violence an inevitable reality for these women (Wright et al., 2003). According to a study by BUILD on the homeless women of Mumbai, around 75% of the respondents claimed to face different forms of violence on daily basis. This data is quite staggering since considering the socioeconomic position of women in general and homeless women in particular majority of the violence goes unreported. Different forms of gender-based violence include domestic violence, physical and verbal abuse, sexual violence, and exploitation, etc. The violence against the homeless women also includes the denial of social security scheme and basic services, forced evictions and displacement, etc. Foul language and verbal abuse by the passerby, police personnel and others are very common form of gender-based violence against the homeless women. Often police personnel uses the muscle power against the homeless people, especially homeless women during forceful evictions. However, majority of these cases go unreported because it has been considered as part and parcel of daily life of the homeless person in India's urban centers.

Further the sexual violence against the homeless women includes eve teasing, molestation, rape, and even survival sex. Homeless women are often compelled for the survival sex in pretext of job and safety from forced evictions and victimization from the authorities. Apart from the sexual exploitation under duress and blackmail, the homeless women especially the young and adolescent girl are forced to live under constant fear of rape and molestation in the hand of locals and fellow homeless men. Their living pattern makes them further vulnerable as in the absence of shelter they often sleep on the pavements and tents which does not have any security against such perpetrators. Similarly, due to lack of washroom in close vicinity they are compelled to walk longer distance in the night too which further aggravates their vulnerability toward sexual violence.

4. Conclusion and recommendations: Achieving gender equality in the streets

Although homelessness is worst form of poverty and deprivation, the homeless people often face stigmatization in the society which makes their survival even more challenging. Despite rise in homelessness across the world due to the urbanization, climate change and conflict, most of the governments across the world does not have specific policy to deal with it. The condition is even worse in the developing economies where homeless people are forced to live under constant invisibilization and victimization. The gender-based inequality, makes the survival for homeless women even more challenging in comparison to the male counterpart. The homeless women are forced to survive in wake up gender-based discrimination. These systematic discrimination against the homeless women manifest in several forms.

Firstly, the occupation segregation of the homeless women does not only perpetuate their economic vulnerability and subjugation but also enforces the feminization of poverty. The report by World Bank states that women earn almost half of the male for similar work due to unrecognized labor and gender-based wage gap. Further, homeless women are mostly engaged in the unskilled, low paying, unorganized, and menial works. Acute shortage of economic opportunities for the homeless women further aggravates their economic vulnerability. Due to stigma attached with homeless women, they don't even get the work as domestic helps.

Secondly, the homeless people are often denied of their accessibility to the basic services like food, water, washroom, education, health, etc. This denial can be comprehended in the context of their policy invisibilization, administrative disengagement, and lack of permanent address and required documents. The homeless women, especially the single homeless women are often subjected to systematic exclusion in the name of not possessing required documents. With denial of basic services like washroom, water, sanitation, shelter, and food, the cases of malnutrition, hunger and psychiatric issues becomes quite common among the homeless women. It becomes even more difficult for the lactating women as they are denied of the access of the public healthcare and are not in condition to afford the private health care.

Thirdly, the rampant gender-based violence makes survival of homeless women more challenging than their male counterpart. Homeless women are subjected to range of violence, starting from the foul language and verbal abuse to physical assault and rape. Among several subgroups of homeless women, the young and minor girls are most vulnerable toward the sexual abuse and predation. Homeless children, living under extreme destitution are also vulnerable toward the child abuse. As the homeless population faces daunting task of survival on daily basis, the homeless women are worst positioned due to their intersectional identity. In this background, there is a need of a complete policy overhaul since the homelessness is not just an economic challenge but a social and humanitarian challenge as it violates the every aspect of right to live with dignity.

4.1 Moving away from linear model

In past, India has experimented with several schemes based on linear model for service delivery to the homeless population like National Urban Livelihood Mission, etc. However, these schemes have had limited instrumentality in dealing with issue of homelessness especially in context of the women. The homeless women are more vulnerable toward the exclusion error. In this backdrop, India needs to experiment with other model of the service delivery like Universal Service Coverage Model or Permanent Address on Temporary Housing. With permanent address on temporary housing scheme, the homeless person can avail the identity document like Aadhaar and also delivery of the services which require permanent address like public distribution system (PDS), Ayushman Bharat, etc.

4.2 From sectoral to comprehensive approach

In the absence of comprehensive policy for the homelessness, India deals with challenges of homelessness through sectoral approach. The limitation associated with sectoral approach is its fragmented instrumentality in resolving the issue of homelessness and associated issues like violence, sheltering, sanitation, etc. Instead of experimenting with different schemes for different needs of the homeless population, a comprehensive policy framework targeting all the concerns and deprivation of homeless population could have better effectiveness.

4.3 Gender blind to gender responsive

Homeless women are subjected to various forms of discrimination in terms of employment opportunity, access to shelter, accessing basic services, and gender-based violence. However, most of the policy initiatives targeted toward poverty and urban poverty lacks specific provisions for the women homelessness.

In this backdrop, it is imperative for the policy makers to shift from gender blind to gender-responsive approach and interventions.

1. Inclusion of homeless women in the policy process to end perpetual invisibilization. By considering temporary address as the permanent residence, government could bring the homeless women in the ambit of social security schemes.
2. Enabling formation of Self-Help Groups and Women Collectives to enhance the bargaining power of homeless women. This will also help in reducing homeless economic and social vulnerability.

3. Gender-specific all season shelter homes must be made mandatory in all the urban centers.
4. Ensuring universal accessibility to gender specific infrastructures like ladies washroom, crèche, etc.
5. Targeted skill development.
6. Establishing one stop center for homeless women near the primary clusters of homeless population.
7. Gender Sensitization of the authorities, especially the Police personnel and municipal workers.

At this stage, as the climate change and conflicts are becoming eminent while the accelerated urbanization is taking place in the global south, the homeless population is expected to register exponential growth in coming decades. The homelessness must be reinterpreted as not just an economic or social problem but as an extreme human rights violation with gendered perspective. A uniform definition of the homelessness along with a transformative framework needs to be developed, so that this global challenge could be addressed in an inclusive manner.

5. Future research

The future research must focus on exploring gendered perspective of homelessness and different dimensions of homelessness. Not only, experience of men and women are different in the streets but even among the women, the plights and vulnerabilities of the single women significantly varies from the women living with the family. Comparative analysis of these two set of subgroups under homeless women could be instrumental in policy process for the single homeless women. Another important area of research could be comparative analysis of homeless women in developed–developing, rural–urban, and between large and cities.

References

2019. Enabling inclusive city for the homeless, indo-global social service society (IGSSS), & organisation functioning for Eytham's (OFFER) New Delhi.

Barker, J. D. (2012). Social capital, homeless young people and the family. *Journal of Youth Studies, 15*(6), 730–743. https://doi.org/10.1080/13676261.2012.677812

Chakravarty, S. R., & Silber, J. (2007). A generalized index of employment segregation. *Mathematical Social Sciences, 53*(2), 185–195. https://doi.org/10.1016/j.mathsocsci.2006.11.003

Greene, J. M., Ennett, S. T., & Ringwalt, C. L. (1999). Prevalence and correlates of survival sex among runaway and homeless youth. *American Journal of Public Health, 89*(9), 1406–1409. https://doi.org/10.2105/AJPH.89.9.1406

Kalia. (2021, November 19). *How the Indian system keeps 'single' women dependent on others.* THE SWDL. https://www.theswaddle.com/how-the-indian-system-keeps-single-women-dependent-on-others. Retrieved from (January 23, 2024).

Kennedy, A. C. (2007). Homelessness, violence exposure, and school participation among urban adolescent mothers. *Journal of Community Psychology, 35*(5), 639–654. https://doi.org/10.1002/jcop.20169

Mayock, P., Bretherton, J., & Baptista, I. (2017). Women's homelessness and domestic violence: (In)visible interactions. In *Women's homelessness in Europe* (pp. 127–154). Ireland: Palgrave Macmillan. https://doi.org/10.1057/978-1-137-54516-9_6

Pleace, N. (2017). Exclusion by definition: The under-representation of women in European homelessness statistics. In *Women's homelessness in Europe* (pp. 105–126). United Kingdom: Palgrave Macmillan. https://doi.org/10.1057/978-1-137-54516-9_5

Prashad. (2020). *Shodhganga@INFLIBNET: Vulnerability abuse and health conditions of homeless people in Delhi.*

Roll, C. N., Toro, P. A., & Ortola, G. L. (1999). Characteristics and experiences of homeless adults: A comparison of single men, single women, and women with children. *Journal of Community Psychology, 27*(2), 189–198. https://doi.org/10.1002/(SICI)1520-6629(199903)27:2<189::AID-JCOP6>3.0.CO;2-M

Salve, P. (2015). *71 million single women, 39% rise over a decade. 71 million single women, 39% rise over a decade.*

Sparks, A., Peterson, N. A., & Tangenberg, K. (2005). Belief in personal control among low-income African American, Puerto Rican, and European American single mothers. *Affilia - Journal of Women and Social Work, 20*(4), 401–415. https://doi.org/10.1177/0886109905279872

Stoner, M. R. (1983). The plight of homeless women. *Social Service Review, 57*(4), 565–581. https://doi.org/10.1086/644139

de Vet, R., Beijersbergen, M. D., Lako, D. A. M., van Hemert, A. M., Herman, D. B., & Wolf, J. R. L. M. (2019). Differences between homeless women and men before and after the transition from shelter to community living: A longitudinal analysis. *Health and Social Care in the Community, 27*(5), 1193–1203. https://doi.org/10.1111/hsc.12752

Wardhaugh, J. (1999). The Unaccommodated woman: Home, homelessness and identity. *The Sociological Review, 47*(1), 91–109. https://doi.org/10.1111/1467-954x.00164

Wright, J. D., Jasinski, J. L., Mustaine, E., & Wesely, J. (2003). *Experience of violence in the lives of homeless persons: The Florida four city study.* ICPSR Data Holdings. https://doi.org/10.3886/icpsr20363

Zugazaga, C. (2004). Stressful life event experiences of homeless adults: A comparison of single men, single women, and women with children. *Journal of Community Psychology, 32*(6), 643–654. https://doi.org/10.1002/jcop.20025

CHAPTER

14

A portrait of poverty in the street of Jakarta, Indonesia: Manusia Karung "Sack People"and their deceptive path to prosperity through compassion

Reza Amarta Prayoga

Research Center for Social Welfare, Village and Connectivity, National Research and Innovation Agency, Jakarta, Indonesia

The arena of the streets is only a stage for the fraud of fake beggars —***Shalih bin Abdullah Al-Utsaim***

1. Introduction

The capital city of Indonesia, Jakarta is the epicenter of government, economic activity, education, and culture. Jakarta's dynamics as the pulse of the nation's life holds a variety of complex problems, one of which is poverty. Jakarta has become a destination of hope for people coming from many regions to try their luck for a better life (Berawi & Miraj, 2023; Hakim et al., 2023). The sparkling facade of massive and constructive development in the nation's capital is tucked away in the marginalized expectations of a holistic life of socioeconomic activity. Not all hopes can be accommodated in this big city. The failure to compete in the contestation and pressure of socioeconomic activities due to lack of capital, skills, and low education levels, has trapped many of the migrants into becoming vagrants, beggars, street performers, and scrap scavengers (Bharoto et al., 2020; Fadri, 2019; Widodo & Galang, 2019).

Thousands of beggars and vagrants invaded Jakarta during February–March 2023, totaling 1631 people, especially in the month of Ramadan (March–April 2023) there is often an increase in the number of seasonal beggars who come to Jakarta to scavenge on the

Homelessness to Hope
https://doi.org/10.1016/B978-0-443-14052-5.00014-8

TABLE 14.1 The number of vagrants and beggars in 2020–21.

Region	Number of vagrants and beggars
South Jakarta	361 people
East Jakarta	381 people
Central Jakarta	508 people
West Jakarta	1017 people
North Jakarta	319 people
Total	**2586** people

From Badan Pusat Statistik DKI Jakarta (2021).

momentum of the Muslim holy month (Yuliani, 2023). This wave of beggary invasion is getting increasingly high as the Eid al-Fitr holiday approaches. Not only in 2023, but in the past 2 years the number of beggars and vagrants spread across several areas in Jakarta has also increased significantly (see Table 14.1).

Jakarta is still perceived as a magnet that is so "sexy" to be visited by the urban population. The lack of capability and sufficient capital leads to their situational choice to seek fortune by begging for the mercy of others. The noticeable increase in the number of beggars and vagrants may be attributed to the deep-rooted generosity within the community. This observation aligns with the 2021 World Giving Index (WGI) data from the Charities Aid Foundation (2021) which ranks Indonesians as the most generous people globally. The profound sense of empathy and the ingrained practice of almsgiving in Indonesian society can be traced back to the influence of the predominant Muslim religion. Within Muslim beliefs, acts of charity, including zakat (almsgiving), waqf (endowments), infaq, and sadaqah (public charity), serve as both a means of purifying wealth and securing blessings in this life and the afterlife (Adzkiya et al., 2023). Yet, this noble spirit of generosity and compassion is, at times, taken advantage of by beggars and the homeless, who seek alms on the streets for personal gain. Living off the compassion of others is used as a weapon for beggars and homeless people to earn a sizable income. Many research findings also justify that begging and vagrancy have mostly become a "profession" in urban areas (Addina & Fuad, 2015; Christiawan et al., 2017; Setiawan, 2020). The way they operate is to "cunningly" exploit the generosity and compassion of others. The tactic of dredging up the compassion and generosity of others is then utilized by them by displaying the pretense of being disabled, holding little children, looking shabby, and utilizing used bottles or cans (Sarmini et al., 2018; Sulastri et al., 2022). Such tactics are sufficient to cover up their cunning on the streets.

Street beggars often employ deceptive tactics (Al-Utsaim, 2022) to exploit the generosity of others, turning it into a lucrative endeavor. For many, it's a strategy to gain substantial income without significant effort. Several cases have highlighted beggars earning hundreds of thousands of rupiah daily, leveraging sympathy to their advantage. Notably, beggars identified as AA (57 years old) and ML (44 years old), who feigned blindness in East Jakarta's Halim Perdanakusuma area, reportedly earned IDR 450 1000 in just 1 day (Ramadhian, 2023). A CNBC Indonesia investigation even revealed certain beggars amassing wealth up

TABLE 14.2 Beggars with fantastic number of wealth.

Initials and age	Hometown	Begging location	Begging's revenues
L (52 years)	Pati, Central Java	On the roadside of Pati Municipality	1 billion IDR in cash
T (60 years)	Malang, East Java	Diponegoro Gas Station, Batu Municipality, East Java	18 million IDR/month in cash
M (64 years)	Padang, South Sumatra	Kebayoran Lama flyover Bridge, South Jakarta	90 million IDR in cash found inside the pocket and plastic bags
S (51 years)	Semarang, Central Java	Red Light intersection yos Sudarso Area, Semarang Municipality, Central Java	140 million IDR in deposits, 16 million IDR in savings, 3 Motorcycles registrations, land certificates, and 400 1000 IDR in cash
W (54 years) dan S (70 years)	Subang, west Java	Pancoran intersection, South Jakarta	25 million IDR in cash

From Anam (2022).

to IDR one billion (Anam, 2022). This investigation underscored five beggars with substantial assets derived solely from their activities on the streets (see Table 14.2).

The COVID-19 pandemic has also affected the poverty situation in urban areas. The phenomenon of the emergence of Sack People begging is like proliferating on the outskirts of town (Alliva, 2020). The Sack People, based on observations, can be said to be a new mode of begging. Usually, they stay on the side of the road carrying a sack, sitting with a shabby look while waiting for compassion and generosity from road users. Interestingly, this Sack People is often identified and associated with waste pickers. In fact, both are functionally very different professions. The waste picker profession is an effort by individuals to find waste to be used as a source of livelihood (Carenbauer, 2021; Malak et al., 2022), while the Sack People profession is the same as a beggar who just carries a sack as a symbol to invite the compassion of others.

While both Sack People and waste pickers carry sacks, they differ mainly in their activity. Waste pickers, searching for recyclables like single-use plastic bottles, are highly mobile and see waste collecting as a livelihood (Sarja, 2020). In contrary, Sack People, essentially beggars with sacks, typically remain stationary in areas with high vehicular traffic, using the sack as a symbol to attract compassion from passersby. This act of begging often arises from situational poverty, where individuals, lacking skills, capital, and education, find themselves economically vulnerable and see begging as their only means of survival (Nugraha, 2021; Roy et al., 2023). Such conditions often stem from limited alternatives for sustenance, relying solely on others' kindness (Alsheikh Ali, 2019; Fadri, 2019). Some, however, exploit this situational begging for personal benefit (Bajari & Kuswarno, 2020).

The growing number of urban beggars, exemplified by groups like the "Sack People," has emerged as a pressing social issue, forming the central focus of this chapter. Jakarta, the center of development, presents an intriguing dichotomy: while being a hub of advancement, it also manifests stark social discrepancies. This includes the marginalization of certain communities, who, due to their lack of competitive skills, economic capital, and limited education, find themselves sidelined in the metropolitan hustle. It is these anomalies in social dynamics that this chapter seeks to probe more deeply. The COVID-19 pandemic, with its devastating consequences like the onset of a fresh poverty wave, including the emergence of the "Sack People," further underscores these disparities (Yuda & Munir, 2023) such developments highlight the widening gap between different social rank, pushing the economically vulnerable into despair. Such developments highlight the widening gap between different social rank, pushing the economically vulnerable into despair. World Bank (2020) report that segments the Indonesians into four expenditure-based social classes: the impoverished class (28 million people spending below Rp354,000 per person monthly), the vulnerable class (61.6 million spending between Rp354,000—Rp532,000), the middle class (53.6 million spending between Rp1,200,000—Rp6,000,000), and the affluent class (3.1 million spending above Rp6,000,000). Tragically, those belonging to the impoverished and vulnerable segments often end up resorting to begging. Furthermore, social inequalities rooted in Jakarta's socio-economic landscape has paved the way for increased poverty (Kaufmann, 2023; Wong et al., 2023). Evidences include the prevalence of slums due to spatial segregation (Wirastri et al., 2023), widespread millennial unemployment resulting from a lack of comprehensive and stable employment opportunities (Rakhmani & Utomo, 2023), and drastic wage disparities (Omar & Inaba, 2020; Palomino et al., 2020). Jakarta's urban setting clearly highlights the significant economic gap, with the differences between the wealthy and the poor affecting daily challenges (Arciniegas, 2021; Janssen et al., 2022). The issue is further complicated by cultural and religious practices. Many view almsgiving as both a sacred act and a duty, which unintentionally encourages the rise of beggars (Adzkiya et al., 2023; Ugwu & Okoye, 2022).

2. Rationale of the study

In light of the findings presented earlier, this chapter delves into the phenomenon of the Sack People in Jakarta, Indonesia. It aims to unfold three key aspects: the prevalence of Sack People on Muslim holy days, especially during Friday prayers; the underlying causes and distinctive characteristics of the Sack People phenomenon; and the strategic use of sacks to attract generosity. These insights set this chapter apart from other studies by offering a nuanced understanding of poverty and begging specific to the Sack People. While past research often equates them with vagrants, this chapter emphasizes the notable differences between the two, as detailed in our findings.

3. Limitations of the study

There are many studies discussing beggars, waste pickers, and vagrants in urban areas, but these studies have limited conceptual explanations of the Sack People as beggars in urban areas. There are not many studies that discuss the existence of the Sack People as a beggar in the city of Jakarta so that the definition of the Sack People is carried out by researchers based on the meaning of the data findings. In addition, the limitation of primary data from informants is also an obstacle in this study, namely the number of informants who avoid direct interaction with researchers because they are considered competitors to get help from others, even though the researchers themselves have disguised themselves as Sack People. This limitation can be further research by expanding the scope of data collection locations, because the Sack People phenomenon is almost common in the big cities throughout Indonesia.

4. Materials and methods

This book chapter employs a qualitative, phenomenological approach to define and interpret the phenomenon of the "Sack People" as beggars. The data exposure is in the form of the definition of the Sack People phenomenon, which is dissected in depth, broadly and thoroughly to find and obtain the truth of the social process of the Sack People as beggars in the city of Jakarta. The expansive city of Jakarta, with its five administrative regions (North, West, East, South, and Central Jakarta), served as the primary data collection sites. Informants were selected through purposive sampling based on specific indicators, such as those carrying sacks for donations and operating at designated times. A total of eight informants, anonymized for confidentiality, contributed to the study. Primary data consisted of in-depth interview transcripts about the Sack People's perspective and detailed field notes from participatory observations. The researcher fully immersed themselves, posing as a Sack Person, ensuring an authentic understanding, and facilitating deep interactions for data gathering. Secondary data was sourced from relevant literature. The data analysis included steps of data reduction, presentation, and conclusion drawing throughout the research process (Denzin & Lincoln, 2017; Miles et al., 2019). Triangulation ensured the validity of the collected data (Denzin, 2012; Rahardjo, 2010).

5. Results and discussion

5.1 The surge of Sack People during the "Friday Blessings"

This chapter delves into the noticeable presence of the "Sack People" predominantly on the significant Muslim day known as "*Jumat berkah*/Friday blessings" On Fridays, these individuals gather in groups, especially on the sidewalks of Jakarta's bustling streets, like commercial hubs, governmental centers, tourist spots, mosques, and markets. This pronounced appearance on Fridays is influenced by the Islamic belief that designates Friday as a special day of virtue. Muslims are often encouraged to allocate a portion of their assets for the needy (Farma & Umuri, 2021), a principle enshrined both in the Quran and the Hadiths. They underscore that a segment of the assets bestowed upon Muslims by Allah SWT inherently belongs to the economically vulnerable, meant to be shared through charitable practices like alms, *infaq*, and *zakat* (Ismail et al., 2022, p. 27–41; Syafiq, 2018). The Quran emphasizes aiding the economically vulnerable—including categories like *fakir miskin*, slaves, and others—as means to foster economic justice (Hidayati, 2022). Notably, the "Sack People" fall under this category, representing those in dire need of help.

Fridays, being abundant with blessings, are days when Muslims intensify their acts of worship and benevolence. Recognized as a day when good deeds, including almsgiving, are believed to be magnified by Allah SWT, the essence of "Friday Blessings" is to embed Islamic virtues and promote empathy (Kumari et al., 2023). Such acts, like alms, are often viewed by many as a religious imperative (Ugwu & Okoye, 2022). In the Indonesian context, "Friday blessings" frequently involve giving out lunch boxes to the poor, especially those living on the streets (Rafi, 2019). Such charitable gestures are widespread across Indonesia, where the majority practice Islam. Distributing packaged rice on this blessed day has evolved as a traditional religious act, especially highlighted during times of crises like the COVID-19 lockdown (Ahmad & Maulida, 2020; Suswanto, 2022). Thus, it is no surprise that on these Fridays, the "Sack People" often receive these donations. As shared by an informant, S, aged 39: "Every Friday becomes a ray of light for those of us battling hunger on the on the streets."

In Islam, almsgiving is more than just charity; it embodies kindness (al-birr), ensures social security (al-ijtima'i), fortifies humanitarian solidarity (al-takaaful al-insani), and serves as a tool for economic empowerment to combat poverty (Rosidi et al., 2023). It aims to bridge the gap of social inequality, especially benefiting those like the Sack People, who often represent the economically vulnerable. Informant MY, a 60-year-old with limited education and skills from West Java, depicts this when he shares his experience from Pondok Indah, South Jakarta: "Friday blessings are a lifeline for us. The generosity of passersby, even those in luxury cars, sharing their rupiahs ensures we survive another day."

> ... It's been 3 years since the pandemic, I've lost my job and only beg on the streets. My life is uncertain, it's tough to feed myself, I don't have a family anymore. In the village (Sukabumi, West Java) I have nothing left. Hoping for the compassion of charity from people in Jakarta is an option to survive. ... ***Interview with Informant MY (60 years old), May 4, 2023.***

On Fridays, many Sack People strategically position themselves in bustling locations across Jakarta, with some traveling to the city specifically for this purpose. These individuals typically station themselves 50–80 m apart along roadsides, ensuring that each has their own space to solicit alms. This informal arrangement, often indicated by white sacks or carts, is rooted in mutual respect and the desire to prevent disputes over donations. Often, these spots are occupied by groups of three to five people, usually connected through familial ties or acquaintanceship. When a potential donor stops near a particular marker, the alms go to the individuals associated with that spot.

The presence of Sack People intensifies in the week leading up to Eid al-Fitr. Rows of them line the streets, hoping for the generosity of passersby. Their numbers grow as Eid draws near, driven by the understanding that Jakarta's Muslim population is particularly inclined to give during this period, amplifying their hopes of receiving alms (see Fig. 14.1). Begging in urban areas has increasingly taken on the character of a business, often veiled in stories of hardship and deceits (Jackman, 2022).

The disclosure by informant MY (60 years old) also confirmed that Friday and the week prior to Eid al-Fitr became the momentum for them to reap the fortune from generous alms. This informant also revealed that instead of going to houses to beg, it is better to sit on the side of the road and will definitely be approached by people who want to give

FIGURE 14.1 Portrait of Sack People on the roadside of Tanah Abang, Central Jakarta.

alms. Furthermore, the results of begging by waiting on the street are greater than going to houses. Besides the proceeds obtained are larger, and there is no need to exhaust themselves walking around. This further explains that there are indications of profit and loss rationality calculated by the Sack People, there is an economic law in each of their actions, where they have calculated the profit and loss (energy, income from alms, and situational certain days of begging). This way of rationality of the Sack People is like an economic calculation, and allegedly reinforces that the Sack People utilize the momentum of the sacredness of religious traditions to reap the benefits of the kindness of others. Thus, findings were obtained that indicated the key to the proliferation of Sack People in Jakarta, namely the momentum of the sacredness of the Friday blessing, the selection of strategic positions or lots, and the calculation of profit and loss.

5.2 Factors contributing to becoming the Sack People

The Sack People begging for money on the streets of Jakarta are a common sight. The characteristic of the Sack People sitting on the side of the road waiting with their sacks filled with the kindness of others is like a hope that is always expected. Living as a beggar just hoping for compassion is a view of a different side of the sparkling massive development of the capital. The existence of the Sack People is a victim of injustice and the implications of uneven development. Sack People are driven to beg just to make ends meet. There is no room for them (Sack People) to be able to live properly due to the high stress of economic needs. Situations that are limited, not having a permanent job, uncertainty and even zero income, no marketable skills, physical disabilities, having no connections, and low levels of education (Suhandi & Erlita, 2021), as well as inherited poverty (generational) (MacDonald et al., 2020), trigger the emergence of the phenomenon of Sack People begging on the streets.

The situation of high competition for life in the capital city of Jakarta makes some people who are limited, finally live at the mercy of others.

Urban street begging often pivots between two narratives: genuine economic despair and deliberate deceit (Jackman, 2022). The prominent factor behind the rise of the Sack People in Jakarta is poverty, a state defined by an individual's or group's inability to sustain basic needs. Economic challenges, intensified by repercussions of the COVID-19 pandemic such as mass layoffs, have pushed many into begging. Informants S (39 years old) and JB (35 years old) exemplify this, having transitioned from jobs in the informal sector to becoming Sack People after facing unemployment. This surge in poverty during the pandemic resonates with findings by Suryahadi et al. (2020). Furthermore, Endrawati (2022), Hill (2021), Renahy et al. (2018) highlight the vulnerabilities of informal sector workers who lack benefits and face rising unemployment due to limited job creation in the formal sector. Macroeconomic challenges and policy missteps synthesis these issues, aggravating poverty and unemployment (Supriyadi & Kausar, 2017), both being primary drivers for the increased begging observed in cities (Hughes, 2017; Kurniawan et al., 2021; Shara et al., 2020).

Secondly, the existence of the Sack People in urban areas arises as a result of inherited poverty. Most Sack People in the field findings are those whose parents already profess to be beggars. Situational poverty is triggered by inheritance or derivative factors due to the past of the previous generation who have long been trapped in poverty (MacDonald et al., 2020). Poverty is inherited from the backward older generation to their children due to limited abilities, lack of assets, and low levels of education so that they do not have the motivation to achieve a better life (Asyifani et al., 2021). As stated by informant UT (30 years old): poor conditions force me to beg, besides having nothing, my parents used to beg, when I was a child, I was often taken by my parents to beg on the street. Begging has become the path of choice for informant UT (30 years old), his parents taught their children to be beggars. Even worse, informant SA (23 years old) stated that he had followed his parents to beg at the crossroads since he was a child, and his parents did not even give him money to go to school. The life of being a beggar is in fact based on hereditary factors from the beggar's own parents (Hasan, 2019; MacDonald et al., 2020).

The rise of the Sack People in Jakarta stems from limited skills, basic education, and weak social ties. Such deficiencies deepen their poverty and isolation (Spicker, 2020, p. 15–34; Suhandi & Erlita, 2021). Most Sack People lack family support and have minimal education, limiting their job opportunities. As Informant S (39 years old) shared, "Being ostracized by my family and lacking skills or education worsened my living conditions."

5.3 Localization and characteristics of the Sack People in Jakarta

The data indicate a pattern in the Sack People's choice of locations on Jakarta's outskirts. They prefer areas with high private vehicle traffic, proximity to shopping hubs, government centers, luxury housing, and places of worship like mosques. Additionally, they choose streets with fewer authority raids on vagrants. Observations identify at least five such prominent locations in Jakarta where Sack People hang out (see Table 14.3).

Informant TR (43 years old) revealed: the right location to hang out should be busy with vehicles and people traffic. It is not uncommon for us to choose around the train station or market. Just sitting on the side of the road, there must be fortune that comes to us. Sitting still

TABLE 14.3 The dispersion of Sack People locations in Jakarta.

Administrative area	Location
North Jakarta	Pluit karang Utara Street, Penjaringan.
West Jakarta	Grogol, Slipi, Palmerah Areas.
East Jakarta	Halim Perdana kusuma, kawasan rawamangun, Pisangan timur, and duren Sawit areas.
South Jakarta	Pondok indah area, fatmawati, kalibata apartment area, and cilandak area.
Central Jakarta	Tanah abang area, kemayoran, and karet Bivak area.

and making shabby gestures can already attract the pity of others. A sack with a little cardboard or used bottles and placed next to each other as they sit is a sign that they need help or alms. The alms from these kind people become our (the Sack People) life savers. Meanwhile, the characteristics that mark the Sack People on the streets of Jakarta are not a difficult matter. Typically carrying a sack and them sitting silently on the side of the road can already be a sign that someone is a Sack People. However, it is not uncommon for people to view Sack People as the same as waste pickers. The fact is that Sack People and waste pickers are different, although there are waste pickers who also turn into Sack People in certain situations.

Waste pickers and Sack People differ in several ways (see Fig. 14.2). Waste pickers are typically characterized by their active movement in seeking used bottles and their use of tools like Ganco. They treat waste picking as an occupation, selling recyclables for income and operating without fixed hours or locations. They often work alone, using sacks to collect items like plastic bottles. On the other hand, Sack People primarily stay in one spot, dressed in

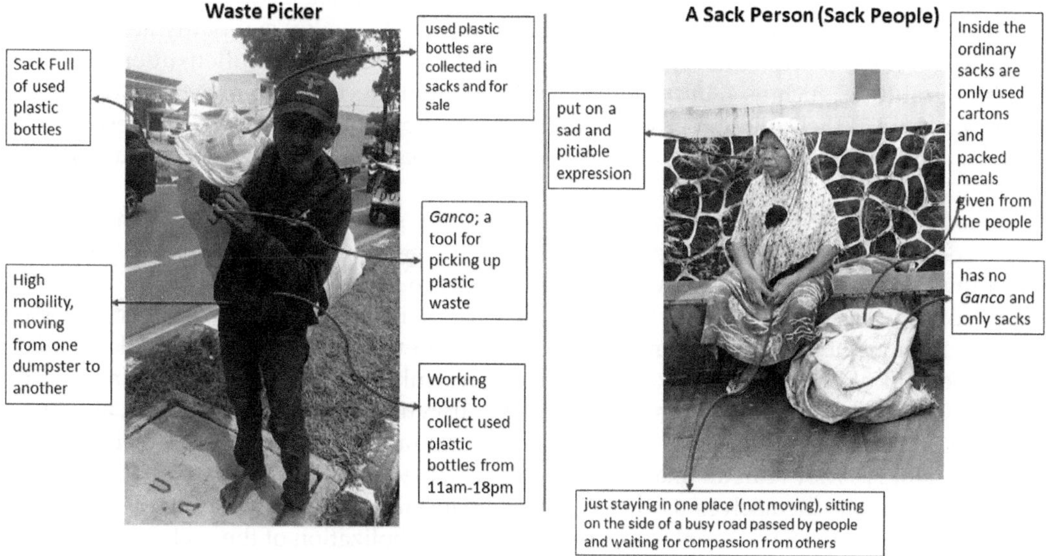

FIGURE 14.2 The difference between a waste picker and a sack person.

tattered clothes to evoke sympathy. They rely on alms, often working in groups and using their sacks to gather donations, be it food or money. Typically found on busy Jakarta streets, these individuals target areas they perceive as populated by empathetic and compassionate individuals.

Just like the assertion of informant TR (43 years old) who views Jakarta as a "a bright space" full of kindness, where the sparkling metropolitan city still has a glimmer of kindness and compassion for those who are poor and have nothing. Informant TR interpreted Jakarta as still holding a space for giving kindness to marginalized people such as Sack People. Informant S (39 years old) stated the same thing, namely that Jakarta is like "*sugar, which will always be swarmed by ants,*" meaning that as long as there are still many people in Jakarta who like to give kindness, Jakarta will remain a magnet for regional people like us to beg. Informant MY (60 years old) emphasized the same thing: as long as there is still kindness from people willing to help, there is a shelter for us to get the blessings of kind people. This is what is extracted from the views of Sack People who feel that being a beggar in Jakarta is promising enough to fulfill their needs. Many Sack People have significant limitations, and often, they turn to begging as a survival strategy due to the challenges of societal structures, such as poverty, unemployment, past convictions, stigmatization, or being ostracized (Griffante, 2020; Rachmah et al., 2023).

5.4 The sack: A symbol of need and a magnet for compassions

The presence of sacks, combined with Sack People's gestures and positioning, serves as an agonizing symbol on Jakarta's roadsides. In areas teeming with activity, these individuals can often be seen seated in rows, their sacks conspicuously placed beside them, signaling their destitute state and silently appealing to passersby for compassion. Unlike active beggars who approach people directly, Sack People adopt a more passive stance, waiting for the kindness of others to come to them in the form of money or meals. The symbolism of their sacks, paired with their prayerful expressions and pleas, effectively evokes sympathy and empathy (Suhandi & Erlita, 2021). One informant, RA (54 years old), would often utter prayerful words of gratitude, such as "*Alhamdulillah ya Allah,*" upon receiving money, reinforcing the symbolic nature of their plight. The combination of their pitiable appearance, prayerful exclamations, and the ever-present sack often draws the compassion of those who encounter them (see Fig. 14.3).

The existence of beggars is often viewed cynically by some people in urban areas. Many people see most beggars as a profession to exploit the generosity of others, even an appeal not to give them something is a way to eliminate them on the streets (Muda et al., 2023). There are many cases of billionaire beggars who earn money just by begging others, beggars can have millions of rupiah in savings or cash, and can even buy houses or vehicles (Anam, 2022). The cynical view sees the existence of beggars only as a pretense tactic to attract the generosity of others in urban areas. According to Abyyu et al. (2023), Sarmini et al. (2018), Sulastri et al. (2022), beggars often adopt a deceptive strategy that leverages powerful symbolic dramaturgy. They craft an image designed to elicit sympathy or pity from potential benefactors, portraying themselves as deserving of financial support. The similarity with this finding shows that the Sack People also beg with the symbolization of the sack as a container to collect the generous donations, putting on a pitiable expression, displaying body gestures

FIGURE 14.3 Sack People on the streets of Jakarta.

of looking down sluggishly, shabby clothes, and barefoot. This symbolization is only to attract sympathy, pity, and exploit the sense of generosity of others to give something to the Sack People. As expressed by informant RA (54 years old), as follows:

> … apart from the sack, begging requires presenting oneself to others as helpless, desperate, in ugly and smelly clothes, barefoot, and if necessary, showing a pitiable face when seen by others. If I don't do this, no one will pity me. … ***Interview with Informant RA (54 years old), April 14, 2023.***

Meanwhile, Jackman (2022), Sarmini et al. (2018), Sulastri et al. (2022) draw attention to the dramaturgical act played out by city street beggars. While these beggars often portray themselves as desperate and physically impaired to earn compassion and donations, many lead comfortable, even luxurious, lives off the sizable incomes from begging. Their displayed symbols and gestures aim to exploit the benevolence of others, often serving personal interests. Consequently, beggars, including the Sack People, have become increasingly prevalent in urban Jakarta, particularly on significant days like Fridays and the lead-up to Eid al-Fitr. On these days, Muslims are especially inclined to generously give alms, seeking blessings and rewards from Allah Almighty (Syu'aibun & Zainarti, 2020).

6. Challenges and solutions

During the process of interpreting data in the field, the issue of the Sack People phenomenon as beggars becomes an urban social problem that is almost similar to other vagrants and beggars. The impact of the seasonal existence of Sack People/beggars like this makes the problem of development inequality and social welfare even more complex. There are two challenges that are noted in the findings, namely first, the rise of Sack People begging cannot be separated from the increasingly difficult socioeconomic conditions after the COVID-19 pandemic. Second, Jakarta is considered an attractive ground to try their luck, whereas migrants who come to Jakarta with no resources, no competence, and low education will be eroded and end up getting involved in social problems.

Solutions that are initiated from the findings in the field, namely first, the need for anticipatory steps in the curative of the Sack People, namely intensive law enforcement from the authorities in the form of strict sanctions against Sack People who are put in order. Second, law enforcement for beggars who are caught receiving assistance and administrative sanctions for illegal aid providers will actually trigger the number of Sack People on the streets. Third, empowerment-based social sequestration should be applied to Sack People to equip themselves so that in the future they can get decent formal jobs. Fourth, the government needs to facilitate guarantees in the form of subsidized food coupons for beggars such as the Sack People on the streets. Fifth, the government needs to promote socialization and ensure the internalization of legal religious philanthropy institutions so as to ensure public trust in philanthropy as a distributor of legitimate and targeted assistance to the poor.

7. Recommendations

This research makes three elementary recommendations, namely

1. **Government**: Recommendations to the government in eliminating beggars such as the Sack People are firm steps with humanist social operations to reduce the seasonal Sack People. This humanist social operation can be carried out by conducting social quarters for Sack People with self-support in the form of business capital, entrepreneurship training, facilitation of formal employment channels protected by the government, and guaranteeing subsidized food coupons.
2. **Science or theoretical**: This research is an effort to produce knowledge in enriching valuable insights into social science, especially regarding the transformation of beggars and poverty. This research also needs further in-depth studies from various perspectives or theories. The involvement of multi-expertise in the study of the Sack People phenomenon will be a recommendation for further research.
3. **Community**: The community needs to put their trust in philanthropic institutions that are officially appointed by the government to distribute aid to those (the poor) who are truly entitled to receive assistance. The BAZNAS (Badan Amil Zakat Nasional) system with the Jakarta bertaqwa, Jakarta cerdas, Jakarta Mandiri, Jakarta Peduli, and Jakarta sadar zakat programs need to be internalized by all levels of society.

8. Conclusions

The Sack People, as beggars, are a portrait of the social problem of poverty in urban areas. The existence of the Sack People is a derivative dimension of the complex problem of poverty in the city of Jakarta. The reality of the Sack People is like mobilizing a new type of beggar. The massiveness of Sack People in Jakarta is a result of the domino effect of the Covid-19 pandemic which has created a new line of poor people. The rise of Sack People as seasonal beggars with certain time movements such as the Muslim holy days on "Friday Blessings" or approaching Eid al-Fitr has become a momentum for beggars in exploiting the generosity of the people in Jakarta. This situational generosity of the people of Jakarta is used by them to gain millions of rupiah just to enrich themselves as beggars.

The community seems to be deprived of a sense of caring for others for the fraudulent practice of enriching themselves from beggars like the Sack People. The Sack People's symbolic management becomes a deceitful tactic in attracting sympathy from alms givers. In fact, the poverty that envelops the Sack People is triggered by the situational scarcity of economic resources, unemployment caused by the wave of massive layoffs, the limitations of the Sack People due to physical disability, the lack of strong support from the social relational network, the low level of education, the absence of job security for income security, and the difficulty in meeting the needs due to the high cost of living to become a circle of poverty that plagues beggars in urban areas.

The reduction of public trust in philanthropic institutions has also strengthened the massive actions of the Sack People as beggars on urban streets. People seem to believe that providing direct assistance to beggars on the streets will be part of easing the burden of Sack People's lives and become a way to eradicate poverty. This belief actually encourages the practice of begging with the symbolization of being the Sack People who desperately need the kindness of others. In fact, the Sack People as beggars have turned into a profession to fulfill their needs, so that the meaning of aiding seems to disappear with the practice of seeking benefits from the generosity of others.

The existence of the "Sack People" as beggars has become a prominent sight on the streets of Jakarta. They often capitalize on the city's inherent generosity and benevolence. Regrettably, many present a misleading façade of hardship. By preying on the kindness, sympathy, and religiosity of the people of Jakarta to give alms, these individuals sometimes misrepresent their circumstances for personal gain. The rise in the number of "Sack People" is driven by the deep-rooted religious conviction and compassionate nature of the people who believe in the duty to help those in need.

References

Abyyu, M. M., Anggraeny, Y., & Hariyanto, V. N. (2023). Dramaturgi kehidupan pengemis alun-alun kabupaten jember. *Journal Sosial Humaniora Dan Pendidikan, 2*(2), 144–153. https://doi.org/10.55606/inovasi.v2i2.1343

Addina, P., & Fuad, F. (2015). Budaya hukum pengemis di DKI Jakarta. *Lex Jurnalica, 12*(2), 146393. https://www.neliti.com/publications/146393/budaya-hukum-pengemis-di-dki-jakarta#cite.

Adzkiya, U., Fittria, A., Wathani, S., Ismail, A. G., Abdullah, R., Zaenal, M. H., & Macmillan, P. (2023). *Islamic philanthropy: Exploring zakat, waqf, and sadaqah in islamic finance and economics* (pp. 1469–7688). Taylor and Francis. https://doi.org/10.1080/14697688.2023.2224395

Ahmad, M., & Maulida, A. H. (2020). Jumat berkah: Community self-help based religious biocultural perpective (RBP) for food reselience to response pandemic in rural ponorogo, East Java. In *Ushuluddin International Conference.* https://vicon.uin-suka.ac.id/index.php/USICON/article/view/287.

Alliva, I. (2020). *Heboh manusia karung, ini kata pengamat sosial.* Gatra. https://www.gatra.com/news-479018-ekonomi-heboh-manusia-karung-ini-kata-pengamat-sosial.html. (Accessed 11 June 2023).

Alsheikh Ali, A. S. A. (2019). Psycho-social determinants of street beggars: A comparative study in Jordan. *Journal of Content, Community and Communication, 10*(5), 96–113. https://doi.org/10.31620/JCCC.12.19/11

Al-Utsaim, S. B. A. (2022). *Mughamarat Al-mutamawwilin baina Al-hajat wa Al ihtiraf (Begging between Necessity and Deception*-Pengemis antara Kebutuhan dan Penipuan*)* (pp. 49–52). Jakarta: Darul Falah.

Anam, K. (2022). *5 pengemis terkaya di RI, Ada yang berharta capai Rp1 miliar! CNBC Indonesia.* https://www.cnbcindonesia.com/lifestyle/20220603131249-33-344102/5-pengemis-terkaya-di-ri-ada-yang-berharta-capai-rp1-miliar. (Accessed 7 May 2023).

Arciniegas, L. (2021). The foodscape of the urban poor in Jakarta: Street food affordances, sharing networks, and individual trajectories. *Journal of Urbanism: International Research on Placemaking and Urban Sustainability, 14*(3), 272–287. https://doi.org/10.1080/17549175.2021.1924837

Asyifani, K., Alauddin, M. A., Herlina, H., & Purnamasari, K. (2021). Solidaritas sosial dalam marginalisasi masyarakat miskin (studi di dusun kentheng kota surakarta). *DIMENSIA: Journal Kajian Sosiologi, 10*(1), 61–75. https://doi.org/10.21831/dimensia.v10i1.41052

Badan Pusat Statistik DKI Jakarta. (2021). *Number of persons with social welfare problems (PMKS) by type and administrative District/City 2019–2021.* https://jakarta.bps.go.id/indicator/27/615/1/jumlah-penyandang-masalah-kesejahteraan-sosial-pmks-menurut-jenis-dan-kabupaten-kota-administrasi-.html.

Bajari, A., & Kuswarno, E. (2020). Violent language in the environment of street children singer-beggars. *Heliyon, 6*(8). https://doi.org/10.1016/j.heliyon.2020.e04664

Berawi, M. A., & Miraj, P. (2023). Towards poverty alleviation for the base of pyramid: Social business model in urban low-cost housings. *Business, Management and Economics Engineering, 21*(1), 169–189. https://doi.org/10.3846/bmee.2023.18822

Bharoto, R. M. H., Indrayanti, I., & Nursahidin, N. (2020). Beggars, homeless, and displaced people: Psycho-social phenomena and the implementation of local government policy. In *International Conference on agriculture, social sciences, education, technology and health (ICASSETH 2019)* (pp. 224–226). https://doi.org/10.2991/assehr.k.200402.052

Carenbauer, M. G. (2021). Essential or dismissible? exploring the challenges of waste pickers in relation to COVID-19. *Geoforum, 120*, 79–81. https://doi.org/10.1016/j.geoforum.2021.01.018

Charities Aid Foundation. (2021). *World giving Index 2021.* https://www.cafonline.org/about-us/publications/2021-publications/caf-world-giving-index-2021. (Accessed 11 October 2023).

Christiawan, P. I., Wesnawa, I. G. A., & Indah, A. R. (2017). Determinasi keberadaan pengemis perkotaan di kecamatan denpasar barat. *Jurnal Ilmu Sosial dan Humaniora, 6*(1). https://doi.org/10.23887/jish-undiksha.v6i1.9711

Denzin, N. K. (2012). Triangulation 2.0*. *Journal of Mixed Methods Research, 6*(2), 80–88. https://doi.org/10.1177/1558689812437186

Denzin, N. K., & Lincoln, Y. S. (2017). *The sage handbook of qualitative research* (5th ed.). Sage Publications, Inc.

Endrawati, D. (2022). Determinants of working poverty in Indonesia. *Journal of Economics and Development, 24*(3), 230–246. https://doi.org/10.1108/JED-09-2021-0151

Fadri, Z. (2019). Upaya penanggulangan gelandangan dan pengemis (gepeng) sebagai penyandang masalah kesejahteraan sosial (PMKS) di yogyakarta. *KOMUNITAS, 10*(1), 1–19. https://doi.org/10.20414/komunitas.v10i1.1070

Farma, J., & Umuri, K. (2021). Filantropi Islam dalam pemberdayaan ekonomi umat. *Jurnal Ekonomi and Perbankan Syariah, 1*(1). https://doi.org/10.37598/jeips.v1i1,%20Mei.953

Griffante, A. (2020). Lazy or diseased? Changing conceptions of beggars and vagrants in the Lithuanian discourse from the end of the nineteenth century to 1940. *Journal of Baltic Studies, 51*(1), 1–15. https://doi.org/10.1080/01629778.2019.1708761

Hakim, A. R., Nachrowi, N. D., Handayani, D., & Wisana, I. D. G. K. (2023). The measuring of urban amenities index and its effect on migration: Evidence from Indonesian cities. *Regional Statistics, 13*(2), 324–351. https://www.ceeol.com/search/viewpdf?id=1161846.

Hasan, M. (2019). *WARISAN PEKERJAAN ORANG TUA (studi deskriptif tindakan sosial anak Pengemis mengikuti pekerjaan orang tuanya mengemis di Kota kediri).* Surabaya: Sosiologi, Fakultas Ilmu Sosial dan Ilmu Politik, Universitas Airlangga. Unpublished content https://repository.unair.ac.id/87557/.

Hidayati, T. W. (2022). Reformulation of the social safety net: A conceptual approach based on qur'anic values. *Analisa: Journal of Social Science and Religion, 7*(1), 1–18. https://doi.org/10.18784/analisa.v7i1.1590

Hill, H. (2021). What's happened to poverty and inequality in Indonesia over half a century? *Asian Development Review, 38*(1), 68–97. https://doi.org/10.1162/adev_a_00158

Hughes, P. (2017). The crime of begging: Punishing poverty in Australia. *Parity, 30*(5), 32–33. https://doi.org/10.3316/informit.017284155363072

Ismail, N., Shafii, Z., Akbar, N., Azid, T., Mukhlisin, M., & Altwijry, O. (2022). Wealth management and investment in islamic settings: Opportunities and challenges. In *Decoding islamic wealth management from qur'anic texts.* Springer. https://doi.org/10.1007/978-981-19-3686-9_2

Jackman, D. (2022). Beggar bosses on the streets of dhaka. *Journal of Contemporary Asia.* https://doi.org/10.1080/00472336.2022.2135580

Janssen, K. M. J., Mulder, P., & Yudhistira, M. H. (2022). Spatial sorting of rich versus poor people in Jakarta. *Bulletin of Indonesian Economic Studies, 58*(2), 167–194. https://doi.org/10.1080/00074918.2021.1876209

Kaufmann, V. (2023). Sociological approaches to mobilities. In J. A. Silva, K. M. Currans, V. V. Acker, & R. J. Schneider (Eds.), *In Handbook on transport and land use: A holistic approach in an age of rapid technological change* (pp. 129–146). UK: Edward Elgar Publishing. https://www.elgaronline.com/edcollchap/book/9781800370258/book-part-9781800370258-14.xml.

Kumari, r., Nurhayati, S., Harmiasih, S., & Yunitasari, S. E. (2023). *Aksara: Jurnal Ilmu Pendidikan Nonformal, 9,* 1067–1074. https://doi.org/10.37905/aksara.9.2.1067-1074.2023

Kurniawan, E., Awaluddin, M., Fitriadi, F., Busari, A., & Darma, D. C. (2021). Contemporary Indonesian GDP: Context of analysis at unemployment, labor force and poor people. *International Journal of Economics and Financial Research, 7*(4), 143–154. https://doi.org/10.32861/ijefr.74.143.154

MacDonald, R., Shildrick, T., & Furlong, A. (2020). 'Cycles of disadvantage'revisited: Young people, families and poverty across generations. *Journal of Youth Studies, 23*(1), 12–27. https://doi.org/10.1080/13676261.2019.1704405

Malak, M. A., Prema, S. F., Sajib, A. M., & Hossain, N. J. (2022). Livelihood of independent waste pickers (tokai) at dhaka city in Bangladesh: Does it incidental choice of them? *The Indonesian Journal of Geography, 54*(1), 92–104. https://doi.org/10.22146/ijg.65461

Miles, M. B., Huberman, A. M., & Saldana, J. (2019). *Qualitative data analysis: A methods sourcebook (fourth).* https://doi.org/10.1177/239700221402800402?journalCode=gjha

Muda, I., Harahap, R. H.i, Amin, M., & Kusmanto, H. (2023). The cost of inaction: A portrait of street beggars in Medan City. *International Journal of Sustainable Development and Planning, 18*(7), 2045–2053. https://doi.org/10.18280/ijsdp.180706

Nugraha, T. R. (2021). Anti-beggar and homeless policy in the context social welfare. *Law Research Review Quarterly, 7*(3), 345–360. https://doi.org/10.15294/lrrq.v7i3.48156

Omar, M. A., & Inaba, Kazuo (2020). Does financial inclusion reduce poverty and income inequality in developing countries? A panel data analysis. *Journal of Economic Structures, 9*(1), 37. https://doi.org/10.1186/s40008-020-00214-4

Palomino, J. C., Rodríguez, J. G., & Sebastian, R. (2020). Wage inequality and poverty effects of lockdown and social distancing in Europe. *European Economic Review, 129,* 103564. https://doi.org/10.1016/j.euroecorev.2020.103564

Rachmah, A. I., Prasetyani, D., & Samudro, B. R. (2023). Analysis of the effect of poverty and unemployment on Indonesian people's welfare during the covid-19 pandemic for 2020-2021. *International Journal of Multicultural and Multireligious Understanding, 10*(4), 213–220. https://doi.org/10.18415/ijmmu.v10i4.4500

Rafi, M. (2019). Living hadis: Studi atas tradisi sedekah nasi bungkus hari jumat oleh komunitas sijum amuntai. *Jurnal Living Hadis, 4*(1). https://doi.org/10.14421/livinghadis.2019.1647

Rahardjo, M. (2010). *Triangulasi dalam penelitian kualitatif. Pascasarjana universitas Islam negeri maulana malik ibrahim malang.* http://repository.uin-malang.ac.id/1133/.

Rakhmani, I., & Utomo, A. (2023). Spiritually surviving precarious times: Millennials in Jakarta, Indonesia. *Critical Asian Studies,* 1–20. https://doi.org/10.1080/14672715.2023.2260382

Ramadhian, N. (2023). *Dinsos jakarta timur temukan uang Rp 450.000 pada pengemis yang pura-pura buta.* https://megapolitan.kompas.com/read/2023/01/19/07504301/dinsos-jakarta-timur-temukan-uang-rp-450000-pada-pengemis-yang-pura-pura?page=all. Kompas.com. (Accessed 5 June 2023)

Renahy, E., Mitchell, C., Molnar, A., Muntaner, C., Ng, E., Ali, F., & O'Campo, P. (2018). Connections between unemployment insurance, poverty and health: A systematic review. *The European Journal of Public Health, 28*(2), 269–275. https://doi.org/10.1093/eurpub/ckx235

Rosidi, Rahman, R., Darusman, D., Ginda, G., Arwan, A., & Antin, T. (2023). Community response to BAZNAS (alms collection agency) as an islamic philanthropic institution in Indonesia. *International Journal of Social Science Research and Review, 6*(4), 33–40. https://doi.org/10.47814/ijssrr.v6i4.981

Roy, A., Mandal, M. H., Sahoo, K. P., & Siddique, G. (2023). Investigating the relationship between begging and homelessness: Experiences from the street beggars of Kolkata, West Bengal, India. *Human Arenas*, 1–23. https://doi.org/10.1007/s42087-023-00330-0

Sarja, S. (2020). Sampah melimpah sebagai sumber kekuatan ekonomi para pemulung. *Madaniyah, 10*(1), 1–14. https://journal.stitpemalang.ac.id/index.php/madaniyah/article/view/4.

Sarmini, M., Musthofa, S., & Sukartiningsih, S. (2018). Scoring sustenance with Simpati capital: The beggar's strategy in getting money. In *1st International Conference on social sciences (ICSS 2018)* (pp. 1583–1586). https://doi.org/10.2991/icss-18.2018.329

Setiawan, H. (2020). Fenomena Gelandangan Pengemis sebagai Dampak Disparitas Pembangunan Kawasan urban dan Rural di Daerah istimewa yogyakarta. *Moderat: Jurnal Ilmiah Ilmu Pemerintahan, 6*(2), 361–375. https://doi.org/10.25157/moderat.v6i2.3218

Shara, A. R. I. D., Listyaningsih, U., & Giyarsih, S. R. (2020). Differences in the spatial distribution and characteristics of urban beggars: The case of the sanglah district in denpasar (Indonesia). *Quaestiones Geographicae, 39*(4), 109–119. https://doi.org/10.2478/quageo-2020-0036

Spicker, P. (2020). *The poverty of nations: A relational perspective.* Bristol University Press. https://doi.org/10.51952/9781447343349.ch001. Poverty.

Suhandi, S., & Erlita, D. (2021). Kemiskinan dan perilaku keagamaan dalam mengungkap simbol keagamaan pengemis. *Ijtimaiyya: Jurnal Pengembangan Masyarakat Islam, 14*(1), 105–132. https://doi.org/10.24042/ijpmi.v14i1.7471

Sulastri, I., Sulaeman, S., Hakim, U. F. R., Zakirman, Novarisa, G., & Ridwan, M. (2022). The dramaturgy communication of beggars in an Indonesian market. *Pertanika Journal of Social Sciences and Humanities, 30*(3), 1299–1317. https://doi.org/10.47836/pjssh.30.3.20

Supriyadi, E., & Kausar, D. R. K. (2017). The economic impact of international tourism to overcome the unemployment and the poverty in Indonesia. *Journal of Environmental Management and Tourism, 8*(2), 451–459. https://doi.org/10.14505/jemt.v8.2(18).18

Suryahadi, A., Al Izzati, R., & Suryadarma, D. (2020). Estimating the impact of covid-19 on poverty in Indonesia. *Bulletin of Indonesian Economic Studies, 56*(2), 175–192. https://doi.org/10.1080/00074918.2020.1779390

Suswanto, B. (2022). The optimization of the nasi bungkus alms movement through the empowerment of micro, small, and medium enterprises in cilebut barat. *Adpebi International Journal of Multidisciplinary Sciences, 1*, 364–373. https://doi.org/10.54099/aijms.v1i1.306

Syafiq, A. (2018). Peningkatan kesadaran masyarakat dalam menunaikan zakat, infaq, sedekah dan wakaf (ZISWAF). *ZISWAF: Jurnal Zakat dan Wakaf, 5*(2). https://doi.org/10.21043/ziswaf.v5i2.4598

Syu'aibun, S., & Zainarti, Z. (2020). *Gerak tanpa titik: Catatan kiprah dan pemikirannya tentang pemberdayaan zakat dan aktualisasi hukum Islam.* FEBI UINSU Press. http://repository.uinsu.ac.id/12196/1/BUKU-SYUAIBUN-OK.pdf.

Ugwu, N. V., & Okoye, K. M. (2022). Begging enterprise: A growing trend among Igbo christians in nsukka urban. *HTS Teologiese Studies/Theological Studies, 78*(4). https://doi.org/10.4102/hts.v78i4.7106

Widodo, W., & Galang, T. (2019). Poverty, evictions and development: Efforts to build social welfare through the concept of welfare state in Indonesia. In *3rd International Conference on globalization of law and local wisdom (ICGLOW 2019)* (pp. 260–263). https://doi.org/10.2991/icglow-19.2019.65

Wirastri, M. V., Morrison, N., & Paine, G. (2023). The connection between slums and COVID-19 cases in Jakarta, Indonesia: A case study of kapuk urban village. *Habitat International, 134*, 102765. https://doi.org/10.1016/j.habitatint.2023.102765

Wong, Z. Z. A., Badeeb, R. A., & Philip, A. P. (2023). Financial inclusion, poverty, and income inequality in ASEAN countries: Does financial innovation matter? *Social Indicators Research, 169*(1), 471–503. https://doi.org/10.1007/s11205-023-03169-8

World Bank. (2020). *Aspiring Indonesia – expanding the middle class.* https://www.worldbank.org/in/country/indonesia/publication/aspiring-indonesia-expanding-the-middle-class.

Yuda, T. K., & Munir, M. (2023). Social insecurity and varieties of family resilience strategies during the COVID-19 pandemic. *International Journal of Sociology and Social Policy, 43*(7/8), 756–776. https://doi.org/10.58671/aswj.v11i1.41

Yuliani, P. A. (2023). *Ribuan pengemis dan manusia gerobak terjaring sejak februari.* Media Indonesia. https://mediaindonesia.com/megapolitan/569011/ribuan-pengemis-dan-manusia-gerobak-terjaring-sejak-februari. (Accessed 3 June 2023).

CHAPTER

15

Effects of homelessness on quality of life and health

Roshan Baa

St. Xavier's College, Ranchi, Jharkhand, India

1. Introduction

Homelessness is the state of having no home, being unsettled, or unhoused which leads one to be on the street, life with brokenness sleeping rough everyday. Homelessness is invisible and easily ignored. It is one of the major social issues worldwide that is poorly impactful on physical and mental health of people. This is a global phenomenon which effects on the quality of life (QoL) and health of the population. It can be defined as the individuals who are unable to have permanent housing and mainly live on the streets are called the homeless people who lack fixed, regular, and adequate nighttime residence, instead living in emergency shelter, transitional houses which are not meant for human habitat. A large number of the population is homeless worldwide in recent times and poverty is the main cause of homelessness. These people live a particular quality of life where they do not have shelter, adequate food and clothes as well. Poverty is arising day by day in the global context and along with that the number of people who are homeless is also increasing which represents the inability of governance and social workers. Almost all the countries have some people who are homeless but some of the countries have the highest number of homeless population. Europe, Asia, Africa, South America, Oceania, and North America are the countries which have the highest percentage of homelessness. Among all of these countries, Nigeria has the highest number of homeless people in the world which is in Africa where ***about 24,400,000 people*** are homeless out of 216 million populations with a rate of **11.3%** (UK Households into Homelessness, 2022). Apart from that, **0.37%** of Europe's population, **8.3%** of Asia's population, **12.35%** of North America's population, **0.5%** of Australia's population is homeless. Therefore, it can be stated that a huge percentage of the population worldwide are suffering from homelessness and that need to be prevented. There are different factors affecting homelessness worldwide which are discussed briefly in this study. In contrast, identifying the impacts of homelessness on QoL and the health of people is the main purpose of this study. It is identified that people might have several physical health issues due to homelessness as they

Homelessness to Hope
https://doi.org/10.1016/B978-0-443-14052-5.00015-X

do not get proper nutrition as well as mental health disorders are identified among the homeless population. Hence, the study includes all the data and information about the impacts of homelessness that are collected through secondary sources. Homelessness needs love, grace, and hope of Gospel.

Homelessness which is one of the biggest problems worldwide and caused by another big social issue that is poverty. Apart from poverty, there are other secondary reasons for the increase of homelessness which are discussed in this section of the study. It is identified that about 100 million people were homeless in 2005 which increased to ***nearly 150 million*** in 2021 (Homeless World Cup 2023, 2023). The number of homeless people is rapidly enhancing and along with that different health issues are also increasing. Hence, not only health issues but also social issues, legal issues, and economic issues might occur due to homelessness. Thereafter, it is identified that ***about* 25%** of homeless people are children in this world (Homelessness Facts, 2022). Poverty is considered as the primary cause of homelessness but there are other reasons as well which leads people to be homeless. It is identified that there are several factors that are increasing poverty worldwide and those are the same determinants of homelessness. Lack of affordability is the main reason for most of the homeless people, which is caused due to ***poverty***. Uncertain events such as COVID-19 pandemic, poor economic system and unemployment are the main causes for increasing poverty (Peterie et al., 2020). It is identified that the number of people who live under the poverty line has increased by **70 *million*** during the pandemic crisis and ***about* 700 *million*** people worldwide are living in extreme poverty in recent times. Along with that, the poverty rate is **9.3%** in 2022 which was **8.4%** in 2019 (Poverty and Shared Prosperity, 2022). Therefore, poverty is increasing among the global population and along with that the inability of people to buy or rent homes is also increasing (Fig. 15.1).

The figure above represents the OECD poverty rates in different countries for understanding homelessness from a global perspective. It can be identified that the United States is at the top of the list with ***about* 18%** OECD poverty rate and along with that Spain has ***about* 16%** and Greece has ***nearly* 15%** poverty rate (Barua et al., 2019). Therefore, it can be stated that poverty is the main reason for increasing homelessness in these countries. On the other hand, ***the increase of unemployment rate*** due to different uncertain events highly contributes to the increase of poverty which is previously mentioned. It is identified that the COVID-19 pandemic has influenced the unemployment rate of people in the world and that led to the enhancement of poverty (Mabhala et al., 2021). Due to this, people became unable to complete their basic needs which includes home expenses and that led them become homeless.

A huge development in the global unemployment rate during the COVID-19 pandemic crisis can be identified in the figure below. In 2019, the unemployment rate was **5.36%** which changed to **6.58%** in 2020 (O'Neil, 2023). Although the unemployment rate has declined in 2021 and 2022, that huge unemployment has poorly contributed to poverty and that has enhanced homelessness. In addition to that, abusive behavior, domestic violence, single parenting, and others are other secondary causes of homelessness. ***Substance abuse*** such as drugs, alcohol is one of the common reasons among youth for homelessness in recent times(Mabhala et al., 2021). It is identified that ***about* 7.1%** of male and **2.2%** of females globally have alcohol abuse which impacts their health and relationships (World Health Statistics 2022, 2022). Hence, substance abuse of drugs and alcohol can be considered as one of the chronic reasons for the increase of homelessness worldwide.

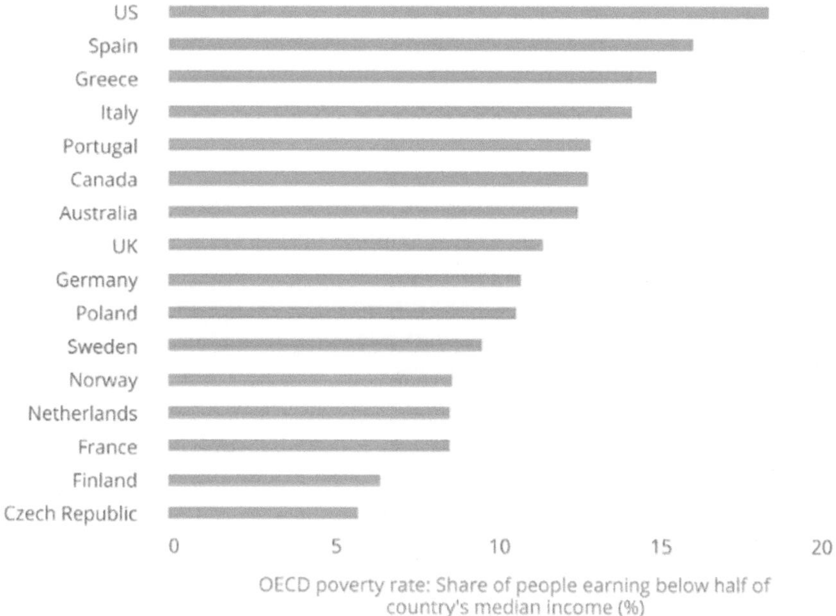

FIGURE 15.1 OECD poverty rates in different Counties. *From Barua et al. (2019).*

Different authors have provided their respective opinions on the causes of homelessness. According to (Zhao, 2023), most of the homeless population are women and children and ***domestic violence*** is the core reason for that. It is identified that most of the homeless women have experienced domestic violence and due to that they left their houses. On the other hand, social norms and conditions are other causes for homelessness. Sometimes the birth of female children is not accepted by families in developing countries and there are some incidents of leaving girl children which makes them homeless. Apart from that, other social norms such as not accepting people from other sexual orientations rather than male and females. Fraser et al. (2019) stated that LGBTIQ+ is a community of people who are not accepted by societies and even their own families which enhances the percentage of homelessness among this particular community. In contrast, ***the high cost of housing*** is another cause for the enhancement of homelessness. The cost of housing is higher than employment which enhances the inability of people to rent or buy houses. The figure below represents the percentage of households that are overburdened by high housing costs. It is noticeable that ***about* 98%** households are overburdened by housing costs in Greece as well as ***almost* 78%** people are overburdened in Denmark (Barua et al., 2019). Similarly, people in all the other countries are overburdened by housing costs and that is a major cause of homelessness. Thus, it can be understood that there are different types of factors affecting homelessness in this world. Homelessness is also poorly impactful on the health of people. As per the perception of Fransham and Dorling

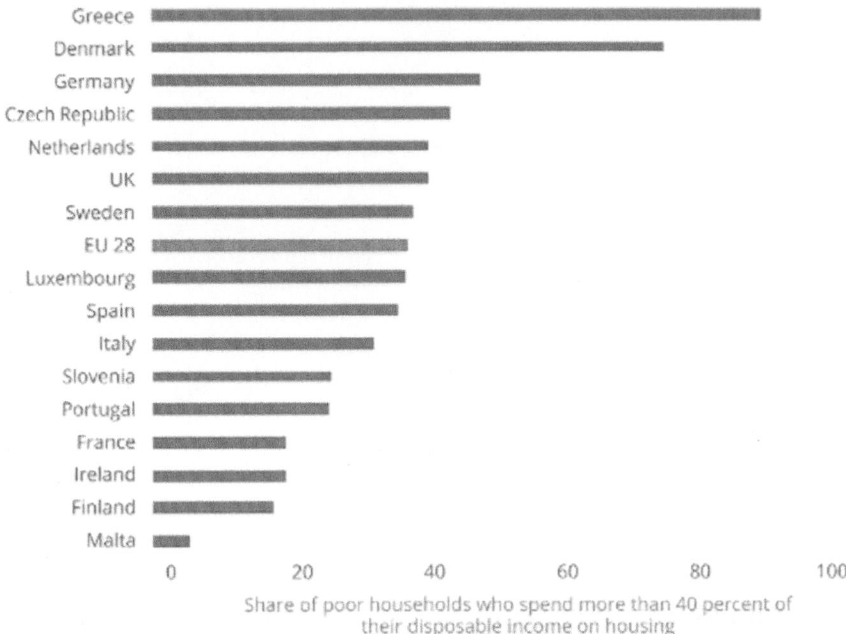

FIGURE 15.2 Housing costs in different countries. *From Barua et al. (2019).*

(2018), poor physical health, mental health, lack of nutrition and other risks are the core effects of homelessness (Fig. 15.2).

On the other hand, it is identified that homelessness is poorly effective on mental health of people. Most of the people who are homeless suffer from anxiety, depression, and stress. These people experience several types of social issues such as sexual abuse and others which enhance mental health disorders among them (Rhoades et al., 2018). Apart from that, physical health issues are quite obvious among people who are homeless. These people are unable to complete their basic needs, even the need of adequate nutrition which led to several diseases such as musculoskeletal disorders, dental problems, respiratory illness and others (Fransham & Dorling, 2018). In addition to that, the chances of chronic diseases such as health damage, brain damage, lungs damage, kidney damage, and others also increase due to improper nutrition. Furthermore, homeless people majorly suffer from infectious diseases such as HIV infection as well as substance abuse also increases among homeless people. According to Gultekin et al. (2020), homelessness is highly impactful on the health of school-age children and youth. Children can suffer from respiratory infections, asthma, ear infections, severe allergies, and others due to homelessness. A brief discussion about the effects of homelessness on physical and mental health of people is presented in this study. It is identified that governments of different countries have recognized homelessness as one of the core issues and taken several initiatives for preventing this issue. "Pradhan Mantri Awas Yojana (PMAY) Scheme," "Housing First and the Affordable Housing Program," "federal housing assistance," "Anjung Singgah," "Reaching Home," and others are some of the governmental initiatives. Hence, participation of governmental authorities in eliminating homelessness in different countries is discussed in this study. On the other hand, it is identified that homelessness enhances different social issues such as rape, sexual exploitation and others. It is identified that increasing homelessness can be a core reason for the decline in literacy rate in future. Therefore, homelessness is poorly effective on society, economy, and other factors and that indicates the requirement of the prevention of homelessness (Gultekin et al., 2020). Along with government authorities, several private organizations are there which contributes to the prevention of homelessness by providing them temporary shelters. The pandemic crisis during 2019–20 has brought a huge growth in unemployment, poverty, and homelessness. Governmental authorities have adopted different strategies and employment rate has increased after the pandemic crisis but some particular initiatives for preventing poverty, violence, and other causes of homelessness are required.

Considering this, some recommendations are presented in this study for improving the condition of societies and eliminating homelessness. Thereafter, all the factors that are influencing the growth of homelessness in a global context are discussed in this study as well as the social effects of homelessness are also presented. Hence, particular methods and tools have been adopted for gathering data and conducting this study and justification for choosing the particular methods are also presented. Secondary sources such as articles, journals, news articles, magazines and others have been used for gathering adequate information about homelessness and the quality of life of the homeless people. It is identified that governmental and nongovernmental authorities are focused on eliminating homelessness without concern about the factors that are influencing the growth of homelessness. Therefore, the authorities need to prevent the causes of homelessness such as poverty, higher costs of housing, domestic violence, and others for faster prevention of homelessness.

2. Materials and methods

2.1 Research philosophy

While investigating a social phenomenon scientific integrity has been observed. The pragmatism, positivism, interpretivism, and realism are assimilated in research process. An interpretivism research philosophy has been selected to use in the following study in order to explore the context of homelessness and its impact on the quality of life. As per the viewpoints of Alharahsheh and Pius (2020), the interpretivism research approach involves a subjective perspective toward a social phenomenon and is concerned with the variables. Therefore, this particular research philosophy has been selected to use as it will help to address the factors influencing homelessness among people as well as the impacts of homelessness.

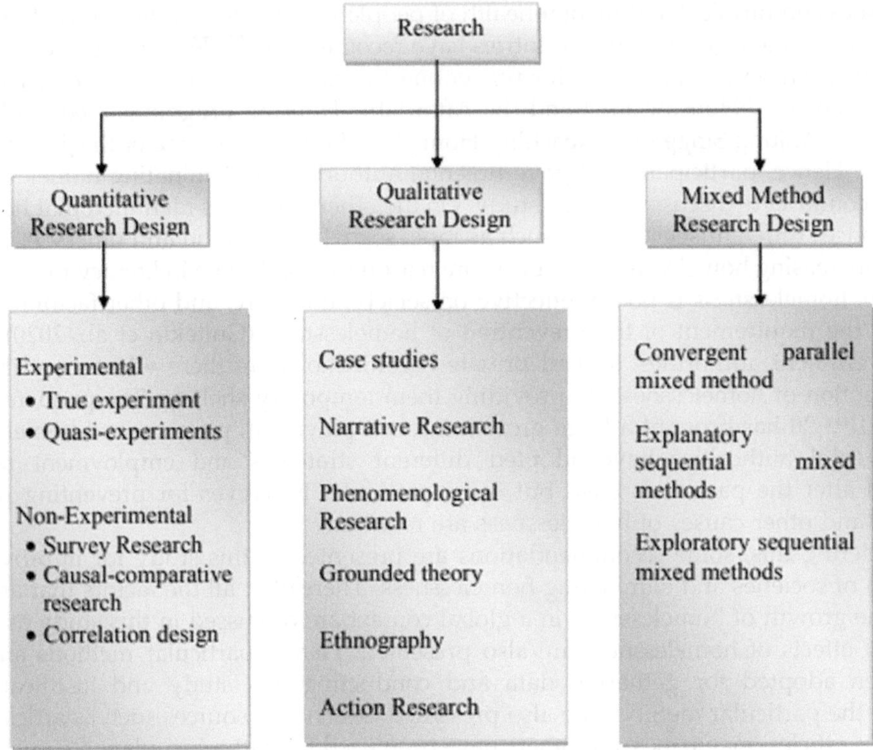

FIGURE 15.3 Types of research design. *From Asenahabi (2019).*

2.2 Research design

Exploratory design has been chosen which has led to qualitative as well as quantitative studies and is regarded as a mixed method (Asenahabi, 2019). Hence, this research design has helped to explore and develop accurate insights associated with homelessness and its impact on the quality of life (Fig. 15.3).

2.3 Research approach

The scientific research approach that has helped to plan the stages leading to collecting, analyzing, and interpreting data and developing potential solutions. There were three types of research approaches, including deductive, abductive, and inductive and in this study, an inductive approach has been selected to use. According to Woiceshyn and Daellenbach (2018), using an inductive approach helped to develop and advance knowledge regarding the research problem. Considering these aspects, an inductive research approach has been selected in the following study to collect relevant data associated with homelessness among people and its social effects and impacts on quality of life.

2.4 Data collection

The relevant and accurate information associated with the research phenomenon by exploring relevant sources have been noted. The primary or a secondary approach is taken for the survey. A secondary data collection method has been dominantly selected for gathering qualitative data associated with homelessness and its social impacts. In order to gather primary data, the research has involved research population directly, while secondary data is already structured and bias-free (Martins et al., 2018). Regarding this, published journals, websites and newspaper articles have been explored for collecting relevant and reliable data associated with the factors influencing homelessness among people and government initiatives.

2.5 Data analysis

Data analysis involved assessing, interpreting, and evaluating the gathered data along with addressing patterns. In order to assess the secondary qualitative data that has been gathered, a thematic analysis has been chosen to conduct and several themes have been developed in this regard. Thematic analysis for analyzing qualitative data (Sundler et al., 2019) is also observed while analysis to understand the factors associated with homelessness and associated issues and impacts.

2.6 Inclusion and exclusion criteria

2.6.1 *Inclusion criteria*

- Reliable and relevant secondary sources, such as websites, newspaper articles, and journals published on or after 2018 have been used for data collection.
- Journals and websites that are free to access have been included in this study.
- Secondary sources published in English language only, have been used for data collection.

2.6.2 *Exclusion criteria*

- Secondary sources that are not free to access have not been used for data collection.
- Journals and websites published in other languages rather than in English have been excluded from data collection.
- Newspaper articles, websites and journals containing relevant information or published before 2018 have been excluded.

2.7 Ethical considerations

During the data collection procedures of this study, all potential research ethics have been maintained and no natural objects have been harmed. Free-to-access secondary sources have been explored and no data has been manipulated during these procedures. A secondary qualitative approach has been followed for collecting relevant data and no person involving the research phenomenon has been asked for disclosing their opinions and experiences.

3. Results and discussion

3.1 Results

3.1.1 *Theme 1: Physical and mental effects of homelessness*

Homelessness can be referred to as a form of social exclusion which can be influenced by several types of factors, such as conflicts with family members or relationship breakdowns. The homelessness can occur due to feeling a shortage of safe or adequate places for living as well as due to extreme poverty. Homelessness can cause severe impacts on the physical and mental health of the people suffering from it. It has been observed that spending a major time of adolescence on the streets can cause Alzheimer's disease as well as other types of brain disorders among homeless adults (Maxmen, 2019). Another major physical effect of homelessness is rapid aging. Consequently, homeless people are more likely to suffer from urinary incontinence, visual impairments and strokes from an early age which indicates that their mortality rate is also high. Homeless people suffer from more physical and mental health issues than people living in homes. A major portion of homeless people suffer from "*Dementia*" and the care service providers of shelters for homeless people claim they lack the infrastructure to treat this disease. According to McCann and Brown (2019), the main issues associated with the physical and mental effects of homelessness are the prevalence of substance use, mental health problems as well as sexual risks. In addition, young people belonging to the LGBTQ+ group have been identified as the most vulnerable group to be affected by these issues. The main causes of death among these people have been identified to be diabetes, heart attacks and cancer. Moreover, it has been observed that poor physical and mental health and homelessness are closely linked and there are several other health issues, such as bronchitis, asthma, blood pressure, pneumonia, epilepsy, liver damage, and wound infections. The unhealthy, irresponsible, and risky lifestyle of people living in stress is the main cause affecting their mental, physical, and emotional health (Fig. 15.4).

Homeless people lack access to sufficient treatment and healthcare facilities which is another vital reason for their poor mental and physical health. Additionally, oftentimes, they receive little or no treatment when required and these aspects influence poor health and various health issues. It has been observed that LGBTQ+ people make up around 20%–40% of homeless people, while they receive less attention and focus (Fraser et al., 2019). Moreover, the prevalence of risky jobs for earning a livelihood is also high among these people who have enhanced infections, wounds, and HIV infections. Homeless people are provided with inadequate access to healthcare as well as shelters, which has influenced high physical as well as mental health risks for these people.

3.1.2 *Theme 2: Impact of homelessness on quality of life*

Homelessness not only impacts the mental and physical well-being of the suffering persons but also affects the quality of life or QoL. It has been observed that homeless people are more likely to attain mental and physical issues, which affects their QoL. As per the opinions of (Tinland et al., 2018), homeless women are more prevalent to develop "post-traumatic stress disorder" or PTSD due to suffering from sexual assaults. Regarding this, homeless women's QoL has been identified to be more highly affected than men living on the streets and this prevalence often leads them to commit suicidal attempts. These women suffer from

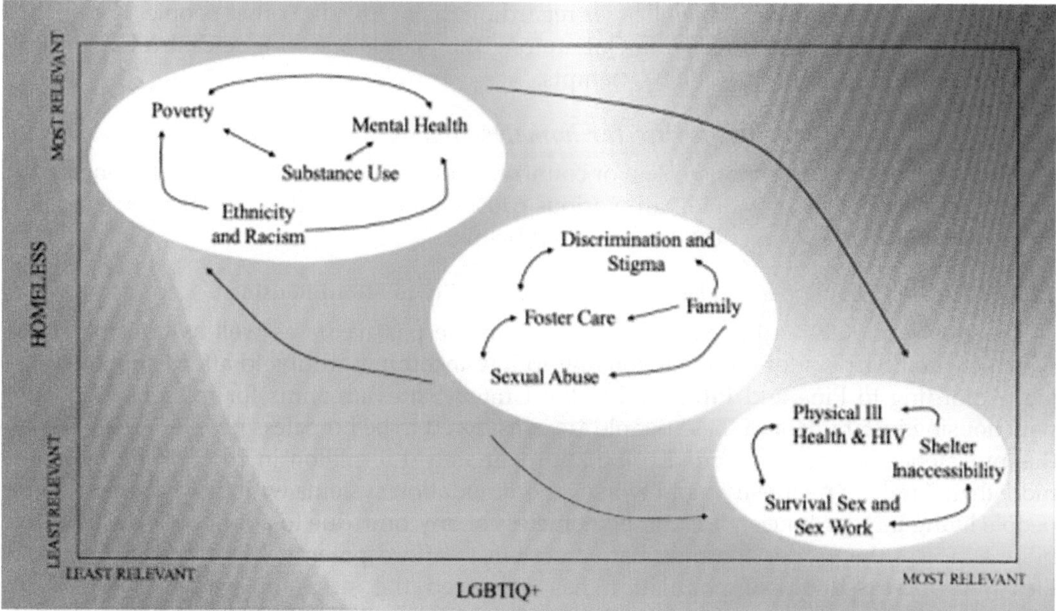

FIGURE 15.4 Experiences of LGBTQ + people as homeless. *From Fraser et al. (2019).*

various mental health disorders including bipolar disorder and schizophrenia and these disorders have a direct impact on their QoL. Poor and insufficient access to essential medications, proper treatment, healthcare, and other basic infrastructure as well as physical and sexual abuse affects the QoL among homeless women and increases suicidal risk. Homeless men face employment issues, healthcare inadequacy, and poverty issues that affect their QoL. Consequently, homelessness not only challenges health and mental well-being but also influences the prevalence of illicit drug consumption among youths. It has been observed that more than half of the young adults living on the streets are habituated to substance usage which affects the QoL of themselves as well as their families. As per the viewpoints of Robinson (2019), governments of different countries have made effective interventions to improve the QoL among homeless people; however, improper management has led to enhance challenges. Moreover, improper management, lack of proper infrastructure and negligence of the responsive persons have worsened the QoL of homeless people in several regions.

The main factors associated with the negative impacts of homelessness on the QoL are abusive incidents, uncertainties, substance consumption, sexual and physical exploitation and associated with traumatic experiences. It has been observed that homelessness affects people living in high-income and low-income, both types of countries and the traumatic experiences affect QoL and well-being (Mejia-Lancheros et al., 2021). Therefore, it can be stated that homelessness consists of a direct connection with negative consequences toward the QoL of the suffering people and government negligence and improper management of available facilities have increased this prevalence. It has been observed that homeless people belonging to indigenous groups face more challenges regarding their QoL than other people (Kidd et al., 2019). Thus, homelessness affects the QoL among the people suffering and also has the potential to lead them toward suicidal attempts.

3.1.3 Theme 3: Influencing factors for homeless people

It has been observed in every region or country, homelessness among people has increased rapidly for many years and there are various causes influencing this prevalence. The main factors influencing homelessness are provided below.

3.1.3.1 Increased population and lack of financial capability to afford housing

The global population has increased rapidly in the past years as well as housing rents, which created barriers for people belonging to low-incoming groups to afford proper housing. According to Piña and Pirog (2019), the United States accounts for millions of people with housing instability and these people are considered to be homeless people. Additionally, this prevalence has been recorded to be approximately 1.5 million in 2015, while there are more than 7 million people living in worse accommodation systems with their families. These people living in poor households can be homeless at any time due to a lack of financial capability to access the proper accommodation system to afford potential housing for their families. The increased global population has enhanced the stress of searching for larger accommodation systems and increased expenses have created destructive barriers, which have eventually led to homelessness.

3.1.3.2 Unemployment and financial risk

Poverty has been identified to be the major cause affecting homelessness among people. Additionally, financial risk and unemployment are also influencing factors associated with poverty which influences homelessness in both developed and developing nations. It has been observed that unemployment is among the major factors influencing and predicting homelessness (Calvo et al., 2018). Additionally, low wages, increasing housing rate, and increasing financial expenses have the potential to influence homelessness among people. Globalization has increased daily expenses and low wages, and unemployment can create barriers for stressed people to rent housing and become homeless.

3.1.3.3 Mental illness and lack of access to essential facilities

It has been observed that nearly half of the homeless youths indulge in substance consumption habits and various types of mental disorders. These practices have eventually influenced homelessness among young adults as well as people of other ages who are suffering from severe mental disorders. It has been observed that the prevalence of mental illness, suicidal behavior and police interactions are more prevalent among homeless people than among people living in homes (Kouyoumdjian et al., 2019). Thus, these are the main factors influencing homelessness among people and it has been observed that these factors also affect the QoL and mental health of homeless people.

3.1.4 Theme 4: Social effects of homelessness

The forces that influence the process of homelessness are reciprocal and complex in nature. Homelessness is a condition when an individual is living without regular accommodation.

People who are homeless cannot continue the living process in a regular and maintainable way. As stated by Bezgrebelna et al. (2021), there are several social forces that can be reasons for occurring homelessness. It includes poverty, mental illness, substance abuse, domestic violence, housing cost, unexpected events, single parenthood, relationship breakdown, and many more. Homelessness is a social problem that causes the irregular lifestyle of people. In many cases, homeless people are considered to have little essentials such as food, money, clothes, shelter, and medical attention. Most homeless people spend their nights uncomfortably on the street or on platforms. In addition to that, many homeless people are also addicted to alcohol or drugs. It has been observed that because of homelessness, the government of an individual country needs to spend more money. Overall, in the world there are several homeless people who are suffering from mental health issues which in most cases lead to traumatic upbringing procedures. In addition to that, it also has become identified that communities with high rates of homelessness are also considered as high rates of crimes. Therefore, homeless people are also identified as vulnerable to the fight against communicable diseases (Pleace & Hermans, 2020). Globally several destructive diseases have been spread with the collaboration of homeless people as they are unable to protect themselves from any kind of negative impact. Sleeping in the street is also an unsafe way to live life and it is also a reason for becoming a health issue among homeless people. In the case of women who become homeless, spending nights on the street or on platforms are not safe and secure. Even they are mostly affected by several social issues which may result in negative experiences.

It is quite natural that all societies are formed based on a particular percentage of the population and among them, few are able to earn high wages to lead life and few become unable and become homeless. In a general aspect, homelessness is not a personal problem but actually, it is a societal issue that is leading to the result of homelessness. There are several societal factors that impact the constantly rising amount of homelessness. As opposed to Yoshioka-Maxwell (2020), homelessness is the failure of society to make sure the living process is successful for an individual. Apart from this, many aged people often become burdens for their family members, and they are thrown out to their houses to the street or old-age homes.

3.1.5 Theme 5: Governmental initiatives for preventing homelessness

In order to secure and balance the population range, the government has adopted several kinds of initiatives for preventing homelessness. As primary prevention, providing employment opportunities, affordable housing, and removing discrimination are the better key tools. In different developing countries, individual governments have taken different initiatives. In India, the Prime Minister has implemented the "***Pradhan Mantri Awas Yojana (PMAY) Scheme***" to reduce homelessness and provide people with essential needs. This is identified as the flagship mission that has been launched on June 25, 2015 to develop urban society (5TH Anniversary of Urban Missions, 2020). Along with this, Canada is also a developing country that has several homeless people, and observing this, the government has initiated "***Reaching Home***" which is considered a community-based program to fund the homeless people. In addition to that, in the UK the government has implemented "***Housing First and the Affordable Housing Program***" to address homelessness and rough sleeping. The "Homelessness Reduction Act" was established on April 3, 2018, in the United Kingdom to take reasonable steps to mitigate all the needs of homeless people (The UK's Homelessness, 2022). Japan is another country that is also trying to control the rate of homeless people by implementing Japan's housing options, examining mental health systems, and strict drug laws. Malaysia is also a renowned country that has undertaken a serious concern regarding the mitigation process of homelessness as the government launched the "***Anjung Singgah***" project in April 2011 (Adib & Ahmad, 2018). It has been implemented in Malaysia basically to provide temporary shelters to needy people to secure them for a period of time.

The United States is another country that is concerned about the homelessness activity of the population. The government of the US has implemented a "***federal housing assistance***" program that is a housing-based solution to mitigate homelessness (State of the Homeless 2022, 2022). In addition to that, to motivate the operators, the "Homeless Assistance Grants" and the "United States Department of Housing and Urban Development (HUD) award" to administer housing and services (Cornes et al., 2018). Following the different aspects of different countries, it has been understood that every country which has social welfare schemes did not face a high level of homelessness.

3.2 Discussion

Several physical and mental effects of homelessness have been identified, including infections, wounds, Alzheimer's diseases, dementia, and other mental disorders. In addition, homeless people are provided with less access to proper medication, treatment and other healthcare facilities which is the main cause of their poor mental and physical health. It has been observed that the prevalence of mental illness due to homelessness is more crucial in low and middle-income countries (Smartt et al., 2019). In addition, a major portion of homeless people belong to the LGBTQ + group and they face more difficulty to access healthcare facilities, which has increased the prevalence of deaths due to heart attacks, diabetes, pneumonia, and cancer. These people suffer from several other physical issues including blood pressure issues, epilepsy, liver damage, asthma, and bronchitis. Homelessness poses a destructive impact on QoL among people and causes severe mental disorders. It has been observed that homeless women suffer more issues regarding their QoL than homeless men and they are more likely to commit suicidal attempts. According to Hoffberg et al. (2018), the suicidal death rate of homeless military veterans in the United States is 81 in every 100,000 deaths. Additionally, the main factors influencing homelessness are severe unemployment and associated financial risk, lack of adequate financial capability for renting a house and mental disorders and substance consumption. Therefore, it can be stated that these are major factors associated with the crucial causes of homelessness that have increased its prevalence and also developed severe mental health issues among the people. Government negligence and improper management of available facilities have also increased the life stress, safety issues and challenges for homeless people. Therefore, proper maintenance of the facilities is highly required as well as ensuring adequate infrastructure for these people. Homelessness is a serious societal issue that causes serious concern all over the world. Due to the absence of a social support system, the experience of homelessness is remaining as a "shadowy edge of society." Homelessness can put an individual in any kind of dangerous situation even if it may cause serious harm. It can lead an individual to depression, loneliness, victimization, and poor health. As per the view of Labella et al. (2019), the activity of homelessness impacts the crime and safety process, as well as also affects the availability of healthcare resources. There are several reasons behind homelessness, and it impacts an individual mind as well as physical wellness.

Poverty, mental illness, substance abuse, domestic violence, housing cost, unexpected events, single parenthood, and relationship breakdown are the reasons behind the increasing number of homeless across the world. Thus, mitigating this societal issue, contrast governments of different countries have taken different initiatives. As opined by Aldridge (2020), all over the world, different developing countries have participated in this journey to secure the world population by implementing effective schemes and programs to improve awareness among homeless people. Implementing all programs and schemes was established based on the requirement of the specific country to balance the population well-being.

4. Limitations of the study

In the present study, the context of homelessness and associated factors as well as its impacts has been discussed. The outcomes of the data analysis procedure have provided significant evidence and insights associated with the mental and physical effects of homelessness. It has been observed that homelessness can be influenced by various types of factors, such as family or relationship bias, poverty, addiction, and abuse. However, the results lack addressing future interventions regarding preventing homelessness among people. In the results section, no future interventions or areas for improvement have been addressed, though the social, physical, and mental effects of homelessness have been addressed. Therefore, effective recommendations will be developed for addressing the limitations and improving the required areas. Additionally, other major limitations are a severe shortage of time and budget that has affected the data collection process and created limitations for utilizing primary data collection methods. Using primary data can provide more accurate insights by directly involving the research population.

5. Recommendations

Following the constantly rising homeless population all over the world it can be mentioned that the government, as well as the community, needs to be more aware of the activities. In this regard, developing a new federal housing framework can help in managing the homelessness process. Thus, poverty is one of the major reasons behind the appearance of homelessness. In this regard to reduce the poverty level, the responsible authority, as well as individual government, needs to take some initiatives that can balance the employment process to make economic stability among the poor. As proposed by Baral et al. (2021), establishing gender equality is one of the greatest elements to reduce the impact of poverty. Along with this, the high cost of housing is another reason for homeless people so in this segment the government needs to implement some schemes to provide shelters to the poor and homeless at a lower rate. Implementing this approach in real life, a coordinated approach is identified as useful as through this the housing cost can be reduced to help the homeless people by having their own shelter. The health issue of homeless people is quite in contrast from the poor people who are directly connected to their homeless states. In the words of (Doshani et al., 2019), the condition of homelessness is identified as a risk factor that influenced health issues. At this stage, homeless people require proper healthcare options to secure their health. The individual government needs to establish healthcare centers in different areas of an individual country. It shall cater to the really needy people such as homeless women, old individuals, or children. As most homeless people are a victim of several unexpected events, it also needs proper healthcare option to deal with mental illness. The COVID-19 pandemic situation is considered one of the serious events that have been unexpected for homeless people.

Improving the medical management process in every individual country can help people to make sense they are under proper observation. They suffered the most as they could not get proper food or shelter to avoid the situation. It needs urgent healthcare solutions to identify health diseases and cure them in time. In addition to that, donating basic items such as food, clothes, and other accessories can be proven a better approach to helping homeless people. According to Magwood et al. (2020), overall, the world people need a stable income level by that they can spend their life away from footpaths. In order to prevent homelessness, it is considered the most effective option that can assist people to make housing affordable. In addition to that, supportive services are necessary for a range of homeless people as that is the reason behind their homelessness. In several cases, it also has been found that homeless people have experienced being chronically mentally ill. They required proper social support systems as well as also a useful healthcare management system to protect them from all the potential issues. Effective discharge planning can be proved as beneficial in this case as it is effective to ensure that these needs are met. Accompanied with supportive services homeless people can become self-reliant which will be a positive approach to modify the ratio of homeless people across the world. As stated by Gibson (2019), due to a lack of proper knowledge many homeless people could not recognize the governmental initiatives. In this context, the government needs to focus on the implementation of several educational programs to make the homeless people aware of the initiatives that the individual government has taken to reduce the rate of homeless people. Domestic violence is also a vital reason behind the appearance of homelessness. In this segment, every individual government needs to take more serious steps to implement the legal act and observe its application to make sure it will sustain to reduce the ratio of homeless people.

6. Conclusions

Homelessness is a bitter feeling that is limited particular people to leading their lives according to their wishes. There are many reasons and factors that are responsible for occurring homelessness on the quality of life across the globe. It has been observed that almost in every country there are some people who are remaining homeless due to some specific reasons. The reasons include poverty, mental illness, substance abuse, domestic violence, housing cost, unexpected events, single parenthood, relationship breakdown, and many more. Among all these reasons, poverty is identified as one of the major reasons by that the number of homeless is gradually increasing day by day overall the globe. In addition to that, the constantly rising cost of everything is also developing the housing cost which made people of being homeless due to not afford the higher cost. Along with this, an uncertain event such as COVID-19 is the influence of unemployment and a poor economic system which are one of the reasons behind increasing homelessness. It has been noticed that after the pandemic situation depending on the unemployment has risen the chances of being homeless have risen enough. In this context, it has been also observed that the number of people who are living under the poverty line has increased a lot after the pandemic. Domestic violence is also responsible for increasing the ratio of homeless people as the act is not applied well in the case of domestic violence. In addition to that, substance abuse such as alcohol consumption and drug addiction are also the reasons behind homelessness among the youth of this recent generation. Thus, the individual government has taken different initiatives to fight against the growth of homelessness. In every country the implementation of government initiatives is different, and it also made the proper initial outcome limited. Moreover, this study can help the raiders to make a clear knowledge regarding the reasons behind homelessness and also can provide effective solutions to make it possible to reduce the negative impact on society and the economy of the entire world.

References

5TH Anniversary of Urban Missions. (2020). https://pib.gov.in/PressReleasePage.aspx?PRID=1634268. (Accessed 18 February 2023).

Adib, N. A. M., & Ahmad, Y. (2018). How effective are the current initiatives in dealing with homelessness in Malaysia? *Journal of Administrative Science, 15*(3).

Aldridge, R. (2020). Homelessness: A barometer of social justice. *The Lancet Public Health, 5*(1), e2–e3. https://doi.org/10.1016/S2468-2667(19)30240-3

Alharahsheh, H. H., & Pius, A. (2020). A review of key paradigms: Positivism VS interpretivism. *Global Academic Journal of Humanities and Social Sciences, 2*(3), 39–43.

Asenahabi, B. M. (2019). Basics of research design: A guide to selecting appropriate research design. *International Journal of Contemporary Applied Researches, 6*(5), 76–89.

Baral, S., Bond, A., Boozary, A., Bruketa, E., Elmi, N., Freiheit, D., Ghosh, S. M., Goyer, M. E., Orkin, A. M., Patel, J., Richter, T., Robertson, A., Sutherland, C., Svoboda, T., Turnbull, J., Wong, A., & Zhu, A. (2021). Seeking shelter: Homelessness and COVID-19. *Facets, 6*, 925–958. https://doi.org/10.1139/FACETS-2021-0004

Barua, A., Frey, R., Suriya, N., & Byrnes, K. (2019). *The homelessness paradox; why do advanced economies still have people who live on the streets?*.

Bezgrebelna, M., McKenzie, K., Wells, S., Ravindran, A., Kral, M., Christensen, J., Stergiopoulos, V., Gaetz, S., & Kidd, S. A. (2021). Climate change, weather, housing precarity, and homelessness: A systematic review of reviews. *International Journal of Environmental Research and Public Health, 18*(11), 5812. https://doi.org/10.3390/ijerph18115812

Calvo, F., Carbonell, X., & Badia, M. (2018). Homelessness and unemployment during the economic recession: The case of the city of Girona. *European Scientific Journal, ESJ, 14*(13), 59. https://doi.org/10.19044/esj.2018.v14n13p59

Cornes, M., Whiteford, M., Manthorpe, J., Neale, J., Byng, R., Hewett, N., Clark, M., Kilmister, A., Fuller, J., Aldridge, R., & Tinelli, M. (2018). Improving hospital discharge arrangements for people who are homeless: A realist synthesis of the intermediate care literature. *Health and Social Care in the Community, 26*(3), e345–e359. https://doi.org/10.1111/hsc.12474

Doshani, M., Weng, M., Moore, K. L., Romero, J. R., & Nelson, N. P. (2019). Recommendations of the advisory committee on immunization practices for use of hepatitis a vaccine for persons experiencing homelessness. *Morbidity and Mortality Weekly Report, 68*(6), 153–156. https://doi.org/10.15585/MMWR.MM6806A6

Fransham, M., & Dorling, D. (2018). Homelessness and public health. *British Medical Journal, 1*(1), 1–5. https://doi.org/10.1136/bmj.k214

Fraser, B., Pierse, N., Chisholm, E., & Cook, H. (2019). LGBTIQ+ homelessness: A review of the literature. *International Journal of Environmental Research and Public Health, 16*(15). https://doi.org/10.3390/ijerph16152677

Gibson, A. (2019). Climate change for individuals experiencing homelessness: Recommendations for improving policy, research, and services. *Environmental Justice, 12*(4), 159–163. https://doi.org/10.1089/env.2018.0032

Gultekin, L. E., Brush, B. L., Ginier, E., Cordom, A., & Dowdell, E. B. (2020). Health risks and outcomes of homelessness in school-age children and youth: A scoping review of the literature. *The Journal of School Nursing, 36*(1), 10–18. https://doi.org/10.1177/1059840519875182

Hoffberg, A. S., Spitzer, E., Mackelprang, J. L., Farro, S. A., & Brenner, L. A. (2018). Suicidal self-directed violence among homeless US veterans: A systematic review. *Suicide and Life-Threatening Behavior, 48*(4), 481–498. https://doi.org/10.1111/sltb.12369

Homeless World Cup 2023. (2023). https://www.homelessworldcup.org/. (Accessed 12 February 2023).

Homelessness Facts. (2022). https://facts.net/homelessness-facts/. (Accessed 9 February 2023).

Kidd, S. A., Thistle, J., Beaulieu, T., O'Grady, B., & Gaetz, S. (2019). A national study of Indigenous youth homelessness in Canada. *Public Health, 176*, 163–171. https://doi.org/10.1016/j.puhe.2018.06.012

Kouyoumdjian, F. G., Wang, R., Mejia-Lancheros, C., Owusu-Bempah, A., Nisenbaum, R., O'Campo, P., Stergiopoulos, V., & Hwang, S. W. (2019). Interactions between police and persons who experience homelessness and mental illness in Toronto, Canada: Findings from a prospective study. *Canadian Journal of Psychiatry, 64*(10), 718–725. https://doi.org/10.1177/0706743719861386

Labella, M. H., Narayan, A. J., McCormick, C. M., Desjardins, C. D., & Masten, A. S. (2019). Risk and adversity, parenting quality, and children's social-emotional adjustment in families experiencing homelessness. *Child Development, 90*(1), 227–244. https://doi.org/10.1111/cdev.12894

Mabhala, M., Esealuka, W. A., Nwufo, A. N., Enyinna, C., Mabhala, C. N., Udechukwu, T., Reid, J., & Yohannes, A. (2021). Homelessness is socially created: Cluster analysis of social determinants of homelessness (SODH) in North West England in 2020. *International Journal of Environmental Research and Public Health, 18*(6), 1–15. https://doi.org/10.3390/ijerph18063066

Magwood, O., Hanemaayer, A., Saad, A., Salvalaggio, G., Bloch, G., Moledina, A., Pinto, N., Ziha, L., Geurguis, M., Aliferis, A., Kpade, V., Arya, N., Aubry, T., & Pottie, K. (2020). Determinants of implementation of a clinical practice guideline for homeless health. *International Journal of Environmental Research and Public Health, 17*(21), 7938. https://doi.org/10.3390/ijerph17217938

Martins, F. S., Cunha, & Serra, F. A. R. (2018). Secondary data in research—uses and opportunities. *PODIUM Sport, Leisure and Tourism Review, 7*(3).

Maxmen, A. (2019). *What are the biological consequences of homelessness?*, 11 https://www.scientificamerican.com/article/what-are-the-biological-consequences-of-homelessness/.

McCann, E., & Brown, M. (2019). Homelessness among youth who identify as LGBTQ+: A systematic review. *Journal of Clinical Nursing, 28*(11–12), 2061–2072. https://doi.org/10.1111/jocn.14818

Mejia-Lancheros, C., Woodhall-Melnik, J., Wang, R., Hwang, S. W., Stergiopoulos, V., & Durbin, A. (2021). Associations of resilience with quality of life levels in adults experiencing homelessness and mental illness: A longitudinal study. *Health and Quality of Life Outcomes, 19*(1). https://doi.org/10.1186/s12955-021-01713-z, 14777525.

O'Neil, A., & Statista. (2023). *Global unemployment rate 2002–2021.* https://www.statista.com/statistics/279777/global-unemployment-rate/. (Accessed 4 February 2023).

Peterie, M., Bielefeld, S., Marston, G., Mendes, P., & Humpage, L. (2020). Compulsory income management: Combatting or compounding the underlying causes of homelessness? *Australian Journal of Social Issues, 55*(1), 61–72. https://doi.org/10.1002/ajs4.79

Piña, G., & Pirog, M. (2019). The impact of homeless prevention on residential instability: Evidence from the homelessness prevention and rapid re-housing program. *Housing Policy Debate, 29*(4), 501–521. https://doi.org/10.1080/10511482.2018.1532448

Pleace, N., & Hermans, K. (2020). Counting all homelessness in Europe: The case for ending separate enumeration of 'hidden homelessness. *European Journal of Homelessness, 3.*

Poverty and Shared Prosperity. (2022). https://www.worldbank.org/en/publication/poverty-and-shared-prosperity. (Accessed 11 February 2023).

Rhoades, H., Rusow, J. A., Bond, D., Lanteigne, A., Fulginiti, A., & Goldbach, J. T. (2018). Homelessness, mental health and suicidality among LGBTQ youth accessing crisis services. *Child Psychiatry and Human Development, 49*(4), 643–651. https://doi.org/10.1007/s10578-018-0780-1

Robinson, T. (2019). No right to rest: Police enforcement patterns and quality of life consequences of the criminalization of homelessness. *Urban Affairs Review, 55*(1), 41–73. https://doi.org/10.1177/1078087417690833

Smartt, C., Prince, M., Frissa, S., Eaton, J., Fekadu, A., & Hanlon, C. (2019). Homelessness and severe mental illness in low- and middle-income countries: Scoping review. *BJPsych Open, 5*(4). https://doi.org/10.1192/bjo.2019.32

State of the Homeless 2022. (2022). https://www.coalitionforthehomeless.org/. (Accessed 29 January 2023).

Sundler, A. J., Lindberg, E., Nilsson, C., & Palmér, L. (2019). Qualitative thematic analysis based on descriptive phenomenology. *Nursing Open, 6*(3), 733–739. https://doi.org/10.1002/nop2.275

The UK's Homelessness Reduction act. (2022), 11 https://www.huduser.gov/portal/pdredge/pdr-edge-international-philanthropic-020822.html.

Tinland, A., Boyer, L., Loubière, S., Greacen, T., Girard, V., Boucekine, M., Fond, G., & Auquier, P. (2018). Victimization and posttraumatic stress disorder in homeless women with mental illness are associated with depression, suicide, and quality of life. *Neuropsychiatric Disease and Treatment, 14*, 2269–2279. https://doi.org/10.2147/NDT.S161377

UK Households into Homelessness. (2022). https://www.greaterchange.co.uk/. (Accessed 15 February 2022).

Woiceshyn, J., & Daellenbach, U. (2018). Evaluating inductive vs deductive research in management studies. *Qualitative Research in Organizations and Management: An International Journal, 13*(2), 183–195. https://doi.org/10.1108/qrom-06-2017-1538

World Health Statistics 2022. (2022). https://www.who.int/news/item/20-05-2022-world-health-statistics-2022. (Accessed 17 February 2023).

Yoshioka-Maxwell, A. (2020). Social work in action: Social connectedness and homelessness amidst a pandemic: Are the social impacts of quarantine on homeless populations being adequately addressed? *Hawai'i Journal of Health and Social Welfare, 79*(11), 329–331.

Zhao, E. (2023). The key factors contributing to the persistence of homelessness. *The International Journal of Sustainable Development and World Ecology, 30*(1), 1–5. https://doi.org/10.1080/13504509.2022.2120109

CHAPTER

16

The quality of life pathways linking homelessness to health: A case of Bangladesh

Md. Emaj Uddin

Department of Social Work, University of Rajshahi, Rajshahi, Bangladesh

1. Introduction

Homelessness associated with social catastrophes (e.g., war, social exploitation, population growth) and natural calamities (e.g., climate change, cyclone, river erosion, drought) is increasing in developed and developing countries (Bretherton & Pleace, 2023; Speak & Tipple, 2009). Despite several measures taken to end homelessness, research shows that homelessness in the developed countries (e.g., United States and European Union) has increased by 20%–40% from 2000 to 2020 (Bretherton & Pleace, 2023; Richards & Kuhn, 2023). In Bangladesh, despite national efforts (e.g., Ashrayan project, social welfare) to reduce homelessness, temporary or chronic homelessness associated with population growth, urbanization, and natural hazards has increased by 20%–30% from 2000 to 2020 (Bangladesh Bureau of Statistics, BBS, 2020). Research during the past several decades has shown that homelessness affects physical and mental health that, in turn, increase morbidity and early mortality among homeless people compared to general population (Chikwava et al., 2022; Frankish et al., 2005; Hashim et al., 2021; Hwang, 2002; Ingram et al., 2023). Although homelessness increases people's physical and mental health risk and mortality, little is known about how homelessness pattern affects physical and mental health across the countries, including Bangladesh (Petrovich et al., 2019; Uddin, 2017). Theory and research suggest that homelessness pattern affects health in many ways, including deteriorating quality of life (QOL; Buccieri et al., 2020; Babyan et al., 2021; Gentil et al., 2019; Krabbenborg et al., 2016). Drawing from QOL theory, extant research reveals that homelessness pattern increases joblessness or unemployment, food insecurity, poor diet, malnutrition, sleeplessness, insecurity, unsafety, social stress, and unhealthy lifestyle among homeless people. In addition, it deteriorates social identity, psychosocial status,

Homelessness to Hope
https://doi.org/10.1016/B978-0-443-14052-5.00016-1

and social images in a community (Laquinta, 2016; Oppenheimer et al., 2018). In turn, these poor QOL pathways may have negative effects on physical, mental, and social health (White et al., 2011). Despite homelessness pattern deteriorates QOL pathways and the domains of health, few studies have explained underlying the poor QOL pathways by which homelessness pattern affects people's physical, mental, and social health over time (Gadermann et al., 2021; LaGory et al., 2001).

QOL is a multidimensional construct which includes both objective (e.g., employment, income, wealth, diet, and nutrition) and subjective (e.g., security, sound sleep, social relation, social identity, social image, and healthy lifestyle) indicators of human life (Hubley, 2014). The quality of life theory suggests that transition into homelessness gradually transforms these positive QOL pathways into negative ones (e.g., joblessness or unemployment, income insecurity, food insecurity, poor diet, malnutrition, sleeplessness, insecurity, identity loss, social stress, and unhealthy lifestyle that, in turn, affect health) (Gadermann et al., 2021; LaGory et al., 2001; McNaughton, 2008, p. 7). Understanding these negative trajectories of QOL pathways through which homelessness pattern affects health (e.g., physical, mental, and social) is informative to social and health researchers, policy-makers, social practitioners, and health service providers to improve health risks associated with reducing homelessness and socioeconomic vulnerability and negative perception about sociopsychological life (Clifford et al., 2019; Gadermann et al., 2021; Hubley, 2014; LaGory et al., 2001). In doing so, a systematic search was conducted in google scholar, Scopus, global health, and the Web of Science to identify relevant literature planned to explain underlying the QOL pathways through which homelessness patterns affect health. The following search terms during January through June 2023 were used: "quality of life pathways," "socioeconomic insecurity," "jobloss," "income insecurity," "food insecurity," "deprivation," "malnutrition," "identity loss," "physical and social threat/unsafety," "sleeplessness," "social stress," "unhealthy lifestyle." In this chapter, narrative review is adopted to analyze relevant theories and literature related to the QOL pathways that help understanding how homelessness patterns affect health. This chapter, however, comprises six sections, with the first introductory section working as a background to the chapter. The second section reviews relevant theories and develops an integrated theoretical framework to understand how homelessness affects health domains. Drawing from the theoretical framework, the third section reviews relevant literature to link homelessness to health. Reviewing relevant literature, the fourth section critically explains the QOL pathways by which homelessness pattern affects health. In the fifth section, we present limitations and direction for future research and policy implications, whereas the sixth section presents concluding remarks.

2. Defining key concepts

This chapter uses three key concepts to understand relationships between homelessness, quality of life, and health. The ways these three concepts have been defined are as follows.

2.1 Homelessness

There is no universally accepted definition of homelessness. In literal sense, the term "homelessness" refers to the lack of a home (Chikwava et al., 2022). In defining homelessness, researchers include three indicators: (i) absence of adequate housing; (ii) lack of privacy; and (iii) lack of legal entitlement to housing (Busch-Geertsema et al., 2016; Chikwava et al., 2022). Despite this, home has several connotations, including a sense of material space and subjectivity (Padgett, 2007). In this sense, homelessness refers to a lack of material space and psychosocial attachment to it. Analyzing homelessness, Sommerville (1992) identified seven signifiers such as shelter, hearth, heart, primary, abode, roots, and paradise. Based on these signifiers, McNaughton (2008, p. 7) defined homelessness as a lack of material shelter, lack of privacy, lack of comfort, lack of citizenship rights, lack of ownership over space, and also merges into subjectivities: Lacking a space to develop intimate relationships; lacking a sense of belonging, and with it a secure sense of identity (p.7). Although the patterns of homelessness vary across the countries or studies, the current study classifies homelessness pattern into three categories: (a) temporary homelessness, (b) episodic homelessness, and (c) chronic homelessness (Kuhn & Culhane, 1998). The operational definition of the patterns of homelessness is given in Table 16.1.

2.2 Quality of life

The term **"quality of life"** refers to the degree to which persons perceive their objective and subjective life in a context. WHO (1998) defined quality of (QOL) as "individuals" perception of their position in life in the context of culture and value systems in which they live, and in relation to their goals, expectations, standards, and concerns. Actually, QOL is a multidimensional construct, encompassing material, social, physical, and psychological well-being. In measuring QOL, it is classified into a material, social, physical, and psychological domain at individual and population level. The operational definition of the indicators of QOL is presented in Table 16.1.

2.3 Health

In simple sense, health refers to the absence of illness or disease within a body. Particularly, health is defined by the World Health Organization (WHO) as "a state of complete physical, mental, and social well-being and not merely the absence of disease or infirmity" (WHO, 1948). This definition has been subject to controversy, in particular as lacking operational value, the ambiguity in developing cohesive health strategies, and because of the problem created by use of the word "complete". Huber et al. (2011) defined health as "the ability to adapt and self-manage in the face of social, physical, and emotional challenges." In this chapter, health is classified into (a) physical, (b) mental, and (c) social condition (see, Table 16.1 in detail).

Patterns of Homelessness.

3. Theoretical framework

Understanding the QOL pathways by which homelessness patterns increase health risks is debatable in social and behavioral science, because several theoretical frameworks over the past several decades have been proposed to explain the critical QOL pathways (Gentil et al., 2019; Hajiran, 2006; Halfon et al., 2018; LaGory et al., 2001; Meleis, 2010; Sirgy, 1986, 2011).

TABLE 16.1 Indicators for homelessness, quality of life, and health.

Key concept	Description
Patterns of homelessness	
Temporary homelessness	Refers to the moment or time when people moving between various forms of temporary shelter such as emergency accommodation, staying with friends or supported accommodation (Busch-Geertsema et al., 2016).
Episodic homelessness	Refer to the frequent shifts between sheltered and unsheltered circumstances among homeless people, ranging from for 3–11 months (Lippert & Lee, 2015).
Chronic homelessness	Refers to people experiencing continuous homelessness for 1 year or more or four or more episodes of homelessness in the last 3 years (Busch-Geertsema et al., 2016).
Indicators for quality of life	
Socioeconomic insecurity	Refers to the uncertainty and absence of education, job, and income by which homeless people suffer from material hardship and starvation.
Physical and social unsafety	Refers to the risky natural and social conditions that lead mild to severe harms for health and adaptation.
Social identity loss	Refers to the identity disturbance or difficulties determining or identifying a person or a group in relation to social network.
Social image degradation	Refers to the loss of image quality in a variety of social contexts, including crime, joblessness, and homelessness.
Social stress	Refers to the situation which threatens one's relationships, esteem, or sense of belonging to a dyad, a group, or a larger social context.
Sleeplessness	Refers to the state of being unable to sleep or sleep disorders over time.
Unhealthy lifestyle	Refers to the persons' choices such as tobacco smoking, illicit drugs, alcohol dependence, physical inactivity, and poor diet that all lead ill health and chronic conditions such as cancer, obesity, diabetes, cardiovascular disease, and premature death.
Health carelessness	Refers to the persons' negligence or failure taking treatment or receiving health services from medical system.
Dimension of health	
Physical health risk	Refers to the overall physical condition of a living organism at a given time. It is the soundness of the body, freedom from disease or abnormality, and the condition of optimal physical well-being (Larson, 1999; WHO, 1948).
Mental health risk	Refers to the state of successful mental functioning resulting in productive activities, fulfilling relationships, and the ability to adapt to change and cope with adversity(WHO, 1948, 1998).
Social health risk	Refers to the ability to form satisfying interpersonal relationships with others, to adapt comfortably to different social situations, and to act appropriately in a variety of settings (Lorig et al., 2003; WHO, 1948).

For example, Hajiran (2006) in economics conceptualizes and focuses only on objective indicators (e.g., standard of living) to understand satisfactions and happiness among persons or community people. Likely, the notion of socioeconomic development theory suggests that economic development and its outcome distribution enhance people's standard of living and social well-being over time (Diener & Diener, 1995; Sirgy, 2011). Personal utility approach by Andrew and Withey (1976) suggests that people of a community evaluate their QOL is high when their life satisfaction is high in all life domains. Likely, bottom-up theory by Diener et al. (1999) suggests that life satisfaction is the result of evaluations made in all life domains, including financial, social, psychological, spiritual and health. Social justice and Max Weberian views are prominent to understand QOL and well-being in a given society. According to social justice (see, Rawls, 1971) and Weberian view (see, LaGory et al., 2001), social structure leads people to homelessness that, in turn, leads them to little life chances (e.g., education, job, and income) and choices (e.g., social support, health care). Overall, these theories argue that although people have rights to home, socioeconomic attainment and health care, people experiencing homelessness have adverse consequences in QOL and health (Sirgy, 2011). The quality of life theory derived from Maslow's developmental perspective suggests that QOL leads people to satisfy their human needs (Sirgy, 1986, 2011). The theory argues that the higher the QOL, people satisfy more their needs, depending on the hierarchy of needs. Although these theories suggest some insights and evidence to understand QOL pathways among general population, they have shortcomings to understand trajectories of the QOL pathways of homelessness linking to health risks.

Home is an objective and subjective space that leads a satisfactory economic, social, psychological, and cultural life (Rivlin, 1990). The QOL theory originated in human development and health research (Halfon et al., 2018; Meleis, 2010) is an important framework that helps understanding home status transition and its consequences throughout the life course. Particularly, the QOL theory during homelessness transition asserts that transition from home ownership to homelessness that may occur at any age leads people to subsequent life risks such as socioeconomic vulnerability and negative perceptions toward life's expectation and goals in a given society (Michalos, 2017a). There are two perspectives in the QOL theory: objective and subjective. The objective perspective suggests that homeless transition increases persons' unemployment, income, and food insecurity. Consequently, homeless persons cannot collect their basic necessities and suffer from meeting human needs (Diener & Diener, 1995; Michalos, 2017b; Sirgy, 2011). The subjective perspective also suggests that homelessness transition increases people's psychosocial risks such as identity loss, social and psychological threats, social stress, unhealthy lifestyle and negative perception, feeling and attitude toward life goals (Lehman et al., 1995; Michalos, 2017a, 2017b; Sullivan et al., 2000). The increased risks in those objective and subjective domains of life over time, in turn, affect bio-psycho-social health in a given context (Beer & Faulkner, 2011; Forrest & Yip, 2013; Halfon et al., 2018). Drawing from the objective and subjective assertions and arguments, some cross-sectional and longitudinal studies have found that chronic homelessness than episodic or temporary homelessness are more likely to deteriorate persons' objective (joblessness, unemployment, socioeconomic insecurity, deprivation, and malnutrition) and subjective (e.g., unsafety, social stress, negative perception toward life, unhealthy lifestyle) QOL and social well-being (Buccieri et al., 2020; Babyan et al., 2021; Gentil et al., 2019; Gadermann et al., 2014, 2021; Hubley et al., 2009, 2014; Krabbenborg et al., 2016; Magee et al., 2019). The increased risks in those domains of QOL are more likely to affect their bio-psycho-social health (Lehman et al., 1995; Sullivan et al., 2000). Gadermann et al. (2014), (2021) found that chronic homelessness than episodic or temporary homelessness negatively influenced health-related objective and subjective QOL that, in turn affected physical and mental health in Canada. Dunn (2002) and Magee et al. (2019) found that homelessness influenced lower qualities of physical (e.g., dilapidated house, open space, poor ventilation, mold and low heat) and social (e.g., feeling of identity, control, friendliness) living spaces, affecting physical and psychological well-being. Intervention studies have found that transition from homelessness to home ownership positively enhances good physical and mental health and well-being via proper treatment, providing healthcare facilities and improving quality of life such as food, dress, social connection, and social support (Kaltsidis et al., 2021; Mejia-Lancheros et al., 2021; Patterson et al., 2013).

The QOL theory and research suggest that health risk development is a crucial aspect of homelessness pattern and its subsequent negative consequences such as poor objective and subjective qualities of life (Michalos, 2017a). Cross-sectional and longitudinal research shows that the patterns of homelessness conceptualized as temporary homelessness, episodic homelessness, and chronic homelessness directly increase socioeconomic vulnerability, food insecurity, malnutrition, as well as psychosocial risks such as identity loss, social stress, social insafety, insomnia) among homeless people. The increased risks of the objective and subjective QOL, in turn, affect their health outcomes such as physical, mental, and social health risks. Although the patterns of homelessness are significantly associated with the dimensions of health risks, little research integrates the objective and subjective QOL pathways that prohibit understanding the effects of the patterns of homelessness on biopsychosocial health risks. Drawing from the QOL theory and relevant research, we broadly conceptualize and integrate both objective and subjective QOL indicators as potential pathways by which the patterns of homelessness increase health risks over time (see, Fig. 16.1). In the current theoretical framework, homelessness is conceptualized and categorized into (1) temporary, (2) episodic, and (3) chronic. The QOL pathways are classified as (1) socioeconomic insecurity, (2) deprivation and malnutrition, (3) physical and social unsafety, (4) sleeplessness, (5) social stress, and (6) unhealthy lifestyle. Health is conceptualized as (1) physical health risk, (2) mental health risk, and (3) social health risk. The sociodemographic background factors are used as covariates or control variables. These variables predictors, mediators, and outcome (also covariates) integrated in a theoretical framework interact in uni- and bidirectional ways that are effective to produce fruitful outcomes in different contexts (e.g., COVID-19, rural and urban areas). This integrated theoretical framework than previous ones, however, may help comprehensive understanding the complex relationships between the patterns of homelessness and health risks via objective and subjective QOL indicators as mechanisms aimed at improving life of the homeless people across the societies and contexts (see, Fig. 16.1).

3.1 Homelessness and health

Theory and research suggests that health risks and its severities depend on the length of time persons being homelessness in a given society (Roca et al., 2019). Although a great deal of research has found a significant relationships between homelessness and health, few studies have directly analyzed the relationships between pattern of homelessness (e.g., temporary, episodic, and chronic) and dimensions of health risks (e.g., physical, mental, and social), as was mentioned in Fig. 16.1 (Fazel et al., 2014; Roca et al., 2019). Based on

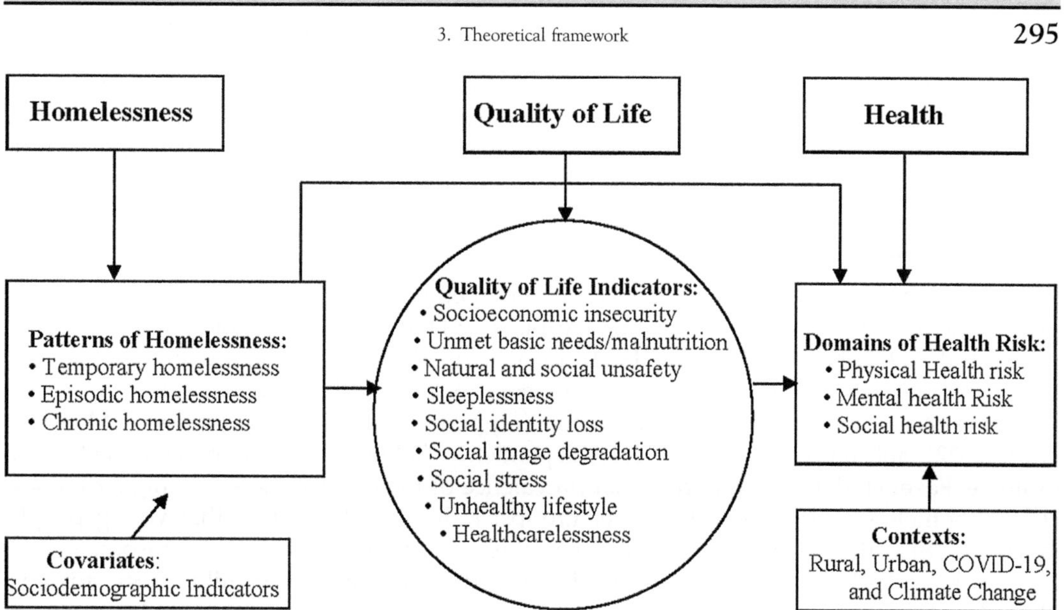

FIGURE 16.1 Conceptual framework.

relevant literature, the following section analyzes the direct relationships of the dimensions of health risk with the pattern of homelessness.

3.1.1 *Homelessness and physical health*

The pattern of homelessness has adverse effects on physical health (Tsai et al., 2019). In developed countries, systematic research and metaanalysis over the past decades reveal that people experiencing chronic homelessness than episodic or temporary homelessness have higher rates of infectious diseases (e.g., pneumonia, fever, tuberculosis, bronchitis, diarrhea, HIV) and chronic health conditions (asthma, hypertension, diabetes, cancer, lung infection, obstructive pulmonary infection), after accounting for sociodemographic characteristics (Richards & Kuhn, 2023; Tsai et al., 2019; Zlotnick et al., 2013; Wiewel et al., 2023). In the United States, based on systematic review, Richards & Kuhn (2023) found that chronic (unsheltered) homeless people have at greater risks of tuberculosis than general population. In Ireland, Ingram et al. (2023) found that homelessness increases the risks of fibrosis, liver infection, and HIV positivity that are the leading causes of immature death among young adults (34%). In developing countries, some studies have found that homeless women have at higher risk in sexually transmitted diseases and unwanted pregnancy (Ray et al., 2001). These chronic physical health conditions, in turn, are consistently associated with comorbidity and premature mortality among chronic homeless people (Richards & Kuhn, 2023). Intervention studies have found that transition from homelessness to supported hosing and health care supports improve physical health conditions among temporary or episodic homeless persons than chronic homeless ones (Tsai et al., 2019; Wiewel et al., 2023). In Bangladesh, although physical health risks (e.g., chronic physical condition, sexually transmitted diseases) associated homelessness patterns are increasing, few studies have directly focused on the relationship between homelessness and physical health condition (Hossain

& Uddin, 2021; Uddin, 2017; Uddin & Ferdous, 2015). Extant research has shown that chronic homeless people than their peer suffer more from chronic and infectious diseases such as STDs, diarrhea, fever, tuberculosis, diabetes, hypertension, and lung infection (BBS, 2020; Uddin, 2017).

3.1.2 *Homelessness and mental health*

The pattern of homelessness has also detrimental effects on mental health and developing psychiatric disorders and hospitalization (Appleby & Desai, 2006; Buckner et al., 1999; Chikwava et al., 2022; Kearns & Smith, 1994). In developed countries, systematic research has found a consistent association of homelessness pattern with mental health disorders, including stress, anxiety, and schizophrenia (Lippert & Lee, 2015), depression (Maestelli et al., 2022), substance abuse, and drug dependence (Johnson & Chamberlain, 2008). For example, Fazel et al. (2014) reported that prevalence of mental disorders was higher among chronic homeless people than their peer groups. Kelly (2020) reported that young people (aged 15–18) with chronic homelessness than their peer with transitional or episodic homelessness have higher rates of mental health disorders, ranging from mild stress, anxiety, alcohol and drug use to severe psychosis, depression, alcohol, and drug dependence. Based on systematic review, Ingram et al. (2023) reported that homelessness is significantly associated with sleep disturbance, self-harm, posttraumatic stress disorders, and illicit drug abuse. In developing countries, several studies have also found similar findings (Smartt et al., 2019; Woan et al., 2013). Intervention studies reveal that housing assistance and long-term mental health services improve mental health disorders among temporary or episodic homeless persons than chronic homeless ones. In Bangladesh, although psychiatric and schizophrenic persons associated with homelessness are increasing, few studies have explained the relationships between homelessness and chronic stress and alcohol dependence (Uddin, 2011a; 2011b).

3.1.3 *Homelessness and social health*

Homelessness also affects social health. Despite medical chronic conditions (e.g., illness, disability), socially healthy persons have an ability to fulfill their goals, to play obligations, to participate in social activities, to maintain social relations with other, and to manage their life with some degree of independence (Lorig et al., 2003). But extant research shows that persistent homelessness decreases a person's psychosocial force and will to fulfill goals in life and detaches from social networks. Consequently, persistent homeless persons suffer from social alienation that affects their social goals and likely social health (Huber et al., 2011). Although the patterns of homelessness significantly affect social health across societies, as well as in Bangladesh, few studies have explained the relationships between homelessness and social health. Although above-mentioned literature review suggests strong relationships between the patterns of homelessness (e.g., temporary, episodic, and chronic) and the domains of health (e.g., physical, mental, and social), we have little knowledge on how the patterns of homelessness affect the domains of health. Review of the intervention studies also suggests that people experiencing homelessness over time have poor improvements in health and less psychosocially adaptive to supportive housing and new socioeconomic environment than their peer with temporary or episodic homelessness. The QOL pathways of homelessness may help an understanding about why and how people experiencing chronic homelessness suffer from health risks than people with other homelessness pattern.

4. Pathways of homelessness to health

Based on the QOLT theory and evidence (Gadermann et al., 2021; Michalos, 2017a,b), the following section describes the QOL pathways such as socioeconomic insecurity, unmet human needs and malnutrition, physical and social unsafety, sleeplessness, social identity loss, social image degradation, social stress, and unhealthy lifestyle by which the patterns of homelessness affect the domains of health.

4.1 Socioeconomic insecurity

Homelessness has adverse effects on socioeconomic security (e.g., education, job, income) and likely human health (Cooke, 2004; Draine et al., 2002; Hatchett, 2004). People experiencing chronic homelessness (PECH) than their peer with temporary or episodic homelessness are at greater risks in socioeconomic insecurity (Poremski et al., 2015; Zuvekas & Hill, 2001). Because PECH than their peer groups suffer more from joblessness or unemployment and income insecurity (Draine et al., 2002; Hatchett, 2004; Pickett-Schenk et al., 2002; Poremski et al., 2015; Zuvekas & Hill, 2001). They also deprive from cash benefits and job supports from social agencies (Riley et al., 2005). In Bangladesh, chronic homelessness associated with river erosion, climate change, settlement expansion, and unequal distribution of land has increased socioeconomic vulnerabilities (40%–70%) than other homelessness pattern (Biswas, 2013; Ghafur, 2004; Islam et al., 2019; Shetu et al., 2016; Uddin, 2017). In turn, the increased socioeconomic vulnerabilities are significantly associated with their severe physical and mental health risks (Biswas, 2013; Kopasker et al., 2018; Uddin, 2011a, 2017).

4.2 Deprivation and malnutrition

PECH than their peers with other types of homelessness are deprived from regular diets and nutrition (Easton et al., 2022; Jones, 2017; Seale et al., 2016). Relevant research shows that PECH than their peer with socioeconomic insecurity or vulnerability are less able to collect material goods (e.g., food, water, clothe, medicine) and pass hungry life over time (Fitzparick & Willis, 2021; Lee & Greif, 2008). They also get less social supports for daily food, water, clothes, and medicine from social networks or agencies (Loftus et al., 2020; Seale et al., 2016; Seligman & Schillinger, 2010). In Bangladesh, about 40%–60% of the PECH than their peers deprive from their survival needs such as food, pure drinking, and nutrition associated with their socioeconomic vulnerability (Biswas, 2013; Islam, 2021; Shetu et al., 2016). When homeless people are deprived from pure drinking water, foods and nutrition over time, their human body and immune system cannot prevent infectious and chronic diseases that increase morbidity and mortality (Dachner & Tarasuk, 2002; Seligman & Schillinger, 2010; Tarasuk et al., 2013). In addition, social deprivation such as dress, occupational prestige, and money damages mental force and social images that ultimately affect health (Crawford et al., 2014; Irwin et al., 2008).

4.3 Physical and social unsafety

PECH who sleep on the street, station, and market places have at greater risks from natural hazards (hot, flood, cyclone) and social threats (social irritation, physical oppression, sexual harassment, rape) than their peers with temporary homelessness (Hsu et al., 2016; Lee & Schreck, 2005). Relevant research shows that PECH usually sleeps on the unprotected places where natural hazards or risks (mosquito, insects, or snake bite) affect their physical health (Lee and Schreck, 2005). In addition, social threats such as irritation, harassment, oppression, and even rape particularly affect young girls or women's physical and mental health (Wenzel et al., 2000). In Bangladesh, several studies have found that chronic homelessness is significantly associated with physical threats (e.g., snake, mosquito or dengue bites) and social risks such as irritation (Biswas, 2013; McNamara et al., 2016). Particularly, homeless young women and children are more likely than men to face sexual harassment, violence, and rape, affecting their health with sexually transmitted diseases or HIV (Helal et al., 2018; Koehlmoos et al., 2009).

4.4 Sleeplessness

Homelessness pattern is significantly associated with level and intensity of sleeplessness (Chang et al., 2015; Gonzalez & Tyminski, 2020). Internationally, research has found that PECH than their peers are more likely to sleep at risky places and suffer more from severe sleeplessness associated with physical and social threats such as hazard space, food insecurity, and depression (Chang et al., 2015; Humphreys & Lee, 2005; Redline et al., 2021). In turn, these severe sleep problems increase threats to normal brain, memory, physical and mental activities. In Bangladesh and internationally, research has found that long duration of sleep problems of the PECH is significantly associated with chronic physical diseases (e.g., diabetes, hypertension, arthritis, and reduced immunity) and mental illness (e.g., anxiety, mood, cognition, or memory disorder) than that of the persons with episodic or temporary homelessness (Chang et al., 2015; Nutt et al., 2008; Shankar et al., 2010; Uddin, 2017). In addition, the sleep disorders also affect social health, including poor performance in job and social activities and adaptation to environment (Redline et al., 2021).

4.5 Social stress

Homelessness creates daily hassles, social anxiety, and social stress. Internationally, relevant research has found that PECH than their peers suffer from daily hassles and social stress related to their vulnerable living conditions (Shinn et al., 2007). Research shows that PECH than their peers suffered more from limited resources for meeting their needs that induce life stressors and psychological distress. In turn, the life stressors partially mediated the relationships between homelessness and social stress. Although social support reduces social stress, PECH than their peers get less social support from social networks (Muñoz et al., 2005; Toro et al., 2008). Consequently, increased social stresses are significantly associated with physical and mental health risks (Muñoz et al., 2005). In Bangladesh, studies by Uddin (2011a, 2017) and others (Islam, 2021) have found that about 60% of the chronic displaced people face more daily hassles and social stresses than their peers that, in turn, negatively influence their health.

4.6 Unhealthy lifestyle

Homelessness also creates unhealthy qualities of life such as poor physical activity or daily exercise and recreation and more alcohol/drug use. Internationally, research has shown that PECH than their peer groups are more likely to engage in unhealthy behaviors such as smoking, drinking, drug, and alcohol use (Clatts et al., 2005; Johnson & Chamberlain, 2008). In Bangladesh, studies by Uddin (2011b) and Hossain and Uddin (2021) have also reported that about 40%–50% of the homeless people are engaged in unhealthy lifestyles such as smoking and local drugs (arrack drinking). These increased unhealthy lifestyles and behaviors are significantly associated with biopsychosocial health risks (Uddin, 2011b, Hossain & Uddin, 2021).

5. Limitations and implications

5.1 Limitations

Based on the QOL theory and literature this chapter develops an integrated theoretical framework to explain underlying objective and subjective QOL pathways of homelessness patterns linking to biopsychosocial health risks (Halfon et al., 2018). In doing so, this chapter uses a systematic narrative review. The systematic review shows that the objective and subjective qualities of life of the PECH are clearly worse than those of the people with episodic or temporary homelessness (Gadermann et al., 2014; Lehman et al., 1995; Sullivan et al., 2000). Particularly, evidence in literature review shows that PECH are more likely than their peer groups to suffer from socioeconomic insecurity and its related deprivation from basic necessities such as water, food, cloth, and medicine. Further evidence shows that PECH are more likely than their counterparts to suffer from physical and social threats and sleep disorders, to maintain unhealthy lifestyles, and to engage in self-harmful practices (Gadermann et al., 2021; Hubley et al., 2014). In turn, these worse qualities of life of the PECH are more likely than their peers to affect their biopsychosocial health across the countries (Gentil et al., 2019; Magee et al., 2019), including Bangladesh (Hossain & Uddin, 2021; Uddin, 2017). Although the theoretical framework and its-related literature help understanding underlying the QOL pathways of homelessness patterns and health risks, evidence on the crucial issue is based on cross-sectional and country-specific research. Future longitudinal research, therefore, may substantiate the theoretical framework to understand the trajectories of the QOL pathways by which the patterns of homelessness negatively influence biopsychosocial health across the countries and contexts.

5.2 Implications

In this chapter, the current theoretical framework and research evidence suggest that the patterns of homelessness increase worse qualities of life in both objective and subjective domains of living conditions that, in turn negatively influence biopsychosocial health (Halfon et al., 2018; Michalos, 2017a,b). The theoretical framework and evidence may help policy-makers and practitioners to understand complex relationships between three domains of life: homelessness, QOL, and health risks. This complex understanding is effective to plan and design social policy and programs to provide (1) material, (2) social, and (3) to psychological, (4) healthcare supports to improve health risks in relation to planned changes in homelessness and objective and subjective QOL. The theoretical framework in this chapter suggests that chronic homeless persons than their peers have vulnerable QOL (e.g., objective and subjective) that, in turn, affects health in both rural and urban contexts, after accounting for socio-demographic factors. Policy-programs implications by stakeholders should (e.g., governments, policymakers, employers, health, and development workers), therefore, should be given priorities on the crucial aspects of the QOL pathways of chronic homelessness and severe health risks. Particularly in Bangladesh, housing policy may improve homeless people's housing. Employers may introduce policy-driven training programs aimed at employing homeless people for earning. The government may also develop national policies and programs, focusing on health needs and QOL of the homeless persons. Health programs may be developed at the national level to use health resources to improve physical, mental, and social health associated with homelessness and QOL in the rural and urban contexts. Although the sectoral policy-program is important, integrated policy-programs are effective to improve health risks associated with improvement in homelessness and QOL.

6. Conclusions

This chapter contributes to the literature through a narrative review of the QOL pathways, demonstrating how these pathways mediate the relationships between homelessness patterns and health risks in places. The theoretical framework developed connects multiple pathways of the QOL of the homelessness patterns to biopsychosocial health risks, signifying a link between the pathways and health risks in place. This chapter implies that investment in the QOL pathways in relation to housing transition and health to enable homeless people to use resources and services to adapt well to any place is worthwhile. Future research should formulate and test theoretical frameworks depicting the link between the QOL pathways, and the homelessness patterns and health risks over time, across places (e.g., rural, urban), and amidst different crises.

References

Andrew, F. M., & Withey, S. B. (1976). *Social indicators of well-being: America's perception of quality life*. New York: Plenum Press.

Appleby, L., & Desai, P. N. (2006). Documenting the relationship between homelessness and psychiatric hospitalization. *Psychiatric Services, 36*(7), 732–737. https://doi.org/10.1176/ps.36.7.732

Babyan, M., Futrell, M., Stover, B., & Hagopian, A. (2021). Advocates make a difference in duration of homelessness and quality of life. *Social Work in Public Health, 36*(3), 354–366. https://doi.org/10.1080/19371918.2021.1897055

Bangladesh Bureau of Statistics. (2020). *Population and housing census report 2020*. Dhaka: Bangladesh Bureau of Statistics Division.

Beer, A., & Faulkner, D. (2011). *Housing transitions the life course: Aspirations, needs and policy*. Bristol: The Policy Press.

Biswas, T. (2013). *Pathways into homelessness and problems of homeless people in Dhaka city*. M. Phil. Dissertation. Dhaka, Bangladesh: National Institute of Preventive and Social Medicine.

Bretherton, J., & Pleace, N. (Eds.). (2023). *The Routledge handbook of homelessness*. New York, NY: Routledge.

Buccieri, K., Oudshoorn, A., Schiff, J. W., Pauly, B., Schiff, R., & Gaetz, S. (2020). Quality of life and mental well-being: A gendered analysis of persons experiencing homelessness in Canada. *Community Mental Health Journal, 56*(8), 1496–1503. https://doi.org/10.1007/s10597-020-00596-6

Buckner, J. C., Bassuk, E. L., Weinreb, L. F., & Brooks, M. G. (1999). Homelessness and its relation to the mental health and behavior of low-income school-age children. *Developmental Psychology, 35*(1), 246–257.

Busch-Geertsema, V., Culhane, D., & Fitzpatrick, S. (2016). Developing a global framework for conceptualizing and measuring homelessness. *Habitat International, 55*, 124–132. https://doi.org/10.1016/j.habitatint.2016.03.004

Chang, H. L., Fisher, F. D., Reitzel, L. R., Kendzor, D. E., Nguyen, M. A., & Businelle, M. S. (2015). Subjective sleep inadequacy and self-rated health among homeless adults. *American Journal of Health Behavior, 39*(1), 14–21. https://doi.org/10.5993/AJHB.39.1.2

Chikwava, F., O'Donnell, M., Ferrante, A., Pakpahan, E., & Cordier, R. (2022). Patterns of homelessness and housing instability and the relationship with mental health disorders among young people transitioning from out-of-home care: Retrospective cohort study using linked administrative data. *PLoS One, 17*(9), Article e0274196. https://doi.org/10.1371/journal.pone.0274196

Clatts, M. C., Goldsamt, L., Yi, H., & Gwadz, M. V. (2005). Homelessness and drug abuse among young men who have sex with men in New York city: A preliminary epidemiological trajectory. *Journal of Adolescence, 28*(2), 201–214.

Clifford, B., Wilson, A., & Harris, P. (2019). Homelessness, health and the policy process: A literature review. *Health Policy, 123*(11), 1125–1132. https://doi.org/10.1016/j.healthpol.2019.08.011

Cooke, C. L. (2004). Joblessness and homelessness as precursors of health problems in formerly incarcerated African American men. *Journal of Nursing Scholarship, 36*(2), 155–160.

Crawford, B., Yamazaki, R., Franke, E., Amanatidis, S., Ravulo, J., Steinbeck, K., Ritchie, J., & Torvaldsen, S. (2014). Sustaining dignity? Food insecurity in homeless young people in urban Australia. *Health Promotion Journal of Australia, 25*(2), 71–78.

Dachner, N., & Tarasuk, V. (2002). Homeless "squeegee kids": Food insecurity and daily survival. *Social Science & Medicine, 54*(7), 1039–1049.

Diener, E., & Diener, C. (1995). Income and quality of life of nations. *Social Indicators Research, 36*, 275–286.

Diener, E., Suh, E., Lucas, R., & Smith, H. (1999). Subjective well-being: Three decades of research. *Psychological Bulletin, 125*, 276–302.

Draine, J., Salzer, M. S., Culhane, D. P., & Hadley, T. R. (2002). Role of disadvantage in crime, joblessness, and homelessness among persons with serious mental illness. *Psychiatric Services, 53*(5), 565–573.

Dunn, J. R. (2002). Housing and inequalities in health: A study of socioeconomic dimensions of housing and self-rated health from a survey of vancouver residents. *Journal of Epidemiology & Community Health, 56*, 671–681.

Easton, C., Oudshoorn, A., Smith-Carrier, T., Forchuk, C., & Marshall, C. A. (2022). The experience of food insecurity during and following homelessness in high-income countries: A systematic and meta-aggregation. *Health and Social Care in the Community, 30*(6), e3384–e3405. https://doi.org/10.1111/hsc.13939

Fazel, S., Geddes, J. R., & Kushel, M. (2014). The health of homeless people in high-income countries: Descriptive, epidemiology, health consequences, and clinical and policy recommendations. *The Lancet, 384*, 1529–1540.

Fitzparick, K. M., & Willis, D. E. (2021). Homeless and hungry: Food insecurity in the land of plenty. *Food Security, 13*, 3–12. https://doi.org/10.1007/s12571-020-01115-x

Forrest, R., & Yip, M. N. (Eds.). (2013). *Young people and housing: Transitions, trajectories and generational fractures*. New York: Routledge.

Frankish, C. J., Hwang, S. W., & Quantz, D. (2005). Homelessness and health in Canada: Research lessons and priorities. *Canadian Journal of Public Health, 96*(Suppl. 2), S23–S29. https://doi.org/10.1007/BF03403700

Gadermann, A. M., Hubley, A. M., Russell, L. B., & Palepu, A. (2014). Subjective health-related quality of life in homeless and vulnerably housed individuals and its relationships with self-reported physical and mental health status. *Social Indicators Research, 116*(2), 341–352.

Gadermann, A. M., Hubley, A. M., Russell, L. B., Thomson, K. C., Norena, M., & Palepu, A. (2021). Understanding subjective quality of life in homeless and vulnerably housed individuals: The role of housing, health, substance use, and social support. *SSM-Mental Health, 1*, Article 1000021.

Gentil, L., Grenier, G., Bamvita, J.-M., Dorvil, H., & Fleury, M.-J. (2019). Profiles of quality of life in a homeless population. *Frontiers in Psychiatry, 10*. https://doi.org/10.3389/fpsyt.2019.00010

Ghafur, S. (2004). Home for human development: Policy implications for homelessness in Bangladesh. *International Development Planning Review, 26*(3), 261–286.

Gonzalez, A., & Tyminski, Q. (2020). Sleep deprivation in an American homeless population. *Sleep Health, 6*(4), 489–494. https://doi.org/10.1016/j.sleh.2020.01.002

Hajiran, H. (2006). Toward a quality of life theory: Net domestic product of happiness. *Social Indicators Research, 75,* 31–43.

Halfon, N., Forrest, C. B., Lerner, R. M., & Faustman, E. M. (Eds.). (2018). *Handbook of life course health development.* Switzerland: Springer.

Hashim, A., Macken, L., Jones, A. M., McGeer, M., Aithal, G. P., & Verma, S. (2021). Community-based assessment and treatment of hepatitis C virus-related liver disease, injecting drug and alcohol use amongst people who are homeless: A systematic review and meta-analysis. *International Journal of Drug Policy, 96,* Article 103342. https://doi.org/10.1016/j.drugpo.2021.103342

Hatchett, B. F. (2004). Homelessness among older adults in a Texas border town. *Journal of Aging & Social Policy, 16*(3), 35–56.

Helal, M. A. A., Chowdhury, M. O. U., & Islam, M. R. (2018). Livelihood pattern of homeless children in Dhaka city: Assessment and reporting. *Asian Profile, 46*(2), 133–150.

Hossain, M. A., & Uddin, M. E. (2021). Education and health: Mediating role of socioeconomic status and health lifestyle among Muslim, Hindu, and Santal men in rural Bangladesh. *Journal of Social Health, 4*(2), 13–29.

Hsu, S.-T., Simon, J. D., Henwood, B. F., Wenzel, S. L., & Couture, J. (2016). Location, location, location: Perceptions of safety and sucerity among formerly homeless persons transitioned to permanent supportive housing. *Journal of the Society for Social Work and Research, 7*(1), 65–88.

Huber, M., Knottnerus, J. A., Green, L., van der Horst, H., Jadad, A. R., Kromhout, D., et al. (2011). How should we define health? *BMJ, 343,* d4163. https://doi.org/10.1136/bmj.d4163

Hubley, A. M. (2014). Quality of life in homeless and hard-to-house individuals inventory (QoLHHI). In A. C. Michalos (Ed.), *Encyclopadia of quality of life and well-being research.* Springer.

Hubley, A. M., Russell, L. B., Gadermann, A. M., & Palepu, A. (2009). *Quality of life for homeless and hard-to-house individuals (QoLHHI) inventory: Administion and scoring manual.*

Hubley, A. M., Russell, L. B., Palepu, A., & Hwang, S. W. (2014). Subjective quality of life among individuals who are homeless: A review of current knowledge. *Social Indicators Research, 115,* 509–524.

Humphreys, J., & Lee, K. (2005). Sleep disturbance in battered women living in transitional housing. *Issues in Mental Health Nursing, 26*(7), 771–880.

Hwang, S. W. (2002). Is homelessness hazardous to your health? *Canadian Journal of Public Health, 93,* 407–410. https://doi.org/10.1007/BF03405026

Ingram, C., Buggy, C., Elabbasy, D., & Perrotta, C. (2023). Homelessness and health-related outcomes in the republic of Ireland: A systematic review, meta-analysis and evidence map. *Journal of Public Health.* https://doi.org/10.1007/s10389-023-01934-0

Irwin, J., LaGory, M., Ritchey, F., & Fitzpatrick, K. (2008). Social assets and mental distress among homeless: Exploring the roles of social support and other forms of social capital on depression. *Social Science & Medicine, 67*(12), 1935–1943.

Islam, M. R. (2021). Drivers of vulnerability and its socio-economic consequences: An example of river erosion affected people in Bangladesh. In G. M. M. Alam, M. O. Erdiam-Kwasie, G. J. Nagy, & W. Leal Filho (Eds.), *Climate vulnerability and resilience in the global south.* Springer.

Islam, T., Azman, A., Singh, P., Ali, S., Akhtar, T., Rafatullah, M., Ismail, N., & Hossain, K. (2019). Socioeconomic vulnerability of river erosion displacees: Case study of coastal villages in Bangladesh. *Indian Journal of Ecology, 46*(1), 34–38.

Johnson, G., & Chamberlain, C. (2008). Homelessness and substance abuse: Which comes first? *Australian Social Work, 61*(4), 342–356.

Jones, A. D. (2017). Food insecurity and mental health status: A global analysis of 149 countries. *American Journal of Preventive Medicine, 53*(2), 264–273.

Kaltsidis, G., Grenier, G., Cao, Z., L'Eperance, N., & Fleury, M.-J. (2021). Typology of changes in quality of life over 12 months among currently or formerly homeless individuals using different housing services in Quebec, Canada. *Health and Quality of Life Outcomes, 19,* 128. https://doi.org/10.1186/s12955-021-01768-y

Kearns, R. A., & Smith, C. J. (1994). Housing, homelessness, and mental health: Mapping an agenda for geographical inquiry. *The Personal Geographer, 46*(4), 418–424.

Kelly, P. (2020). Risk and protective factors contributing to homelessness among foster care youth: An analysis of the National Youth in Transition Database. *Children and Youth Services Review, 108*(1). https://doi.org/10.1016/j.childyouth.2019.104589

Koehlmoos, T. P., Uddin, M. J., Ashraf, A., & Rashid, M. (2009). Homeless in Dhaka: Violence, sexual harassment, and drug use. *Journal of Health, Population and Nutrition, 27*(4), 452–461.

Kopasker, D., Montagna, C., & Bender, K. A. (2018). Economic insecurity: A socioeconomic determinant of mental health. *SSM Population Health, 15*, 184–194.

Krabbenborg, M. A. M., Boersma, S. N., van der Veld, W. M., Vollebergh, W. A. M., & Wolf, J. R. L. M. (2016). Self-determination in relation to quality of life in homeless young adults: Direct and indirect effects through psychological distress and social support. *The Journal of Positive Psychology, 12*(2), 130–140. https://doi.org/10.1080/17439760.2016.1163404

Kuhn, R., & Culhane, D. P. (1998). Applying cluster analysis to test a typology of homelessness by pattern of shelter utilization: Results from the analysis of administrative data. *American Journal of Community Psychology, 26*(2), 207–232. https://doi.org/10.1023/A:1022176402357

LaGory, M., Fitzpatrick, K., & Ritchey, F. (2001). Life chances and choices: Assessing quality of life among the homeless. *The Sociological Quarterly, 42*(4), 633–651.

Laquinta, M. S. (2016). A systematic review of the transition from homelessness to finding a home. *Journal of Community Health Nursing, 33*(1), 20–41.

Larson, J. S. (1999). The conceptualization of health. *Medical Care Research and Review, 56*, 123–136.

Lee, B. A., & Greif, M. J. (2008). Homelessness and hunger. *Journal of Health and Social Behavior, 49*(1), 3–19.

Lee, B. A., & Schreck, C. J. (2005). Danger on the streets: Marginality and victimization among homeless people. *American Behavioral Scientist, 48*(8), 1055–1081. https://doi.org/10.1177/0002764204274200

Lehman, A. F., Kernan, E., DeForge, B. R., & Dixon, L. (1995). Effects of homelessness on the quality of life of persons with severe mental illness. *Psychiatric Services, 46*, 922–926.

Lippert, A. M., & Lee, B. A. (2015). Stress, coping, and mental health differences among homeless people. *Sociological Inquiry, 85*(3), 343–374. https://doi.org/10.1111/soin.12080

Loftus, E., Lachaud, J., Hwang, S., & Mejia-Lancheros, C. (2020). Food insecurity and mental health outcomes among homeless adults: A scoping review. *Public Health Nutrition, 24*, 1766–1777.

Lorig, K. R., Ritter, P. L., & Gonzalez, V. M. (2003). Hispanic chronic disease self-management: A randomized community-based outcome trial. *Nursing Research, 52*, 361–369.

Maestelli, L. G., Anderson, S. M. S., Cintia, de A.-M. P., Julio, T., Antonio, V., & Castaldelli-Maia, J. M. (2022). Homelessness and depressive symptoms: A systematic review. *The Journal of Nervous and Mental Disease, 210*(5), 380–389.

Magee, C., Norena, M., Hubley, A. M., Palepu, A., Hwang, W., & Gadermann, A. (2019). Longitudinal associations between perceived quality of living spaces and health-related quality of life among homeless and vulnerably housed individuals living in three Canadian cities. *International Journal of Environmental Research and Public Health, 16*, 1–12.

McNamara, K. E., Olson, L. L., & Rahman, M. A. (2016). Insecure hope: The challenges faced by urban slum dwellers in Bhola slum, Bangladeesh. *Migration and Development, 5*(1), 1–15.

McNaughton, C. (2008). *Transitions through homelessness*. UK: Palgrave Macmillan.

Mejia-Lancheros, C., Woodhall-Melnik, J., Wang, R., Hwang, S. W., Stergiopoulos, V., & Durbin, A. (2021). Associations of resilience with quality of life levels in adults experiencing homelessness and mental illness: A longitudinal study. *Health and Quality of Life Outcomes, 19*(74). https://doi.org/10.1086/s12955-021-01713-z

Meleis, A. I. (Ed.). (2010). *Transitions theory: Middle range and situation specific theories in nursing research and practice*. New York: Springer.

Michalos, A. C. (2017a). *Connecting quality of life theory to health, well-being and education*. Switzerland: Springer.

Michalos, A. C. (2017b). *Development of quality of life theory and its instruments*. Switzerland: Springer.

Muñoz, M., Panadero, S., Santos, E. P., & Quiroga, A. (2005). Role of stressful life events in homelessness: An intragroup analysis. *American Journal of Community Psychology, 35*(1/2), 35–47.

Nutt, D., Wilson, S., & Paterson, L. (2008). Sleep disorders as core symptoms of depression. *Dialogues in Clinical Neuroscience, 10*(3), 329–336. PMID: 18979946.

Oppenheimer, S. C., Nurius, P. S., & Green, S. (2018). Homelessness history impacts on health outcomes and economic and risk behavior intermediaries: New insights from population data. *Families in Society: The Journal of Contemporary Social Services, 97*(3), 230–242.

Padgett, D. (2007). There is no place like (a) home: Ontological security among persons with a serious illness in the United States. *Social Science & Medicine, 64*, 1927–1936.

Patterson, M., Moniruzzaman, A. K. M., Palepu, A., Zabkiewicz, D., Frankish, C. J., Krausz, M., & Somers, J. M. (2013). Housing first improves subjective quality of life among homeless adults with mental illness: 12 months findings from a randomized controlled trial in vancouver, British columbia. *Social Psychiatry and Psychiatric Epidemiology, 48*(8), 1254–1259. https://doi.org/10.1007/s00127-013-0719-6

Petrovich, J. C., Hunt, J. J., North, C. S., Pollio, D. E., & Roark, M. E. (2019). Comparing unsheltered and sheltered homeless: Demographics, health services use and predictors of health services use. *Community Mental Health Journal, 56*, 271–279. https://doi.org/10.1007/s10597-019-00470-0

Pickett-Schenk, S. A., Cook, J. A., Grey, D., Banghart, M., Rosenheck, R. A., & Randolph, F. (2002). Employment histories of homeless persons with mental illness. *Community Mental Health Journal, 38*, 199–211.

Poremski, D., Distasio, J., Hwang, S. W., & Latimer, E. (2015). Employment and income of people who experience mental illness and homelessness in a large Canadian sample. *Canadian Journal of Psychiatry, 60*(9), 379–385.

Rawls, J. (1971). *A theory of justice*. Cambridge: Harvard University Press.

Ray, S. K., Biswas, R., Kumar, S., Chatterjee, T., Misra, R., & Lahiri, S. K. (2001). Reproductive health needs and care seeking behavior of pavement dwellers of Calcutta. *Journal of the Indian Medical Association, 99*, 142–145.

Redline, B., Semborski, S., Madden, D. R., Rhoades, H., & Henwwod, B. F. (2021). Examining sleep disturbance among sheltered and unsheltered transition age youth (TAY) experiencing homelessness. *Medical Care, 59*, S182–S186.

Richards, J., & Kuhn, R. (2023). Unsheltered homelessness and health: A literature review. *AJPM Focus, 2*(1). https://doi.org/10.1016/j.focus.2022.100043

Riley, E. D., Moss, A. R., Clark, R. A., Monk, S. L., & Bangsbarg, D. R. (2005). Cash benefits are associated with lower risk behavior among the homeless and marginally housed in San Francisco. *Journal of Urban Health: Bulletin of the New York Academy of Medicine, 82*(1), 142–150.

Rivlin, L. G. (1990). The significance of home and homelessness. *Marriage & Family Review, 15*, 39–56.

Roca, P., Panadero, S., Rodríguez-Moreno, S., Martin, R. M., & Vázquez, J. J. (2019). The revolving door to homelessness. The influence of health, alcohol consumption and stressful life events on the number of episodes of homelessness. *Annals of Psychology, 35*(2), 175–180. https://doi.org/10.6018/annalsps.35.2.297741

Seale, J., Fallaize, R., & Lovegrove, J. (2016). Nutrition and homeless: The underestimated challenge. *Nutrition Research Reviews, 29*(2), 143–151.

Seligman, H. K., & Schillinger, D. (2010). Hunger and socioeconomic disparities in chronic disease. *New England Journal of Medicine, 363*(1), 6–9.

Shankar, A., Syamala, S., & Kalidindi, S. (2010). Insufficient rest or sleep and its relation to cardiovascular disease, diabetes and obesity in a national, multiethnic sample. *PLoS One, 5*(11).

Shetu, M. S. R., Islam, M. A., Rahman, K. M. M., & Anisuzzaman, M. (2016). Population displacement due to river erosion in Sirajganj district: Impact on food security and socio-economic status. *Journal of Bangladesh Agriculture University, 14*(2), 191–199.

Shinn, M., Gottlieb, J., Wett, J. L., Bahl, A., Cohen, A., & Ellis, D. B. (2007). Predictors of homelessness among older adults in New York City: Disability, economic, human and social capital and stressful events. *Journal of Health Psychology, 12*(5), 696–708.

Sirgy, M. J. (1986). A quality-of-life theory derived from Maslow's developmental perspective: 'Quality is related to progressive satisfaction of a hierarchy of needs, lower order and higher. *The American Journal of Economics and Sociology, 45*, 329–342.

Sirgy, M. J. (2011). Theoretical perspectives guiding QOL indicator projects. *Social Indicators Research, 103*, 1–22.

Smartt, C., Prince, M., Frissa, S., Eaton, J., Fekadu, A., & Hanlon, C. (2019). Homelessness and severe mental illness in low- and middle-income countries: Scoping review. *BJPsych Open, 5*(e57), 1–8.

Sommerville, P. (1992). Homelessness and the meaning of home: Rooflessness or rootlessness? *International Journal of Urban and Regional Research, 16*(4), 529–539.

Speak, S., & Tipple, G. (2009). *The hidden millions: Homeless in developing countries*. London: Routledge.

Sullivan, G., Burnam, A., Koegel, P., & Hollenberg, J. (2000). Quality of life of homeless persons with mental illness: Results from the course-of-homelessness study. *Psychiatric Services, 51*, 1135–1141.

Tarasuk, V., Mitchell, A., McLaren, L., & McIntyre, L. (2013). Chronic physical and mental conditions among adults may increase vulnerability to household food insecurity. *The Journal of Nutrition, 143*(11), 1785–1793.

Toro, P. A., Tulloch, E., & Qullette, N. (2008). Stress, social support and outcomes in two probability samples of homeless adults. *Journal of Community Psychology, 36*(4), 483–498.

Tsai, J., Gelberg, L., & Rosenheck, R. A. (2019). Changes in physical health after supported hosing: Results from the collaborative initiative to end chronic homelessness. *Journal of General Internal Medicine, 34*(9), 1703–1708. https://doi.org/10.1007/s11606-019-05070-y

Uddin, M. E. (2011b). Cross-cultural social stress among muslim, hindu, santal and oraon communities in rasulpur of Bangladesh. *International Journal of Sociology & Social Policy, 31*(5/6), 361–388.

Uddin, M. E. (2017). Family demographic mechanisms linking of socioeconomic status to subjective physical health in rural Bangladesh. *Social Indicators Research, 130*(3), 1263–1279.

Uddin, M. E., & Ferdous, Z. (2015). Men's condom use or non-use and health risk: A comparative study on premarital, marital and extramarital men at rajshahi city, Bangladesh. *Advances in Applied Psychology, 1*(2), 128–134.

Uddin, M. E. (2011a). Relationship between social stress and arrack drinking pattern: A cross-cultural comparison among Muslim, Hindu, Santal and Oraon arrack drinkers in Rasulpur of Bangladesh. *International Journal of Sociology & Social Policy, 31*(5/6), 335–360.

Wenzel, S. L., Koegel, P., & Gelberg, L. (2000). Antecedents of physical and sexual victimization among homeless women: A comparison to homeless men. *American Journal of Community Psychology, 28*, 367–390. https://doi.org/10.1023/A:1005157405618

White, C. R., Gallegos, A. H., O'Brien, K., Weisberg, S., Pecora, P. J., & Medina, R. (2011). The relationship between homelessness and mental health among alumni of foster care: Results from the Casey young adult survey. *Journal of Public Child Welfare, 5*(4), 369–389. https://doi.org/10.1080/15548732.2011.599754

Wiewel, E. W., Zhong, Y., Xia, Q., Beattle, C. M., Brown, P. A., & Rojas, J. F. (2023). Homelessness and housing assistance among persons with HIV, and associations with HIV care and viral suppression, New York City 2018. *PLoS One*. https://doi.org/10.1371/journal.pone.0285765

Woan, J., Lin, J., & Auerswald, C. (2013). The health status and street children and youth in low- and middle-income countries: A systematic review of the literature. *Journal of Adolescent Health, 53*, 314–321.

World Health Organization. (1998). *Program on mental health: WHOQOF user manual. division of mental health and prevention of substance abuse*. Geneva: WHO. who/mnh/mph/98.4.rev.1.

World Health Organization. (1948). *The constitution of World health Organization*. Geneva: WHO.

Zlotnick, C., Zerger, S., & Wolfe, P. B. (2013). Health care for the homeless: What we have learned in the past 30 years and what's next. *American Journal of Public Health, 103*(Suppl 2), S199–S205.

Zuvekas, S. H., & Hill, S. C. (2001). Income and employment among homeless people: The role of mental health, health and substance abuse. *The Journal of Mental Health Policy and Economics, 3*(3), 153–163.

SECTION IV

Homelessness, governancy policy and sustainable solution

CHAPTER

17

Exploring the model of Urban Innovation District as a solution for housing need in Neoliberal era: Case of Amravati, India

Sampreeth Inteti[1] and Vibhore Bakshi[2]

[1]Department of Architecture, School of Planning and Architecture, Vijayawada, India; [2]IIT Roorkee, School of Planning and Architecture Bhopal, Department of Urban and Regional Planning, Bhopal, Madhya Pradesh, India

1. Introduction

Innovation Districts can be defined as organizations engaged with one other along with anchor institutions to stimulate innovation and entrepreneurship, with active assistance from local governments in enabling efficient infrastructure, enticing services under the aegis of various policy initiatives (Gadecki et al., 2020). The pragmatics of Innovation district involves anchor institutions such as university labs which is further determined by relational geography of various knowledge firms (Hariharan & Biswas, 2020; Huggins, 2008). Infrastructure, redevelopment funds, and supporting land use with certain zoning restrictions are major impetus to the proposed spatial planning framework that encourages innovation-based mechanisms (Yigitcanlar, 2011; Yigitcanlar & Velibeyoglu, 2008). In this chapter, we have tried to study the concept of an Urban Innovation District and its conceptual relevance as impetus to housing solution for Amaravati, which has recently emerged as a new capital in India. The chapter has three sections. In the first section, the conceptual study of an Innovation District, its need, and its characteristics in a city's development will be explored through qualitative research on the concept of Innovation district, Innovation ecosystem, its key actors, and organizations. Furthermore, various regional theories are studied from the perspective of identifying the relevance of Innovation District in imparting housing solutions' also embarks on various model cases of Innovation District majorly Boston,

Homelessness to Hope
https://doi.org/10.1016/B978-0-443-14052-5.00017-3

Detroit, and St. Louis. In the second section of chapter, the Site area of Amaravati is undertaken with respect to the need for setting up an Innovation district in the city with emphasize on Infrastructure and business environment for studying the innovation landscape of Amaravati. According to State Economic Policy of Amravati, entrepreneurs and start-up businesses play a critical role in urban and metropolitan job creation, and innovation districts may aid this trend in a number of ways. This can be particularly helpful in the case of Amaravati as the city currently has a lot of underdeveloped lands with attractive investment policies by the state government which can be used to connect with entrepreneurs who would in turn help in boosting the economy and provide jobs to the residents (APCRDA, 2016). In the final stage of book chapter, broad policy and regulatory-based interventions are explored that would aid in setting up an Innovation district in Amaravati in purview of housing provisions through zoning mechanisms. This is achieved through a thorough synthesis of the various qualitative literature obtained from the secondary data. Spatial assessment is done for Amravati in terms of identifying thematic areas for boosting innovation in relevance to 2050 Amravati masterplan. Notably, Amravati describes the case of "Political Will" of state government as a knowledge capital that enables housing provisions by private entities for different thematic zones of as per zoning regulations of masterplan 2050. The study undertakes assessment of current ICT enabling ecosystem with broadband penetration and mobile users as per recent estimates of state government which are accountable for fostering innovation ties of Amravati.

1.1 Emerging housing need in Neoliberal period

Housing Issues in India has proliferated in last few decades, with unprecedented growth in urbanization levels; resulting in polarization of incomes and contributing to other externalities (Tiwari & Rao, 2016). In neoliberalism period, post 1990s, advent of privatization of housing and urban development has resulted in significant reduction of public housing size, and the state government lack the fiscal resources and capacity to expand the sector (Tsenkova & Polanska, 2014). Era of Neoliberalism has shifted the approach of government from enablers to facilitator by strengthening institutional capacity, by inducing flexibility in protocols for private enterprises for housing supply, and delegating the power and authority to local and state administration without much resource allocation (Mazer, 2019). However, the flagship initiative of "Pradhan Mantri Awaas Yojana" under aegis of the Prime Minister Housing Scheme envisions to facilitate the construction of two crore houses by 2022 is in the second phase of enabling new houses for the states. The idea of "Housing for all" ensures that the welfare state focuses on government support for the inseparable commodities of land and housing through direct and indirect aid for the targeted beneficiaries. Furthermore, there is a need to embark on the well-known emerging approach of Innovation-led development, which may render some relief to the impending housing issues in India (Biswas & Bakshi, 2022). The agony of housing shortage in India is not new; it dates back to much before the refugee crisis after India's partition during independence (Hariharan & Biswas, 2021).

1.2 Nested approach for innovation district for emerging housing need

The rationale for "Nested Approach" underpins on the various interlinkages that are essential for Innovation District to suffice market needs. The nested approach for Innovation-led development Fig. 17.1 highlights on the important pillars for formulation of policy framework for housing need. The nested approach of Innovation-based development interlinks firm, location, functional, and spatial attributes. The significance of Innovation district proliferates the idea of housing by ascertaining policy interventions in the form of clustering; and following the global economic connections (Lawrence et al., 2019). In nutshell for Innovation-based development, there is an identified cluster for housing in proximity to knowledge industries, certainly for greenfield development; incorporated in the projected masterplans; this book chapter tries to envisage the applicability of the fuzzy concept of Innovation district in the case of recently emerging knowledge capitals of Amravati, Andhra Pradesh, India as emerged recently.

1.3 Process of knowledge-based development

Silicon Valley depicts the change in knowledge diversification to knowledge specialization from the period of 1960s to the 1990s. The knowledge firms are propelled by public policies, ancillary industries, and market dynamics (Zingler & Schumpeter, 1940). The knowledge-based urban development framework focus on the facets of sociocultural development, Enviro-Urban development, Institutional development, and Economic Development

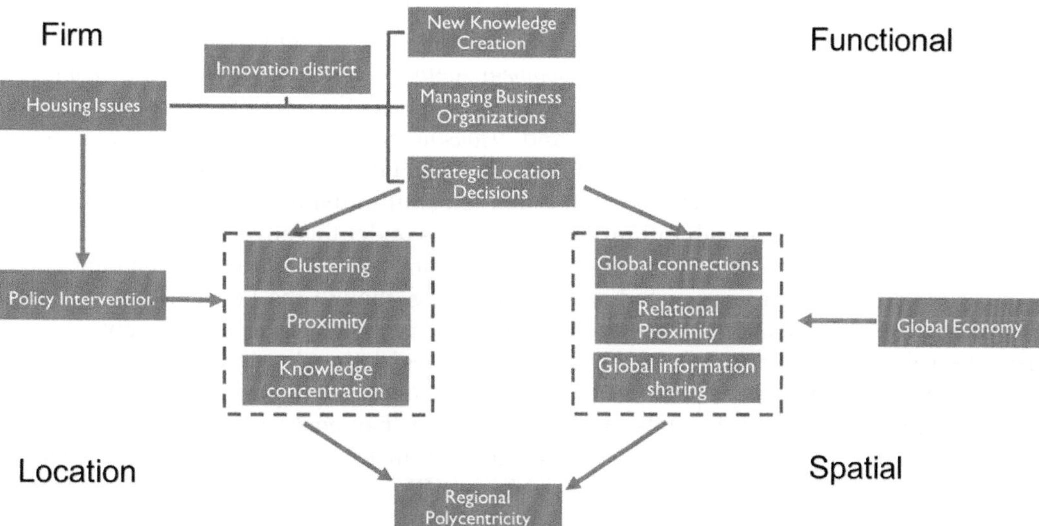

FIGURE 17.1 Nested approach for housing policy needs for innovation-led development. *Adapted from literature pertaining to Innovation-based development for Housing needs.*

(Yigitcanlar, 2011). Location established a pivotal role in the knowledge economy (Asheim et al., 2005). Globally, the literature ascertains the immense role of Boston, Medellin, Philadelphia, Sydney, and Rotterdam in transforming the urban fabric for innovation districts. The innovation district process deciphers a regenerative role for underperforming Abadi areas; with the intent of transforming them into knowledge capitals however with political will (Yigitcanlar & Velibeyoglu, 2008). Knowledge firms are well established in cluster form and possess interlinkages with other industries (Gadecki et al., 2020). The innovation district concept is not new, many researchers are attuned to various regional theories that decipher the significant role of knowledge and innovation in urban development (Bakshi & Biswas, 2023). Poverty and Income polarization has increased over period of time with rising housing inequality in recent years for metropolitan cities in India (Ahmad, 2015). Various knowledge theories are described with housing propositions as mentioned in Table 17.1.

1.4 Growth trajectory of housing policies in India

The landscape of housing development in India has witnessed tumultuous change post independence (Gopalan & Venkataraman, 2015). The policies and programs to boost housing in India is not new, the efforts for enabling efficient housing mechanism came in limelight around 1950s; when there was emerging need to settle the refugees post partition. The 1rst Fiveyear plan emphasized up on creating the housing for refugees and settling them in various colonies. However, the housing shortage has increase in last 5 decades as the population has grew many folds in India; which has a huge impact on the land market for housing. As revealed by 12th Five year plan the Housing Shortage in India has increased drastically. Various schemes and policy motifs have restructured the housing governance in India with imperative on economic and political manifestations. There are two reforms for housing development regarded as pre-reform and post-reform period. The Pre-Reform phase covers periods 1 and 2, namely "*Housing as a Development Activity*" and "*Housing as an Economic Activity*." This phase covers the 1st–7th Five-Year Plans. The Post-reform phase covers periods 3 and 4, namely "*Housing as a Market Good*" and "*Housing as a Commodity*." This phase covers the 8th to 12th Five-Year Plans and beyond, whereas the Post-reform phase covers Periods 3 and 4 of the Housing Policy Timeline with a marked shift in the role of the State as a facilitator, post liberalization. The 8th to 12th Five-Year Plans under this cover the key policy shifts between 1992 to the current period. The 12th Five-Year Plan was marked by change in the name of Planning Commission in India by NITI Aayog, though the Policy Timeline covers the period post 2015 under the Post-Reform phase. This phase reiterated that the government plays the role of a facilitator by creating an enabling legislative, legal, and financial framework for private sector participation. This phase envisages better prospects for housing in terms of shifting the role of the State as facilitator rather than enabler through the establishment of various policy initiatives like the National Slum Development Programs (NSDP), Jawaharlal Nehru National Urban Renewal Mission (JNNURM), and the Pradhan Mantri Awas Yojana (PMAY).

TABLE 17.1 Timeline of Regional theories of Innovation-led development with housing propositions.

Serial No.	Regional theories	Timeline	Author	Knowledge and innovation propositions	Housing propositions
1.	Creative destruction theory	1942	Joseph Schumpeter	Innovation replaces the capitalist model of development; well evident in the United States and other parts of the globe	Enabling housing for the innovation workers
2.	Core peripheral model	1962	John Freidman	Interdependencies between core and periphery; knowledge firms propel spatial development in peripheral areas	Enabling housing and industry connections from core-periphery
3.	Cluster proposition	1966	Michael porter	Knowledge industries are always perceived in the cluster; these clusters include the interlinkages with other ancillary industries in the cluster	Industries and inter −intra housing linkages for cluster
4.	Concept of technopolis, space of flows	1994	Manuel Castells and peter Halls	Knowledge firms depend on virtual networks; digitalization and big data	Technology and space; housing for the high tech, advance production workers
5.	Regional advantage	1995	Anna Lee Saxenian	Regional factors play a crucial role in the location of industries	Regional location for housing clusters and their distance with the parent industries
6.	Network Cities and Globalization	2001	Sasskia Sassen	Global connections between the countries; information flows, export, and import determine the economic significance	Policy impetus on housing for advanced industries
7.	Creative class	2005	Richard Florida	Creative people play a crucial role in wealth generation	Housing for creative occupations; informal and formal workers
8.	Knowledge framework of urban development	2013	Yigitcanlar	Essential sectors for knowledge-based development	Relevance of housing for knowledge clusters in Australia
9.	BCI Global	2020	Business Continuity Institute	Deciphers the role of talent, education, technology, mobility, and market dynamics determine the knowledge base development.	Housing in triple helix model

Data from various literature pertaining to Regional Theories of Knowledge-based development.

2. Research methodology

2.1 Research need in the Indian context

There are a few questions to embark on the research chapter as mentioned below in Table 17.2. The emerging research needs for Innovation District in enabling knowledge-based development can be identified for Indian cities from the attributes like global percentage share estimates by Niti Aayog in the research and development sector for India which is just 0.68% of the overall country's GDP. UNDP, global innovation report card 2021 reveals that India is a moderate performer since it lags in the attributes of preliminary education, Higher education, ICT, Economy, and Enabling environment in comparison to the overall average.

2.2 Aim, research questions, and objectives

This chapter aims to understand the relevance of Innovation Districts (IDs) as an impetus to housing issues in India; by identifying the manifestations of Innovation District through literature; and making these work for one of the newer developed cities of Amravati, Andhra Pradesh, India. However, the manifestations of Innovation district (IDs) is limited to developed nations; developing Innovation district (IDs) require humongous efforts for developing a knowledge capital, and the political motifs and policy frameworks can determine the growth trajectory. In the context of developing nation, these manifestations face enormous challenges. Therefore, there is a need to interlink the concept to context; in juxtaposition to the spatial plan for the city; that envisions the future growth trajectory. To ascertain the research process, there are three objectives; the first objective identifies the key aspects from the literature on Innovation Districts (IDs); the second objective includes an assessment

TABLE 17.2 Research questions and objectives.

Serial No.	Research questions	Objectives of the research	Inclusions
1.	What are the key factors for innovation district (IDs)?	To identify the key factors for innovation district (IDs) for enabling housing provisions	Anchor models; role of institutions; major key players for innovation district (IDs) for imparting solution to housing issues
2.	How different cities have successfully followed the growth trajectory of innovation district?	To assess the pragmatics of innovation district (IDs) in the case of Boston, Detroit, St. Louis, etc.	Parameters for innovation district; case studies of cities which have successfully adopted the principles of innovation district (IDs)
3.	How does Amravati, Andhra Pradesh possess the potential to be one of the innovation districts in India?	To identify the potential of innovation district (IDs) in relevance to Amravati Masterplan 2050	Infrastructure, business environment, human capital, and quality of ICT infrastructure

Research questions and objectives.

of successful manifestations of Innovation District (IDs). Third objective focus on the potential for Amravati, Andhra Pradesh, India from the perspective of the Innovation District. In the nutshell, the research chapter tries to establish "How the concept of Innovation district can change the housing landscape of Amravati" through certain interventions from masterplan 2050?

3. Literature review

Literature for Innovation district highlights the advent of knowledge-based development globally. It is vital to assess how innovation holds the key to the growth and prosperity of regions. There is a need to strengthen innovative capacities at regional scale; supported by knowledge exchange leading to knowledge spill over. Major regional theories of Creative Destruction (Schumpeter, 1950); "Cluster proposition" by Michael Porter around 1960s and role of technology in "space of flows" by Saxenian in 2001; furthermore the propagation of philosophy of Creative class by Richard Florida in 2005 reflects on the principles of how "creative economy" can contribute to the development of the city. There is a need to promote the twin concept of Innovation from the perspective of knowledge-enabled development; that may facilitate housing for knowledge workers and have provisions for housing needs in the near future (Asheim et al., 2005). The approach of Innovation-based-economy focusses on the revival of downtrodden urban areas in cities by providing jobs and induces influx of working population (Gadecki et al., 2020). Cities like Bangalore and Hyderabad already have huge tech firms that have spatially segregated IT Corridors that are scattered around the city (Hariharan & Biswas, 2020). These cities already face a lot of issues related to housing and traffic issues which are persistent due to a lot of planning failures (Sanyal, 2018). Innovation districts aid in solving these issues by mixing commercial and residential spaces forming a cluster (Kim & Ben-Joseph, 2013). There is a need to re-evaluate the current strategies for Innovation district as problem-solver for housing issues in India. Therefore, this chapter undertakes the approach for studying the impact of Urban Innovation District on housing landscape of a city. Innovation districts help cities and metropolitan regions grow jobs in ways that are both disruptive economic forces and take use of their particular economic position (Bruce Katz, 2014). In innovation areas, entrepreneurs act a big engine of economic growth and job creation.

3.1 Key factors and approaches for an innovation district

As determined from the literature, the key factors for Innovation districts (IDs) include real estate developers, advanced research institutions, anchor companies, social networking programmers, economic cultivators like accelerators and incubators, politicians; mayors, and local governments, philanthropic investors, advanced medical campuses, managers of research campuses (Rissola et al., 2019). In the "anchor plus" model it is evident that large-scale mixed-use is evident around the downtown areas of central cities. However, the process of Innovation districts (IDs) is triggered by the anchor institutions, entrepreneurs, and digital networks (Sara Lawrence, 2019). Mobility, land values, and proximity to downtown areas determine the establishment of industries in the Innovation District. Synthesis of Literature highlights the emergence of parameters for Innovation District (ID) at the Regional level and National Level as described in Table 17.2. The predominant key factors for the Innovation District include Real estate developers, Advanced research institutions, anchor companies, virtual networks, politicians, investors, etc. Spatial characteristics of Innovation districts (IDs) include Dense mixed-use space, open spaces, spaces for experimentation, community participation, digital technology, and strong ties (Cohen, 2019).

3.2 Global approaches to innovation

This section of the chapter includes the global frameworks for innovation districts. To ascertain the development of Innovation, there are various indicators to embark on housing perspective (Table 17.3).

3.3 Case studies from global context: Boston, Detroit, and St. Louis

This section of the chapter inculcates the applicability of "Urban Innovation district in knowledge-based development with impetus on housing," approaches undertaken by the global cities of Boston, Detroit, and St. Louis (Boston Planning & Development Agency, 2022). Furthermore, in subsequent sections of the chapter, Amravati, one of the newer greenfield cities of India is assessed from the potential of the Innovation District (Amravati, 2050).

TABLE 17.3 Global approaches, parameters, and Housing Imperatives for Innovation District (IDs).

Serial No.	Global approaches	Parameters	Housing imperatives
1.	Bloomberg Innovation Index 2021	High tech companies, research, and development, patents, postsecondary education, research personnel, manufacturing per capita	Housing facilitation around high-tech clusters
2.	EIU's Global Innovation Ranking	Innovation output (performance), innovation input (enablers), quality of local research, infrastructure research and development, innovation environment, etc.	Housing in proximity to anchors of helix model
3.	Massachusetts's Innovation Economy	Economic impact, research, technology, business development, capital, talent, etc.	Housing in innovation district with talent pool and creative occupations
4.	Global Innovation Index	Innovation index (potential and performance), innovation input index, and innovation output index	Housing governance for enabling significant contribution of innovation-based development
5.	India Innovation Index	Human capital, investment, knowledge workers, business environment, safety legal environment, knowledge output, knowledge diffusion, etc.	Housing as essential pillar for enabling a good business environment

Various approaches for Innovation-based development with impetus on housing.

3.3.1 *Boston Innovation District: Creating clusters of Innovation*

Boston Innovation District represents the classical case of Innovation-led economic revitalization. The case of Boston depicts a political inclination for transforming 1000 acres of south Boston into an atmosphere that supports innovation, cooperation, and entrepreneurship along with housing creation for the workers (Cohen, 2019). Policy impetus for creating Innovation districts (IDs) through cluster of houses and public infrastructure. Boston has created many jobs from 2010 onwards through collaborations. Plans for 12,000 additional residential units have been authorized in 15% of the District's housing would be affordable. Housing, and another 15% going to education (Boston Planning & Development Agency, 2022). These collaborations include business, technology, law, education, creative economy, and many knowledge exchanges. In future, many organizations tend to shift to Boston; which may diversify the economy with 2500 jobs. Boston re-development process is envisaged from the map; which has impetus on creating a cluster of exhibition space, development proposal, brewery, world trade Center, etc. Various manifestations for transforming Boston into innovation capital, can be looked from the map. Two essential components of this invention were affordability and networking-oriented shared areas.

3.3.2 *Detroit Innovation District: Reviving the economy through innovation-cluster with impetus on housing development*

The Detroit innovation district project envisage to create a central development project of 4.3 square miles; that includes central business district, midtown and Corktown, including downtown area and housing space for Innovation and knowledge-based workers (Bruce Katz, 2014). Majorly, the emerging need of innovation district was evident amidst 2008 economic collapse; therefore to prevent such unforeseen circumstances; the manifestation of Innovation district can help to diversify economic base through clustering at the Center with housing provision; and with impetus on housing allocation as a prerequisite for masterplans. The interlinkage model is beneficial for economy of Detroit. Regional advantage of Detroit has a potential of automobile manufacture clusters, business networks, proximity to world renowned institutions, Michigan State, the University of Michigan, and Wayne State, renowned prestigious academic clusters. While the population of Detroit was on verge of extinction, there was a need to revive the need of the Detroit Innovation District potential area accounted for 3.1% of the city's land area, little more than 3% of the city's population (22,018), 52% of the city's employment base, and 9% of the city's commercial establishments (4700) (Bruce Katz, 2014). The Detroit Innovation district allocates housing spaces for the innovation and knowledge workers.

3.3.3 *St. Louis innovation district: Boosting knowledge ties with housing provisions*

Cortex Innovation community is regarded as commercial home with 200 acre hub of biological and technical science research, with emerging knowledge trade-offs. Access to public transit is enhanced in proximity to Washington University. The district inculcates humongous potential; in 14 years it has attracted 3800 IT employees. Cortex officials estimate that their master plan will cost $2.3 billion to construct and will include approximately 4.5 million square feet of mixed-use development and 13,000 technology jobs. It is one of the region's most important innovation campuses, created and chosen to help high-growth startups and existing innovative firms with an intent to expand. It was founded by Washington University in St. Louis, Saint Louis University, the University of Missouri—St. Louis, BJC Healthcare, and the Missouri Botanical Garden (Bruce Katz, 2014). Cortex is a well-known innovation center that serves a variety of local and international technological industries. Four of the anchor institutions put up $29 million in equity in 2002 to help fund the property acquisition for the new location. Cortex was able to purchase restricted properties inside the area while determining the future use of each land parcel furthermore it demarcated the vertical housing units for Cortex Innovation district.

3.4 Collating the characteristics of innovation districts

The three specific cases of Boston ID, Detroit ID and Cortex IC are explored in Table 17.4 through dimensions like the geographic location, the predominant regional economic structure and trajectory, the organizational ownership of the innovation district initiative, and the degree of anchor institution engagement and leadership's three innovation districts range in size from 187 acres in Cortex to 2750 acres in Detroit whereas Knowledge city of Amaravati has the highest area allocated of 8547 Acres. All are sited within a few miles of the central business district, or, in the case of Detroit, encompassing the central business district. The mayor's office in Boston spearheaded and directed the program, while a consortium of

TABLE 17.4 Characteristics of innovation districts (Boston, Detroit and St. Louis).

Serial No.	Parameters of innovation district	Boston	Detroit	St. Louis
1.	**Name of area**	Seaport (Boston Innovation District)	Detroit Innovation district	Cortex Innovation district
2.	**Year**	2010	2014	2010
	Physical delineation	1000 acres	2750 acres	187 acres
3.	**City population**	4.3 lakhs	3.5 lakhs	2.2 lakhs
4.	**Salient feature of (IDs)**	Start-ups	Start-ups and firm networking	Start-ups and academic institutions
5.	**Anchor leaders**	Public (city)	Nonprofit community and anchor institutions	Anchor institutions
6.	**Industrial strength**	Innovation and knowledge sharing	Traditional manufacturing	Traditional manufacturing
7.	**Entrepreneurial mix**	Large and small firms	Large firms	Large firms
8.	**Funding**	Private sector investment	Organizations and private sector	Government, organizations, and private sector
9.	**IDs envision**	Urban restructuring	Foster interregional connections	Community development

Collating the observations from Innovation Districts (Boston, Detroit and St. Louis).

anchor institutions in St. Louis pushed it and the nonprofit community in Detroit supported it with significant anchor institution engagement. In Amaravati, the State government took the initiative step in creating knowledge city which is focused on creating a world class knowledge impact hub. These four cities' innovation areas cater to both existing and start-up firms. The funding strategy used in Boston ID and Amaravati knowledge city was to leverage public resources to catalyze private sector investment. The Detroit Innovation District does not manage programmatic funding itself, but rather operates based on resources available for component aspects such as small business development and targeted industry clusters. In St. Louis, it was the substantial initial contributions from the surrounding research institutions that furnished Cortex with the resources to begin acquiring land and developing structures (Drucker et al., 2019).

3.5 Critical issues in the model approaches

In the previous section of the chapter there are certain approaches from the cases of Boston, Detroit and St. Louis that entails to the different nature of development prospects. From the beginning, the Innovation District's goal include flexible housing alternatives. The government policies for Boston reflect on lowering building costs, inculcating innovative housing mechanisms, ensuring affordability, and incorporating new designs. As per the policy regimes for Boston, the Innovation District manifests to increase housing stock by 53,000 units which enables housing for lower, middle, higher income groups and students. Many administrator have strongly contested the idea of "Innovation District for Boston" few have argued that the model of Innovation district in Boston is still in fancy stage as it is still not brought in to lime light by the city administrators; furthermore these innovation strategies are more politically stated as we comprehend it from the literature and ground reality, the implementation of housing allocation through various policy regimes in Innovation cluster is not evident in full swing; however these clusters have experienced more polarisation of workers; thus increasing social fragmentation. Moreover on a positive side; there are certain actions by the government that promises to ensure equitable distribution of amenities for inhabitants. However, out of many complexities in enabling the effective mechanisms for housing need, the adoption of Housing choice bill in Boston for empowering cities to build a strong housing ecosystem for expanding housing subsidies to all communities with immediate urge of private-sector investment in imparting housing solutions has proven to be successful model in ensuring more inclusivity of residents by various grants and schemes. (Cohen, 2019). However, the case of Detroit case reflects on innovation-based development that yields good paying jobs to the city residents and maximized the investments. However, the emergence of Innovation-based development during the global financial crisis have proven to be a stopgap measure to impart economic benefits until propelled by market forces, and it was too rigid in enabling housing provisions, there needs to be certain flexibilities in the Innovation policy of Detroit that foster housing inclusions of all the workers contributing in innovation economy (Mazer, 2019). The period of 2002–10, there was a lack of formalized workers, however to combat with the global recession the Cortex shifted strategy to live-work-play model by diversifying the portfolio; the major challenge emerged amidst the timeline was the affordable housing and enabling housing for the cortex workers since the diversification process required land; there were certain limitations for the private entities who were dictating the housing land markets to enable the housing in the same premise since "location" was a big question in St. Louis Cortex model. However, the intent of all three models pondered on generating revenue through creativity and Innovation, but these case-models were reluctant to address housing issues in much detail. Certainly, there are few takeaways that describe these as successful examples of Innovation model. In this chapter, an attempt is made to demystifies the successful cases with certain critiques to reflect on Amravati, India as a successful Innovation district model for imparting solution to housing issues.

4. Potential of Amravati as innovation district

In this section of chapter, there are few observations that demystifies the classical case of Amravati as Innovation district; however, Amaravati is the state capital of Andhra Pradesh in India. It is situated on the banks of the Krishna River in Guntur District. Because of its strategic location between the two major metropolitan centers of Vijayawada and Guntur in the Capital Region, Amaravati was chosen as the future capital of Andhra Pradesh (8600 sq. km). The township was built with the cooperation of 24,000 farmers who pooled 35,000 acres of land in India's largest-ever land pooling based on agreement (APCRDA, 2016). The Amaravati Capital City spans 25 villages in three mandalas in the Guntur district (Thulluru, Mangalagiri, and Tadepalli), with a total area of 217.23 km^2. There are around one lakh people residing in 25 villages in the Capital City region, totaling about 27,000 households. The closest cities are Vijayawada (30 km) and Guntur (18 km). The nearest railway station is Tadepalli's KC Canal, while the nearest airport is Gannavaram, 22 km distant. The metropolis will cover 217 km^2 and cost $553.43 billion, with the state government providing just 126 billion over 8 years from April 2018 to March 2026. Amravati has enormous potential of Innovation and Knowledge-based development (Fig. 17.2).

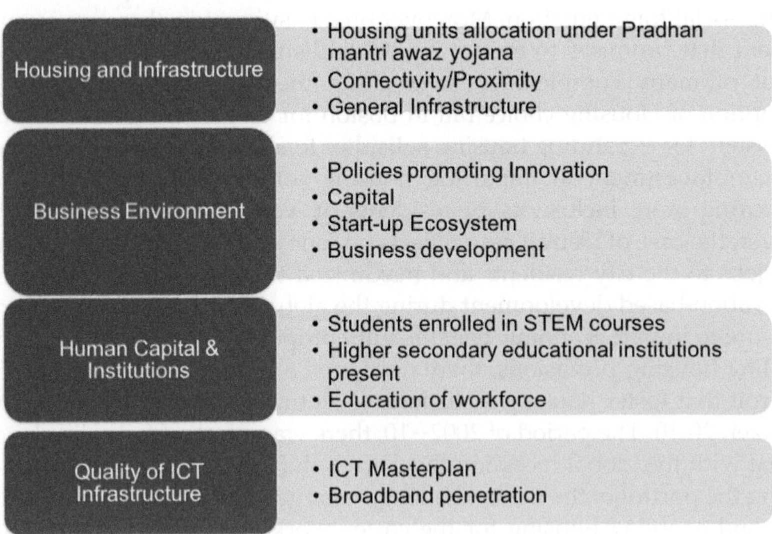

FIGURE 17.2 Innovation potential of Amravati Region potential of Amravati as innovation capital. *Identified factors for Amravati as Innovation district.*

4.1 Political realization of ID from masterplan 2050

Under the aegis of N. Chandrababu Naidu, Andhra Pradesh's 13th Chief Minister Amaravati is envisioned as India's first people-centric smart city, constructed on sustainability and livability principles, and as the world's happiest city (Amravati, 2050). To embark on, various initiatives for manifesting Amravati as an "Innovation District," with knowledge exchanges, there are certain considerations:

- World class infrastructure—The road network for enhancing better connectivity will be increased to 600 km of Road Network by 2050 Jobs and Homes for all-Expected
- Resident population by 2050 will be 5.5 million, and Amravati will house 2.5 million jobs by 2050
- Clean and green–Amravati Masterplan envisions reserving nearly 20% of the area for integrating blue-green spaces
- Quality of life- Parks and public facilities within 5–10-minute walking distance
- Efficient resource management—Flood resistant city toward net zero discharge
- Identity and heritage—Heritage and tourism network using roads, metro, and waterways

Within the capital city, the Concept Plan envisions nine themed development cities. These cities were designed with complementary functions in consideration. Within the capital city, each of these cities will be a focus of activity with a distinct purpose and employment generators' concept plan of the Amravati region is envisioned with nine thematic areas of development as per statutory provision under APCRDA, 2016; which may trigger employment. To ascertain, the manifestation of the "Innovation District" by 2050, there is a government complex at the core of 5.5 km by 1 km in proximity to knowledge city, Electronic city, health city, Finance city, green city, media city, sports city, justice city, etc. The manifestations of thematic areas will play a crucial role in deciphering the ideology of the Innovation District and shall be a problem solver for current housing issues (Fig. 17.3).

4.2 Housing provisions in impetus to masterplan Amravati, 2050

Residential projects envisage to be clustered into townships, according to the plan. Each township is sought to be constructed by utilizing a hierarchical distribution of people, land uses, open spaces, and infrastructure in line with the township model. Principles explained above are adopted in Preparation of Master Plan of capital City, Amaravati. The following zoning districts are proposed for the Capital city:

- Residential (R1, R2, R3, R4) for housing needs
- Commercial (C1, C2, C3, C4, C5, and C6) for different zones
- Industrial (I1, I2, I3) area for enabling knowledge and Innovation
- Open Spaces and Recreation (P1, P2, P3) for residents in adjacent to housing
- Institutional Facilities (S1, S2, S3) around the housing clusters
- Infrastructure Reserve (U1, U2) for projected population

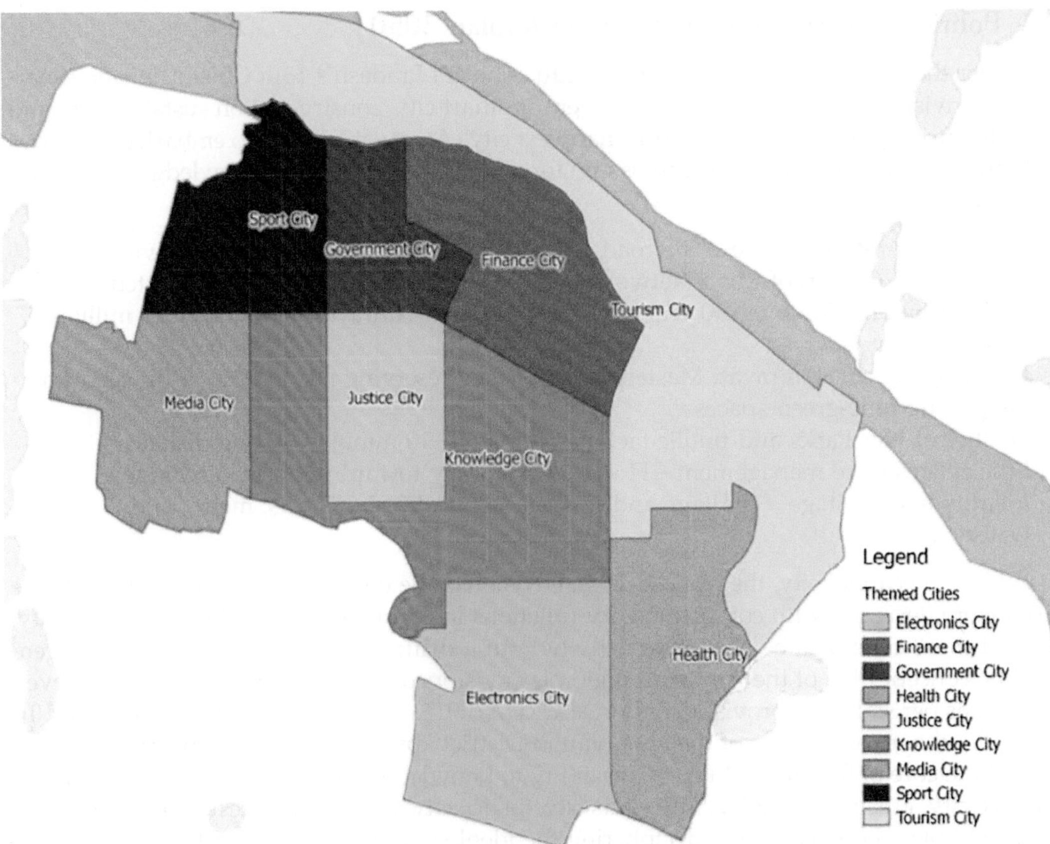

FIGURE 17.3 Amravati concept plan 2050 concept plan of Amravati—masterplan 2050 *Amravati Concept Plan adapted from Amravati Masterplan 2050.*

4.3 Business environment in Amravati

The Concept Plan of Amaravati envisions nine themed development cities within the Capital city. Complementary functions were taken into account when designing these cities (APCRDA, 2016) Each of these cities will be a hub of activity inside the capital city, with a particular mission and job sources. South of the Justice and Finance cities lies Knowledge City. The city comprises around 3459 ha and is home to a university, many institutions, and a knowledge precinct. With a mission to "transform Amaravati Knowledge City (AKC) and the Capital Region into a world-class knowledge impact hub by supporting knowledge production for human growth." This project is being constructed in two stages. The government plans to establish four universities, one skills university, one entrepreneurship development institute, 25 R&D businesses, and 60 start-ups in the first phase, which will employ 15,000 people. It aspires to expand to nearly eight universities, one Skills University, one entrepreneurship Development Institute, boost 350 research and development businesses, and more than a 1000 start-ups in the second phase, employing 150,000 people.

4.4 Overview of startup ecosystem in Amravati: Potentials

A well-organized startup Ecosystem is present in Andhra Pradesh that emphasizes in particular incubation and institutional assistance. The state government's startup policy, which is effective through 2020, was introduced in 2014 under the ITE&C Department. The state hopes to develop a world-class technological startup ecosystem in the region that will contribute to greater knowledge, income, and employment in society, by supporting entrepreneurship and a culture of innovation. The State's Startup policy is based on five pillars:

- Shared infrastructure
- Accelerators/incubators
- Human capital
- Incentives for startups
- Funding

The objectives of the policy start-up policy are aimed at creating at least one homegrown billion-dollar technology start-up from this region, developing one million square feet of incubation space, incubate 5000 companies and start-ups, mobilize venture capital fund of fund worth INR 1000 crore for innovation, Establish 100 incubators/accelerators. The State's start-up policy is assessed on seven pillars by the Start-up Ranking 2019 by the Ministry of Commerce and Industry. Looking at the state's performance concerning the best-performing state in that pillar, at 14% the State got the highest under the Awareness and Outreach pillar. Under simplifying regulations pillar it scored 10% and 3% under the Easing public procurement pillar. However, it ranked the lowest under Incubation facilitation, Seed funding, and Venture funding support. Although mentioned in the Startup policy of the state, the three pillars lacked implementation. Despite substantial efforts to boost the startup ecosystem, the programs aimed at "Venture Funding Support," "Seed Funding Support," and "Incubation Support" need to be improved because outcomes in these pillars have been marginal.

4.5 Potential of human capital and academic institutions for Amravati

The potential of knowledge trade-offs and exchanges can be witnessed by an article published by story research on Amaravati that entails funding for emerging start-ups and anchor companies Among the Tier-II cities in India, it stands in second place by nearly 113 million dollars dedicated to boosting start-up culture. The AP Innovation Society which was established in the city by the Government of Andhra Pradesh is one of the anchor enablers for the innovation district, for booming start-up hub. Currently, there are two Incubator cells in this region, one is the Tadepalli Incubation Center inside the Innovation Society and the other is the SHRISTI-SRM Incubation Center which is housed inside the SRM University of Amaravati. According to the Statistics Handbook 2013–14, there are 248 higher secondary institutions present in the AP Capital region. Out of these, 63 are Engineering colleges that specialize in STEM courses and 22 are ITI colleges that teach Skill development courses to the students. Inside the Knowledge City of Amaravati, there are four major Universities present which are National Institute of Design,

SRM University Amaravati, Vellore Institute of Technology, and AMRITA University. Overall, there are 80,232 students enrolled in the 248 colleges in the Capital region. Out of this, there are 17,888 students enrolled in non-STEM courses, whereas 62,344 students are enrolled in STEM courses specializing in Engineering, Science, and technology. This shows that 72% of total students are pursuing engineering and other related disciplines. There are five colleges per 10,000 people in towns. To understand the knowledge ecosystem; Amravati has the potential for knowledge sharing and networking as we see prestigious institutions in the proximate distance (Fig. 17.4).

4.6 ICT masterplan of Amravati

The ICT master plan for Amaravati is envisioned to construct a city data center and Information Infrastructure Network through hierarchical construction during the project design period of vision 2050. Key objectives of the ICT infrastructure plan envisage to include end-to-end internet connection of 15–20 Mbps, and institutions and companies with 1 Gbps to 10 Gbps with data protection Center for the users.

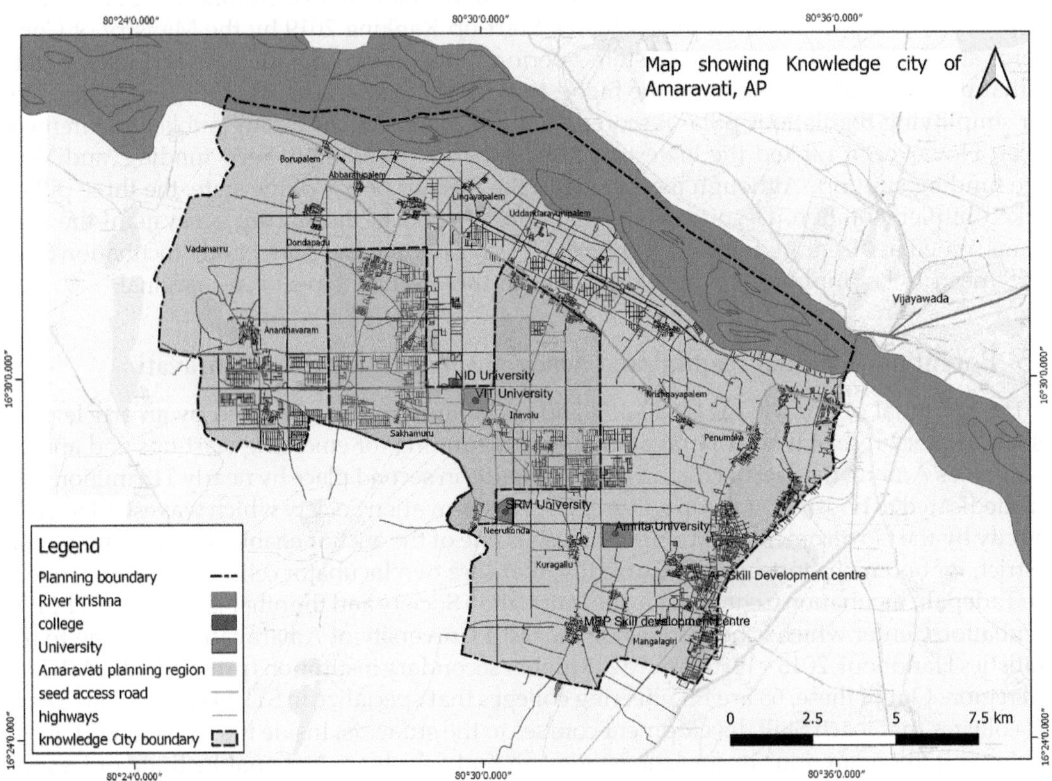

FIGURE 17.4 Knowledge city of Amravati with prestigious institutions Presence of prestigious academic institution in Amravati. *Author generated Map showing the Knowledge City of Amaravati & Universities.*

4.6.1 *Broadband penetration and number of mobiles*

The Amaravati Capital city and Vijayawada Urban and Rural regions have 2.6%–8.0% of households with broadband Connectivity. This is due to the increased efforts by the State Government in launching the AP Fibernet which is aimed at providing cheaper Internet even to the rural areas of the State. Also, Digital India has indirectly impacted the increase of Smartphone users in this region using Mobile data revealed by Andhra Pradesh Regional Development Authority 2022, there is a regional variation in Internet users for Amravati; however, this is significantly lower as the average broadband penetration rate in India is at 51%. Although in terms of wireless connectivity, Andhra Pradesh is the second highest as far as Internet subscriptions are concerned with over 58.65 million subscribers. Amaravati and Vijayawada regions also have the highest share of Mobile users with households using mobiles ranging from 55.1% to 70%.

5. Conclusion

This research chapter dwells on "What contextualization's can be thought in India's context for enabling successful dissemination of this fuzzy concept for Indian cities to resolve housing issues for greenfield cities?" however, the timeline is still visionary. The Innovation ecosystem involves a synergistic relationship between Universities, Research institutes, Companies, Startups, and the local government in combined form. Clustering and knowledge exchange among the various actors leads to knowledge spillovers thus spurring Innovation's footprints as evident from few model cases of Detroit, Boston, and St. Louis. The dichotomy of District is evident at neighborhood scale that generates options for pedestrian friendly atmosphere with combination of diverse mix of land uses and activities, with a blend of "live-work-play" atmosphere that is conducive to the lifestyles of the innovation workforce. These Innovation districts adhere to three models which includes "anchor plus model," "the re-imagined urban areas model," and "the urbanized science park model." The Concept of an Innovation district has proven to be successful for developed economies, although it poses certain challenges in developing nations. However, the impact of Innovation district embarks on economic development of a region is still in a very conceptual stage; the diversity and spatial fragmentation of administration functions depicts a complex procedure; guided by political motifs although the process of adopting this concept in the case of India pose several challenges "geographically," "politically," and "economically," although it promises the idea of enabling efficient housing as witnessed by the literature review evident in third section of chapter; but questions on the fact that "who will implement the concept in India and how"? The concept of Innovation district reflects on efficient approach for enabling knowledge capital with adequate housing needs for future as witnessed by the case of green-field towns through Amravati Masterplan 2050. Implementation procedures for the Innovation district for India needs to be backed up by statutory provisions and certain policy regimes.

References

Ahmad, S. (2015). Housing poverty and inequality in urban India. *Economic Studies in Inequality, Social Exclusion and Well-Being*, 107–122. https://doi.org/10.1007/978-981-287-420-7_6

Amravati, Masterplan. (2050). (Amravati Masterplan).

APCRDA. (2016). *Background on master plan - Amaravati.* APCRDA.

Asheim, B., Coenen, L., Vang, J., & Moodyson, J. (2005). *Regional innovation system policy: A knowledge-based approach* (Vol 13). Centre for Innovation.

Bakshi, V., & Biswas, A. (2023). Global narratives of knowledge and innovation-based development. In U. Chatterjee, N. Setiawati, & S. Sarkar (Eds.), *Urban ccommons, future Smart cities and sustainability.* Springer Geography. Springer, Cham. https://doi.org/10.1007/978-3-031-24767-5_1

Biswas, A., & Bakshi, V. (2022). *Relevance of cooperative housing in a neoliberal era in India.* NCHF. https://nchfindia.net/Sep-2022.pdf#page=37.

Boston Planning and Development Agency. (2022). Boston Planning & Development Agency. http://www.bostonplans.org/business-dev/initiatives/innovationboston/overview#:~:text=The%20Innovation%20District, innovation%2C%20collaboration%2C%20and%20entrepreneurship.

Bruce Katz, J. (2014). *The rise of innovation districts: A new geography of innovation in America.* Brookings.

Cohen, A. (2019). *The development of BOSTON'S innovation district: A case study of cross-sector collaboration and public entrepreneurship the intersector project.*

Drucker, J. M., Kayanan, C. M., & Renski, H. C. (2019). Innovation Districts as a Strategy for Urban Economic Development: A Comparison of Four Cases. *Center for Economic Development Technical Reports.* In press https://scholarworks.umass.edu/cgi/viewcontent.cgi?article=1193&context=ced_techrpts, 2019.

Gadecki, J., Afeltowicz, L., Anielska, K., & Morawska, I. (2020). How innovation districts (do not) work: The case study of Cracow. *European Spatial Research and Policy, 27*(1), 149–171. https://doi.org/10.18778/1231-1952.27.1.07

Gopalan, K., & Venkataraman, M. (2015). Affordable housing: Policy and practice in India. *IIMB Management Review, 27*(2), 129–140. https://doi.org/10.1016/j.iimb.2015.03.003

Hariharan, A. N., & Biswas, A. (2020). A Critical review of the Indian knowledge-based industry location policy against its theoretical arguments. *Regional Science Policy and Practice, 12*(3), 431–454. https://doi.org/10.1111/rsp3.12257

Hariharan, A. N., & Biswas, A. (2021). Global Recognition of India's knowledge-based industry evolution through empirical analysis. *Journal of the Knowledge Economy, 12*(3), 1399–1423. https://doi.org/10.1007/s13132-020-00673-x

Huggins, R. (2008). The evolution of knowledge clusters: Progress and policy. *Economic Development Quarterly, 22*(4), 277–289. https://doi.org/10.1177/0891242408323196

Kim, M., & Ben-Joseph, E. (2013). *Spatial qualities of innovation districts: How third places are changing the innovation ecosystem of kendall square.*

Lawrence, S., Hogan, M., & Brown, E. (2019). *Planning for an innovation district: Questions for practitioners to consider.* RTI Press. https://doi.org/10.3768/rtipress.2018.op.0059.1902

Mazer, K. (2019). Making the welfare state work for extraction: Poverty policy as the regulation of labor and land. *Annals of the American Association of Geographers, 109*(1), 18–34. https://doi.org/10.1080/24694452.2018.1480929

Rissola, G., Bevilacqua, C., Monardo, B., & Trillo, C. (2019). *Place-based innovation ecosystems Boston-Cambridge innovation districts (USA).* https://doi.org/10.2760/183238

Sanyal, B. (2018). A planners' planner: John Friedmann's quest for a general theory of planning. *Journal of the American Planning Association, 84*(2), 179–191. https://doi.org/10.1080/01944363.2018.1427616

Sara Lawrence, M. (2019). *Planning for an innovation district: Questions for Practitioners to consider.* RTI International.

Schumpeter, J. A. (1950). *Capitalism, socialism and democracy.* Allen and Unwin.

Tiwari, P., & Rao, J. (2016). Housing markets and housing policies in India. *SSRN Electronic Journal.* https://doi.org/10.2139/ssrn.2767342

Tsenkova, S., & Polanska, D. V. (2014). Between state and market: Housing policy and housing transformation in post-socialist cities. *Geojournal, 79*(4), 401–405. https://doi.org/10.1007/s10708-014-9538-x

Yigitcanlar, T. (2011). Position paper: Redefining knowledge-based urban development. *International Journal of Knowledge-Based Development, 2*(4), 340–356. https://doi.org/10.1504/IJKBD.2011.044343

Yigitcanlar, T., & Velibeyoglu, K. (2008). Knowledge-based Urban development: The local economic development path of Brisbane, Australia. *Local Economy, 23*(3), 195–207. https://doi.org/10.1080/02690940802197358

Zingler, E. K., & Schumpeter, J. A. (1940). Business cycles: A theoretical, historical, and statistical analysis of the capitalist process. *Southern Economic Journal, 6*(4). https://doi.org/10.2307/1053493

CHAPTER

18

Urbanization as a remedy for homelessness: An analysis of the positive effect of peri-urban development

Rafia Anjum Rimi and Tasnuva Rahman

Rajshahi University of Engineering & Technology (RUET), Department of Urban and Regional Planning, Rajshahi, Bangladesh

1. Introduction

Globally, urban regions are increasingly faced with the difficulty of coping with highly dynamic metropolitan growth. Urbanization is one of the most enabling processes in this rapid growth because most people live in urban areas after 2007 (Vlahov & Galea, 2002). Urbanization initially accelerates the city's growth, benefiting the development of job markets and infrastructure. Sequentially job opportunities work as a pull factor in large cities driving and migration, aggravating the problem of homelessness. Urbanized areas attract rural people whose major goal is to find better-paying work (Akinluyi & Adedokun, 2014). Migration to cities has resulted in a significant loss of urban land, increasing property prices beyond the reach of middle- and low-income people (Brennan & Richardson, 1989). In developing countries with rapid urbanization, the problem of inadequate accommodation for the population, particularly those with low incomes, is one of the greatest obstacles to economic development and the population's welfare. Lack of security, poverty, social neglect, poor health, and lack of education were identified as some of the effects of homelessness (Jasni et al., 2022). The issue of homelessness remains a pressing concern in urban areas across the globe. As cities grapple with the challenges of rapid urbanization, finding effective solutions to alleviate homelessness has become a critical endeavor. When properly managed, urbanization can offer opportunities for economic growth, improved living conditions, enhanced social well-being, as well as reduce homelessness. In different regions of the world, development

https://doi.org/10.1016/B978-0-443-14052-5.00018-5

has different characteristics, ranging from new suburban development on the edges of large cities to the infill of rural areas caused by the expansion of villages and towns surrounding cities (Schneider et al., 2015). These areas often called peri-urban areas combine both urban and rural land (Alexander Wandl et al., 2014). According to authors such as Adell (1999), peri-urban development develops in areas between the metropolis and the countryside. These locations represent a territory comprised of a variety of features, the majority of which are the result of urban center activities, and which are located near and on formerly agricultural land. This emphasis also discusses the merging of urban and rural areas, at least in terms of the closer coexistence of agricultural and nonagricultural activities in urban regions. Rural areas are better connected to urban agglomerations, and many rural households engage in peri-urban activities. Improved transportation infrastructure and increased urbanization have facilitated rural access to labor and product markets, making it easier for rural households to purchase and sell commodities (Kedir et al., 2016). Considering that peri-urban development can alleviate pressure on city centers, individuals associated with primary services such as farming, agriculture, and animal husbandry can access urban amenities without migrating to the CBD central business District. Bangladesh has experienced rapid urban population expansion, as well as rural-to-urban migration (Hasan, 2022). Unemployed and disaster-affected people migrate to urban areas for a better life and face the challenges of housing affordability. In Bangladesh, the housing sector is characterized by an excess of housing stock for upper-income groups and a dearth of affordable housing for the vast majority of middle- and lower-income population groups (Shams et al., 2014). Rajshahi, like many other regions experiencing urbanization, has embarked on various development projects to stimulate economic growth and foster regional development. Among these projects, constructing the bypass road in Kismat Kukhandi, a peri-urban area of Rajshahi City Corporation, has been a significant catalyst for change. This road development has had far-reaching impacts on the region's overall development, influencing various socio-economic aspects, including housing conditions. Bangladesh has not worked on peri-urban development, mostly they have worked on various development projects, positive effects of peri-urban development on urbanization have not been explored in a sense. On the basis that the study aims to explore the potential positive effects of peri-urban development to alleviate homelessness. By analyzing the key characteristics of peri-urban development, this study seeks to shed light on how this development has potential to address the complex challenges associated with homelessness.

The objective of the study is to compare the development of housing conditions in the years 2018 and 2022 and establish the reason behind the development. This study also focuses on the many factors of peri-urban development that support its potential beneficial effects. These dimensions include the provision of affordable housing, and access to various factors like road network development, economic condition, and employment. The findings of this study have significant implications for policymakers, urban planners, and other parties involved in addressing the issue of homelessness. This study provides valuable insights into the positive effects of peri-urban development on reducing homelessness by analyzing changes in housing conditions over the specified time. Understanding the positive effects of urbanization in peri-urban areas can guide the design and implementation of innovative housing models and social support systems, ultimately leading to more effective and sustainable approaches to addressing homelessness.

2. Limitations of the study

While conducting the research in a relatively small peri-urban area close to the primary urban area of Rajshahi City Corporation, not all urbanization factors were thoroughly investigated. In this context of peri-urban development, factors such as population growth, the establishment of industries, government policies, and social influences were not examined in detail. For example, the study did not delve extensively into the impact of rapid population growth on the demand for housing, infrastructure, and services, which are key drivers of urbanization and the growth of urban areas. Similarly, the role of industrialization and the transition from agrarian to manufacturing and service-based economies, as well as the influence of government policies and regulations on urbanization trends, were not comprehensively investigated. In addition, social factors such as altering lifestyles, aspirations, and cultural influences, which are influenced by factors such as education, healthcare facilities, entertainment, and social opportunities, were not thoroughly examined. In pursuance of a higher quality of life, these factors frequently drive people to migrate to urban areas. Due to the study's limited scope and concentration on a particular peri-urban area, these multidimensional factors of urbanization were not investigated in depth. It is essential to acknowledge that these factors should have been included in the analysis in order to provide a more complete comprehension of the urbanization dynamics in the region.

3. Methodology

3.1 Study area profile

The study area which is selected for the study named Kismat Kukhandi, Budhpara, and Rajshahi. This place is situated in Rajshahi ZI, Rajshahi DIV, Bangladesh. Its geographical coordinates are 24° 23′ 0″ north, 88° 40′ 0″ east and its original name is Kismat Kukhandi (Google Map, n.d.) Kismat Kukhandi is a village under Paba Upazilla, Hariana Union. Hariana Union is directed East from the Paba Upazila center. Its total volume is 23,062 Acres or 93.33 square Km. The total housing of this Union is 5952. Hariana Union has 16 Mouza and the Mouza are divided into 11 administrative Ward. Kismat Kukhandi is one of them. Total population of Kismat Kukhandi is 1532 (Bangladesh Population and Housing Census, 2011). It is a low-income area. The condition of the people and housing capacity is very poor. The monthly income of people in this area is 5-20k. All are agriculture dependent. The local people of this area lead their area by farming and fishing and some are involved in handicraft works. There is no high school or college in this area. A primary school and Madrasha where the children go for education. Kismet Kukhandi is 4.26 Km away from Hariana Union. It is just in the middle between Rajshahi bypass and the Dhaka-Rajshahi highway. Rajshahi bypass is 5.89 km north and Dhaka-Rajshahi the highway is 3.53 km south of our study (n.d.). Our study area Kismat Kukhandi is surrounded by Bangladesh Agricultural Research Institute, Katakhali peaking power plant, Rajshahi Jut Mills and Rajshahi Sugar Mill. The minor road to access Kismat Kukhandi is just beside the Fruit station and Bangladesh Rice Institute (Fig. 18.1).

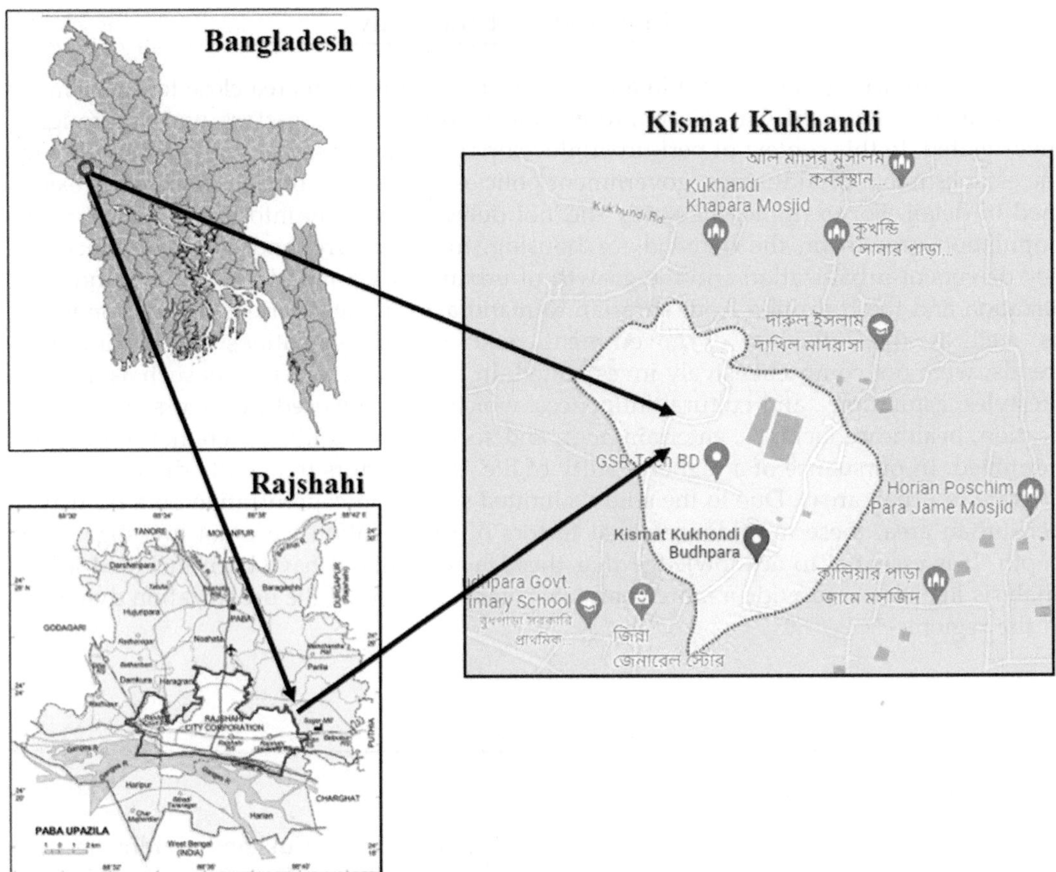

FIGURE 18.1 Study area.

Kismat Kukhandi is approximately 19 km away from the main CBD which is Shaheb Bazar crossing Katakhali. The total area of Kismat Kukhandi is 298 acres. Most of the lands are agricultural covered with crops and mustard fields. A primary survey was done in the study area to have a basic knowledge about it. A preview of the observed notable points about the study area is that most of the area is surrounded by green field agricultural land, about 7 Brick kilns are seen in Kismat Kukhandi, but 4 Brick kilns are situated in the mouza boundary area in 2018 which has increased know about 10 Brick kilns. One masjid and one madrasa are situated here. The maximum building was one-storied and in the building process in 2018 and most of those buildings has raised to two-story in 2022, other houses are made of soil, tin, and bamboo fence, and water bodies are used for fishery.

3.2 Data collection

The data for this study was collected through a questionnaire survey, utilizing primary data. The sample size was determined based on a 90% confidence level and a 10% confidence interval, resulting in a sample size of 68. The survey employed a simple random sampling technique, which is a statistical method where each individual in the population has an equal and independent chance of being selected. In simple random sampling, the selection of one individual does not affect the selection of others, ensuring that every member of the population has an equal probability of being chosen. This approach guarantees that all possible samples have an equal likelihood of being selected, thus providing a fair and unbiased means to obtain a representative sample from the larger population. The survey was conducted in the same area using the same questionnaire in both 2018 and 2022, allowing for a comparative analysis of the housing conditions and the factors influencing urbanization over time. The study ensures continuity and consistency in data collection by employing a consistent methodology across the two survey periods.

3.3 Statistical analysis

Understanding the relationships between various factors and their effects on various phenomena requires statistical analysis (Destiny, 2017). Particularly, correlation analysis offers valuable insights into the strength and direction of associations between variables. It has been used significantly in research to analyze the relationships between indicators and outcomes (Schober & Schwarte, 2018). Numerous studies have utilized correlation analysis to investigate the relationships between various factors. The correlation analysis is conducted with statistical software (SPSS), which calculates correlation coefficients such as Pearson Co-Relation. R-values are used to assess statistical significance by determining the strength and direction of the relationships between variables (Turok & McGranahan, 2013). The correlation coefficients are interpreted within the context of the research objective, considering their magnitude and significance. The information is collected at the data collection level, the value is put into Excel, then the final correlational values have been found. The findings are reported using tables, graphs, or visualizations to present the correlation coefficients and their significance. There is a discussion of the implications of the results, providing insights based on the research objective and specific domain.

3.4 Multicriteria decision-making technique

In the questionnaire survey conducted, various factors such as housing conditions, education facilities, and waste disposal facilities were taken into account to assess the development of the area. However, it was important to identify the most influential factor among them and validate it by consulting experts in the field of urbanization. To incorporate their opinions and views, the well-regarded and highly appreciated multi-criteria decision-making method known as the analytical hierarchy process (AHP) was utilized. In the analysis of the survey outcomes, specific factors were identified and defined as alternatives for pairwise comparisons within the AHP framework. The experts were asked to evaluate these alternatives using a scale of 1–9, considering the relative importance of the factors. Matrices were developed for each expert to capture their judgments. Additionally, the experts were given scores based on their involvement and expertise in the study. Through this process, the factor that received the highest score was determined as the most influential factor among the considered criteria. By employing the AHP methodology, which combines expert opinions and mathematical calculations, the study aimed to provide a reliable and informed identification of the factor that had the greatest impact on the development of the area under investigation.

4. Result and discussion

This study examines the changes in homelessness from 2018 to 2022, focusing on factors such as occupation, income, land values, housing conditions, socioeconomic status, and transportation infrastructure. The development of a well-structured road network has significantly contributed to the reduction of homelessness. It enhances economic opportunities, improves income levels, and expands housing options in peri-urban areas. This leads to better socio-economic upliftment, enabling individuals to maintain their housing and meet basic needs. Additionally, the availability of efficient and affordable transportation options helps individuals access essential services, healthcare, and education without relocating to city centers, reducing homelessness risk in urban center. Eventually, the development of a robust road network contributes to the reduction of homelessness by improving the quality of life for individuals and families.

4.1 Demographic condition

According to BBS, census 2011, Kismat Kukhadi has 380 households where 1531 people live, as 2022's census result is not published yet, but through the survey, it is found that the population has increased multiple times till this time. As this peri-urban area is situated on the border of the Rajshahi city corporation area, this area is basically agriculture based. But the construction of roads has shifted people to other secondary and tertiary occupations too Fig. 18.2.

The presented graph illustrates the changes in basic occupations within the area, highlighting the significant ones. However, it should be noted that the mentioned occupations are not exclusive or limited to those stated. In 2018, a higher number of individuals were engaged in farming and day laborer roles, but these numbers decreased by 2022. The phenomenon of decreased homelessness can be subject to critical analysis by evaluating the evolving dynamics of many employment. It is worth mentioning that occupations such as rickshaw pullers, van pullers, day laborers, and auto-rickshaw drivers have experienced a rise within the corresponding timeframe. The observed change can be related to the improvement of the transport infrastructure, which has enabled persons to pursue employment opportunities in these sectors. Moreover, the data collected from survey participants has revealed an increased accessibility of transit alternatives in comparison to earlier time periods. As a result, the necessity for those residing in peripheral areas migrate toward the urban core has decreased. The shift in job patterns can be perceived as a favorable advancement within the framework of mitigating homelessness. A growing number of career options that do not necessitate a permanent place of residence have the potential to offer economic stability to individuals who may otherwise face the risk of homelessness. Furthermore, the enhanced availability of mobility can facilitate individuals in their pursuit of job or use of services without necessitating a move to metropolitan areas. Consequently,

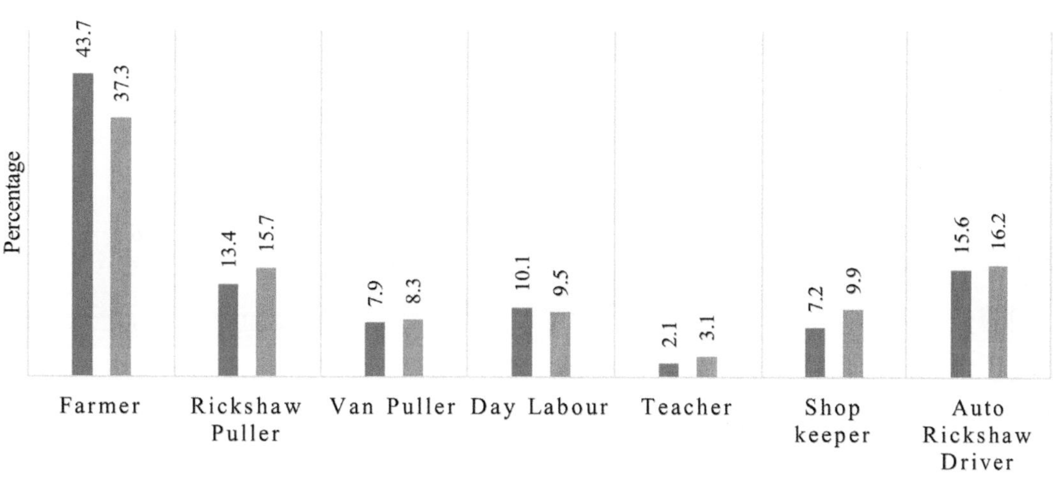

FIGURE 18.2 Occupation.

this can contribute to the preservation of stable living conditions for those residing on the outskirts. The decline in homelessness can be partially linked to the evolving dynamics of work and transit choices.

4.1.1 Income

Analyzing income ranges is crucial for understanding the overall socioeconomic condition of individuals residing in a specific area, as it reflects the improvement in their lives. The provided graph illustrates the changes in income between 2018 and 2022. It reveals a decrease in the percentage of people earning below 7000 per month, dropping from 38.1% in 2018 to 19.9% in 2022. Conversely, there has been a significant increase in the percentage of people earning between 7000 and 15,000, rising from 27.4% in 2018 to 38.2% in 2022. Furthermore, the income range of 15,000–22,000 also experienced growth, indicating a steady increment between the 2 years (Fig. 18.3).

These shifts in income ranges can be attributed to the development of people's occupations and their transition toward better social conditions. The upward trend in income signifies an improvement in the overall economic well-being of the population within the area under study.

4.2 Land value

The value of land undergoes significant changes with the development of adjacent road networks, serving as a key indicator of overall area development. Improved road

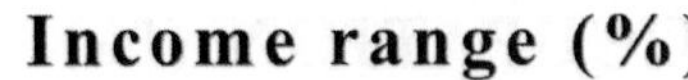

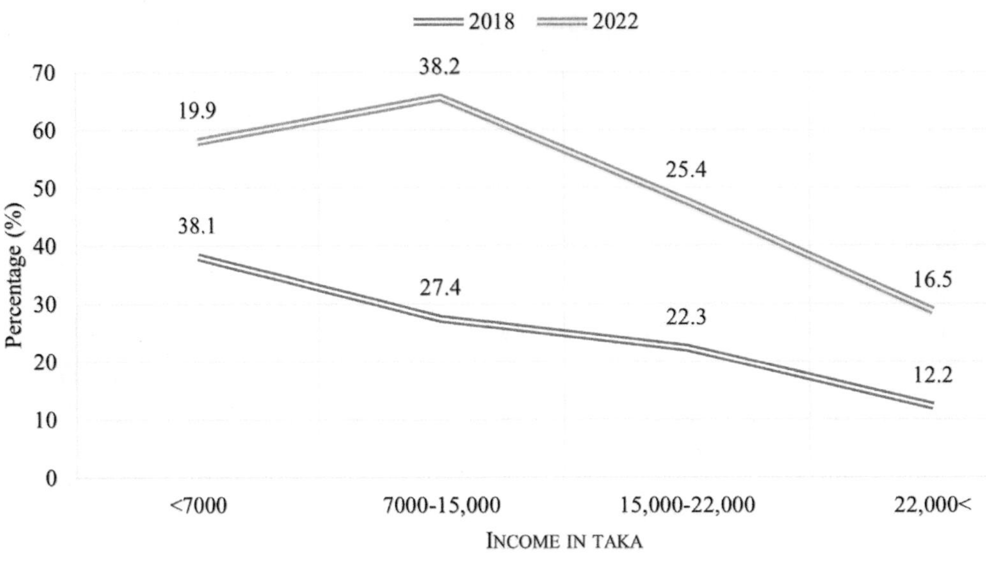

FIGURE 18.3 Income range in percentage.

networks have an immediate impact on land values. The provided graph depicts the land value per katha (a unit of measurement) in 2018 and 2022. Findings from the survey reveal that before the construction of the bypass road, the land price was 4.5 Lacs per katha in 2018. However, after the completion of the bypass road, the land price soared to approximately 16 lacs per katha in 2022. Moreover, land prices have continued to increase steadily over time (Fig. 18.4). This significant rise in land value has had a profound impact on people's lives. With the improved road network, individuals from remote villages are migrating to the Kismat Kukhandi area, purchasing land, and constructing buildings. Simultaneously, residents who own land are selling small parcels of their property to establish businesses or enhance their existing houses. This surge in land value has brought about transformative changes in the socioeconomic landscape, shaping the aspirations and opportunities for both newcomers and long-time residents of the area which created housing for more people in affordable rent or land price and reduced homelessness.

4.3 Housing conditions

4.3.1 Housing material

Housing material plays a crucial role in determining an area's housing development level. Therefore, analyzing changes in housing materials is an important factor for understanding the area's overall development. In 2018, a significant number of houses were constructed using soil, bamboo, and tin materials. However, in 2022, there has been a noteworthy increase of approximately 19% in houses constructed with concrete (Fig. 18.5).

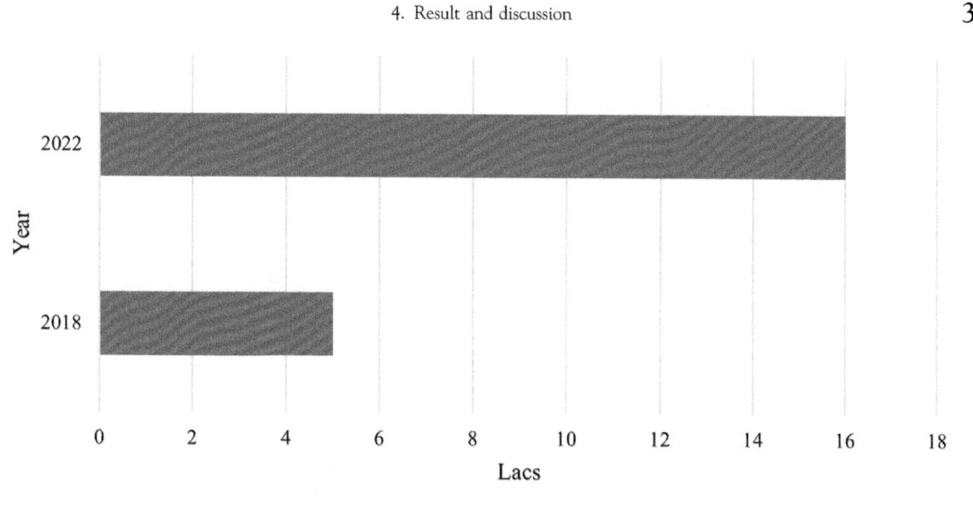

FIGURE 18.4 Land value (lacs per katha).

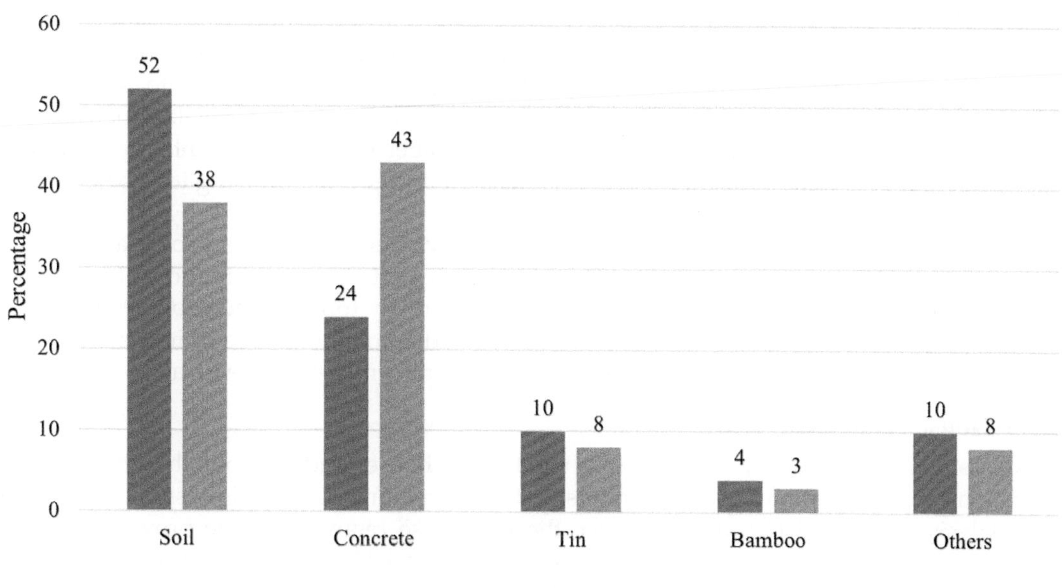

FIGURE 18.5 Housing material (%).

This shift toward concrete construction indicates an improvement in housing facilities within the area. The increased availability of housing options has contributed to a reduction in the problem of homelessness. With the transition to more durable and permanent housing structures, people now have access to improved housing facilities, which positively impacts their quality of life and provides greater stability and security.

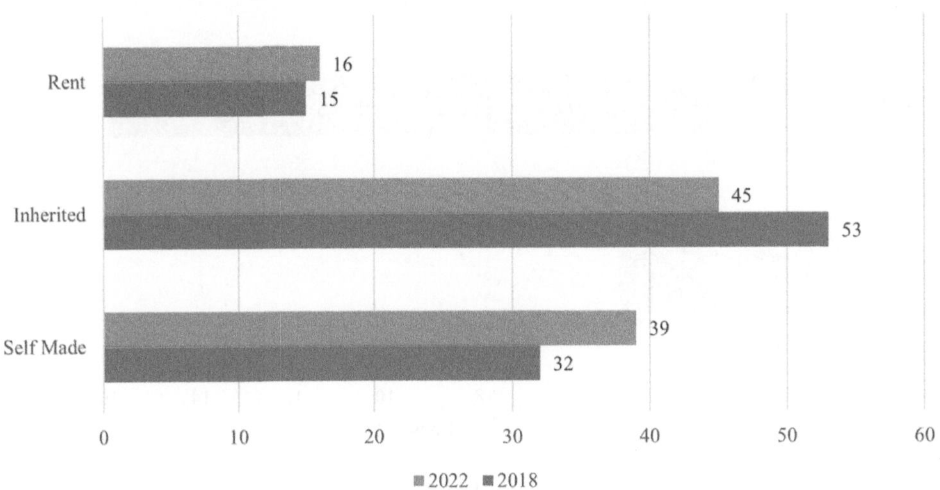

FIGURE 18.6 Ownership of the houses.

4.3.2 *Ownership of house*

To analyze homelessness, it is essential to examine the increase in house ownership. The provided graph illustrates the ownership of houses, offering insights into self-made, inherited, and rented houses. The data reveals that there has been a notable increase of approximately 7% in self-made houses, while the number of rented and inherited houses has decreased. This trend indicates that individuals are increasingly moving toward owning their own houses (Fig. 18.6).

The shift toward self-made houses signifies a positive development as people are taking proactive steps to secure their housing. This shift plays a crucial role in reducing homelessness as individuals attain greater stability and security through homeownership. By becoming homeowners, people are less reliant on rented accommodations or inherited properties, which leads to a decrease in the homelessness problem within the area.

4.3.3 *House renovating years*

Assessing how frequently people can renovate their houses and improve their household conditions is an important aspect of understanding development. To gain insights into this, the analysis considered the duration since the last house renovation. The provided graph highlights the following findings: In 2018, approximately 40% of people had not renovated their houses for over 10 years, whereas in 2022, this percentage decreased to 25%. Within the 5- to 10-year range, 25% of people had renovated their houses in 2018, which slightly decreased to 22% in 2022. Similarly, in the 2- to 5-year range, 20% of people had renovated their houses in 2018, while this figure decreased to 15% in 2022 (Fig. 18.7).

However, the most significant observation is that only 15% of people had renovated their houses within the past 2 years in 2018, whereas in 2022, this percentage increased to 38%. This indicates that a larger proportion of people had the opportunity to develop and improve their housing conditions within a span of less than 2 years. These findings demonstrate a positive trend of increased house renovation, reflecting the improved capacity of individuals to

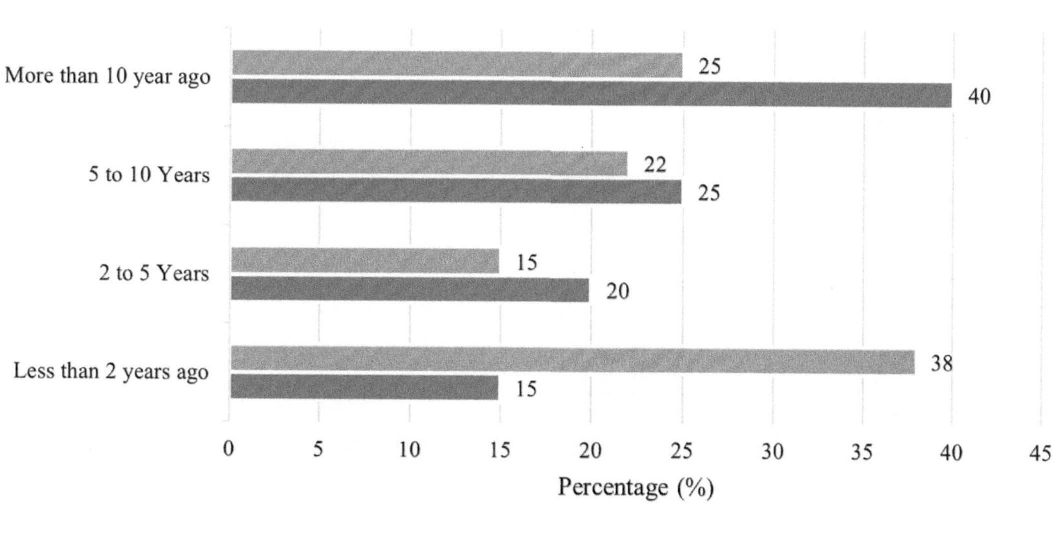

FIGURE 18.7 Housing renovation timeline.

enhance their household conditions over time. The ability to renovate houses more frequently indicates progress in the development of the area and contributes to an overall improvement in the quality of life for its residents.

4.3.4 *Road width in front of house*

The width of the roads in front of each house has undergone significant expansion. In 2018, approximately 25% of houses had a road width of less than 5 feet, while 35% had roads widened between 5 and 10 feet. Additionally, 7% of houses had roads widened between 11 and 20 feet, and 5% enjoyed road widths exceeding 20 feet (Fig. 18.8).

In 2022, there was a noticeable shift in these numbers. The percentage of houses with road widths less than 5 feet decreased to 38%, indicating a significant improvement. Houses with road widths between 5 and 10 feet increased to 42%, reflecting a larger proportion of houses benefitting from widened roads. Furthermore, the percentage of houses with road widths between 11 and 20 feet experienced an increase of approximately 12%, demonstrating further progress. Lastly, the proportion of houses with road widths exceeding 20 feet also increased to 8%. These changes in the front road width of houses signify a positive development in the area's infrastructure. The expansion of road widths provides residents with improved accessibility and convenience, enhancing transportation options and overall connectivity.

4.4 Socio-economic facilities

4.4.1 *Education*

The two-pie chart describes the education opportunities in the Kismat Kukhandi area between 2018 and 2022 (Figs. 18.9 and 18.10).

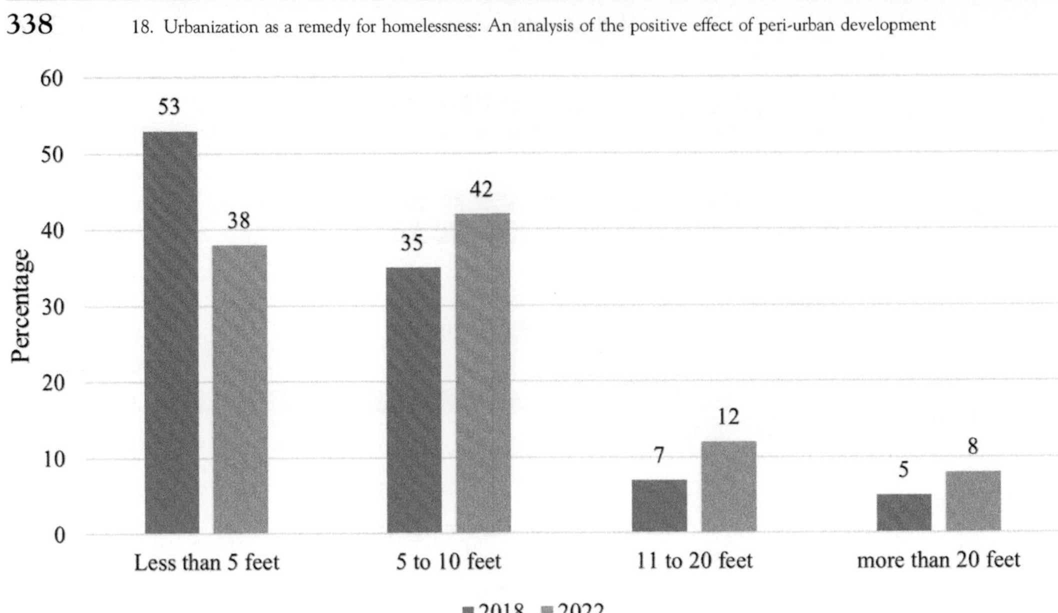

FIGURE 18.8 Road width in front of house.

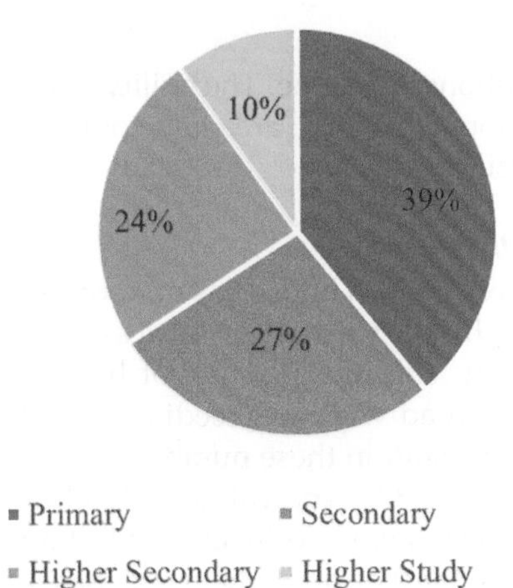

FIGURE 18.9 Education opportunity (2018).

In 2018, the primary education category accounted for 39% of the pie, indicating that a substantial portion of the educational opportunities were awarded to primary education. Secondary education was in second with 27%, constituting a significant but significantly smaller amount. Higher secondary education accounted for 24% of the opportunities. Finally, higher education opportunities accounted for 10% of the total. In 2022, the proportion of the pie chart representing primary education has decreased to 30%, indicating a decrease relative to 2018. Secondary education reached marginally 28%, remaining at a similar level. Higher secondary education remained constant at 27% of the population. The most notable increase occurred in the higher education category, which increased from 10% to 15%.

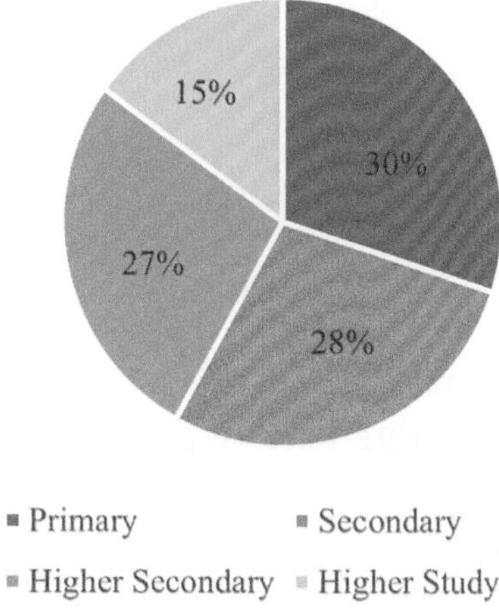

FIGURE 18.10 Education opportunity (2022).

4.4.1.1 Waste disposal

In the Kismat Kukhandi area, a central waste disposal system was absent, but in 2022, there was an increase in awareness regarding waste disposal systems, resulting in a higher adoption rate compared to 2018 (Figs. 18.11 and 18.12).

Most households relied on disposing of their waste in Mudholes, which accounted for 37% of the pie. Then, open spaces were utilized for waste disposal, comprising 29% of the distribution. It is worth noting that 18% of households dispose of their waste in water bodies, which poses a significant environmental concern as it adversely affects water quality. Lastly, a smaller segment of 7% resorted to disposing of waste on the roadside. These findings emphasize the evolution of waste disposal practices in the Kismat Kukhandi area over time. While more people have embraced waste disposal systems, there is still a need to address improper disposal in water bodies, roadside, and open spaces to protect the environment and preserve water quality.

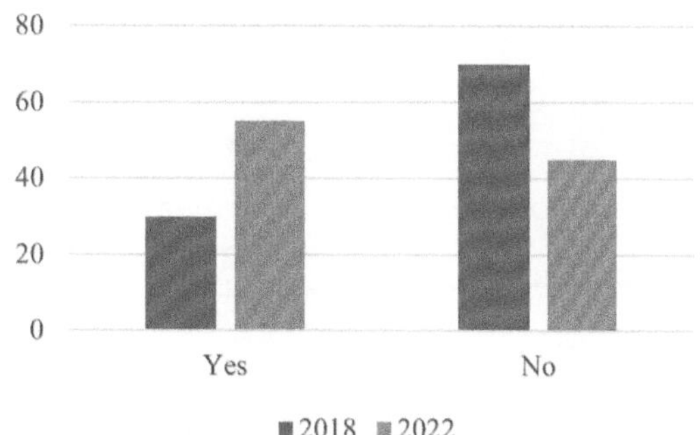

FIGURE 18.11 Waste disposal system availability.

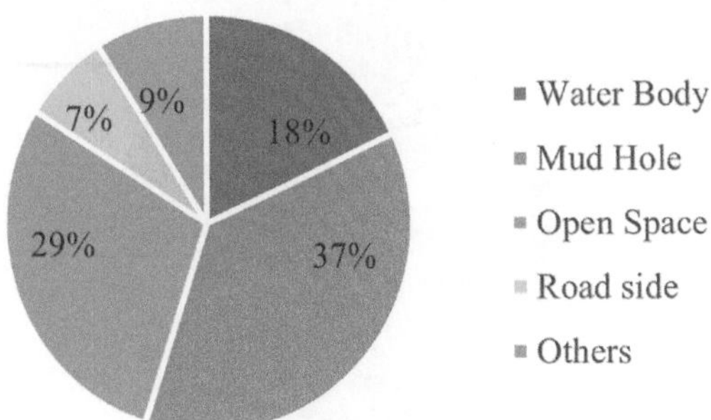

FIGURE 18.12 Waste dispose.

4.5 Transportation facilities

4.5.1 Vehicle availability

The graph illustrates the various modes of transportation utilized for commuting purposes, particularly for traveling to the CBD at Shaheb Bazar, as well as for work-related purposes. In 2018, most individuals relied on vans for commuting and transporting raw materials. Additionally, a significant portion of the population used rickshaws, autos, and bicycles for personal transportation. By the year 2022, due to improvements in road infrastructure, there was a notable shift in transportation patterns. Rickshaws and autos emerged as the preferred modes of travel, while the usage of vans experienced a decline. Furthermore, the utilization of bicycles decreased, while motorcycles witnessed an increase in popularity as a means of transportation (Fig. 18.13).

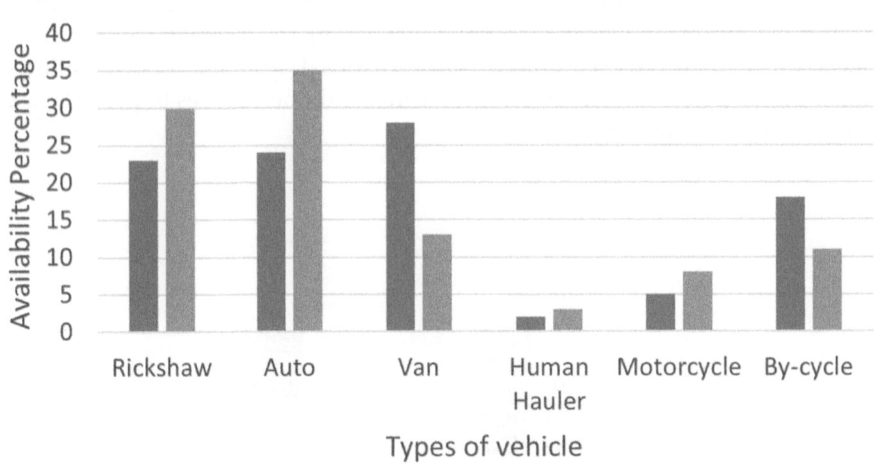

FIGURE 18.13 Available vehicle.

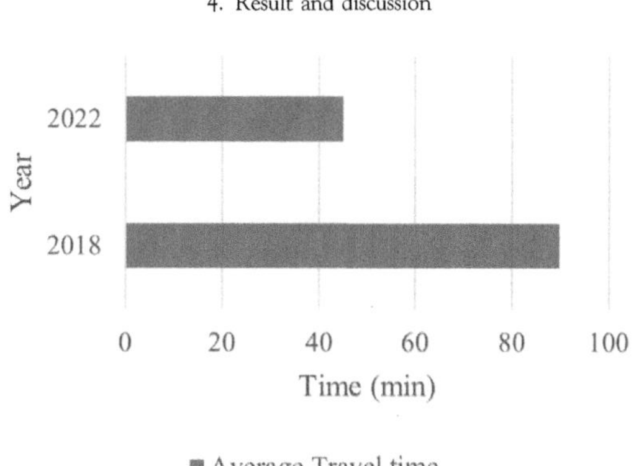

FIGURE 18.14 Average travel time.

4.5.2 *Travel time & travel frequency*

The Kishmat Kukhandi area is located approximately 9 km away from the CBD at Shaheb Bazar. In 2018, the average travel time from Kishmat Kukhandi to the CBD was 90 min. However, this travel duration significantly decreased to 45 min in 2022, because of notable improvements in the road network (Figs. 18.14 and 18.15).

Moreover, there has been an increase in the frequency of travel to the CBD in 2022. In 2018, most individuals visited the CBD around 2–4 days per week, with a significant portion visiting less than 2 days per week. A smaller proportion of people visited the CBD on a daily basis. In contrast, travel patterns have shifted in 2022. People now visit the CBD more frequently, with a majority going there daily for work purposes and returning home afterward. Consequently, fewer individuals visit the CBD about 4–6 days per week or less than 2 days per week compared to the previous years.

4.6 Correlation between factors

The graph shows the relationships between four factors: Road network, Land value, Employment, and Housing condition. The correlation values indicate the strength and

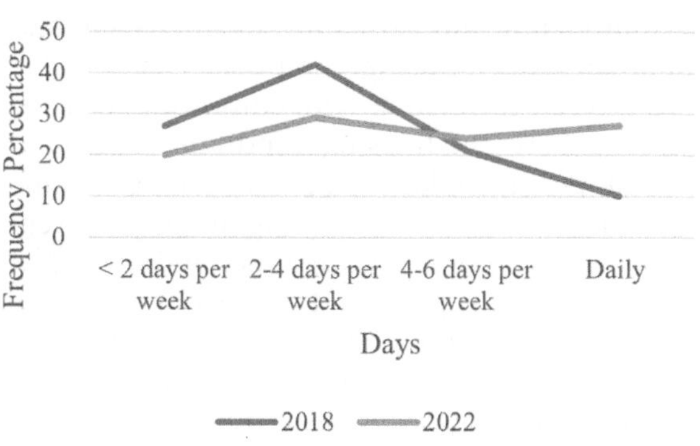

FIGURE 18.15 Travel frequency.

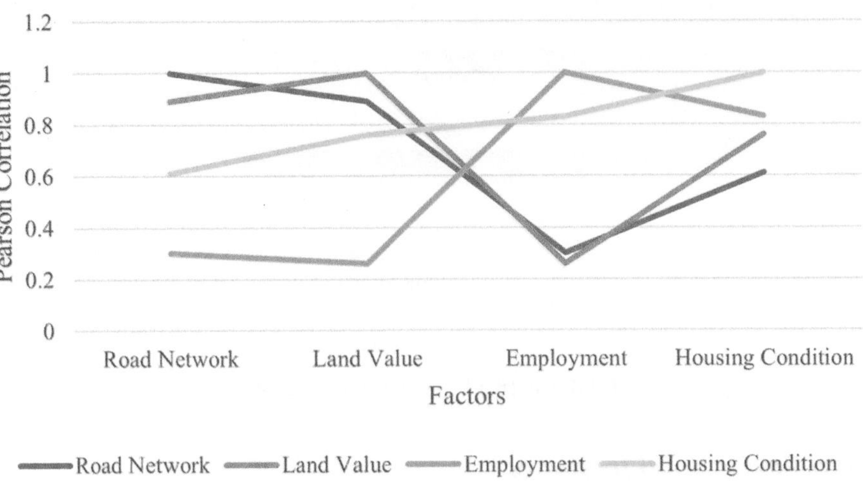

FIGURE 18.16 Correlation between factors.

direction of the relationships between these factors and highlight the interplay between these factors (Fig. 18.16).

The road network significantly influences land value, housing conditions, and employment conditions. The correlation value of 0.89 indicates a strong positive relationship between the road network and land value. It indicates that areas with better road networks tend to have higher land values. This could be because good transportation access and connectivity provided by well-developed road networks make the land more desirable and convenient for various purposes like transportation, commerce, and development. The relationship between road networks and employment yields a weak positive correlation, with a value of 0.3. This indicates that areas with better road networks may exhibit slightly higher employment rates. The presence of sound road infrastructure can enhance access to job opportunities, easing commuting for individuals and facilitating the transportation of goods and services for businesses. Furthermore, the correlation value of 0.61 between the road network and housing conditions indicates a moderately positive relationship. This indicates that better road networks tend to boast improved housing conditions. Enhanced road networks augment accessibility to housing areas, attracting infrastructure, maintenance, and development investments, thereby contributing to better housing conditions. Land value also correlates with housing conditions but has a weak correlation with employment, while employment shows a strong positive relationship between housing conditions and land value, 0.26. This correlation value suggests a weak positive relationship between land value and employment. It implies that higher land values may have slightly higher employment rates. The land value and housing condition correlation value is 0.83, which indicates that higher land values tend to have better housing conditions. Higher land values often indicate desirable locations with good amenities, infrastructure, and quality of life, which can contribute to better housing conditions. Moreover, employment demonstrates a strong positive correlation with both housing conditions and land value, as evidenced by a correlation value of 0.83. This represents that higher employment rates tend to enjoy better housing conditions. The availability of job opportunities and increased economic activity stimulate investments in housing, leading to enhanced housing conditions. Ultimately, these relationships provide insights into how these factors are interconnected and can help understand patterns and make informed decisions about future development.

4.7 Analytical hierarchy process

In order to evaluate the factors of urbanization and identify the key influencer in shaping the city's growth and addressing the homelessness problem, a multicriteria decision-making technique called AHP was employed. The study engaged professionals from three prominent sectors within the urban context of Rajshahi, namely the Rajshahi City Corporation (RCC), Rajshahi Development Authority (RDA), and Urban Planner & Lecture of Rajshahi University of Engineering & Technology (RUET). Each professional was assigned a weightage based on their expertise and experience in dealing with various urban decisions and challenges. Through pairwise comparison matrices involving the perspectives of RDA, RCC, and RUET, weights were determined for each organization. RDA, which plays a pivotal role in urban planning and granting building permissions within Rajshahi, obtained the highest weightage of 48%. RCC received a weightage of 30%, followed by RUET with 22%. These weightages were assigned based on the level of influence each organization exerts in the urban context. Considering these organizations as the main criteria, the study defined subcriteria including land value, road network, employment, and housing conditions. The final scores of these factors are depicted in the provided graph (Fig. 18.17).

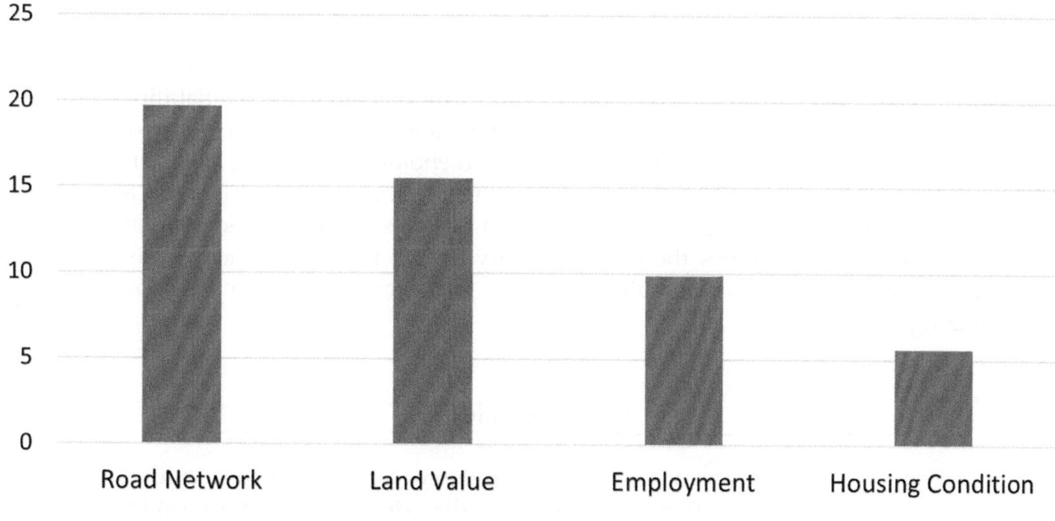

FIGURE 18.17 AHP score of factors.

According to the graph, it is evident that the road network emerges as the most influential factor in driving housing development and, consequently, reducing the problem of homelessness.

5. Findings and recommendations

After analyzing various factors of peri-urban area of Rajshahi from 2018 to 2022 from questionnaire survey, it is found that this development in peri-urban area has played a significant role in addressing the housing needs of vulnerable communities. There has been a noticeable improvement in the availability of affordable housing options. The role of peri-urban development in improving access to various factors that contribute to enhanced housing conditions. The development of road networks has facilitated better connectivity between urban and peri-urban areas, making it easier for residents to access job markets, educational institutions, healthcare facilities, and other essential services. This improved infrastructure has positively influenced the overall living standards in the peri-urban region. Additionally, the development of road networks has not only facilitated better access to essential services but has also had a substantial influence on economic growth. The increased connectivity has attracted investments and businesses, resulting in the creation of new job opportunities. This, in turn, has provided individuals with greater income-generating prospects, thereby reducing the risk of homelessness and poverty within the peri-urban area. The expansion of peri-urban regions has been instrumental in stimulating economic growth and fostering a favorable environment for employment. However, it is worth noting that the escalating land values in these peri-urban areas have had some adverse effects, particularly on the housing affordability of the middle and lower classes. As land values continue to rise, it becomes increasingly challenging for these population segments to access affordable housing options. It is essential to continue and expand efforts to create more affordable housing options for middle- and lower-income groups. This can be achieved through partnerships between government agencies, private developers, and nonprofit organizations. Innovative financing mechanisms and supportive policies can further facilitate the availability of affordable housing in peri-urban areas. Also, policymakers and urban planners should prioritize collaborative planning processes that involve key stakeholders, including local communities, NGOs, and private sector entities. By engaging all relevant parties, the planning and implementation of peri-urban development projects can be more inclusive and sustainable. Sustainable development improves the overall quality of life, minimizes the negative impacts on the environment, and reduces vulnerability to future challenges, such as homelessness in urbanizing regions.

6. Conclusion

In urbanization, the challenge of homelessness emerges as a critical concern that demands immediate attention and innovative solutions. Peri-urban development has been identified as a potential remedy, and this study has shed light on its significant role, particularly in enhancing road networks. The comprehensive analysis conducted in the Kismat Kukhandi area underscored the pivotal influence of road network development on various aspects of urban life, including housing, employment, and access to essential amenities.

Because of road network development, the Kismat Kukhandi area developed employment opportunities and reduced the urban migration rate, which helped to reduce the homelessness problem. The recognition of the paramount importance of road infrastructure development as a primary focus for policymakers offers a tangible pathway toward addressing the intricate challenges associated with homelessness within urban settings. The findings of this study emphasize the need for a more expansive exploration of these critical factors in peri-urban areas. This is why it is mentioned in the title that urbanization as remedy for homelessness. Through this statement, effective urbanization in peri-urban region has been focused. By undertaking comprehensive studies in larger contexts, policymakers can gain deeper insights into the complex dynamics of urbanization and homelessness, facilitating sustainable, inclusive, and far-reaching city planning strategies.

With the indispensable role of peri-urban development, specifically the emphasis on fortifying road networks, it becomes increasingly evident that strategic decision-making forms the cornerstone for driving positive change within urban environments. By prioritizing peri-urban development, cities can aspire to create a more sustainable, inclusive, and supportive ecosystem for their inhabitants. This concerted effort is essential for mitigating homelessness and nurturing dynamic, resilient, and thriving urban landscapes that prioritize the well-being and advancement of all individuals. Thus, through thoughtful and comprehensive planning, urban centers can strive to foster a more equitable and prosperous future for their communities and reduce the homelessness problem.

Acknowledgments

The study acknowledges the invaluable contributions of town planner Azmeri Ashrafi from Rajshahi Development Authority, town planner Bony Ahsan from Rajshahi City Corporation, and lecturer Waresul Hassan Nipun from Rajshahi University of Engineering & Technology. These individuals and organizations have greatly enriched the understanding of the subject matter and contributed significantly to the research process. The study's success is attributed to the cooperation and assistance provided by the Rajshahi Development Authority, Rajshahi City Corporation, and RUET. The dedication, professionalism, and willingness to share expertise have significantly enhanced the quality and comprehensiveness of the findings.

References

Adell, G. (1999). Theories and models of the peri-urban interface: A changing conceptual landscape. *Development Planning Unit.*

Akinluyi, M. L., & Adedokun, A. (2014). Urbanization, environment and homelessness in the developing world: The sustainable housing development. *Mediterranean Journal of Social Sciences, 5*(2), 261–271. https://doi.org/10.5901/mjss.2014.v5n2p261

Alexander Wandl, D. I., Nadin, V., Zonneveld, W., & Rooij, R. (2014). Beyond urban-rural classifications: Characterising and mapping territories-in-between across Europe. *Landscape and Urban Planning, 130*(1), 50–63. https://doi.org/10.1016/j.landurbplan.2014.06.010

Bangladesh Population and Housing Census. (2011). http://203.112.218.65:8008/WebTestApplication/userfiles/Image/National%20Reports/SED_REPORT_Vol-4.pdf.

Brennan, E. M., & Richardson, H. W. (1989). Asian megacity characteristics, problems, and policies. *International Regional Science Review, 12*(2), 117–129. https://doi.org/10.1177/016001768901200201

Destiny, O. (2017). Quantitative research methods : A synopsis approach. *Arabian Journal of Business and Management Review, 33*, 1–8.

Google Map. https://www.google.com/maps.

Hasan, M. R. (2022). *Urban planning in Bangladesh: Challenges and opportunities*. The Business Standard.

Jasni, M. A., Hassan, N., Ibrahim, F., Kamaluddin, M. R., & Che Mohd Nasir, N. (2022). The interdependence between poverty and homelessness in Southeast Asia: The case of Malaysia, Indonesia, Thailand, and Singapore. *International Journal of Law, Government and Communication, 7*(29), 205–222. https://doi.org/10.35631/ijlgc.729015

Kedir, M., Schmidt, E., & Wagas, A. (2016). *Pakistan's changing demography: Urbanization and peri-urban transformation over time*. International Food Policy Research Institute (IFPRI.

Schneider, A., Chang, C., & Paulsen, K. (2015). The changing spatial form of cities in Western China. *Landscape and Urban Planning, 135*, 40–61. https://doi.org/10.1016/j.landurbplan.2014.11.005

Schober, P., & Schwarte, L. A. (2018). Correlation coefficients: Appropriate use and interpretation. *Anesthesia & Analgesia, 126*(5), 1763–1768. https://doi.org/10.1213/ANE.0000000000002864

Shams, S., Shohel, M. M. C., & Ahsan, A. (2014). Housing problems for middle and low income people in Bangladesh: Challenges of Dhaka Megacity. *Environment and Urbanization ASIA, 5*(1), 175–184. https://doi.org/10.1177/0975425314521538

Turok, I., & McGranahan, G. (2013). Urbanization and economic growth: The arguments and evidence for Africa and Asia. *Environment and Urbanization, 25*(2), 465–482. https://doi.org/10.1177/0956247813490908

Vlahov, D., & Galea, S. (2002). Urbanization, urbanicity, and health. *Journal of Urban Health, 79*(1), S1–S12. http://www.springerlink.com/(amnxz545o4qr0f55b4vq0l55)/app/home/journal.asp?referrer=parent&backto=linkingpublicationresults,1:119977,1.

CHAPTER

19

Meta-analysis of housing policy in India (1990–2022)

Somenath Halder

Kaliachak College, Faculty, Department of Geography, Malda, West Bengal, India

1. Introduction

Between the late 1960s and early 1970s, three basic needs of humans, "roti, kapra, aur makan," (meaning: food, clothes, and housing), attracted the attention of the political economy of India (Bharti & Divi, 2019). While this *chapter is* mainly concerned with the *third* item, i.e., housing. Several co-concepts have emerged into the SDGs or part of the development issue of housing, e.g., affordable housing, housing for the poor, urban housing, and housing for the homeless (Akinwande & Hui, 2022; Baqutaya et al., 2016; Force, 2008; Ganga Warrier et al., 2019; Gangani et al., 2016; Jadhav et al., 2022; Killemsetty et al., 2022; Sethi, 2017; Tiwari et al., 2016, pp. 83–139), and so on. From Fig. 19.1, it is clearly observable that during several sets of calendar year, the policy, like "housing," is targeting specified economic groups and regions. For instance, before 1956, subsidized housing for classified groups like industrial workers, economically weaker sections, or lower income groups, plantation workers, slum clearance, and improvement had been activated. During 1957–69, the fashion of housing policy shifted toward the beneficiary of the middle-income groups, rental housing for state employees, village housing, and land acquisition and development initiative(s). On the other side, in 1971, an attempt was made to homeless workers in rural areas, urban slums (1972), and sites and service schemes (1980). Thereafter, in several calendar years, many announced schemes dedicated directly toward the Indian housing policy or policy of affordable shelters, *such as* the Indira Awas Yojana (1985), Night Shelter Scheme for Pavement Dwellers (1990), National Slum Development Program (1996), Two Million Housing Program (1959), and Valmiki Ambedkar Malin Basti Awas Yojana (2000). After 2000, several concise approaches to solve the problem of a secured "roof above the head" have been enacted with greater precision. For example, Pradhan Mantri Gramodaya Yojana (2001), Jawaharlal Nehru National Urban Renewal Mission (2005), Pradhan Mantri Adarsh Gram Yojana (2009–10), Rajiv Awas Yojana (2011), and Pradhan Mantri Awas Yojana (2015) are the

Homelessness to Hope
https://doi.org/10.1016/B978-0-443-14052-5.00019-7

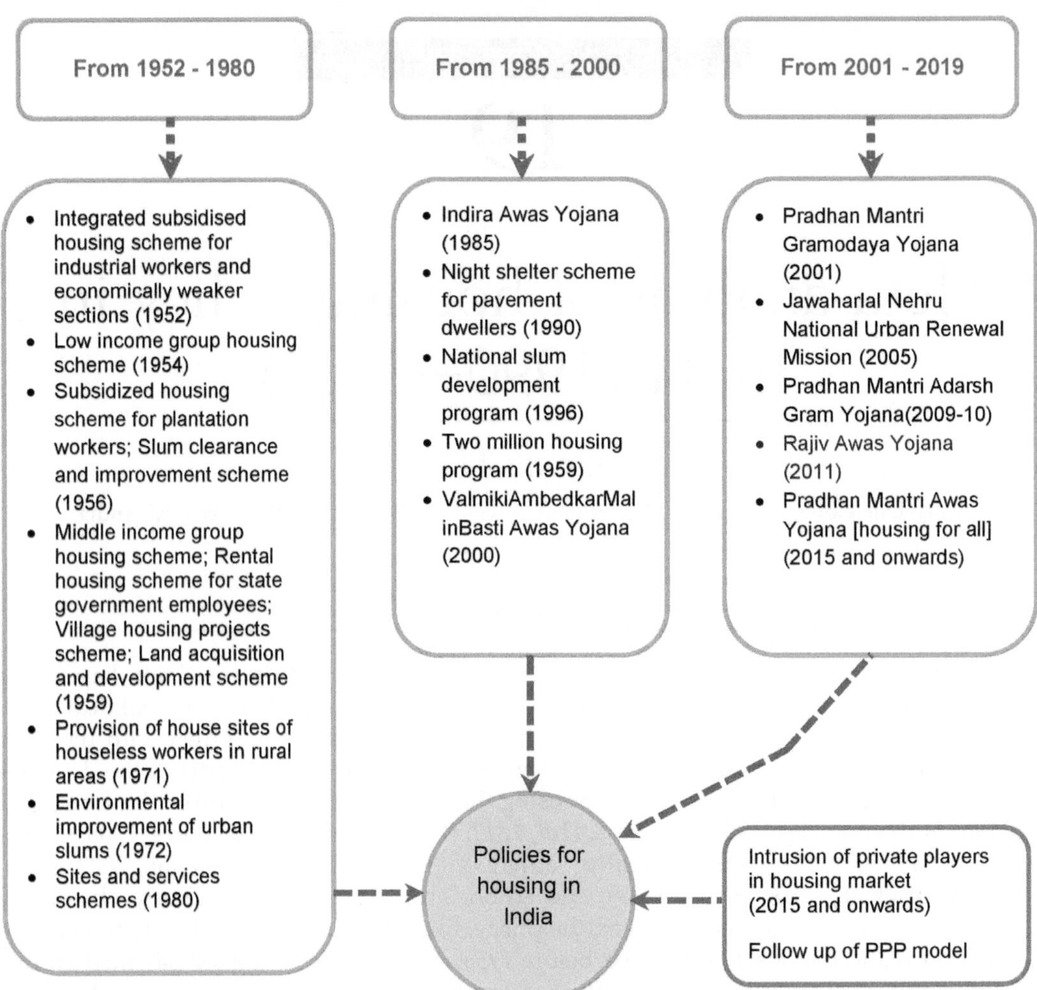

FIGURE 19.1 Changing the phase of housing policy in India after the 1950s. **Source**: *Developed by author,* **after** *Chatterjee (2020), Gohil and Gandhi (2019).*

worthy some schemes (MoHUPA, 2009; Ministry of Urban Development and Poverty Alleviation, 2015; Ministry of Housing and Urban Poverty Alleviation (MoHUPA) Pradhan Mantri Awas Yojana—Housing for All (Urban) Scheme Guidelines, Government of India, 2016; Pradhan Mantri Awas Yojana (Urban)—Housing for All; Credit Linked Subsidy Scheme for EWS/LIG, 2017; Pradhan Mantri Awas Yojana (Urban)—Housing for All; Credit Linked Subsidy Scheme for EWS/LIG, Operational Guidelines, Government of India, 2017; Gohil & Gandhi, 2019). Amidst the chaos of bilateral drifts of theoretical endeavors, area-centric or theme-centric execution, the entrance of private players, and the follow-up of private—public partnership (PPP) are the generous effects (Fig. 19.1). Beyond the above-mentioned periodical statements of a gradual progression toward the secured shelter of Indian citizens, this *chapter* encompasses three major objectives: (1) Seeking the development of studies relating to housing policy in India; (2) indicating toward the heterogeneity assessment on the theme of housing policy; (3) observing the meta-analysis and meta-regression in reframing welfare policies for affordable housing, and conferring the experiences and make statements for futuristic guidance under the prevalence of housing policy research.

The use of "meta-analysis" in *social science* for synthesizing any relevant theme is a newly endeavored approach (Field & Gillett, 2010). Previously, *this* technique has been dominantly applied in medical science, virology or other branches of human physiology, and pure science (Guzzo et al., 1987; Hunter & Schmidt, 1991; L'Abbe et al., 1987). On an instant mood, generalized question(s) may arise "What is meta-analysis?" and "What is the significance of meta-analysis?" The summarized definition of "meta-analysis" refers to a well-acknowledged method of quantitative (and the most systematic) literature review by using the standard statistical methods and combining the outcomes of independent research questions and estimated values of means and variances on the specific targeted research problems in several empirical research works (Bonett, 2009; Davies & Crombie, 1998; Greco et al., 2013; Hedges, 1992; Hunter & Schmidt, 1991). When put forward to the next level, "meta-analysis" give some logical reply(ies) not only on simple significant effect but also the direction and magnitude of effect-size about the prime research question. Ad-hoc significance relies on its hybrid techniques of agglomerations to produce synthesized knowledge acquired through multiple qualitative and quantitative means of studies (Bailar III, 1997; Crits-Christoph, 1992; Durlak, 1995; Mengist, Soromessa, & Legese, 2020a, 2020b). Side by side, the current *chapter* focuses on the boarded welfare issue of secured shelter for the homeless in the name of "housing policy" in the Indian subcontinent.

The forthcoming sections are designed as follows. The just-after subsection deals with the significance of the presently raised issue of housing policy in a developing country like India. The subsequent *section* (Section 2) is merged for the generalized understanding of the model of the *chapter*, discussing data sources and methodological inputs. In Section 3, based on the selected objectives, the derived results cover the advancement of welfare studies (underlining housing policy) in India between 1990–2022, incorporated methods, the scale of the study, assessment style, correlated issues of housing policy, quality control, and purpose of research. The subsequent sections (Sections 4 and 5) deal with limitations, recommendations, and conclusive remarks.

1.1 Rationale of the raised agenda: housing for all

Similar to other basic human needs, a secure and affordable shelter is indispensable. Under the Global South, many nations still need to provide *this* basic amenity to their underprivileged citizens. However, according to the United Nations (2008) facilitating affordable housing for needy citizens comes under the prime targets of SDGs (sustainable development goals). More surprisingly, the importance of family-wise ownership of a house got enlightened during the COVID-19 pandemic or when a severe total lockdown and social isolation were enacted worldwide (Anderson et al., 2021; Luthar et al., 2021). Therefore, these are the very lines that encouraged studying state of the art in the case of housing policy in a developing country like India after post-economic reform (after the 1990s). It is well acknowledged (almost) that the current pandemic has changed everything (Halder & Paul, 2021). Therefore, for the sake of an upgraded systematic literature review (SLR), this *chapter* advocates the meta-analytical research of housing policy in India after economic reform.

2. Materials and methods

2.1 Data

In view of SLR and the study's objectives, the tertiary data source has been used in the true sense of the term. At this stage, it is necessary to clarify "what is a tertiary database." Tertiary databases are the sort of assembled information in the form of numerical data, statistics, flow-diagram, maps, and plates, which would be examined through standard statistical tools to reach the finest target(s), and concludes. The current *chapter* has used a tertiary database from standard publishing channels, *like* Google Scholar, Scopus/ScienceDirect, ResearchGate, and Academia.edu. To reach the prefixed objective in this *chapter*, the sum of studies included and excluded are as follows: [a] Google Scholar (included: $N=48$; excluded: $N=34$), Scopus/ScienceDirect (included: $N=8$; excluded: $N=16$), ResearchGate and Academia.edu (included: $N=13$; excluded: $N=28$). Lastly, the meta-analysis and meta-regression have incised 69 research studies that dealt with the (broader) focused aspect of housing policy in a developing country like India. The consequent section and subsection emphasize the central methodological part.

2.2 Methodology

It has been generously observed that, for systematic review and meta-analysis (SRMA), especially in the case of medical research, frames like PICOT and ECLIPSE7 (Gogtay & Thatte, 2017) have been applied for synthetization. To reach the targeted outcome of this *chapter*, this current piece of academic endeavor (meta-analysis) has been applied well-accepted frame like PSALSAR (delAmo et al., 2018; Mengist et al., 2019). The term "PSALSAR" refers to P = Protocol, S = Search, AL = Appraisal, S = Synthesis, A = Analysis, and R = Report (Table 19.1). On the other side, Table 19.1 helps understand the research frame incorporated for meta-analysis and meta-regression. After a rigorous review of the past literature belonging to the old mode or classical mode of literature review, a considerable research gap has been found in the arena of housing policy. Subordinately, to reach the detailed "state of the art" dealing with the *housing policy in India*, several standard doorways *like* Scopus/ScienceDirect, Google Scholar, ResearchGate, and Academia.edu have been chosen for the research time-frame during 1990–2022. Table 19.2 guides the criteria of selection and de-selection of research focused on Indian housing policy. In the section above, it has been mentioned that the availability of academic research works highlighting the agenda, like the Indian housing policy, is limited, and as a suitable solution, a well-recognized virtual (or online) medium is used for SLR. The title, abstract, and keywords are the vital search keys, and inclusion of available research works. On the contrary, the criteria of de-sections are replica work (or not-original), gray literature, self-publishing literature, monograph (in

TABLE 19.1 Framework of meta-analysis.

Frame	Stages	Products	Approaches
PSALSAR frame	Protocol	Defined the objective, scope, and extent	Exclusively focusing on housing as a major welfare policy in a country like India
	Search	Define the search strategy	Searching strings (major gateways: Google Scholar, Scopus/ScienceDirect, ResearchGate, etc.)
		Search studies	Search databases (mainly published research articles and chapters of the books)
	Appraisal	Selecting studies	Determining the inclusion and exclusion criteria
		Quality assessment of selected studies	Quality criteria
	Synthesis	Data extraction	Template oriented extraction
		Data categorization	Categorize the data based on the iterative definition and prepare it for further analysis
	Analysis	Data analysis	Quantitative categories, description, and narrative analysis of the tabulated and rectified data
		Result and discussion	Based on the analysis, show the trends, identify the gap(s), and make the comparison
		Conclusion	Developing conclusion and recommendation
	Report	Report writing	PRISMA methodology
		Preparing for publication	Summarizing the report result for the larger audience

Source: Developed by author (delAmo et al., 2018; Mengist et al., 2019).

the book format), editorial letters, and so forth. Hence, finally, those works that did not concentrate on the directed theme of "housing policy in India" and are inaccessible full-text literature reviews have been excluded. In Fig. 19.2, the roundup methodological layout can be seen.

To pursue the meta-analysis and meta-regression on the topic of housing policy (in India), a code sheet has been prepared based on the chosen criterion (*see* Table 19.3). In the first column of the code sheet, only the surnames of the first or corresponding author, with a year of publish; have been recorded as a statement. While in the case of research works having more than a single author, "et al." is added. As this *chapter* exclusively deals with the housing policy concerned a country like "India," the second criterion (study area) is not included in the quantitative review analysis, a common factor for all included studies. The third criterion, like a source of the database used, has two subcriterions, and the inputted values are binary (0 or 1). The following criteria are set with the subjective quantification of the quality of each subtype (1 for *good*, 2 for *better*, and 3 for *best*). Significantly, criterion like "quality control" is newly adopted (which shows the citation frequency). After a rigorous cross-checking of the tabulated data and required modification of prearranged information, this systematic thematic review goes through critical statistical analysis. This quantitative review adopted

TABLE 19.2 Criteria of inclusion for meta-analysis.

Criteria	Decision
Selection of the research work depends on prefixed keywords contained as a whole or at least in title, keywords, abstract	Inclusion
Chapter in an edited book, published in a peer-reviewed publication	Inclusion
Research article/book chapter written in the English language	Inclusion
Research works that present piece(s) of evidence on India or any provinces of the same	Inclusion
Published material in the shape of a monograph or limited accessibility	Exclusion
Self-published material like a dissertation or without peer-reviewed	Exclusion
Research works that are not accessible in full format	Exclusion
Papers that are not primary/original research	Exclusion
Research work that got published before 1950	Exclusion

Source: Developed by author (Mengist et al., 2019; Mengist et al., 2020a, 2020b).

standard methods *like* descriptive statistics, odd ratio, and bilinear regression. For final data representation and computation, several recommended tools *like* MS Excel and IBM SPSS (Version 29.0.1.0(171)) have been used to show the meta-analysis results through *forest plot, funnel plot, Galbraith plot*, and *bubble plot*. Lastly, a critical comparative assessment is made with the best possible alternative suggestions for future studies; in this way, the *chapter* has its synthesized shape.

3. Results and discussion

3.1 Prelude discussion on "housing policy"

Since independence, India, as a nation that belongs to the Global South, has suffered from so many problems (Mohanty, 2018; Prashad, 2014; Thomas-Slayter, 2003). Among them, the agenda of secured shelter (or, say, housing) is recently capable of drawing the attention of the government as well as the private sector. The problem of accommodation for the homeless is ever-growing due to the ever-increasing population (Manomano, 2023), massive number of urban immigrants (Jha & Kumar, 2016), option deficiency of easy and cheap financing for house construction in rural India (Hardoy & Satterthwaite, 2014, pp. 1–374), unavailability of residential lands in already overcrowded slums (Dear & Wolch, 2014; Jha & Kumar, 2016), and so on. After economic reform, several housing policies in India partially relate to the homeless, but the number of homeless people gradually increased. Meanwhile, the indirect influencing causes behind the rising number of homeless are malfunctioning housing policy (Burt, 1991; McChesney, 1990), complexity in implementation in rural and urban areas, the lacuna in financing transparency (before 2014) (Bauman & Bordoni, 2014; Khan, 2022,

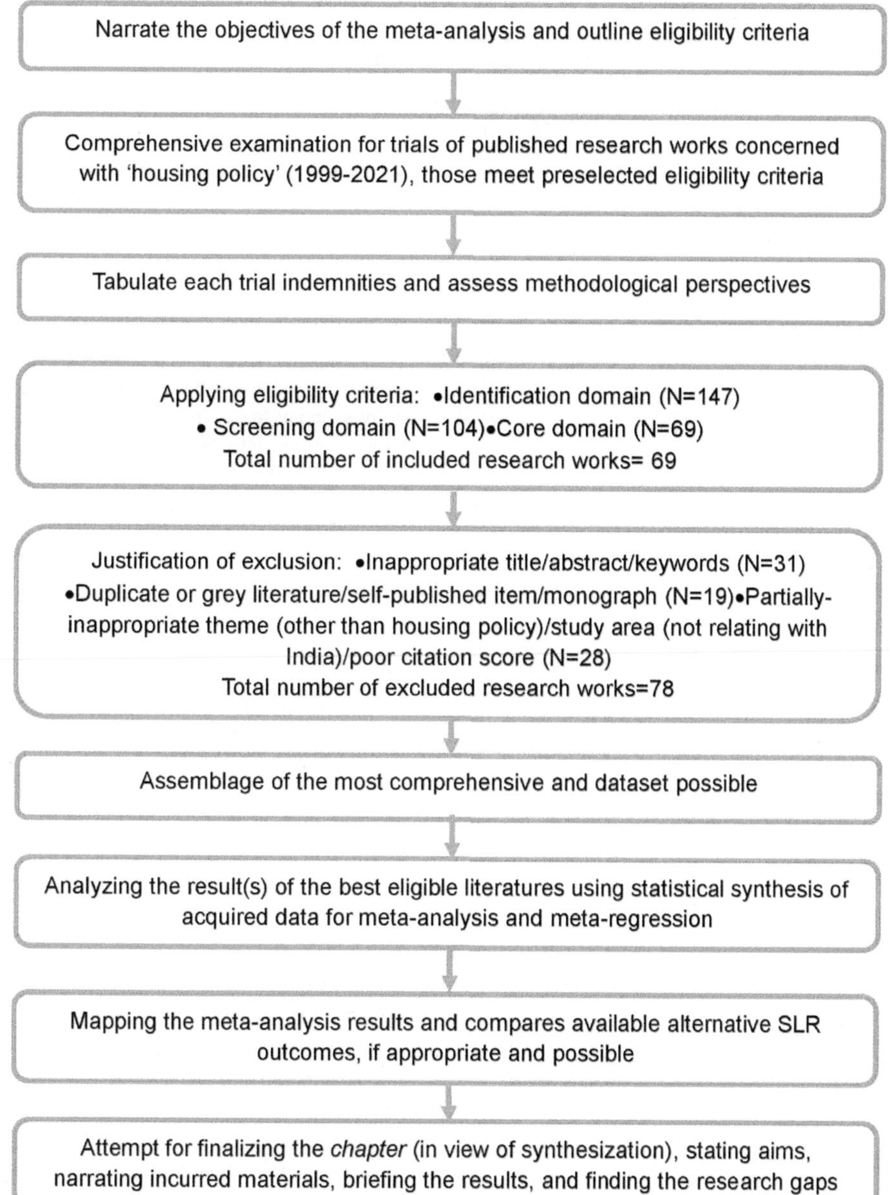

FIGURE 19.2 Flow diagram displaying the methodological profile. **Source**: *Developed by author,* **after** *Akobeng (2005), Tawfik et al. (2019), Mengist et al. (2020a, 2020b).*

pp. 229–258), backwardness and unawareness of poor and underprivileged sections in availing housing scheme (Kozel & Parker, 2003; Singh, 2011), unstructured rental policy (for the middle-income group) especially in big cities (Soederberg, 2018), and so forth. However,

TABLE 19.3 Statement of selected criteria.

No.	Criteria	Chosen categories	Justification
1.	Author(s) and year of publication	Between 1990 January to 2021 December	Those research works published before 1990 were excluded
2.	Fixed study area	India	As this study is focused on Indian housing policy, thus other than Indian scenario-oriented works (e.g., South Asian, Global South, World, etc.) are discarded
3.	Data sources	Primary data	Any kind of data collected through sampling method(s) (e.g., field data, surveys, interviews, or census data)
		Secondary data	The source of secondary data which are not collected by the author(s) but collected through an available set of information (e.g., remote-sensed data, socioeconomic data, and mixed sources of databases like regional/national statistics)
4.	Method	Qualitative analysis	Use available or rare kinds of literature
		Quantitative analysis	Use of descriptive statistical tools for analysis
		Comparative analysis	Use of any cross-comparative assessment at any specified level (i.e., regional, national, or global)
5.	Scale of study	Micro level	Rural or urban
		Regional	State level, cluster of states
		National	As the entire country level
6.	Focusing the (small and seminegligible) precise agendas with housing policy	Rural and urban housing problems	Studies not only discussing food security and welfare but also touching the issue of hunger and starvation
		Slum housing	Studies not only discussing housing in slum areas and welfare but also touching the issue of homelessness
		Role of banking and insurance sectors	Studies not only discuss the role of the banking and insurance sectors in allocating funds but also touch the issue like welfare guarantee
		Gaps between framed policy and implementation	Studies highlighting gaps between framed policy and implementation and welfare but also touching the issue of insecure life
		Unfulfilled SDGs	Studies not only discuss housing policy and welfare but also touch the broader issue like sustainable development goals
7.	Quality control	Average cited score	Number of cited research under each study
8.	Roundup purpose	Contribution toward knowledge	Works especially contributed toward the world of knowledge for any further development of the concept

TABLE 19.3 Statement of selected criteria.—cont'd

No. Criteria	Chosen categories	Justification
	Critical evaluation	To develop reframed policy or examining the existing method(s) and reformation of policy
	Management option	To endorse management option(s) for suitable implementation of accessible resources
	Policy implementation	To suggest new or modified policies for the eradication of problems concerned with the crisis of secured shelter

Source: Developed by author (Mengist et al., 2020a, 2020b).

during meta-analysis (1990–2022), a shortfall has been noticed in the case of research studies on rental policy in large urban areas or metropolitan cities of India. In short, the scenario of a secured home for the homeless is a major concern not only for the researchers but also for the government and policymakers.

Before entering into the core part of a meta-analysis, it is significant to study the state-of-art published research on the specific theme, i.e., housing policy. This current *chapter* has decided to choose 1990 to 2022 as a time frame and India as a study region. After a rigorous (available) literature review shows that before 1998 and from 2000 to 2004, a minimal amount of research has been concentrated on *housing policy in India* (e.g., 4.35% and 5.80%, respectively). While during 2015–19, the highest proportion (43.48%) of research has been focused on the *said* theme, followed by 2020–22 (23.19%), 2005–09 (13.04%), and 2010–14 (10.14%). Conversely, most of the research on the consulted topic is qualitative (about 73%), while above 80% of the study used secondary data, and nearly 76% of works are stacked to country-level assessment. The following subsections are endorsed for a more critical analysis of the highlighted theme.

3.2 Housing policy in India: A heterogeneity test

Although the meta-analysis falls under the group of *Systematic Literature Review*, it differs a lot due to its analytical precision, mode of data processing, level of authentication, and many more. In this *chapter*, 69 studies have been included, and each study is treated as singular. This section is primarily focusing on heterogeneity test. The term "heterogeneity" explains the nature of departure or dissimilarity among studies and the end products concerned with any singular or multidimensional theme or concept(s) (Sandercock, 2011; Sedgwick, 2015). More clearly, in view of the standard synthesizing of a sizable number of empirical research in the shape of a single academic research under any specific theme, heterogeneity or statistical heterogeneity have unconditional importance. It also helps to study its χ^2 value, I^2 value, z value, and P-value for accurate data outcome (Fletcher, 2007). Here, the overall size of the meta-analysis is 69. Meanwhile, particularly in the case of "effect-size" selection, the "fixed-effects size" model has been chosen as ideal. And the reason is: when a meta-analysis deals with a particular mono-theme, it would be a better choice to incorporate "fixed effects size" (Barili et al., 2018; Tufanaru et al., 2015). The forest plot diagram (Fig. 19.3)

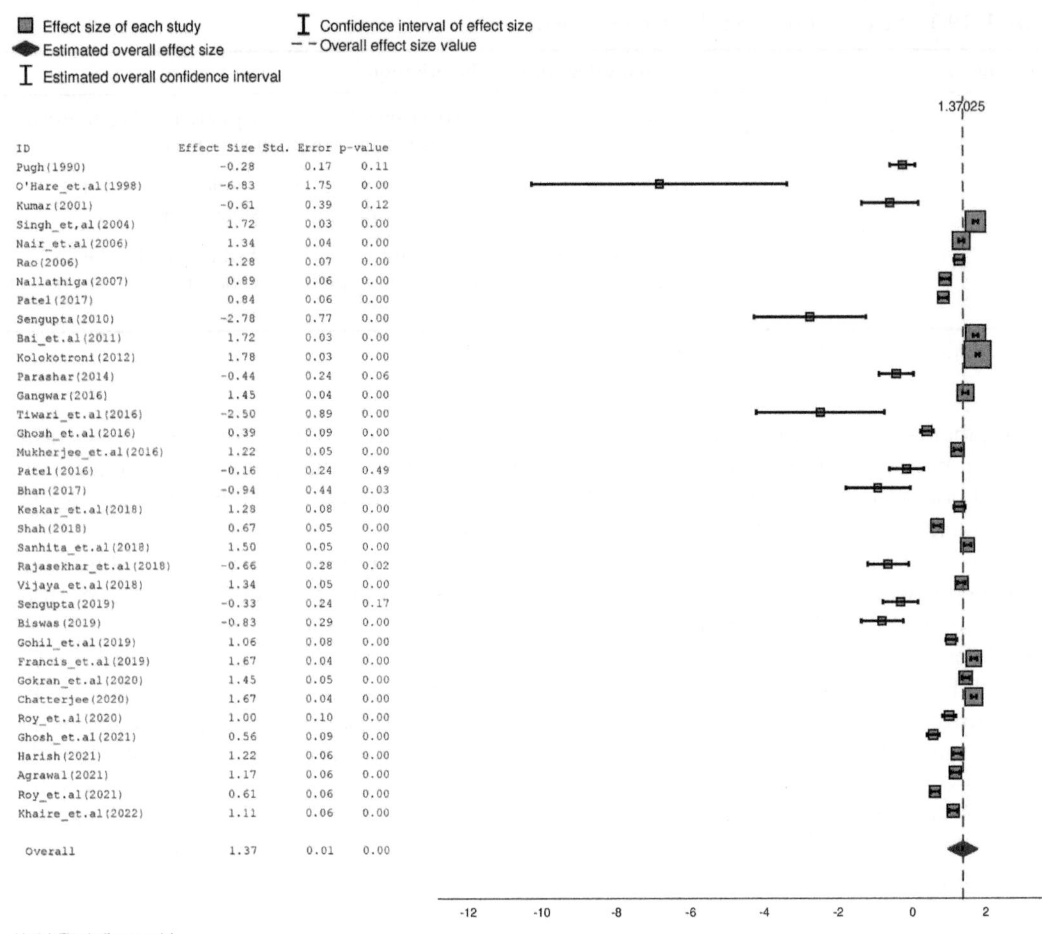

FIGURE 19.3 Forest plot displaying heterogeneity forest plot displaying heterogeneity.

depicts the overall effect size of the study, χ^2 value, I^2 value, z value, and P-value. As (Borenstein et al., 2010) suggested, this *chapter* applied "fixed effects size" to analyze the statistical heterogeneity and other correlated indices. The overall effect size (z) is 135.78, which should be considered rational. The overall (statistical) heterogeneity of housing policy demonstrates that the Chi-squared value is 51.73, which shows remarkable heterogeneity. According to (Higgins et al., 2003; Higgins & Thompson, 2002), I^2 statistic represents the "percentage of variation across studies." At the same time, this meta-assessment corresponds to the I^2 value is 0.98, which proves its immense significance level. The summed-up effect-size value of this meta-analysis is 1.37. On the contrary, the P-value is generally used for obtaining the "means of converting meta-analysis results to defined test statistics which are expressible as a function of the estimates of the βs and σs" (Chyou, 2012; Greenland et al., 2016). In this study, the P-value is 0.0001, replicating the greater significance level of the meta-analysis (at 95% confidence level).

3.3 Meta-analysis: testing dissimilarity and precision

This part mainly includes two types of meta-analysis: *Funnel Plot* and *Galbraith Plot*, for testing the dissimilarity and precision of the topic "Housing Policy in India." A funnel plot is used as a standard meta-cartographic means to represent a scatter plot of the "effect estimates" (consulting any definite theme), individual research studies made on some specific measure. Whereas the outcome of the means as mentioned earlier (funnel plot) differs based on the size of (each) empirical work or precision meticulously integrated into the meta statistical analysis (MSA). Under the funnel plot diagram, the y-axis (vertical axis) shows the "standard error," and the x-axis (horizontal axis) displays the "mean difference" (Sterne & Egger, 2001; Sterne et al., 2011). Side by side, under the fundamental (triangular) geometric shape, i.e., triangular funnel, the scattering pattern of computed and plotted data (belonging to each study) placed either the bottom left-hand corner or upper-head part directly relies on smaller or larger studies (Begg & Berlin, 1988; Sterne & Egger, 2001). From Fig. 19.4, it is revealed that there is (nearly) a moderate symmetry, which ultimately indicates moderately higher publication biases. Additionally, the above-stated remark becomes logical when this study includes a 95% pseudo-confidence interval. In detail, in the upper left-side part of the funnel, some studies have negative standard error (below 0.5), and upper right-side part, some studies have slightly higher standard error (ranges between 0–1). Contrastingly, three studies (published in 1998, 2015, and 2010, respectively) have higher standard errors (ranges from 1.7 to 3.1). Briefly, it would not be wrong to say that the angular part of the left side of the funnel shows the *P*-value is less than 0 or with a negative value, and here null hypothesis is rejected. On the opposite side, some studies have a *P*-value >.01 to 5.1, where it can be stated that the null hypothesis would not be rejected completely.

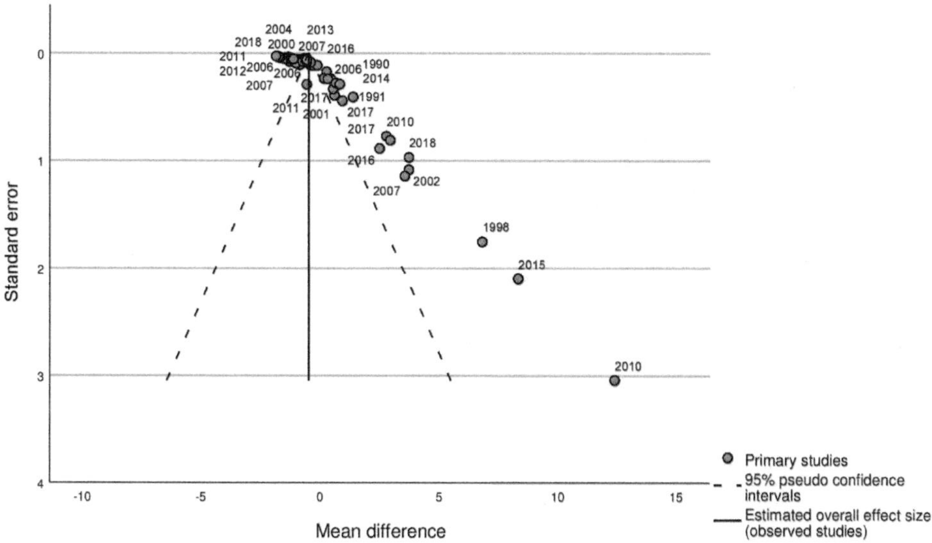

FIGURE 19.4 Funnel plot showing publication biases on the issue of housing policy.

Simultaneously, on the right portion of the funnel, three studies (mentioned previously) replicated the case of complete publication biases.

According to (Colditz et al., 1994), the *radial plot* (commonly known as Galbraith Plot) is widely accepted for displaying the roundup story of MSA. This established method is purposively applied to show the "unweighted regression of z-scores on the inverse standard error with the intercept constrained to zero" (Galbraith, 1988, p. 272). As a generalized and ad-hoc content, *this* method displays the outlier in the effect size or effect estimates. The overall picture of the outlier in the effect-size on the underlined issue, *like* housing policy in India, is shown in Fig. 19.5. From the said diagram (Fig. 19.5), it is clearly observable that a small number of empirical researches came under within the broken line of the 95% confidence interval region; where a major portion of research works remain beyond the mentioned region. Moreover, the calibrated regression line displays a negative trend. There is a common agreement, as well as understanding, that 95% of the published research works should be plotted under two defined broken lines, or, *say*, "confidence interval region." In the case of the present *chapter* (regarded as a singular and synthesized study), as a major proportion of published works remains beyond the defined area, it certainly confers precision. In short, a much smaller proportion of the research works closer to the y-axis have lower precision, while the precision increases toward the x-axis, where comparatively a higher proportion of research is plotted.

3.4 Meta-regression of Indian housing policy

Contemporarily, "meta-regression" is considered a modern and robust method for drawing conclusive remark(s) under the objective of synthetization the already attempted

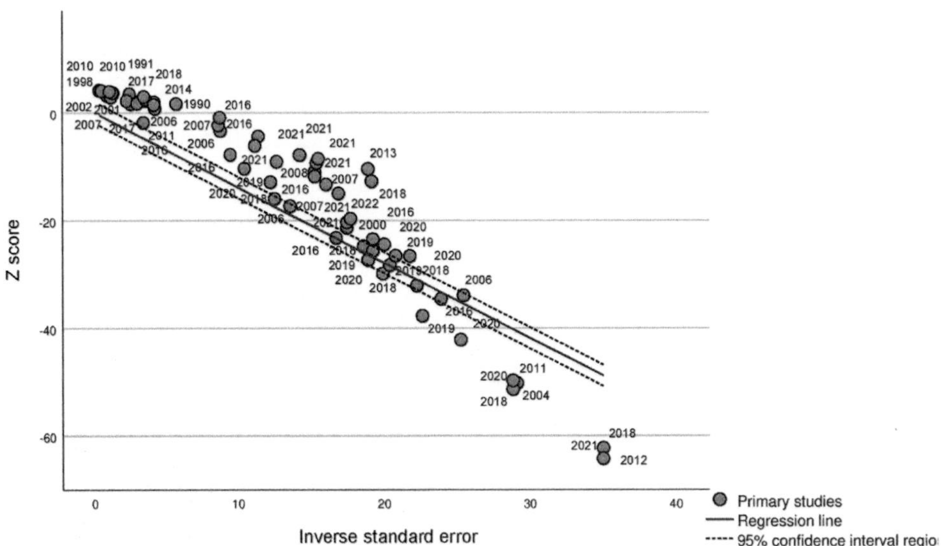

FIGURE 19.5 Galbraith plot showing research precision on the issue of housing policy.

outcomes presented by various academic types of research, particularly on the singular (or specific) themes. The concerned method solely depends on "effect size," i.e., fixed effect size (FES) or random effect size (RES). To pursue a linear relationship between study-generated covariates and the effect-size among diversified disciplines and subdisciplines, a scientific modus operandi "meta-regression" has been used for model creation (Fagerland, 2015; Jakubowski et al., 2015, pp. 697–725). The application of the said method, nowadays, intrudes into the sphere of social science, though it originated from the arena of pure science. In other words, the prime merit of "meta-regression" is like that its usage for forecasting the core objectives (e.g., combination, comparison, and synthetization) of the already selected outcomes from manifold types of academic research and for destined the ultimate aim(s) (Bonett, 2009; Thompson & Higgins, 2002). A cartographic expression like "Bobble Plot" is usually adopted to show the product of meta-regression. Based on a moderator variable, a scatter plot is functionally plotted (in respect of the "x" and "y" axis) depending on the effect-size of each preselected study. In the same meta-regression model, any moderator variable is measured through the x-axis, and the observed effect-sizes or effect estimates are on the y-axis. As a result, an auto-generated regression line has come up from the *said* model. Additionally, it can (also) be remarked that in this type of cartographic expression of meta-regression, the (software) generated points (headed by each selected research work) are characteristically perpetuated with diversified sizes to replicate their precision or weightage in the model (Thompson & Higgins, 2002; Viechtbauer, 2010). The concerned *chapter* has chosen the criterion of "citation" of each selected study as moderator of meta-regression (represented on the horizontal axis). At the same time, the "effect-size" of each included study has been shown on the vertical axis (or y-axis) (Fig. 19.6). As per the nature of the current endeavor, *here*, the "fixed effect-size model" and *t*-distribution (estimation based) have been adopted to show 95% confidence level. The final outcome shows the ultimate negative trend of the regression line. The summary regression line depicts a sharp decline in accordance with the x-axis, where the effect-size is very nominal. On the contrary, where the effect-size is (comparatively) higher, the studies have a meager citation impact; exceptionally, two studies (on housing policy) have a lower effect-size (less than −0.2) but with comparatively moderately higher citation (40.5–50.5). In short, it would be fitting to comment that, tagged with the 95% confidence interval, the individual studies' effect size (concerned with India's housing policy) has increased over time but with lower citations. *Op cit*, in the case of older studies, the citation impacts are higher instead of lower effect-size, which is a natural phenomenon.

4. Limitations of the study

On behalf of the (real) development and/or welfare of the poor (or underprivileged) and homeless people, the synthetization of policy matters the current research endeavor (which should also be considered as novel one) concerning the agenda of secured shelter, along with the format of welfare policy, the targeted MSA has some limitations too. As per predecided materials and methods, this *chapter* has not included any studies before the 1990s. Furthermore, in accordance with the predecided line and length, this *chapter* only covers India (with its Judiciary State and Union territories), whereas in special cases, intra-country cross-

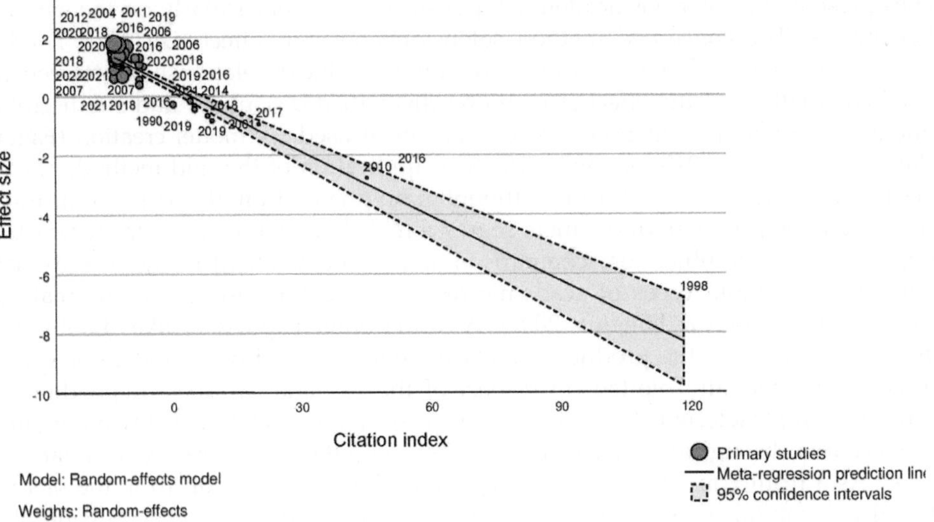

FIGURE 19.6 Bubble plot showing the meta-regression on the issue of housing policy aims.

comparisons have been allowed. The off-listed works are given as follows: (a) monograph, (b) self-published dissertation (without peering), and (c) nonpeer-reviewed academic writings. Conversely, the present *chapter* is built upon certain preselected objectives and methodological outfits. Thus, the *chapter* has been unable (partially) to focus beyond this framed issue, *like* housing policy.

5. Conclusion

There are so many hegemonies in terms of methodological inputs, selection of study area (micro to macro), research scale, database, and the subthemes covered under the housing policy of India during 1990–2022. After a rigorous and systematic literature review, it can be said that nearly 40% of the research covers the sub-theme of secured homes for the slum-dwellers, and about 58% and 32% of works focused on the subtheme of housing for the homeless in urban and rural areas, respectively. Side-by-side, a considerable research gap is found in the case of affordable rental housing, especially in urban areas for immigrants from the middle-income group. The following section also pinpoints the main findings and suggestions for next-generation researchers. Among long-listed welfare policies, "secured shelter," which protects a household from (nearly) every evil (natural calamities or manmade misact), is a prime need of human beings. This *need* seems more crucial for homeless people and immigrants, especially in a country like India. As this *chapter* is purposively concerned with previously conducted housing policy-oriented standard research endeavors across the nations, the overall result of MSA helps find the gaps as well as new suggestive avenues of further research. This study also proves a plodding progression of academic studies, on this particular theme, between the years 1990–2022. The overall heterogeneity of this meta-analysis is 51.47. Correspondingly, this study also reveals good precision, authenticity, and validation of the previously attempted research on Indian housing policy. After a thoughtful churning, few recommendations are given as follows. (a) In the future, studies on housing policy should be done on cross-country-based comparative analysis with the help of advanced methodology and models. (b) The upcoming research should focus on areal data-oriented products and, in the case of drawing recommendations, should address micro-regional crisis management for a secured hosting. (c) The next generation of studies on housing for the homeless should incur stakeholders' perceptions regarding the status of the ongoing housing welfare schemes and predict the solutions. Apart from this, it has also been observed that there is a significant research gap in the section on secured housing policy for the middle-income group and the reframing of laws on rental housing (in urban areas) for immigrants. Cohesively, if any research persuasion of the given recommendations has been incorporated, even partly, the aim of this *chapter* might be fulfilled.

Acknowledgments

The author would like to thank those anonymous reviewers whose constructive comments have made this *chapter* acceptable to a global audience. Secondly, the authors would like to show courtesy to those previous authors/researchers who contributed much in the field of housing policy, shelter management, and other welfare study, which encouraged a lot to conduct of this meta-analysis. The author would also thank the online sources and tools for the execution of the study.

Funding information
This study was not funded by any authority (Government, Private, or any Society in India or Abroad).

Conflict of interest
The author declares there is no potential conflict of interest concerning the research, authorship, and/or publication of this chapter.

References

Akinwande, Timothy, & Hui, Eddie C. M. (2022). Housing supply value chain in relation to housing the urban poor. *Habitat International*. ISSN: 01973975, *130*, 102687. https://doi.org/10.1016/j.habitatint.2022.102687

Akobeng, A. K. (2005). Understanding systematic reviews and meta-analysis. *Archives of Disease in Childhood*. ISSN: 14682044, *90*(8), 845–848. https://doi.org/10.1136/adc.2004.058230

Anderson, S., Parmar, J., Dobbs, B., & Tian, P. G. J. (2021). A tale of two solitudes: Loneliness and anxiety of family caregivers caring in community homes and congregate care. *International Journal of Environmental Research and Public Health*. ISSN: 16604601, *18*(19). https://doi.org/10.3390/ijerph181910010

Bailar III, J. C. (1997). The promise and problems of meta-analysis. *New England Journal of Medicine, 337*(8), 559–561. https://doi.org/10.1056/NEJM199708213370810

Baqutaya, S., Ariffin, A. S., & Raji, F. (2016). Affordable housing policy: Issues and challenges among middle-income groups. *International Journal of Social Science and Humanities*. ISSN: 20103646, *6*(6), 433–436. https://doi.org/10.7763/IJSSH.2016.V6.686

Barili, F., Parolari, A., Kappetein, P. A., & Freemantle, N. (2018). Statistical primer: Heterogeneity, random- or fixed-effects model analyses? *Interactive Cardiovascular and Thoracic Surgery*. ISSN: 15699285, *27*(3), 317–321. https://doi.org/10.1093/icvts/ivy163

Bauman, Z., & Bordoni, C. (2014). *State of crisis*. John Wiley & Sons.

Begg, Colin B., & Berlin, Jesse A. (1988). Publication bias: A problem in interpreting medical data. *Journal of the Royal Statistical Society: Series A*. ISSN: 09641998, *151*(3), 419. https://doi.org/10.2307/2982993

Bharti, M., & Divi, S. (2019). Housing policies: Common goals and diverse approaches to social housing in India and other countries. *PDPU Journal of Energy and Management, 4*(1), 41–49.

Bonett, D. G. (2009). Meta-analytic interval estimation for standardized and unstandardized mean differences. *Psychological Methods*. ISSN: 1082989X, *14*(3), 225–238. https://doi.org/10.1037/a0016619

Borenstein, Michael, Hedges, Larry V., Higgins, Julian P. T., & Rothstein, Hannah R. (2010). A basic introduction to fixed-effect and random-effects models for meta-analysis. *Research Synthesis Methods*. ISSN: 17592879, *1*(2), 97–111. https://doi.org/10.1002/jrsm.12

Burt, M. R. (1991). Causes of the growth of homelessness during the 1980s. *Housing Policy Debate*. ISSN: 2152050X, 2(3), 901–936. https://doi.org/10.1080/10511482.1991.9521077

Chatterjee, A. (2020). Housing the urban poor: Understanding the policy shifts. In P. R. Choudhury, & A. Narayana (Eds.), *Land in India: Issues and debates* (pp. 26–34). India Land and Development Conference (ILDC) 2020.

Chyou, P. H. (2012). A simple and robust way of concluding meta-analysis results using reported P values, standardized effect sizes, or other statistics. *Clinical Medicine and Research*. ISSN: 15546179, *10*(4), 219–223. https://doi.org/10.3121/cmr.2012.1068

Colditz, G. A., Brewer, T. F., Berkey, C. S., Wilson, M. E., Burdick, E., Fineberg, H. V., & Mosteller, F. (1994). Efficacy of BCG vaccine in the prevention of tuberculosis: Meta-analysis of the published literature. *JAMA, the Journal of the American Medical Association*. ISSN: 15383598, *271*(9), 698–702. https://doi.org/10.1001/jama.1994.03510330076038

Crits-Christoph, P. (1992). The efficacy of brief dynamic psychotherapy: A meta-analysis. *American Journal of Psychiatry*. ISSN: 0002953X, *149*(2), 151–158. https://doi.org/10.1176/ajp.149.2.151

Davies, H. T. O., & Crombie, I. K. (1998). Getting to grips with systematic reviews and meta-analyses. *Hospital Medicine*. ISSN: 14623935, *59*(12), 955–958.

Dear, M. J., & Wolch, J. R. (2014). *Landscapes of despair: From deinstitutionalization to homelessness* (Vol 823). Princeton University Press.

Durlak, J. A. (1995). Understanding meta-analysis. *Reading and understanding multivariate statistics*, 319–352.

Fagerland, M. W. (2015). Evidence-based medicine and systematic reviews. *Research in medical and biological sciences: From planning and preparation to grant application and publication* (pp. 431–461). https://doi.org/10.1016/B978-0-12-799943-2.00012-4, 9780128001547.

Field, A. P., & Gillett, R. (2010). How to do a meta-analysis. *British Journal of Mathematical and Statistical Psychology*. ISSN: 00071102, *63*(3), 665–694. https://doi.org/10.1348/000711010X502733

Fletcher, John (2007). What is heterogeneity and is it important? *BMJ, 334*(7584), 94–96. https://doi.org/10.1136/bmj.39057.406644.68, 0959-8138.

Force, H. L. T. (2008). *Affordable housing for all. Ministry of housing and poverty alleviation, Government of India*.

Galbraith, R. F. (1988). Graphical display of estimates having differing standard errors. *Technometrics*. ISSN: 15372723, *30*(3), 271–281. https://doi.org/10.1080/00401706.1988.10488400

Ganga Warrier, A., Tadepalli, Pavankumar, & Palaniappan, Sivakumar (2019). Low-cost housing in India: A review. *IOP Conference Series: Earth and Environmental Science, 294*(1), 012092. https://doi.org/10.1088/1755-1315/294/1/012092, 1755-1307.

Gangani, M. G., Suthar, H. N., Pitroda, J., & Singh, A. R. (2016). A critical review on making low cost urban housing in India. *International Journal of Constructive Research in Civil Engineering (IJCRCE), 2*(5), 21–25. https://doi.org/10.20431/2454-8693.0205004

Gogtay, N. J., & Thatte, U. M. (2017). An introduction to meta-analysis. *Journal of the Association of Physicians of India*. ISSN: 00045772, *65*(October), 78–85. http://www.japi.org/october_2017/12_sfr_an_introduction_to_meta_analysis.pdf.

Gohil, J., & Gandhi, Z. H. (2019). Pradhan Mantri Awas Yojana (PMAY) scheme—An emerging prospect of affordable housing in India. *International Research Journal of Engineering and Technology, 6*(12), 2546–2550.

Greco, T., Zangrillo, A., Biondi-Zoccai, G., & Landoni, G. (2013). Meta-analysis: Pitfalls and hints. *Heart Lung and Vessel, 5*(4).

Greenland, S., Senn, S. J., Rothman, K. J., Carlin, J. B., Poole, C., Goodman, S. N., & Altman, D. G. (2016). Statistical tests, P values, confidence intervals, and power: A guide to misinterpretations. *European Journal of Epidemiology*. ISSN: 15737284, *31*(4), 337–350. https://doi.org/10.1007/s10654-016-0149-3

Guzzo, R. A., Jackson, S. E., & Katzell, R. A. (1987). Meta-analysis analysis. *Research in Organizational Behavior, 9*(1), 407–442.

Halder, Somenath, & Paul, Sourav (2021). Long-term vision of development in post COVID-19 era: A normative theorem in world perspective. *Ensemble*. ISSN: 25820427, *SP-1*(1), 30–34. https://doi.org/10.37948/ensemble-2021-sp1-a004

Hardoy, J. E., & Satterthwaite, D. (2014). *Squatter citizen: Life in the urban third world* (pp. 1–374). Argentina: Taylor and Francis. https://doi.org/10.4324/9781315070193, 9781134157389.

Hedges, Larry V. (1992). Meta-analysis. *Journal of Educational Statistics, 17*(4), 279–296. https://doi.org/10.3102/10769986017004279, 0362-9791.

Higgins, J. P. T., Thompson, S. G., Deeks, J. J., & Altman, D. G. (2003). Measuring inconsistency in meta-analyses. *British Medical Journal*. ISSN: 09598146, *327*(7414), 557–560. https://doi.org/10.1136/bmj.327.7414.557

Higgins, J. P. T., & Thompson, S. G. (2002). Quantifying heterogeneity in a meta-analysis. *Statistics in Medicine*. ISSN: 02776715, *21*(11), 1539–1558. https://doi.org/10.1002/sim.1186

Hunter, J. E., & Schmidt, F. L. (1991). Meta-analysis. *Advances in Educational and Psychological Testing: Theory and Applications, 28*. https://doi.org/10.1007/978-94-009-2195-5_6

delAmo, I. F., Erkoyuncu, J. A., Roy, R., Palmarini, R., & Onoufriou, D. (2018). A systematic review of augmented reality content-related techniques for knowledge transfer in maintenance applications. *Computers in Industry, 103*, 47–71. https://doi.org/10.1016/j.compind.2018.08.007

Jadhav, B., Phadke, R., & Patil, S. (2022). Study of PMAYU India: Affordable housing for all. *ECS Transactions*. ISSN: 19385862, *107*(1), 17789–17796. https://doi.org/10.1149/10701.17789ecst

Jakubowski, Jacek, Kusy, Michał, & Migut, Grzegorz (2015). *The Poland medical bundle* (pp. 697–725). https://doi.org/10.1016/b978-0-12-411643-6.00031-4, 9780124116436.

Jha, M. K., & Kumar, P. (2016). Homeless migrants in Mumbai: Life and labour in urban space. *Economic and Political Weekly*. ISSN: 00129976, *51*(26–27), 69–77. http://www.epw.in/system/files/pdf/2016_51/26-27/Homeless_Migrants_in_Mumbai_0.pdf.

Khan, Mohammad Aslam (2022). *Mega risks, social protection, and sustainability* (pp. 229–258). https://doi.org/10.1007/978-3-031-14088-4_9, 978-3-031-14087-7.

Killemsetty, N., Johnson, M., & Patel, A. (2022). Understanding housing preferences of slum dwellers in India: A community-based operations research approach. *European Journal of Operational Research*. ISSN: 03772217, *298*(2), 699–713. https://doi.org/10.1016/j.ejor.2021.06.055

Kozel, V., & Parker, B. (2003). A profile and diagnostic of poverty in Uttar Pradesh. *Economic and Political Weekly, 38*(4), 385–403.

L'Abbe, K. A., Detsky, A. S., & O'Rourke, K. (1987). Meta-analysis in clinical research. *Annals of Internal Medicine*. ISSN: 00034819, *107*(2), 224–233. https://doi.org/10.7326/0003-4819-107-2-224

Luthar, S. S., Ebbert, A. M., & Kumar, N. L. (2021). Risk and resilience during COVID-19: A new study in the Zigler paradigm of developmental science. *Development and Psychopathology*. ISSN: 14692198, *33*(2), 565–580. https://doi.org/10.1017/S0954579420001388

Manomano, T. (2023). Inadequate housing and homelessness, with specific reference to South Africa and Australia. *African Journal of Development Studies (formerly AFFRIKA Journal of Politics, Economics and Society)*. ISSN: 26343630, *13*(1), 93–110. https://doi.org/10.31920/2634-3649/2022/v12n4a5

McChesney, K. Y. (1990). Family homelessness: A systemic problem. *Journal of Social Issues*. ISSN: 15404560, *46*(4), 191–205. https://doi.org/10.1111/j.1540-4560.1990.tb01806.x

Mengist, Wondimagegn, Soromessa, Teshome, & Legese, Gudina (2020a). Ecosystem services research in mountainous regions: A systematic literature review on current knowledge and research gaps. *Science of the Total Environment*. ISSN: 00489697, *702*, 134581. https://doi.org/10.1016/j.scitotenv.2019.134581

Mengist, Wondimagegn, Soromessa, Teshome, & Legese, Gudina (2020b). Method for conducting systematic literature review and meta-analysis for environmental science research. *MethodsX*. ISSN: 22150161, *7*, 100777. https://doi.org/10.1016/j.mex.2019.100777

Mengist, W., Soromessa, T., & Legese, G. (2019). Ecosystem services research in mountainous regions: A systematic literature review on current knowledge and research gaps. *Science of the Total Environment, 702*, 134581. https://doi.org/10.1016/j.scitotenv.2019.134581

Mohanty, M. (2018). Inequality from the perspective of the global south. *The Oxford Handbook of Global Studies*, 211–228. https://doi.org/10.1093/oxfordhb/9780190630577.013.42, 9780190630577.

MoHUPA, Ministry of Housing and Urban Poverty Alleviation. (2009). *India: Urban poverty report, Government of India.*

Ministry of Housing and Urban Poverty Alleviation (MoHUPA). (2016). *Pradhan Mantri Awas Yojana; Housing for all (Urban) scheme guidelines, Government of India.*

Ministry of Housing and Urban Poverty Alleviation, MoHUPA. (2017). *Pradhan Mantri Awas Yojana (Urban)—Housing for all; Credit linked Subsidy scheme for EWS/LIG.* . (Accessed 21 May 2021).

Ministry of Urban Development and Poverty Alleviation, MoUDPA. (2015). *Mission overview, Jawaharlal Nehru national urban renewal mission, Government of India.* http://jnnurm.nic.in/. (Accessed 1 March 2022).

Pradhan Mantri Awas Yojana (Urban)—housing for all; Credit linked subsidy scheme for EWS/LIG, operational guidelines, Government of India. (2017). *Ministry of Housing and Urban Poverty Alleviation (MoHUPA.*

Prashad, V. (2014). *The poorer nations: A possible history of the Global South.* Verso Books.

Sandercock, Peter (2011). The authors say: 'The data are not so robust because of heterogeneity'—So, how should I deal with this systematic review? *Cerebrovascular Diseases, 31*(6), 615–620. https://doi.org/10.1159/000326068, 1015-9770.

Sedgwick, P. (2015). Meta-analyses: What is heterogeneity? *BMJ, 350*(1). https://doi.org/10.1136/bmj.h1435, 1756-1833, mar16 h1435-h1435.

Sethi, Harshleen Kaur (2017). Affordable housing in India*. *International Journal of Engineering Research, V6*(06). https://doi.org/10.17577/IJERTV6IS060375, 2278-0181.

Singh, P. (2011). We-ness and welfare: A longitudinal analysis of social development in Kerala, India. *World Development.* ISSN: 0305750X, *39*(2), 282–293. https://doi.org/10.1016/j.worlddev.2009.11.025

Soederberg, S. (2018). The rental housing question: Exploitation, eviction and erasures. *Geoforum.* ISSN: 00167185, *89*, 114–123. https://doi.org/10.1016/j.geoforum.2017.01.007

Sterne, J. A. C., & Egger, M. (2001). Funnel plots for detecting bias in meta-analysis: Guidelines on choice of axis. *Journal of Clinical Epidemiology.* ISSN: 08954356, *54*(10), 1046–1055. https://doi.org/10.1016/S0895-4356(01)00377-8

Sterne, J. A. C., Sutton, A. J., Ioannidis, J. P. A., Terrin, N., Jones, D. R., Lau, J., Carpenter, J., Rucker, G., Harbord, R. M., Schmid, C. H., Tetzlaff, J., Deeks, J. J., Peters, J., Macaskill, P., Schwarzer, G., Duval, S., Altman, D. G., Moher, D., & Higgins, J. P. T. (2011). Recommendations for examining and interpreting funnel plot asymmetry in meta-analyses of randomised controlled trials. *BMJ, 343*(1). https://doi.org/10.1136/bmj.d4002, 0959-8138, jul22 d4002-d4002.

Tawfik, G. M., Dila, K. A. S., Mohamed, M. Y. F., Tam, D. N. H., Kien, N. D., Ahmed, A. M., & Huy, N. T. (2019). A step by step guide for conducting a systematic review and meta-analysis with simulation data. *Tropical Medicine and Health.* ISSN: 13494147, *47*(1). https://doi.org/10.1186/s41182-019-0165-6

Thomas-Slayter, B. P. (2003). *Southern exposure: International development and the Global South in the twenty-first century.* Kumarian Press.

Thompson, S. G., & Higgins, J. P. T. (2002). How should meta-regression analyses be undertaken and interpreted? *Statistics in Medicine.* ISSN: 02776715, *21*(11), 1559–1573. https://doi.org/10.1002/sim.1187

Tiwari, Piyush, Rao, Jyoti, & Day, Jennifer (2016). *Housing development in a developing India* (pp. 83–139). https://doi.org/10.1057/978-1-137-44610-7_4, 978-1-137-44609-1.

Tufanaru, C., Munn, Z., Stephenson, M., & Aromataris, E. (2015). Fixed or random effects meta-analysis? Common methodological issues in systematic reviews of effectiveness. *International Journal of Evidence-Based Healthcare.* ISSN: 17441609, *13*(3), 196–207. https://doi.org/10.1097/XEB.0000000000000065

United Nations. (2008). Department economic and social affairs sustainable development goals. In *Millennium development goals report 2008.* https://www.un.org/en/development/desa/publications/millennium-development-goals-report-2008.html. (Accessed 3 March 2023).

Viechtbauer, Wolfgang (2010). Conducting meta-analyses in R with the metafor package. *Journal of Statistical Software, 36*(3). https://doi.org/10.18637/jss.v036.i03, 1548-7660.

CHAPTER

20

Homelessness in the context of extreme poverty: Social policy from Indonesia

Habibullah Habibullah

National Research and Innovation Agency (BRIN), Jakarta, Indonesia

1. Introduction

The issue of homelessness has a strong relationship with the phenomenon of poverty, causes homeless people to live on the streets, become beggars, and disrupt public order. The government often carries out policing of the homeless, but it is often considered inhumane and does not respect human rights. Homelessness is often associated with a confluence of factors, including limited financial resources, compromised physical health, inadequate educational attainment, heightened susceptibility to harm, restricted access to resources, and unsafe living conditions (Ganti et al., 2022). Homelessness is closely related to social exclusion and urban poverty (Lee & Schreck, 2005). Previous research shows evidence of a relationship between poverty, exclusion, and homelessness (Decker, 2004; Edgar & Doherty, 2001). Homelessness is a problem in some countries, but not all countries have exact data on the number of homeless people. Each country has a different definition, so it cannot be compared between countries. Based on OECD data, 36 countries provide estimates of the number of homeless people, with the highest homeless estimates in the United States 580,466 (2020), the United Kingdom 289,800 (2020), Germany 337,000 (2018), France 141,500 (2012), Canada 129,127 (2016), Australia 116,427 (2016), and Brazil 101,854 (2015) (OECD, 2021). Indonesia does not report homelessness data in the OECD report. However, it is estimated that in 2015 there were 18,599 homeless people (Kuntari & Hikmawati, 2017), and in 2019, there were an estimated 77,500 homeless people in Indonesia. Currently, there is no official data published by official government agencies on the number of homeless people in Indonesia (Yusuf et al., 2022). The official statistical agency of Indonesia, Central Statistics of Indonesia (BPS), solely provides information on the count of homeless locations, which has been reported as 516 (Badan Pusat Statistik Indonesia, 2022). In Indonesia, the

Homelessness to Hope
https://doi.org/10.1016/B978-0-443-14052-5.00020-3

condition of being homeless is referred to as "Gepeng," which refers to the acronym of the terms "Gelandangan" (which refers to homelessness) and "Pengemis" (meaning beggar). In addition to "Gelandangan," the label against homeless people in Indonesia is called "Tuna Wisma" the term "homeless," derived from old Javanese, literally means "no (tuna) house (wisma)" (Speak, 2012). In 1980, the Indonesian government began recognizing homelessness as a social problem that must be addressed through intervention. The Indonesian government deals with the problem of homelessness and beggar through preventive, repressive, and rehabilitative efforts by issuing Government Regulation No. 31 of 1980 on Combating Homelessness and Begging. Homelessness is a complex issue in big cities in Indonesia, such as Jakarta, Surabaya, Bandung, Medan, Semarang, and Makassar.

What are the underlying causes of homelessness in metropolitan areas? Indonesia's major urban centers appeal to a broad demographic, encompassing individuals with limited educational attainment, health resources, and vocational training. Homelessness in Indonesia is attributed to the disparity in development between urban and rural regions (Setiawan, 2020). Development in Indonesia prioritizes urban areas, so there are more job opportunities in urban areas than in rural areas. One of the distinctive features of the homeless phenomenon in Indonesia and other developing countries is the emergence of homelessness due to development inequality between urban and rural areas. The general problem of homelessness is closely related to the problems of order and security that disrupt order and security in urban areas. The increasing number of homeless people is suspected of providing opportunities for security and order disturbances. Security and order disturbances disrupt development stability, and the national goal of realizing a just and prosperous society cannot be achieved (Iqbali, 2008). Several previous studies have successfully provided an understanding of the types of homelessness, credible demographic estimates, and homelessness and coping strategies. With conceptual, theoretical, and methodological advances, research literature provides a more complete understanding of homelessness. There is such agreement among experts on the macro- and micro-level underlying causes of homelessness. Research on community, media, and government responses to and efforts made by homeless people to mobilize themselves has also been conducted.

However, research on homelessness in Indonesia is still rarely conducted or published internationally. In the Scopus database using the keywords homelessness and Indonesia, there are only 29 documents in the form of journal articles, books, and conference proceedings. None of the two–nine documents specifically discuss homelessness and extreme poverty in Indonesia. The theme of research on homelessness in Indonesia is related to street children (Black & Farrington, 1997a, 1997b; Suzanne Speak, 2005), life on the streets (Karsono, 2020; Stodulka, 2015), homelessness related to natural disasters (Doocy et al., 2013), particularly the Tsunami in Aceh (Ashkenazi & Shemer, 2005; Du et al., 2012; Lee et al., 2015; Morrow & Llewellyn, 2006; Santos-Reyes et al., 2017; Santos-Reyes et al., 2014), and themes related to mental health (Daugherty et al., 2020; Stratford et al., 2014; Sukma & Irawati, 2021).

2. The rationale of the study

Indonesia is making significant strides toward achieving the Sustainable Development Goals (SDGs) target of eradicating extreme poverty by 2030, with a new goal of achieving it by 2024. Extreme poverty is a global issue that is determined by various indicators, one of which is homelessness—a characteristic that is prevalent in Indonesia. Despite this, there is currently no published research on the topic of homelessness in the context of extreme poverty and the policies that Indonesia has in place to address it. Historically, research on homelessness in Indonesia has focused on disaster-induced homelessness and street children. This section will delve into the study of homelessness and extreme poverty in Indonesia, as well as the policies that are being implemented to tackle these issues. This social policy is expected to serve as a blueprint for other developing countries facing similar challenges and contribute toward the resolution of homelessness and extreme poverty.

3. Materials and methods

This study aims to review homelessness in the context of extreme poverty and the solutions offered by Indonesia can be seen in Fig. 20.1.

A literature review is the research method used in this study. To explain all aspects being investigated and support the current study direction, the literature review highlights pertinent theoretical and research findings (Bowden, 2022). The process undertaken in a literature review is to identify relevant materials, synthesize the search results (tables, graphs, and narratives), and make conclusions about the research theme. The goals of a literature review are to determine what has already been accomplished, to facilitate consolidation, to expand on earlier work, to summarize said work, to prevent unnecessary repetition, and to spot any omissions or gaps in said work (Grant & Booth, 2009). Data was collected from online databases, namely Scopus and Google Scholar. No date limit for data search, resulting in information on homelessness and extreme poverty in Indonesia. The data search resulted in 235 articles that could be included in this review. Sixty-one of t is final total relates to information overload in such homelessness and extreme poverty. In addition to literature searches from Scopus and Google Scholar, this research material is in the form of regulations, policies and technical reports, and various official sources of information from official Indonesian Government institutions such as the Indonesian Ministry of Social Affairs and the Central Statistics Agency as well as news from online media. The data from this study were examined using thematic analysis. The main objective of the qualitative descriptive method of thematic analysis (Vaismoradi et al., 2013) is to find, analyze, and communicate themes and patterns in data (Braun & Clarke, 2006).

4. Results and discussion

Homelessness is often defined as difficulty in owning a home due to poverty. Although not all homeless people are poor, some wealthy people become homeless due to floods, fires, and landslides (Lee et al., 2010). There are various definitions of homelessness. Different academics, institutions, countries, and levels define different definitions of homelessness. Homelessness is an individual problem that focuses on individual characteristics and behaviors. At the individual level, homeless individuals do not have definite, regular, and suitable night residences (Palmer et al., 2023). A person who spends the night in a long-term motel, campground, public space, shelter, car, or without permanent residence is considered

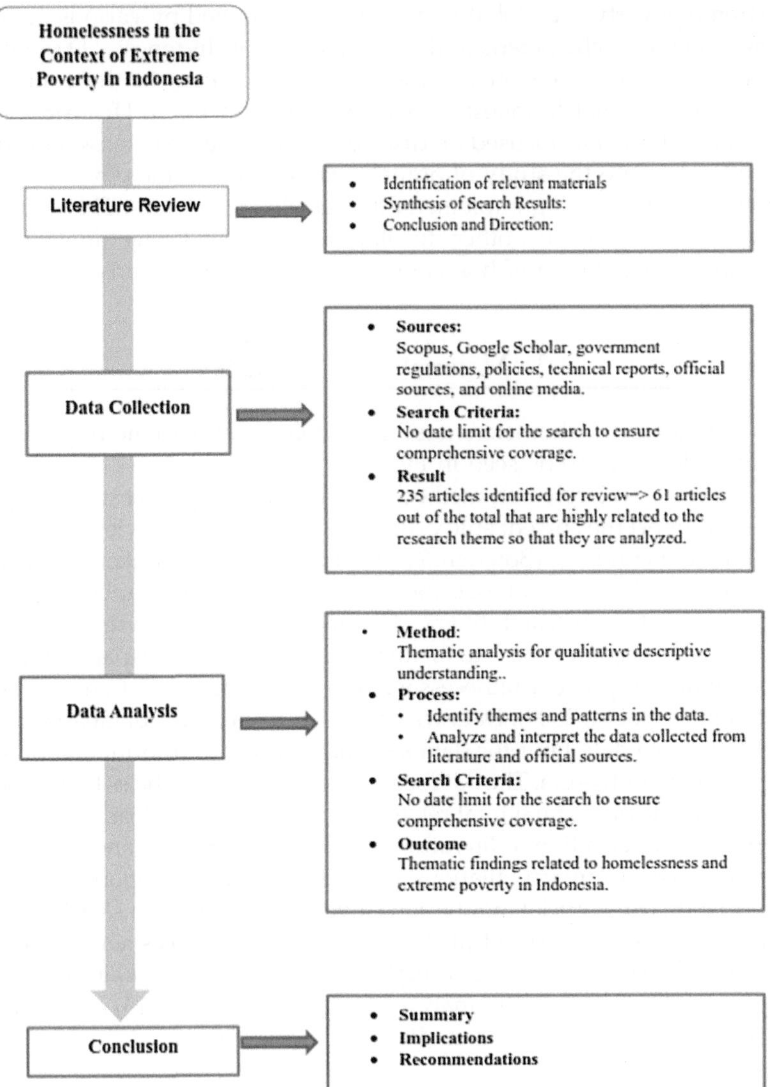

FIGURE 20.1 Methodological flow diagram. Homelessness in the context of extreme poverty: Social policy from Indonesia.

homeless (Parrott et al., 2022). Schutt and Garrett (2013) stated that spending the night at a bus station does not constitute homelessness if it occurs while someone is on their way to a vacation destination. Living for a few months with friends or family for a few months does not constitute homelessness for an adult if the arrangement is preferable to other regular housing alternatives; they become homeless if there is no alternative when eviction from a regular residence or search for housing is fruitless (Schutt and Garrett, 1992). Speak (2005) states that when a person does not have a safe and adequate place to live, they rely on emergency accommodation or live in a long-term social institution. Living in a cramped house can be defined as homelessness. Pauly et al. (2014) state that homelessness is not just an individual, family, or group homelessness problem, but a combination of individual, systemic, and structural factors. Homelessness as a structural problem focuses on social and economic structures, such as the job market, social assistance programs, and the housing market. Systemic behavioral factors involve power and access to resources that are important to the homeless. The complexity of homelessness has led to several disciplines conducting studies on homelessness, such as economics, geography, sociology, psychology, criminology, law, anthropology, and social work.

1. Macroeconomics states that the causes of homelessness include excess demand for housing, economic conditions leading to unemployment and poverty, drug epidemics, and other factors (Lee et al., 2010). In other words, structural homelessness falls into this category, which links the causes of homelessness to greater forces such as housing market conditions, poverty, and unemployment (Bramley & Fitzpatrick, 2018; Nwokah et al., 2017). Four factors contribute to homelessness: poverty, unaffordable accommodation, lack of social services and support, and individual disabilities (Sullivan, 2023).
2. Geography examines spatial patterns of homelessness, studying where the homeless population is concentrated and how urbanization and geography affect the availability of resources for the homeless (Lee & Price-Spratlen, 2004).
3. Psychology focuses on the mental health of homeless individuals, including issues such as depression, anxiety, and drug and narcotic abuse. Psychology also explores coping and intervention mechanisms.
4. Sociology links homelessness with social stratification. Homelessness occupies the lowest position when viewed from the perspective of social stratification. On the other hand, sociology emphasizes one's position in society. If they have few social connections, migrate frequently, or are heavy drinkers, single men living in cheap hotels and lodging houses are considered homeless. Homeless people are not affiliated with social structure networks. This suggests that homeless people experience social exclusion or marginalization (Edgar et al., 1999).
5. Social work relates to homelessness as a deviation in people's behavior. Therefore, efforts are needed to return homelessness to the community and their families, as well as to prevent the phenomenon of homelessness through various welfare programs.
6. Criminology examines the relationship between homelessness and crime, studying the victimization of homelessness and any criminal activities they may engage in due to their vulnerable circumstances.
7. Law science studies aspects of homeless law, including the right to housing, the civil liberties of homeless individuals, and the implications of antihomelessness laws and policies.
8. Anthropology studies the cultural and social factors contributing to homelessness by examining how communities perceive and interact with homeless individuals.

Some institutions responsible for homelessness define homelessness as follows:

1. The US Department of Housing and Urban Development defines *homelessness* as lacking a fixed, regular, and adequate nighttime residence (US Department of Housing and Urban Development, 2021). A household without shelter falls within the scope of shelter.

Homelessness is simply the inability to obtain permanent housing when such housing is desired. Homelessness is a symptom of extreme poverty but an even more extreme condition of deprivation—the absence of a place called home. Spending a night in a bus station does not constitute homelessness if it occurs while one is route to a vacation destination; homelessness occurs if it follows an eviction from a usual place of residence or an unfruitful search for housing. Staying for several months with friends or family for several months does not constitute homelessness for adults if the arrangement is preferred over other regular housing alternatives; it becomes homeless if there are no alternatives (Schutt & Garrett, 1992, pp. 1–20).

2. The European Federation of National Organizations Working with the Homeless (FEANTSA) (Nicholas & Bretherton, 2013) created ETHOS: the European Typology of Homelessness and Housing Exclusion, as well as a shorter version, ETHOS Light, to specify data gathering on homelessness. The typology of homelessness according to ETHOS Light is as follows: (1) People living on the streets or in public spaces need a fixed place of residence; (2) people living in emergency shelters; (3) people living in accommodations for the homeless; (4) people living in institutions: e.g., institutions and people in prisons who have no place to live before release; (5) people living in nonconventional dwellings due to lack of housing; (6) people living temporarily in conventional housing with family and friends due to a lack of housing.
3. The Government of Indonesia defines homelessness as people who live in a state that is not under the norms of a decent life in the local community, do not have a permanent place of residence and work in a specific area and live to wander in public places (Pemerintah Republik Indonesia, 1980). Some local governments, such as Yogyakarta Special Region Province, Palu City, Makassar City, and Pangkal Pinang City, establish the criteria for homeless people through their regional regulations. According to the regional government, homeless are people with criteria such as: (1) Not having an identity card (KTP); (2) not having a definite or fixed place of residence; (3) not having a fixed income; and (4) not having plans for the future of their children and themselves. The government of Indonesia considers that homelessness is not following the norms of the life of the Indonesian nation based on Pancasila (five principles, the Indonesian state philosophy) and the UUD 1945 (The 1945 State Constitution of the Republic of Indonesia), so efforts to overcome homelessness are needed. The government of Indonesia carries out countermeasures against homelessness through preventive, repressive, and rehabilitative efforts.

4.1 Homelessness in developing countries

In developing countries, the problem of homelessness is more complicated than in developed countries. Homelessness in Europe is associated with migration between countries, such as Italy and Sweden (Giansanti et al., 2022). In contrast, in developing countries, urbanization is often caused by a lack of affordable housing, poverty, and inadequate social support systems. Some of the key factors and challenges associated with homelessness in developing countries are as follows:

1. Urbanization, developing countries often experience rapid urbanization as people migrate from rural areas to cities for better economic opportunities. This can lead to

overcrowding in cities and inadequate housing facilities, resulting in homelessness. Some developing countries that have greater urban population changes than rural areas include the Philippines, Indonesia, South Africa, and Iran (Yamashita, 2017).

2. In developing countries, poverty is the leading cause of homelessness. Poor people often cannot afford decent housing, forcing them to live in slums or on streets.
3. Lack of affordable housing. The gap between housing demand and supply causes housing prices to rise, making it difficult for low-income individuals and families to find a decent place to live in. Decent housing, which is unaffordable for low-income communities, can be found in Mexico and Angola. Some countries have built affordable homes for low-income people, but the new problem of difficult transportation and infrastructure has created new transportation costs.
4. Developing countries are often more vulnerable to natural disasters and conflict. These events can displace many people, leaving them homeless and without adequate resources. Some developing countries are more affected by environmental disasters and climate changes than others, causing mass destruction of homes and the loss of lives. For example, cyclones in Bangladesh and Southern India caused hundreds of thousands of people to lose their homes. The 2004 tsunami in Indonesia that left more than 1.5 million people homeless.
5. Limited access to social services such as healthcare, education, and job training can perpetuate cycles of poverty and homelessness. Homeless people often do not receive social welfare programs, because they are not registered as permanent residents.
6. Mental health and substance abuse are prevalent among the homeless population, including in developing countries.
7. Discrimination and Stigmatization: Homeless individuals, especially women and children, often face discrimination and stigmatization from society. This can make it difficult for them to access support services and reintegrate into the mainstream society.
8. Limited Government Resources: Developing countries often have limited resources to address homelessness. Government budgets may not allocate sufficient funds to social welfare programs, including housing and support services.
9. Lack of Legal Protection: Homeless individuals in developing countries may lack legal protection and rights. They may not have access to legal representation, making it difficult to defend their housing and social service rights.
10. Nonprofit nongovernmental organizations (NGOs), and nonprofit groups are important in providing shelter, food, and support services to the homeless in many developing countries. However, their resources are often limited and they cannot fully cope with the scale of the problem.

4.2 Conventional handling of homelessness

Homelessness and begging are two of the urban social problems in Indonesia that have emerged since 1980. The Indonesian government has made various efforts to ensure that homeless people and vagrants do not appear in public spaces. Since the issuance of Government Regulation Number 31 of 1980 concerning Handling Homelessness and Begging, which regulates the handling of homeless people and beggars, it has been carried out in a repressive, rehabilitative, and preventive approach.

4.2.1 *Repressive approach to homelessness*

The Indonesian government conducts repressive efforts against homelessness through organized efforts, whether through institutions or not, to eliminate homelessness and prevent its spread in society. The repressive approach aims to reduce the visibility of homeless people, especially in public places such as streets, city parks, shopfronts, under bridges, and riversides. Repressive efforts are carried out through raids. A joint team between the Civil Service Police Unit (Satpol PP), the police, and the social service carries out the raids. Homelessness is a criminal problem, so for centuries, many European countries have developed strategies and policies related to vagrancy and homelessness, using punitive and repressive approaches as the primary intervention (Beier & Ocobock, 2008). This policy became widespread and international, combining relief and repression. 19th-century society viewed homeless people as criminals (Beier & Ocobock, 2008). A vagrant is someone who does not have (1) a fixed place of residence, (2) the means to support themselves, or (3) gainful employment, according to the Belgian penal code from 1867. Those detected in such conditions are imprisoned (Maeseele et al., 2014).

Homelessness caught in the raids will be accommodated in temporary shelters for assessment. The Indonesian government provides temporary shelters for homelessness caught in raids. Unlike in the United Kingdom, third-sector organizations provide as much as 95% of temporary shelters (DeVerteuil et al., 2009). In Indonesia, the government fully provides temporary shelters, especially local ones. In temporary shelters for homelessness, an assessment will be conducted. The results of the assessment will determine the next steps for homeless people. Actions against homelessness that have been assessed include Homelessness, who are often raided and are criminals will be handed over to the court for criminal law. Homeless people who still have the potential to be rehabilitated will be channeled into social care institutions. Homeless people who experience health problems will be given health services first. Homelessness is returned to their parents, family, or homeland. Homeless people are released under certain conditions. Social and security policies have a thin line between repressive approaches because the main goal is to disperse marginalized people and keep them away from public places (Bonnet, 2009).

4.2.2 *Rehabilitative approach to homelessness*

The rehabilitative approach to homeless people is a follow-up to the repressive approach. Homeless people caught in the raid are temporarily accommodated in shelter institutions and rehabilitated in social care institutions. The social rehabilitation process of homelessness is carried out with the assumption that homeless people are a low-class group and often fall into deviant behavior and face undesirable situations. This behavioral deviation must be treated or re-educated so that it does not interfere with the social order of society. The main focus of rehabilitative efforts lies in the condition of homelessness, especially to make changes or improvements to conditions that are not expected or considered problematic into conditions that meet expectations or applicable social standards. The main assumption of rehabilitative efforts is that homelessness has the potential to return to normal conditions. If the assumption is that the reality inherent in homelessness is a condition that cannot be changed, then rehabilitative efforts are pointless.

Social rehabilitation efforts for homelessness are carried out in ways (Fadri, 2019)

1. Social care institutions care for homelessness by giving them a place to live with a full infrastructure and multiple families. The Ministry of Social Affairs and several provincial governments in DKI Jakarta, West Java, Central Java, and East Java manage social care centers for the homeless in Indonesia.
2. The social cottage environment (Liposos), namely the homeless system, prioritizes living together in the social sphere as it befits community life.
3. Temporary shelters, a temporary handling of homelessness before getting permanent housing, is a transition from life on the streets to a designated residence.
4. Community settlement is a form of handling homelessness by providing permanent housing in a certain location. The placement of homelessness in the community relations system is carried out when homelessness is ready to live side by side with the community. Community settlement occurs after homelessness receives social guidance and skills training through a social care service system.
5. Transmigration: handling homeless people with a transmigration system, namely by sending homeless outside the region and the island so that population density and high labor competition are no longer a problem for homelessness. Homelessness who has been moved to rural areas or even returned to their home villages are given counseling and an understanding of the contribution and motivation of businesses that can be done in the village so that the thought of living and residing in the city as a homeless person is not the only effort to fulfill their daily needs.

Selecting social rehabilitation methods for homelessness must be done carefully; otherwise, it will be charity and tend to create dependence for homelessness on government programs. The dependence of homelessness on government programs means the state must provide more social welfare funds. An ideal form of rehabilitative effort is handling homeless problems oriented toward capacity building. Capacity building is carried out utilizing social guidance, mental guidance (psychological and spiritual), physical guidance (exercise and health care), and skills guidance (carpentry training, salted egg making and other culinary skills, batik skills, sewing, catfish cultivation, and agriculture). Handling homelessness rehabilitation should not be momentary, temporary, or sporadic but sustainable. For this reason, institutionalized social service action is needed through institutions with patterned and continuous activities.

4.2.3 *Preventive approach to homelessness*

Rehabilitative approaches are efforts to handle homelessness with a focus on the condition of homelessness; thus, it is an effort to change and improve so that the problem of homelessness can be resolved. Meanwhile, preventive approaches have focused on the fact that the homeless problem has not yet occurred but has the potential to occur. Preventive approaches target individuals, groups, or communities that are not yet homeless. At the individual level, controlling or directing the socialization process can prevent the possibility of becoming homeless. At the group level, prevent efforts mainly focus on social groups that are considered to have the potential to become homeless, such as street children and people living in illegal and slum areas. People who are vulnerable to becoming homeless live in illegal places (under bridges, river sides, railroad sidings), which can be evicted or repressed by the government at any time. At the community level, preventive efforts can be made by creating a conducive community life and providing opportunities for people to develop themselves, including access to decision-making, various social services, and employment opportunities. The Indonesian government and the community make various efforts to prevent the problem of homelessness, starting with conducting social counseling for vulnerable groups and providing family-based social assistance for poor families that could become homeless. Providing housing assistance and the provision of rental flats, as well as various skills training and business capital, for individuals or families who have the potential to become homeless.

4.3 Multidimensional and extreme poverty

Poverty has many dimensions, and it can also involve various other challenges beyond the economy, such as limited capabilities, marginalization, discrimination, and poor health. FAO conceptually distinguishes between absolute and relative poverty. Absolute poverty is a condition where the basic needs for a decent life, both food and nonfood-related, are unmet. In contrast, relative poverty refers to a person's position relative to average income (FAO, 2021). According to Chambers (2014), poverty is a broad concept encompassing social and moral aspects, ranging from the inability to satisfy basic needs and improve conditions to lacking business. Therefore, poverty is multi-dimensional. Muti-dimensional poverty can be divided into four forms (Chambers, 2014).

1. Absolute poverty occurs when an income falls below the poverty line or is insufficient to satisfy their basic needs, which include food, clothing, shelter, health care, and education.
2. Relative poverty is a condition caused by development policies that have not reached all people, resulting in income inequality, or it can be said that a person has lived above the poverty line but is still below the ability of the surrounding community.
3. Cultural poverty refers to the problem of a person's or group's attitude caused by cultural factors, such as a reluctance to attempt to improve one's life, laziness, wastefulness, and a lack of creativity despite assistance from outsiders.
4. Structural poverty is an impoverished situation caused by limited access to resources that occurs in a socio-cultural and socio-political system that does not support poverty alleviation but often promotes the growth of poverty.

Oxford Poverty and Human Development developed the Multidimensional Poverty Index (MPI) Initiative using information from 10 indicators arranged in three equally weighted dimensions: health, education, and standard of living (Mulya et al., 2021) MPI includes

1. Two health indicators: Nutrition and infant mortality.
2. Two indicators of education: years of schooling and school attendance.
3. Six indicators of living standards: fuel for cooking, sanitation, drinking water, electricity, housing, and assets.

In 2022, according to the 2022 Global Multidimensional Poverty Index, poverty exists in 111 countries, with 1.2 billion people, or 19.1%, living in multidimensional poverty (UNDP and OPHI, 2002).

The government, the private sector, and the community all have a role in contributing to reducing poverty (Setiawan et al., 2023). Eliminating poverty and hunger globally by 2030 is a major goal of sustainable development and a commitment of all countries. Therefore, a standard measure of global poverty elimination is needed. The elimination of poverty in this context is the elimination of extreme poverty. The measure of extreme poverty uses the concept of absolute poverty, which is poverty that can be compared across countries and over time. The SDGs use the absolute poverty indicator, which means that people experiencing extreme poverty have a daily income of less than $1.25 (Taufiq, 2022). In difference to the poverty reduction target based on the SDGs, the World Bank defines the extreme poor as those who fulfill their daily needs with no more than USD 1.9 PPP (Purchasing Power Parity). Purchasing Power Parity is a unit of price that has been adjusted so that the value of currencies in different countries can be compared with one another (The World Bank, 2022). Based on this poverty line, extreme poverty globally decreased in 2015 by 10.1% or 740 million to 8.6% or 656 million people in 2018. The COVID-19 pandemic disrupted the achievement of eliminating extreme poverty. Data from the Department of Economic and Social Affairs of the United Nations states that there was an increase in poverty during the Covid-19 pandemic in 2019 (8.3%) to 9.2% in 2020. As a result, multiple actions are required to combat extreme poverty and the consequences of the Covid-19 pandemic (The World Bank, 2020).

4.4 Extreme poverty in Indonesia

Globally, extreme poverty occurs when people's well-being is below the extreme poverty limit of USD 1.9 (Purchasing Power Parity). Extreme poverty was measured using consistent measures of absolute poverty across countries and times. In 2022, the extreme poverty rate in Indonesia is 2.04% or 5.59 million people. This extreme poverty rate is lower than the national poverty rate based on data from the National Socioeconomic Survey (Susenas), released regularly by the Central Statistics Agency, which was 9.54% in March 2022. The Central Bureau of Statistics (BPS) in Susenas calculates the national poverty rate based on the inability to achieve a minimum level of basic needs, including food and nonfood needs. BPS measures poverty based on the basic needs of the population. Indonesian society generally defines *extreme poverty* as very poor. The poor as regulated by Law Number 13 of 2011 concerning Handling the Poor. The poor are defined as people who have no source of livelihood at all and have a source of livelihood but cannot meet the basic needs that are appropriate for their lives of themselves and their families. The definition of the poor is much more complicated than the measurement of extreme poverty with purchasing power parity. To implement Law Number 13 of 2011, the Indonesian government issued Decree Number 146/HUK/2013 of the Minister of Social Affairs of the Republic of Indonesia, addressing the formulation of criteria and data collecting for the poor and disadvantaged. Based on the decree, the poor are categorized as registered and unregistered. The registered poor and disadvantaged are households that have the following criteria:

1. They do not have a source of income or have a source of income but are unable to meet basic requirements.
2. Have expenditures that are primarily used to fulfill fundamental food needs.

3. Unable to seek medical treatment or having difficulty doing so, excluding Puskesmas (Public health center) and government-subsidized treatments.
4. They cannot afford to purchase clothing annually for each household member.
5. Can send their children to school only through the junior high level.
6. Contains walls made of bamboo or wood in poor condition or of low quality, as well as unplastered walls.
7. The floor is in poor condition and comprises soil, wood, cement, and ceramic.
8. A palm fiber, thatch, tiles, zinc, or asbestos roof is in subpar condition or quality.
9. Does not use electricity for residential building illumination or use electricity without a meter.
10. A small house's floor space per individual is less than 8 square meters.
11. Using water sources from springs, rivers, rainwater, or other unprotected drinking sources.

Homeless persons fall within the category of unregistered poor, according to the Decree of the Minister of Social Affairs of the Republic of Indonesia Number 146/HUK/2013 concerning the Determination of Criteria and Data Collection of the Poor and Disadvantaged. Homeless people are categorized as unregistered poor because, in 2013, the Indonesian population had not all registered and received a Population Identification Number (NIK). NIK is a single identity number in Indonesia, and homeless people are the most populous population who do not have NIK. Homeless people who usually have irregular residence and often move around have not been registered and have an Identity Card (eKTP) and NIK. As a result of not having a population identity, homeless people often do not get social protection from the Indonesian government, even though homeless people are vulnerable groups and need social protection. Getting social protection from the government requires having a population identity. Therefore, officers acquire data on their identities when they raid homeless individuals. If they do not have NIK, the Population and Civil Registry Office assists them in obtaining it. The Decree of the Minister of Social Affairs of the Republic of Indonesia Number 146/HUK/2013 concerning the determination of criteria and data collection for the poor and the incapable is no longer relevant to the current conditions, so the Government of Indonesia issued the Decree of the Minister of Social Affairs of the Republic of Indonesia Number 262/HUK/2022 concerning poor criteria. The main reason for the change in poverty criteria is that 99.21% of the Indonesian population already has a national single identity in the form of a (Nomor Induk Kependudukan: NIK) and an electronic identity card (Kartu Penduduk Elektronik: eKTP). There are 272,229,372 Indonesians, only 0.8% of whom do not have an eKTP. Those who do not have an eKTP are scattered in Papua and West Papua. Meanwhile, homeless people in big cities in Indonesia have been recorded and have NIK and eKTP.

Therefore, based on the Decree of the Minister of social affairs of the Republic of Indonesia number 262/HUK/2022, the criteria for the poor are used to detect initial poverty conditions as part of handling the poor. The poor criteria for detecting early poverty conditions need shelter/day-to-day residence. If someone does not have a lace to live daily, they are immediately categorized as poor. This initial criterion of the poor is identical to the homeless, namely, not having a place to live daily.

However, if a person has a shelter to live in daily or is not considered homeless, further detection is carried out with criteria including

1. Head of household or caretaker of the head of household who is not working
2. Feeling worried about not being able to eat in the past year
3. Expenditure on food needs is greater than half of the total expenditure
4. No expenditure on clothing during the last 1 year
5. The dwelling has mostly dirt and or stucco floor
6. The dwelling has mostly bamboo, wire, wooden planks, tarpaulin, cardboard, unplastered walls, thatch, or zinc wall
7. Do not have their toilets or use community toilets
8. The lighting source comes from electricity with a power of 450 V or no electricity.

Based on the criteria for the poor set by the Government of Indonesia, housing is essential in determining a person to be poor in addition to employment and fulfillment of food and clothing needs. Because to overcome poverty, it is very necessary to fulfill housing needs.

4.5 Solutions from Indonesia for integrated approach to ending homelessness and extreme poverty

Indonesian President Joko Widodo ordered the 2030 SDGs target to end extreme poverty be moved to 2024. Eliminating extreme poverty requires the right policy framework and interventions as a program to accelerate the elimination of extreme poverty; the government allocates a social protection budget of IDR 479.1 trillion (USD 31 million) in 2023. Based on Presidential Instruction Number 4 of 2022 concerning the Acceleration of Extreme Poverty Eradication, 22 ministries, 6 Institutions, and Local Governments (governors/regents/mayors) are tasked with implementing measures within the scope of their responsibilities and authorities to accelerate the process of alleviating extreme poverty. The Indonesian government established three main activities that are expected to address extreme poverty.

1. Social assistance and subsidies, namely, groups of programs/activities, reduce the expenditure burden of the extreme poor.
2. Community empowerment increases the income of the poor.
3. Build basic service infrastructure to reduce the number of poor areas.

Policies and programs to help reduce the expenditure burden of poor and vulnerable groups through various subsidized social assistance, social security, and social safety nets. Programs implemented related to social assistance and subsidies include conditional cash transfers (Family Hope Program), basic food programs (BPNT), and scholarships (Smart Indonesia Card) (Habibullah et al., 2023). The Contribution Assistance Program through the Healthy Indonesia Card with National Health Insurance (JKN) is related to social security. Homeless people have not obtained various government programs in extreme poverty alleviation in the form of social assistance and subsidies, empowerment programs for income, and decent housing services. With the integration of handling homelessness and extreme poverty, homeless people receive these services. Homeless people receive conditional cash assistance (PKH), food assistance, and health insurance. Likewise, economic empowerment programs and skill improvement as provisions for earning income are obtained by homeless people. Skill training is tailored to the potential possessed by homeless people and is not uniform in its type of training; for example, homeless people are interested in farming, so agricultural training is provided. The variety of skills training is often uniform, even though it does not match the potential possessed by people who experience homelessness, so the skills acquired cannot be applied. Skill training should be tailored to opportunities. For example, when there is an opportunity to work in a cleaning service, training in cleaning services is provided. Vocational training that matches potential and opportunities provides job opportunities for the homeless to earn income. Income adequacy is essential for securing housing (Deverteuil, 2005). Without adequate income, affordable housing, and employment, families find it difficult to break the cycle of poverty and homelessness (Pauly et al., 2014). The Government of Indonesia is integratively handling homelessness and extreme poverty by establishing homelessness as a key poverty criterion. Perceptions and definitions of homelessness have significant implications for how homelessness is explained and policies are implemented (Olufemi, 2002). An overview of the integrative approach to addressing homelessness and extreme poverty can be seen in Fig. 20.2 below.

Previously, homelessness and poverty were addressed separately. Handling homelessness was done through conventional approaches: repressive, rehabilitative, and preventive. The conventional approach is not effective and efficient; homeless people who receive the conventional program will return to being lost and return to the streets after the program is completed. On the other hand, poverty reduction programs only target registered poor households, and homeless people are one of the unregistered populations, so they do not get poverty reduction programs from the government. The determination of homelessness as the main criterion of poverty makes homelessness the main target of poverty reduction programs. Various poverty alleviation programs require beneficiaries to be registered as residents of Indonesia and have a resident identification number. Therefore, homeless people in Indonesia have been registered with a national identification number online, making it easier to identify them when they are found or caught in raids. The Indonesian government to facilitate the distribution of social protection programs uses a population identification number and often homeless people do not get social assistance, especially during the Covid-19 pandemic (Susantyo et al., 2023) because they do not have a NIK are not considered residents of the area where the homeless are located. This identification is crucial for further interventions in handling homelessness. Homeless people who have been identified will receive

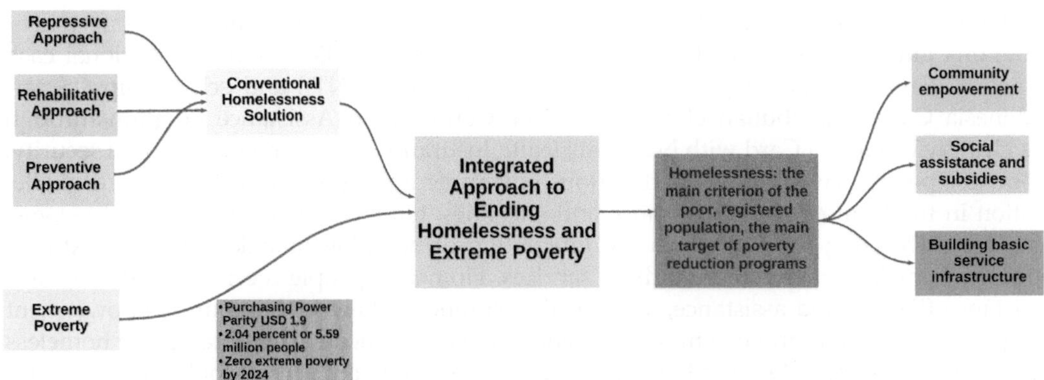

FIGURE 20.2 Integrated approach to ending homelessness and extreme poverty.

integrative treatment so they do not become homeless again and are free from extreme poverty. The integrative approach to handling homeless people is carried out by placing homeless people from both repressive raids and preventive activities in temporary shelters or social care institutions. Activities carried out in social care institutions are in the form of skills training following their potential. The Indonesian government provides affordable homeless housing by building flats (Rusunawa). The rental flats have 93 units in three locations: Bekasi City, East Jakarta City, and Surakarta City (Kumparan, 2022), flats for homeless people in Bekasi City can be seen (Fig. 20.3).

In contrast to the conventional policy of handling homeless people with a rehabilitation approach in social care institutions that does not apply rental fees, this flat applies a rental fee of IDR. 10,000/month (USD 0.76) to create a sense of responsibility for those formerly homeless. Regarding the need for adequate housing for the homeless, the Government of Indonesia provides affordable housing by building flats (Rusunawa). Rental flats have 93 units in 3 locations: Bekasi City, East Jakarta City, and Surakarta City (Kumparan, 2022), homeless flats in Bekasi City can be seen. Unlike the conventional policy of handling homelessness through a rehabilitation approach in social institutions that does not apply rental fees, this flat applies a rental fee of Rp. 10,000 per month (USD 0.76) to create a sense of responsibility for those who were previously homeless. The construction of rental flats in strategic places facilitates the mobility of homeless people to make a living because they are close to the transportation network. The application of cheap rent to be affordable for the poor and with the rental system is expected not to cause the transfer of home ownership from low-income groups to middle- and high-income groups, which often occurs in developing countries. When viewed from the distribution of rental flats built by the government, which are only 93 units, it is still minimal. The number of homeless people in Indonesia is estimated

FIGURE 20.3 Rental flats for homelessness in Bekasi City. *From Kumparan (2022).*

to be 77,500. Likewise, rental apartment locations are distributed only in Bekasi City, East Jakarta, and Surakarta City. Large cities with pockets of homelessness, such as Surabaya, Makassar, and Medan, do not yet have special rental flats to accommodate the homeless.

The Indonesian government has integrated the reduction of homelessness with extreme poverty; it is an effective and efficient activity because it addresses two problems with one intervention. However, the government's intervention is not necessarily in accordance with the needs of the homeless. Based on the characteristics of homelessness, what is done by the Government of Indonesia is to only deals with homeless people who are unable to have a decent place to live and eliminate the appearance of homeless people in public places so that it does not seem shabby. However, not all homeless people in public places face poverty problems. Some homeless people found on the street are individuals who experience mental disorders and have problems with their families. Hence, they leave home or make a living on the street by busking and begging, who earn far more than working, such as factory workers, working in shops, and restaurants, or trying to be independent, such as being traders.

5. Limitations of the study

This research only uses literature studies from previous research results, mass media coverage, regulations, and policies of the Government of Indonesia, not from field data collection. The limited previous research has caused this study to be poorly supported by quantitative data describing homelessness and extreme poverty in Indonesia. Therefore, field research on extreme poverty and homelessness in Indonesia is needed to continue this research. In addition, further studies are required on extreme poverty in other countries, especially developing countries, to determine the causal relationship between extreme poverty and homelessness.

6. Recommendations

This chapter recommends addressing homelessness and extreme poverty simultaneously. The Indonesian government has set homelessness as a key criterion for people experiencing extreme poverty, making it a prime target of poverty alleviation programs. The designation of homelessness as a critical indicator of extreme poverty has the consequence that homeless people receive the same interventions as people experiencing extreme poverty. Homeless people receiving these poverty reduction programs must first be registered and have a resident identification number to facilitate identification. Providing social welfare services to the homeless on par with people experiencing poverty is a state intervention that does not discriminate against the homeless. There is still a need to expand integrated interventions for homelessness and extreme poverty, which are not limited to specific locations. Establishing homelessness as an important criterion for poverty alleviation and combining homelessness with poverty can be applied in countries with problems caused by extreme poverty, especially in developing countries. However, homelessness reduction interventions such as poverty reduction should be carried out jointly between the government, the private sector, and the community (Setiawan et al., 2023).

This chapter recommends further research on the appropriateness of homelessness policies as a critical indicator of extreme poverty, especially from the perspective of homelessness. Government policies toward homelessness by equating forms of intervention with the poor may not necessarily be compatible with the needs of the homeless. The solutions offered by the Government must be sustainable to address the problem of homelessness. Further research on the relationship between homelessness and extreme poverty using quantitative methods in different countries, especially developing countries, is of great interest.

7. Conclusions

Homelessness and extreme poverty are two social problems that are interconnected. Homelessness can result in extreme poverty, and extreme poverty can often lead to homelessness. By using the national single identity system to identify and categorize homelessness, the government can better control the problem. When homelessness is recognized as a criterion for poverty, homeless people can receive the same rights as other citizens, including poverty alleviation programs that they may have not previously qualified for due to being considered criminals. While poverty reduction interventions can be effective in addressing homelessness that is caused by poverty, the issue of homelessness itself is multifaceted. There are many reasons why someone may become homeless, and simply addressing poverty may not solve the problem entirely. Nonetheless, efforts to alleviate poverty are still important steps in the right direction toward tackling homelessness.

Acknowledgments

The researcher would like to thank Prof. Dr. M. Alie Humaedi, M.Ag., M.Hum, as Head of the Research Center for Social Welfare, Villages and Connectivity of the National Research and Innovation Agency, Indonesia, for guiding the writing of this article.

References

Ashkenazi, I., & Shemer, J. (2005). Tsunami - The death waves. *Harefuah, 144*(3), 154–159. https://www.scopus.com/inward/record.uri?eid=2-s2.0-21244453959&partnerID=40&md5=943d9c23b2de563c44a2c460edff6897.

Badan Pusat Statistik Indonesia. (2022). *Statistik Potensi Desa Indonesia 2021*. Jakarta: Badan Pusat Statistik. Village Potential Statistics of Indonesia 2021 https://www.bps.go.id/publication/2022/03/24/ceab4ec9f942b1a4fdf4cd08/statistik-potensi-desa-indonesia-2021.html. (Accessed 2 June 2023).

Beier, A. L., & Ocobock, Paul (2008). *Cast out: Vagrancy and homelessness in global and historical perspective*. Athens: Ohio University Press.

Black, B., & Farrington, A. P. (1997a). *Promoting life for Indonesia's street children* (Vol 45, pp. 10–11). AIDSCAP. AIDSlink: Eastern, Central & Southern Africa https://www.scopus.com/inward/record.uri?eid=2-s2.0-0031139058&partnerID=40&md5=575c6daf9506121b6de688cfb112cbb9.

Black, B., & Farrington, A. P. (1997b). Preventing HIV/AIDS by promoting life for Indonesian street children. *Aidscaptions, 4*(1), 14–17. https://www.scopus.com/inward/record.uri?eid=2-s2.0-0031161706&partnerID=40&md5=dfc1bf57415f4cf600d192a99b8cdf2d.

Bonnet, Francois (2009). Managing marginality in railway stations: Beyond the welfare and social control debate. *International Journal of Urban and Regional Research, 33*(4), 1029–1044. https://doi.org/10.1111/j.1468-2427.2009.00920.x

Bowden, Vicky R. (2022). Types of reviews – Part 3: Literature review, integrative review, scoping review. *Pediatric Nursing, 48*(2), 97–100. https://www.proquest.com/openview/ca2d215cb9c9e5b61424acf3d7ade4e9/1?pq-origsite=gscholar&cbl=47659.

Bramley, G., & Fitzpatrick, S. (2018). Homelessness in the UK: Who is most at risk? *Housing Studies, 33*(1), 96–116. https://doi.org/10.1080/02673037.2017.1344957

Braun, V., & Clarke, V. (2006). Using thematic analysis in psychology. *Qualitative Research in Psychology, 3*(2), 77–101. https://doi.org/10.1191/1478088706qp063oa

Chambers, R. (2014). *Rural development: Putting the last first* (pp. 1–246). Taylor and Francis. https://doi.org/10.4324/9781315835815

Daugherty, B., Warburton, K., & Stahl, S. M. (2020). A social history of serious mental illness. *CNS Spectrums, 25*(5), 584–592. https://doi.org/10.1017/S1092852920001364

Decker, Pascal De (2004). Dismantling or pragmatic adaptation? On the restyling of welfare and housing policies in Belgium. *European Journal of Housing Policy, 4*(3), 261–281. https://doi.org/10.1080/1461671042000307297

DeVerteuil, Geoffrey, May, Jon, & von Mahs, Jürgen (2009). Complexity not collapse: Recasting the geographies of homelessness in a 'punitive' age. *Progress in Human Geography, 33*(5), 646–666. https://doi.org/10.1177/0309132508104995

Deverteuil, Geoffrey (2005). The relationship between government assistance and housing outcomes among extremely low-income individuals: A qualitative inquiry in Los Angeles. *Housing Studies, 20*(3), 383–399. https://doi.org/10.1080/02673030500062350

Doocy, S., Daniels, A., Dooling, S., & Gorokhovich, Y. (2013). The human impact of volcanoes: A historical review of events 1900-2009 and systematic literature review. *PLoS Currents*. https://doi.org/10.1371/currents.dis.841859091a706efebf8a30f4ed7a1901

Du, Y. B., Lee, C. T., Christina, D., Belfer, M. L., Betancourt, T. S., O'Rourke, E. J., & Palfrey, J. S. (2012). The living environment and children's fears following the Indonesian tsunami. *Disasters, 36*(3), 495–513. https://doi.org/10.1111/j.1467-7717.2011.01271.x

Edgar, Bill, & Doherty, Joe (2001). Supported housing and homelessness in the European Union. *European Journal of Housing Policy, 1*(1), 59–78. https://doi.org/10.1080/14616710110036418

Edgar, B., Doherty, J., & Mina-Coull, A. (1999). *Services for homeless people: Innovation and change in the European Union*. Bristol: The Policy Press.

Fadri, Zainal (2019). Upaya penanggulangan gelandangan dan pengemis (GEPENG) sebagai penyandang masalah kesejahteraan sosial (PMKS) di Yogyakarta. *Komunitas, 10*(1), 1–19. https://doi.org/10.20414/komunitas.v10i1.1070

FAO. (2021). *Ending poverty and hunger by the context strategic investments for achieving SDG 1 and SDG 2 examples of strategic investments for poverty reduction*. FAO. https://www.fao.org/3/i7556e/i7556e.pdf. (Accessed 2 June 2023).

Ganti, Mery, Yusuf, Husmiati, Wismayanti, Yanuar Farida, Setiawan, Hari Harjanto, Susantyo, Badrun, Konita, Ita, Budiarti, Menik, & Sulubere, Muhammad Belanawane (2022). The issues and social economic potentials of urban marginal groups in Indonesia. In *Proceedings of the International Conference on Sustainable Innovation on Humanities, Education, and Social Sciences (ICOSI-HESS 2022)* (pp. 246–259). https://doi.org/10.2991/978-2-494069-65-7

Giansanti, Enrico, Lindberg, Annika, & Joormann, Martin (2022). The status of homelessness: Access to housing for asylum-seeking migrants as an instrument of migration control in Italy and Sweden. *Critical Social Policy, 42*(4), 586–606. https://doi.org/10.1177/02610183221078437

Grant, Maria J., & Booth, Andrew (2009). A typology of reviews: An analysis of 14 review types and associated methodologies. *Health Information and Libraries Journal, 26*(2), 91–108. https://doi.org/10.1111/j.1471-1842.2009.00848.x

Habibullah, H., Yuda, T. K., Setiawan, H. H., & Susantyo, B. (2023). Moving beyond stereotype: A qualitative study of long-standing recipients of the Indonesian conditional cash transfers (CCT/PKH). *Social Policy and Administration*. https://doi.org/10.1111/spol.12946

Iqbali, Saptono (2008). Studi Kasus Gelandangan – Pengemis (Gepeng) Di Kecamatan Kubu Kabupaten Karangasem. *Jurnal Piramida, 4*(25). https://garuda.kemdikbud.go.id/documents/detail/1342917.

Karsono, S. (2020). Flâneur, popular culture and urban modernity: An intellectual history of new order Jakarta. *Asian Studies Review*, 1–19. https://doi.org/10.1080/10357823.2020.1784092

Kumparan. (2022). *Pemerintah Buat Rusun untuk Eks Gelandangan, Uang Sewanya Rp 10.000 per Bulan*. https://kumparan.com/kumparanbisnis/pemerintah-buat-rusun-untuk-eks-gelandangan-uang-sewanya-rp-10-000-per-bulan-1zq7ojLV06F/full. (Accessed 18 June 2023).

Kuntari, Sri, & Hikmawati, Eni (2017). Melacak akar permasalahan gelandangan pengemis (gepeng). *Media Informasi Penelitian Kesejahteraan Sosial, 41*(1), 11–26. https://garuda.kemdikbud.go.id/documents/detail/2151404.

Lee, B. A., Tyler, K. A., & Wright, J. D. (2010). The new homelessness revisited. *Annual Review of Sociology, 36*, 501–521. https://doi.org/10.1146/annurev-soc-070308-115940

Lee, Barrett A., & Schreck, Christopher J. (2005). Danger on the streets: Marginality and victimization among homeless people. *American Behavioral Scientist, 48*(8), 1055–1081. https://doi.org/10.1177/0002764204274200

Lee, Barrett A., & Price-Spratlen, Townsand (2004). The geography of homelessness in American communities: Concentration or dispersion? *City & Community, 3*(1), 3–27. https://doi.org/10.1111/j.1535-6841.2004.00064.x. (Accessed 14 October 2023)

Lee, C., Du, Y. B., Christina, D., Palfrey, J., O'Rourke, E., & Belfer, M. (2015). Displacement as a predictor of functional impairment in tsunami-exposed children. *Disasters, 39*(1), 86–107. https://doi.org/10.1111/disa.12088

Maeseele, Thomas, Roose, Rudi, Bouverne-De Bie, Maria, & Roets, Griet (2014). From vagrancy to homelessness: The value of a welfare approach to homelessness. *British Journal of Social Work, 44*(7), 1717–1734. https://doi.org/10.1093/bjsw/bct050

Morrow, R. C., & Llewellyn, D. M. (2006). Tsunami overview. *Military Medicine, 171*(10 Suppl.), 5–7. https://doi.org/10.7205/milmed.171.1s.5

Mulya, Carunia, Dwi, Firdausy, & Budisetyowati, Andayani (2021). Variables, dimensions, and indicators important to develop the multidimensional poverty line measurement in Indonesia. *Social Indicators Research*. https://doi.org/10.1007/s11205-021-02859-5

Nicholas, Pleace, & Bretherton, Joanne (2013). *Measuring homelessness and housing exclusion in Northern Ireland a test of the ETHOS typology*. Feantsa. https://www.york.ac.uk/media/chp/documents/2013/measuring_homelessness_and_housing_exclusion_in_northern_ireland.pdf. (Accessed 2 June 2023).

Nwokah, E. E., Becerril, S., Hardee, W. P., & Brito, E. (2017). Play with homeless and low-income preschoolers: University student experiences with service learning. *International Journal of Play, 6*(1), 53–77. https://doi.org/10.1080/21594937.2017.1288397

OECD. (2021). *HC3.1 Homeless population [estimates]* (pp. 1–12). OECD Affordable Housing Database. http://oe.cd/ahd.

Olufemi, Olusola (2002). Barriers that disconnect homeless people and make homelessness difficult to interpret. *Development Southern Africa, 19*(4), 455–466. https://doi.org/10.1080/0376835022000019455

Palmer, Alyssa R., Piescher, Kristine, Berry, Daniel, Dupuis, Danielle, Heinz-Amborn, Britt, & Masten, Ann S. (2023). Homelessness and child protection involvement: Temporal links and risks to student attendance and school mobility. *Child Abuse & Neglect, 135*, 105972. https://doi.org/10.1016/j.chiabu.2022.105972

Parrott, K. A., Huslage, M., & Cronley, C. (2022). Educational equity: A scoping review of the state of literature exploring educational outcomes and correlates for children experiencing homelessness. *Children and Youth Services Review, 143*. https://doi.org/10.1016/j.childyouth.2022.106673

Pauly, Bernie, Wallace, Bruce, & Perkin, Kathleen (2014). Approaches to evaluation of homelessness interventions. *Housing, Care and Support, 17*(4), 177–187. https://doi.org/10.1108/HCS-07-2014-0017. . (Accessed 10 June 2023)

Pemerintah Republik Indonesia. (1980). *Peraturan Pemerintah Republik Indonesia Nomor 31 Tahun 1980 Tentang Penanggulangan Gelandangan dan Pengemis* (pp. 1–13). Peraturan Pemerintah. https://peraturan.bpk.go.id/Home/Details/66630/pp-no-31-tahun-1980.

Santos-Reyes, G. S., Gouzeva, T., Santos-Reyes, J. R., Cepin, M., & Bris, R. (2017). Preliminary results on historical data on homelessness and post-earthquake disaster emergency shelter. In *Safety and Reliability – Theory and Applications – Proceedings of the 31st European Safety and Reliability Conference* (pp. 2525–2532). https://doi.org/10.1201/9781315210469-320

Santos-Reyes, J., Santos-Reyes, G., & Gouzeva, T. (2014). Earthquakes and homelessness: A review of historical data. In *Homelessness: Prevalence, impact of social factors and mental health challenges* (pp. 259–276). Nova Science Publishers, Inc. https://www.scopus.com/inward/record.uri?eid=2-s2.0-84922307621&partnerID=40&md5=c7fa5927fbabf5ba402ac53a7cc9b111.

Schutt, R. K., & Garrett, G. R. (2013). *Responding to the homeless: Policy and practice*. Springer Science & Business Media.

Schutt, Russell K., & Garrett, Gerald R. (1992). *The problem of homelessness*. Springer Nature. https://doi.org/10.1007/978-1-4899-1013-4_1

Setiawan, Hari Harjanto, Yuda, Tauchid Komara, Susantyo, Badrun, Sulubere, Muhammad Belanawane, Ganti, Mery, Habibullah, Habibullah, Sabarisman, Muslim, & Murni, Ruaida (2023). Scaling up social entrepreneurship to reduce poverty: Exploring the challenges and opportunities through stakeholder engagement. *Frontiers in Sociology, 8*. https://doi.org/10.3389/fsoc.2023.1131762

Setiawan, H. (2020). Fenomena Gelandangan Pengemis Sebagai Dampak Disparitas Pembangunan Kawasan Urban dan Rural di Daerah Istimewa Yogyakarta. *Moderat: Jurnal Ilmiah Ilmu Pemerintahan, 6*(2), 361–375.

Speak, S. (2012). *Alternative understandings of homelessness in developing countries* (Doctoral dissertation).

Speak, Suzanne (2005). Relationship between children's homelessness in developing countries and the failure of women's rights legislation. *Housing, Theory and Society, 22*(3), 129–146. https://doi.org/10.1080/14036090510034581

Stodulka, T. (2015). Emotion work, ethnography, and survival strategies on the streets of Yogyakarta. *Medical Anthropology: Cross Cultural Studies in Health and Illness, 34*(1), 84–97. https://doi.org/10.1080/01459740.2014.916706

Stratford, A., Kusuma, N., Goding, M., Paroissien, D., Brophy, L., Damayanti, Y. R., Fraser, J., & Ng, C. (2014). Introducing recovery-oriented practice in Indonesia: The Sukabumi project — An innovative mental health programme. *Asia Pacific Journal of Social Work and Development, 24*(1–2), 71–81. https://doi.org/10.1080/02185385.2014.885210

Sukma, N. M., & Irawati. (2021). Development of physical and mental abilities in fulfilling children's rights in the program of 'Kampung Anak Negeri' in Surabaya. *Indian Journal of Forensic Medicine and Toxicology, 15*(1), 213–216. https://doi.org/10.37506/ijfmt.v15i1.13407

Sullivan, A. A. (2023). What does it mean to be homeless? How definitions affect homelessness policy. *Urban Affairs Review, 59*(3), 728–758. https://doi.org/10.1177/10780874221095185

Susantyo, B., Habibullah, H., Irmayani, N. R., Erwinsyah, R. G., Nainggolan, T., Sugiyanto, S., Rahman, A., Arifin, J., As'adhanayadi, B., & Nurhayu, N. (2023). Social cash assistance for food security during a disaster: Lesson learned from Indonesia. *IOP Conference Series: Earth and Environmental Science, 1180*(1), 012047. https://doi.org/10.1088/1755-1315/1180/1/012047

Taufiq, Nuri (2022). Penciri Kemiskinan Ekstrem di 35 Kabupaten Prioritas Penanganan Kemiskinan Ekstrem. *Seminar Nasional Official Statistics, 2022*(1), 895–904. https://doi.org/10.34123/semnasoffstat.v2022i1.1258

The World Bank. (2022). *Fact sheet: An adjustment to global poverty lines*. The World Bank. https://www.worldbank.org/en/news/factsheet/2022/05/02/fact-sheet-an-adjustment-to-global-poverty-lines. (Accessed 2 June 2023).

The World Bank. (2020). *World Bank Indonesia investing in people social protection for Indonesia's 2045 vision*. https://www.worldbank.org/en/country/indonesia/publication/investing-in-people-social-protection-for-indonesia-2045-vision. (Accessed 20 June 2023).

UNDP and OPHI. (2002). Global MPI 2022 — Unpacking deprivation bundles to reduce multidimensional poverty. *Unpacking deprivation bundles*. https://ophi.org.uk/global-mpi-report-2022/. (Accessed 2 June 2023).

US Department of Housing and Urban Development. (2021). *The 2020 annual homeless assessment report to congress*. https://www.huduser.gov/portal/sites/default/files/pdf/2020-AHAR-Part-1.pdf. (Accessed 2 June 2023).

Vaismoradi, Mojtaba, Turunen, Hannele, & Bondas, Terese (2013). Content analysis and thematic analysis: Implications for conducting a qualitative descriptive study. *Nursing and Health Sciences, 15*(3), 398–405. https://doi.org/10.1111/nhs.12048

Yamashita, Akio (2017). *Rapid urbanization in developing Asia and Africa* (pp. 47–61). https://doi.org/10.1007/978-981-10-3241-7_3

Yusuf, Husmiati, Setiawan, Hari Harjanto, Ganti, Mery, Wismayanti, Yanuar Farida, Susantyo, Badrun, Konita, Ita, Sulubere, Muhammad Belanawane, & Budiarti, Menik (2022). Access to basic needs for marginalized groups in Indonesia: A case study of the homeless. In *Proceedings of the International Conference on Sustainable Innovation on Humanities, Education, and Social Sciences (ICOSI-HESS 2022)* (pp. 237–245). Atlantis Press. https://doi.org/10.2991/978-2-494069-65-7

CHAPTER

21

Housing intervention for people who are homeless in Indonesia: Combining institutional and community-based approaches

Hari Harjanto Setiawan[1] *and Yanuar Farida Wismayanti*[2]

[1]National Research and Innovation Agency (BRIN), Research Center for Social Welfare, Villages and Connectivity, Jakarta, Indonesia; [2]National Research and Innovation Agency (BRIN), Research Center for Public Policy, Jakarta, Indonesia

1. Introduction

Homelessness is a complex problem caused by social and economic factors. Factors causing homelessness include poverty, physical and mental health, housing problems, dysfunctional families, and drug addiction. This is a challenge for most countries in the world (Mago et al., 2013). Data from the United Nations Department of Economic and Social Affairs shows that as many as 1.6 billion people worldwide live with inadequate living conditions, and around 15 million people are forcibly evicted yearly. A number of these people have no certainty of life and become homeless. About 3 million homeless people are in Indonesia, and some 28,000 are in Jakarta. As many as 77,500 people spread across many big cities throughout Indonesia in 2019. The problem of homelessness in Indonesia is a welfare problem at the national level. The homeless live in conditions that do not follow the standard of living of the local community, do not have a permanent place of residence, and work in an uncertain business (Regulation of the Government of the Republic of Indonesia Number 31 of 1980). The cause of homelessness is the condition of the village that is not prosperous, so villagers go to the city without education and work skills. Job opportunities are limited, so they become homeless. Elements that are interrelated with the problem of homelessness are structure, system, early intervention, prevention of evictions, and stabilization of houses (Dej et al., 2020). Homeless people in Indonesia have unique characteristics compared to other countries.

Homelessness to Hope
https://doi.org/10.1016/B978-0-443-14052-5.00021-5

They live without having a permanent and decent place to live. Most homeless live under bridges, railroad tracks, rivers, and on the outskirts of shops. The homeless do not have regular jobs such as looking for cigarette butts, pulling carts, and looking for used goods. In Indonesia, the life of the homeless is always associated with beggars because they usually work as beggars in public places such as bus terminals, train stations, and shop fronts. Homeless people are a social problem that disturbs people's lives. Some stereotypes of homeless people include disturbing order, beauty, decency, cleanliness, peace, and sexual exploitation (Saewyc et al., 2021).

In Indonesia, homelessness is a social problem that must receive serious attention. Homeless people in urban areas have shown contradictory situations and conditions. The development that has been carried out has brought social problems in the form of homeless people. The ease of making money in big cities in Indonesia has become a unique attraction for migrants from outside the region without bringing sufficient skills and education to try their luck. Better, want something instant, and have low resilience capabilities. The relatively large number of homeless people in Indonesia is the government's responsibility to be present in overcoming the problem of homeless people (Murphy & Tobin, 2012). The central and regional governments focus on reducing the social problems of the homeless through various programs. Counseling is done to make the homeless aware of living a better life. The government also provides skills training to improve skills so that they can live a better life. Follow-up coaching is routine coaching that is carried out to monitor daily activities after completing social rehabilitation. The government is also trying to build a place to live for them. The transmigration program or moving residents to other islands still has land for planting (Lastiwi & Badruesham, 2017). In line with this right, the Indonesian Government has committed to implementing the SDGs (Sustainable Development Goals) Agenda that focuses on ending poverty in all its practices everywhere. The government's vision has main characteristics: the present state, building from the boundary, and the Mental Revolution. This aims to guarantee the quality of human life, including community productivity, and realize economic independence by driving strategic sectors of the domestic economy, especially for marginalized groups, including the homeless. The government and the private sector have launched many service programs for the homeless. However, the problem of homelessness still arises, especially in big cities such as DKI Jakarta, Surabaya, Semarang, Bandung, and Makassar. Homeless people in these big cities will burden the central, provincial, and city governments. This research is essential to find the best intervention for homeless people. This study also describes the assistance carried out by social workers from the stage of identifying problems, motivating, and connecting with the resource system. The ultimate goal of service to the homeless is social functioning. Homeless people who have received services can live in society. Beneficiaries can recognize their problems and can provide their solutions so that survival can be maintained. Hopefully, this research will help formulate policies to solve the social problems of homeless people (Khayyatkhoshnevis et al., 2020).

The problem of homelessness is very complex, including health, education, employment, and housing. The prevalence of mental illness among the homeless population contributed to homelessness (Brittany Bingham et al., 2019). The problem of homelessness is also associated with high levels of chronic and acute diseases and increased mortality rates (Alexander et al., 2023). The low education of people without homes impacts the acquisition of employment and housing. This problem requires various approaches to solve it. Thus, the government has developed two approaches: an institution-based approach and a community-based one. The institutional-based approach in the Indonesian context is that homeless people are placed in temporary shelters (Panti in Bahasa) as a dormitory to provide health services, education, work skills, and job distribution. Homeless people are given guidance and services, including training for various periods, from 6 months to 1 year in the dormitory, depending on the services' assessment and progress. After completing the training, they will be returned to the community to live independently. The second approach is community-based, where people experiencing homelessness are given services and social assistance in a specific location through local community organizations. Both approaches have their respective advantages and disadvantages. This chapter will describe implementing these two approaches to perfect the housing program. The government provided the building with 200 families in the dormitory area. Homeless people participate in development programs in dormitories but live and socialize in housing. This combined approach is more humane and creates a safe social environment for people experiencing homelessness. This approach prioritizes human values and rights to restore social functions in society. This approach is considered capable of improving the welfare of homeless people because housing is a basic need that must be fulfilled for all people. This chapter answers three main problems: (1) What is the condition of homelessness in Indonesia? (2) How can housing be developed by combining institutional and community-based approaches? (3) How is housing provided for homeless people implemented? (4) How do social workers in the housing sector provide social guidance for homeless people? The novelty of this paper is the form of intervention practice for homelessness by combining the two previous approaches. This approach is essential, and it will be helpful in inspiring other countries to develop suitable and more humane approaches to homelessness.

2. The rationale of the study

2.1 Institution-based approach

Institution-based social rehabilitation of homeless people is placing homeless people in a particular institution as a shelter that aims to function socially. The government forms institutions that are part of the Ministry of Social Affairs with regulations issued by the Minister of Social Affairs. The government provides shelters that have complete facilities such as health services and skills training. Homeless people put in shelters are given motivation by social workers to change their lives so that they can live in society normally. They live in a shelter for around 6 months with a social rehabilitation program with planned stages. However, this approach requires huge costs because it covers all their living expenses (Anggriana & Dewi, 2016).

Homeless people are also given skills to open a business or get job opportunities. Beneficiaries are expected to have the chance to work in the formal sector based on the skills that they have acquired in social rehabilitation institutions. Skills development at this institution includes carpentry skills, salted egg production, culinary skills, batik (traditional printing) skills, sewing, catfish cultivation, and agriculture. Those who are interested and meet the criteria can take part in the empowerment program individually or in groups (Aykanian et al., 2020). This rehabilitation system faces several problems: limited skills to compete for work. In addition, the graduates generally fare below labor market standards.

Mental and social development methods are also provided in the institutional approach. They are forming a homeless person into a strong person by providing mental, psychological, and spiritual guidance. Cognitive development should be the main focus of rehabilitation programs with an institutional approach. In practice, mental development still refers to the transfer of knowledge only. Social workers' role is crucial to increasing awareness because mental awareness must be prioritized. Ethics, functions, social responsibilities, and group dynamics are given to live in society (Temesváry & Drilling, 2022). Physical guidance is also provided when they get a mental issue, which is vital for their activities. The physical advice provided includes sports and health services.

After finishing the institution's social rehabilitation program, homeless people will be returned to society. Continuous guidance when leaving the institution is essential. It must be routinely monitored to ensure that beneficiaries can socialize in the community. Every beneficiary is guaranteed to be part of society. They can be accepted in society and maintain norms. After returning to live in the community (Nasution et al., 2021), they are still accompanied by social workers to provide consultations if there are problems. Consultative aims to resolve issues and better develop conditions by utilizing society's potential. Social workers place social welfare services on homeless people not as objects but as subjects who will improve themselves.

The institution-based approach has several limitations that must be corrected. Homeless accommodation is often unable to meet the diverse needs of individuals. These needs range from physical and mental health services to social support and access to education and employment opportunities. The health service models have proven inadequate in addressing the complex needs of homeless people (Konduru, 2019). Integrating healthcare delivery within homeless shelters effectively increases access to healthcare for individuals experiencing homelessness (Alexander et al., 2023). Therefore, there is an increasing need for practical approaches that integrate housing to improve the well-being of homeless people (Zabkiewicz, 2012).

2.2 Community-based approach

Community-based social rehabilitation for homeless people is a service model carried out at the community level to increase public awareness using community resources (Bessel, 2019). The community-based approach aims to return the homeless to their places of origin or homes in groups at specific locations. This program places homeless people in authentic community life. This program is carried out when they are ready to live as part of society with the skills acquired during training. The community-based approach is an effort to empower the community's capacity to identify, analyze, and take initiatives to solve existing problems independently. The community-based approach is to increase the capacity of the community and seek to reduce the vulnerability of individuals, families, and society at large (Fowler et al., 2019). Changes in the homeless community are carried out to deal with environmental problems. Community-based programs use a reality-based approach. The relatively simple and easy-to-implement method is aimed at the grassroots level so people can make positive changes for the better. The target of this program is the homeless who live in vulnerable areas and are willing to accept change. This method emphasizes internal community-based program planning rather than external factors. The approach used is a bottom-up approach, not a top-down one. The potential threat is not outside but within the social system. Reducing the level of threats and risks should be part of development considerations. Community-based programs will only be optimally successful with partnership and high participation. The components of society, government, and institutions/NGOs are stakeholders who must participate. Strengthening partnerships and participation, in this case, is not only directed at providing funds, materials, and manpower. The partnership also on planning, implementation, monitoring, and evaluation (Mosley, 2021). Stakeholders must be involved in program sustainability. Strengthening partnerships and participation encourages communication, coordination, and cooperation from various disciplines and professions. Professionals invited to cooperate with community-based homelessness include community development workers, health practitioners, economists, medical, social workers, counselors, and teachers. Each profession has different expertise but can complement each other.

Community-based programs are expected to reduce the vulnerability of homeless people by empowering community capacities. The increasing uncertainty of environmental, physical, social, economic, and political situations makes homeless people very vulnerable to the dangers and impacts of these situations. This requires a lot of effort to empower the homeless through community organizing in social and economic awareness, environmental awareness, education/training, etc. Empowerment of the homeless is needed in decision-making, planning, and policy-making. Community-based empowerment of the homeless is also needed so that the community can control inputs, processes, outputs, and sustainability. Through a community-based approach, the main goal is for homeless people to live together with the community (Bhattacharya & Priya, 2022). A community-based homeless management program is a program that does not only focus on short-term needs but is more than that. It must also be oriented to the long term. All supporting elements, such as strategies, approaches, models, instruments, and methods, must be institutionalized from generation to generation. So that this program can maintain, maintain, and develop the programs that have been implemented. Program sustainability means that the community takes responsibility independently when homelessness-related problems are in their environment. The government, private sector, NGOs, academics, and all elements of society have a social responsibility to continue working together to create an independent society to overcome the problem of homelessness (Sinatra & Lanctot, 2016). With its authority, the government makes programs and policies that support realizing these goals.

2.3 Housing: A mix of community and institutional-based

The government's attention to the homeless is manifested by building housing that starts in 2021 and will be completed in 2023. This housing for the homeless combines institutional and community-based intervention approaches. This housing is a safe and attractive vehicle to enter community life. The housing location is in a homeless training institution, so the training for them is better. Coaching is carried out informally to help the homeless overcome problems and find alternative solutions. This housing will reshape the attitudes and behavior of homeless people by the values and norms prevailing in society (Fitzpatrick & Stephens, 2014). The main goal is to return homeless people to life in society. Homeless people can carry out their roles as members of society and comply with prevailing values and norms. The right of the homeless to get a place to live from the perspective of human rights is the same as the people's right to get a place to live O'Shaughnessy (O' Shaughnessy et al., 2021). The right is related to a person's life, it's also connected with the housing issue. International human rights standards for adequate housing refer to services, resources, facilities, and infrastructure availability. The word adequate also implies fulfilling the principles of affordability, habitability, and accessibility. The state's responsibility in housing and settlement development in the context of human rights must be seen in availability, affordability, and sustainability. These three aspects are the principles of fulfilling the rights of the homeless from a human rights perspective. The implementation of state responsibilities has been regulated, starting from the Constitution to various regulations in the field of housing and settlements.

The fact is there are still many who do not enjoy home and become homeless. It is estimated that around eight million households in Indonesia do not yet have decent housing. Most of those who do not occupy proper housing are the urban poor or low-income homeless groups. The urban poor living as homeless people is a social fact that cannot be covered up and ignored. They work in the informal sector, such as porters in traditional markets, collecting used goods and beggars. Homeless people are more common in urban areas than in rural areas. Homeless people arise as a result of development disparities in villages and cities. Thus, not a few groups of individuals flocked to the city to try their luck. The problem of homelessness originates from economic limitations, so its handling requires cooperation from various parties (Calderón-Villarreal et al., 2022). People of all nationalities universally recognize the right to adequate housing for the homeless. Every state has several obligations related to housing. Housing for the homeless in Indonesia is a collaboration between the Ministry of Social Affairs and the Ministry of Public Works and Public Housing to allocate the budget. The Ministry of Social Affairs provides land and guidance for the homeless. Meanwhile, the Ministry of Public Housing allocates funds for housing construction. Housing is built in the location of a homeless service center with an institutional-based and community-based approach. Meanwhile, the Ministry of Public Works and Public Housing builds housing, provides funds, and builds housing for the homeless. Combining these two approaches is expected to solve the problem of homelessness in Indonesia. Every citizen, including the homeless, has the right to meet their housing needs. The service concepts framework for the homeless can be seen in Fig. 21.1.

3. Materials and methods

This study uses a mixed methods approach, which combines quantitative and qualitative methods. The combined research method is an approach in research that combines or links quantitative and qualitative research methods (Harrison et al., 2020). It includes a philosophical foundation, using both quantitative and qualitative approaches and combining them in research. The purpose of this method is to obtain more comprehensive information. A mixed methods approach is a better way to research vagrant vulnerabilities. The quantitative approach will describe the characteristics of homeless people in Indonesia, including housing, education, income, health, and identity. A qualitative approach is used to obtain in-depth information about social services carried out in housing for the homeless. The combined research method can eliminate the weaknesses in the quantitative and qualitative methods.

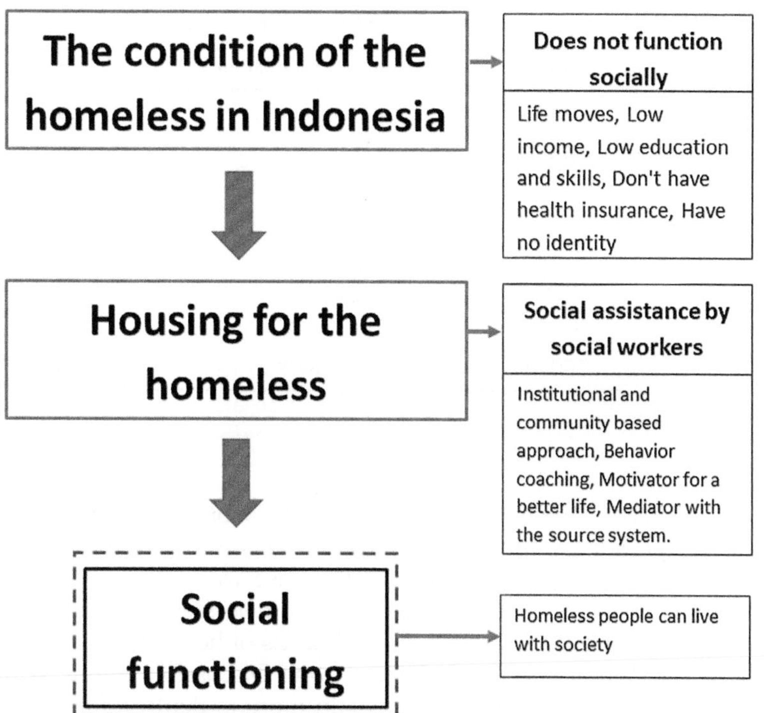

FIGURE 21.1 Conceptual framework of housing services for the homeless: housing is a social service that aims for the social functioning of the homeless in society.

3.1 Population and sample

The population of this study is homeless people who receive guidance from the Ministry of Social Affairs. Sampling was purposive; the researcher determined the sample by determining unique characteristics according to the research objectives. The reason for using purposive sampling is that the distribution of beneficiaries of social services for the homeless is concentrated in big cities. Locations represent each area in the six major cities with institutions for the homeless. Samples are taken proportionally because the population is divided into several groups. Then each group is taken separately. The population is in six provinces: DKI Jakarta, West Java, East Java, Central Java, North Sumatra, and South Sulawesi. The total population is 3578, with a confidence level of 95% an alpha value of 0.5, and a total sample of 306 respondents. The sample proportions for each province are shown in Table 21.1.

3.2 Data collection

Data collection in this study began with documentation and literature studies to gather further information regarding homelessness. Supporting documents in diaries, activity reports, and photos are also required to support the information. In addition, this study also

TABLE 21.1 The proportion of sample of each province.

No.	Province	Population	Sample
1	DKI Jakarta	1681	254
2	West Java	579	88
3	Central Java	310	47
4	East Java	400	61
5	South Sulawesi	458	70
6	North Sumatra	250	38
	TOTAL	**3678**	**558**

used several data collection methods, namely: (1) Questionnaires; this study used an e-questionnaire using the Survey Monkey application. This e-survey was distributed to two groups, namely institutional managers, who in this case were represented by the Head of the Institution, and beneficiary groups, namely the homeless. (2) In-depth interviews were conducted to learn more about the conditions and needs of homeless people and solve their problems (Deterding & Waters, 2021). (3) Focus Group Discussion (Sim & Waterfield, 2019) conducted to process data triangulation and obtain more in-depth information from stakeholders about social rehabilitation services for the homeless. Quantitative data collection was assisted by enumerators whom researchers had trained. On average, an enumerator conducts interviews with two to four people daily. It took each enumerator two to 4 days to complete the interview. The enumerators were selected based on the condition that social workers work in homeless service institutions. The consideration is that the enumerator knows the situation and condition of the beneficiaries. The research team chose enumerator locations to carry out their duties objectively to avoid bias.

3.3 Data processing and analysis

The data obtained through the data collection process is analyzed in several stages. The first stage will be the processing of quantitative data from the survey results using the SPSS application, including (Stockemer, 2019) (1) editing, checking all lists of questions filled out by respondents; (2) coding, making symbols or signs in the form of numbers for research respondents' answers; and (3) tabulation, compiling and calculating data from coding results which will then be presented in tabular form. Furthermore, the data processing results will be analyzed to obtain descriptive data. Qualitative data processing was carried out by examining all data from research instruments, such as documents, notes, results of FGDs and interviews, recordings, and results of observations. Thus, the data were analyzed based on the thematic analysis approach.

4. Results and discussion

4.1 The condition of the homeless in Indonesia

Homelessness is a social problem often found in big cities in Indonesia. Homeless people appear because of one of the negative impacts of the development. People's living needs are increasing, but the available jobs are insufficient. The ease of making money in big cities is an attraction for villagers. They come without adequate education and skills. A lazy attitude and wanting something instantly make them homeless (Anggriana & Dewi, 2016). Dealing with the problem of homeless people requires social workers to explore the lives of homeless people. Building a trusting relationship is necessary to know the needs and potential of homeless people[(Phillips et al., 1988). The importance of the assessment is to find out the potential and social welfare resources needed (Cheng & Yang, 2010). Homelessness is a social problem that needs attention from the Indonesian government. The homeless are often stigmatized as lazy, unwilling to work, mentally weak, and having physical or psychological disabilities. In addition, there are also external factors, including social, cultural, economic, educational, environmental, religious, and geographical location. This paper examines the problem of homelessness as multidimensional and multilevel. So, the resulting intervention ideas can solve the problem completely (Somerville, 2013). An overview of the condition of homeless people in Indonesia can be seen from various aspects, including housing, education, health, income, and identity. Most of the homeless people in Indonesia come to live in big cities with their families, such as spouses and children. They only have an opportunity to work in the informal sector due to no permanent residence. This study shows that 28% of respondents received services in an orphanage. Furthermore, 22% of homeless people mention living in a rented room in a slum area. While the other 17% live in a relative's house. The homeless also admit to living nomadically under bridges or on store terraces. While others stated that they had lived in an empty building. They make a living as scavengers or beggars to survive. Some live in rented houses in slum areas close to garbage dumps, riverbanks, under bridges, and locations that pose a safety hazard. Conditions like this must be given access to basic housing needs and job assistance (Maryatun & Muhammad Taftazani, 2022). They arrive in big cities with inadequate educational backgrounds and skills. They have less than a junior high school education. Homeless people are a vulnerable group and do not have access to a decent life. The types of skills possessed are pretty varied. As many as 25% have skills in making food, 9% in handicrafts, 5% in farming and animal husbandry, 4% in automotive and sewing, and 2% in trade, driving a car, salon, and others. The number of unskilled homeless people is still relatively high, namely 42%. Implementation of education and skills will be a solution for the homeless. The low education of homeless people will affect their children's education, so policy changes must be made for a better life (Rafferty, 1995). Fifty percent of the homeless have an average weekly income ranging from IDR 100,000–IDR 300,000. They have an average expenditure that is higher than income. Homeless people with an income between IDR 300,000–IDR 700,000 is 18%. However, homeless people who spend between IDR 300,000–IDR 700,000 show a figure higher than the average weekly income of 26%. The most used expenditure is to meet daily needs. This makes their lives vulnerable to debt. If this condition continues without intervention, their lives will still be categorized as poor. The condition is getting worse because most of the homeless people are in debt. Most of the beneficiaries stated that they did not have any savings. 80% of the homeless do not have savings, and only 20% have savings. However, more than half of the beneficiaries admit to having debts. This condition requires increased income through Productive Economic Business so the homeless can live independently (Huber et al., 2022).

Most of the homeless in Indonesia do not have health insurance. Only 16% of homeless people have health insurance. The health in question does not include mental health. Many homeless people experience mental disorders. One study revealed the importance of recreation and mental health for the homeless (Chan et al., 2014; Chan et al., 2014). Mental health issues are faced by a person who has a problematic experience. This situation should be necessary for homeless people who have mental problems. Using an addictive substance is also a wide problem faced by the homeless to cope with life. Many homeless people do not have an identity as much as 13%. Their identity awareness is still low because many homeless people feel that identity is not essential. The local government is trying to reach out to provide ID cards to them because of the conditions for accessing social assistance from the government. Resident identity is very important for all citizens, including homeless people. The importance of this population document in achieving human rights for all citizens (Abdullah et al., 2018). Citizens receiving social assistance must have a resident identity as proof of citizenship.

4.2 Critical perspectives on intervention

There are two main approaches to addressing homelessness: Institution-based and community-based approaches. Institution-based approaches focus on implementing public policies and structural changes to address the root causes of homelessness (Pierse et al., 2019). This approach recognizes that homelessness is often caused by broader systemic problems such as unemployment, unaffordable housing, and inadequate social welfare programs. Instead, a community-based approach involves local engagement and collaboration to provide immediate support and services to individuals experiencing homelessness. This approach recognizes the importance of engaging local communities, traditional institutions, and leadership in addressing homelessness and designing solutions to meet specific community needs. Homelessness is associated with various adverse social, economic, and health impacts (Byrne et al., 2020). People experiencing homelessness often interact with multiple government-funded systems of care, including emergency shelters, health services, mental health, substance use disorder treatment, and the criminal justice system, providing various points to address housing, health care, and other social needs. An institution-based approach to homelessness involves implementing public policy and structural changes to address the root causes of homelessness, such as unemployment and unaffordable housing. This approach focuses on system-level change, such as creating affordable housing initiatives, strengthening social welfare programs, and addressing systemic issues contributing to homelessness (Baxter et al., 2019). Ensuring the effectiveness and inclusiveness of efforts to address homelessness is essential. Additionally, it is crucial to recognize the impact of systemic factors on individuals and communities experiencing homelessness, particularly those that intersect with race, culture, and mental health (Crawford et al., 2022). By leveraging both institution-based and community-based approaches to homelessness, we can better address the complex needs of individuals and communities. An institution-based approach involves working with government agencies, nonprofit organizations, and service providers to develop and implement policies, programs, and services that address homelessness at a systemic level. This approach addresses structural factors contributing to homelessness, such as unemployment rates, lack of affordable housing, and cuts to welfare programs.

On the other hand, a community-based approach emphasizes the participation and empowerment of individuals and communities in finding solutions to homelessness. This approach recognizes the expertise and knowledge of communities impacted by homelessness and aims to involve them in the decision-making process. By utilizing both approaches, we can ensure that interventions are comprehensive, responsive to individuals' and communities' unique needs and experiences, and involve indigenous institutions, knowledge, and leadership. This collaborative approach is critical in creating sustainable solutions and reducing the stigma and discrimination surrounding homelessness (Ngo & Turbow, 2019).

4.3 Housing program for the homeless

The housing development is a combination of two approaches, namely institutional and community-based. Development is carried out in the orphanage complex to empower and develop productive economic businesses in collaboration with the Ministry of Social Affairs and the Ministry of Public Works and Public Housing. It is hoped that housing will become a place to empower nonincome people and reduce urban slum settlements. The Ministry of Public Works and Public Housing is building housing as livable places for the homeless in Jakarta and Bekasi worth Rp 86.6 billion. Beneficiaries included in the category of homeless people are scavengers, beggars, cart people, abandoned parents, and other social problems. Housing will be used as a decent standard of living for the homeless by the Social Rehabilitation Center and the Ministry of Social Affairs by first participating in social rehabilitation, job training, and entrepreneurship coaching. The construction of residential houses was marked by laying the first stone in mid-February 2021. Residential houses were built in two places, namely the Bambu Apus Orphanage Complex, Cipayung, East Jakarta, with an area of 1932 m2, and the Balai Karya Pangudi Luhur Complex in Bulak Kapal, East Bekasi, with an area of 3880 m2. Each flat is built on five floors consisting of 108 units of type 24 to accommodate 428 people. The two housing complexes have infrastructure, parking facilities, clean water, sanitation, and electricity networks. In addition, each room is also equipped with a bed, cupboard, table, and chairs. Regarding space utilization, the ground floor is used as a management, function, and disabled room. The cost of constructing the two houses and their complementary facilities is estimated at Rp 86.6 billion. Residential development as a livable place for the homeless is part of the One Million Houses Program through the Directorate General of Housing. Since its launch on April 29, 2015, the achievements of the One Million Houses Program have increased gradually; in 2015–19, 4,800,170 houses were built, and in 2020 there were 965,217 houses or as many as 9799 units. Housing for the homeless that has been completed can be seen in Fig. 21.2.

The big cities identify with the development and social problems as an impact, including increasing the number of homeless and case Study beggars due to limited jobs, housing, and lack of education (Fitri, 2019). The government tries to manage the population of homeless

FIGURE 21.2 Housing for the homeless in Bekasi, West Java, Indonesia: The Ministry of Public Works and Public Housing and the Ministry of Social Affairs have built flats for the homeless working in the informal sector. The flat has been built in Bekasi city under the name "Pangudi Luhur integrated center."

people and beggars through police operations, however; their numbers never decrease and even tend to increase. The ease of making money in big cities is the main attraction for migrants from outside the area without bringing adequate skills and education. Laziness, reluctance to improve life, wanting something instant, and low resilience ability (Anggriana & Dewi, 2016). Causes of being homeless include poverty, geographic origin, and social psychology (Samari & Groot, 2023). The government has made various efforts to overcome the problem of homelessness. Through an institutional approach, the government established an orphanage to foster homeless people. The institutional approach is the responsibility of the government at the provincial level. Meanwhile, the community-based approach is the government's responsibility at the district level. The community also plays a role in overcoming this problem by establishing social welfare institutions. Institutions established by the community must obtain permission from the government. The two approaches are still not optimal because they work with each other. A combined approach is needed to solve this problem.

This study also found several factors regarding their reasons for becoming homeless and beggars (Kuntari & Hikmawati, 2017), economic reasons or poverty, geographical constraints in their area of origin, and psychological and sociocultural factors to be factors that must be considered. Because their income from begging is relatively high, they are reluctant to look for other jobs. The government has made various efforts to overcome this problem. Building halls, houses belonging to the central and regional governments, and Social Welfare Institutions. Maximum social rehabilitation efforts because homeless people and beggars tend to return after receiving guidance at the rehabilitation center. This study provides recommendations to the government regarding housing for homeless and beggars. This effort is a form of social rehabilitation, providing strict sanctions for those who return to homelessness (Merlindha & Hati, 2017). The government and society have implemented programs to overcome the problem of homelessness. There are still many other problems related to homelessness in Indonesia. If this problem is not resolved, it becomes the government's responsibility as the party responsible (Suradi et al., 2020). The government is trying to carry out social rehabilitation to change their living habits. Changing habits from expecting help to doing good. So that they can function socially in society. Social entrepreneurship is one of the programs to address this problem. Meanwhile, another function is to make people aware of the importance of the rehabilitation program because it is related to daily life, to help fulfill their basic needs, develop their potential, and help them to be able to behave normatively. Welfare is the ideal that everyone wants to achieve. Efforts to achieve prosperity cannot run smoothly, but various obstacles exist.

The state of homelessness and begging shows they are not functioning socially. They function socially when they can access basic needs (Husmiati Yusuf et al., 2022). They also have to carry out their duties and responsibilities as members of society. The government helps them basically to solve their problems. The lives of homeless and beggars who are socioeconomically vulnerable need special programs to change their lives (Mery Ganti et al., 2022). They live without a place to live because they make a living, do not have jobs at home, and face psychological problems that require them to leave their homes. Their presence in public places raises pros and cons for the community. Some think they interfere with the view (Baillergeau, 2014). One of the limitations in accessing needs is due to the low level of education (Alberton et al., 2020). They cannot work in the formal sector. They also do not have a permanent residence. Various factors cause them to become a group that is not accepted by society. Low education level and limited work skills affect their income level. They choose to be homeless and beggars because they are not engrossed in work. The state is said to have failed to provide the rights of its citizens if there are still homeless people roaming in public places. This group is vulnerable and discriminated against. They do not have access to adequate education. They are called a vulnerable group because they do not have a permanent place to live. Some of them live in carts and move from place to place. They do not have ID cards or official identities. Their children cannot go to school because they do not have birth certificates. Under these conditions, their lives need social protection from the state. The social function of this group is disrupted to meet needs and carry out life tasks.

4.4 Social assistance by social workers

Centra Pangudi Luhur is a place that provides services to homeless people so that they can live appropriately in society. This institution provides services by conducting social rehabilitation for the homeless with various coaching. Service aims to improve the quality of physical and spiritual knowledge and skills. Development programs include physical development, mental development, and skills development (Rutenfrans-Stupar et al., 2019). The beneficiaries are people who experience social problems, such as homeless people. The stages of social rehabilitation are carried out in various stages, namely (1) Approach, (2) Assessment, (3) Intervention Plan, (4) Problem solving, (5) Settlement, (6) Termination,

(7) Further Development. The initial approach stage was carried out using socialization and identification. This stage also motivates the homeless at reception (Misje, 2020). The assessment stage carried out by the Pangudi Luhur Center for Homeless People is the disclosure and understanding of the problems experienced by homeless people by listening, collecting, analyzing, and formulating problems, needs, potentials, and resources that can be utilized in social rehabilitation services. The intervention plan stage is the stage of planning activities to be carried out in socially rehabilitating socially assisted members. The results of the assessment stages are the main provisions used in making intervention plans for social rehabilitation. The stages of problem-solving carried out by the Pangudi Luhur Center for beneficiaries are the implementation of a problem-solving plan in the form of physical, mental, social, and skills development. The activities provided in each coaching are intended so beneficiaries can be independent and play an active role in community life. Physical guidance is given to beneficiaries in various activities related to physical health. The mental guidance given to socially assisted members is given in various activities related to mentality and spirituality, such as religious activities and providing motivation (Edidin et al., 2012). Social guidance is provided to assisted members in various activities that can improve the assisted members' social skills, such as group dynamics activities, tours, and outbound activities. Skills guidance is given to beneficiaries in various activities that make beneficiaries have skills such as welding skills guidance, car mechanics, motorcycle mechanics, graphic design, carpentry, sewing, soybean processing, salon and cosmetology, and food processing. Apart from that, there are also extracurricular activities that socially assisted residents can participate in, such as farming, fishing, and music.

The resocialization stage is the stage of getting used to beneficiaries in work. At this stage, beneficiaries carry out work-learning practices. These stages are carried out in collaboration with agencies that follow the skills guidance in the Pangudi Luhur Center. The termination stage is the stage of stopping the provision of social rehabilitation services. At this stage, the Pangudi Luhur Center will provide basic food packages so that social assistance members can continue to live in the community. The subsequent development is strengthening the independence of service recipients after receiving social rehabilitation services. Sentra Pangudi Luhur saw the efforts made by beneficiary alumni and then provided business guidance. Skills guidance is coaching provided to support beneficiaries to have the provision for living independently (Sisselman-Borgia, 2021). Skills development in the Pangudi Luhur Center includes welding, car mechanics, motorcycle mechanics, graphic design, carpentry, sewing, soybean processing, salon and cosmetology, and food processing. These skills are easy to apply and match market needs. Skill guidance time is 6 months, supported by adequate learning media. One beneficiary can take one skill guidance by the career goals to be developed. Skill guidance is given based on the results of the first assessment. The skills guidance provided follows the beneficiaries' abilities which is a solution to the problems experienced. Guidance on food processing skills is guidance given to help beneficiaries acquire knowledge and skills for their survival. Guidance is assisted by instructors who have culinary and business skills. Coaching is carried out by providing material and practice of making food. Guidance on food processing skills has been able to help beneficiaries improve their quality of life (Barton et al., 2022). This skills guidance has produced alumni who have succeeded in changing their lives, including: (1) some beneficiaries can produce and sell bread; (2) some can produce and sell cakes; (3) some work in restaurants and catering; and (4) some can live decently and send their children to school.

5. Limitations of the study

The research data does not yet show national data because the sample for this study was taken from a population of homeless people who receive social services in orphanages and community-based services. This chapter is limited to answering problems in five regions and cannot be generated for all regions of Indonesia. The uniqueness of society and unstable social situation are challenging in this study on how to make a standard measurement of to solve the homeless problems. Thus, this approach might be applied in other areas if the situation and condition are the same.

6. Recommendations

Homeless people's indifference to their social environment is one of the factors inhibiting services in apartments. This is the most challenging factor to resolve. If homeless people no longer care about their environment, then it can be said that any efforts made by various parties will not work optimally. Awareness is the key to empowering the homeless. With awareness, homeless people have the desire to change and innovate in change. By looking at the characteristics of homeless people, what needs to be instilled is an awareness of wanting to live in society. Housing, which is a combination of institutional and community-based approaches, will involve many stakeholders. This is an important part of social intervention in carrying out its goals. Stakeholders can be found anywhere, especially in community activities, so the aim of social intervention cannot be separated from the existence of the most influential figures. With stakeholders in community-based activities, developing intervention objectives will be very necessary. However, not all stakeholders will have a positive impact on the intervention objectives. The stakeholders in intervention against homelessness which are the key to its success are the local community who accept homeless life after undergoing activities in housing. Interventions with homeless people require intensive assistance from social workers. Mentoring is an activity that means coaching, teaching, and directing, which has the connotation of mastering, controlling, randomized, and controlling. Assistance is an effort to involve homeless people in developing their potential to achieve a better quality of life. The main role of homeless companions is as a motivator, namely providing stimulation and encouragement to beneficiaries so that they are able to develop their potential. The second role is as a mediator who connects potential homelessness with existing resource systems in the community.

7. Conclusions

The problem of homelessness has become a national social welfare problem. Homeless people in Indonesia do not have access to housing, education, income, health, and identity. Most homeless people have limited access to social services such as social assistance, health and education services, employment, and housing assistance. In addition, support from social institutions, both government and private, are often not able to provide full support due to limited human resources, facilities, and infrastructure, as well as complex regulations and policies. Some people consider homeless people as people who have a negative image, disturb public order, beauty, decency, cleanliness, taking care of themselves, and peace. The problem of homelessness is socio-cultural, such as the inability to follow the rules of social life so that they become increasingly vulnerable and marginalized in society. Housing for the homeless, built by the Indonesian government, combines institutional and community-based approaches. Residents of this housing will be assessed to receive skills training by their career choices. After having the skills will be channeled into the world of work. Optimizing the service approach to the homeless involves many stakeholders. Cooperation between ministries started with the Ministry of Social Affairs and the Ministries of Public Works and Public Housing. The central government and local governments work together according to their respective roles. Collaborating with the community and the business world in the process of empowering the homeless will strengthen the interventions that have been implemented. The involvement of the business world, especially in the company's labor distribution. Homeless people need intensive assistance after entering housing due to changes in new conditions. The homeless are a vulnerable group because of limited access to social services and opportunities for advancement. Homeless assistance requires special competence because the problems faced differ from society. Increase the number of professionals, especially social workers, to meet the needs of assistance in homeless housing. Social workers are professionals who deeply understand the problem of homelessness. Empathy and genuine commitment from the parties involved will contribute to the success of this program. The stages of assistance must be carried out appropriately to improve the social functioning of the homeless.

Acknowledgment

Both authors were the main contributors who contributed to data collection and writing. We would like to thank the Research Center for Social Welfare, Village and Connectivity and Research Center for Public Policy for supporting the writing of this book chapter.

References

Abdullah, Husni, U., Mataram, H. A., & Manusia. (2018). Pentingnya dokumen kependudukan sebagai wujud hak asasi manusia. *Prosiding PKM-CSR, 1*, 1744–1753. https://www.prosiding-pkmcsr.org/index.php/pkmcsr/article/view/84.

Alberton, Amy M., Angell, G. Brent, Gorey, Kevin M., & Grenier, Stéphane (2020). Homelessness among Indigenous peoples in Canada: The impacts of child welfare involvement and educational achievement. *Children and Youth Services Review, 111*, 104846. https://doi.org/10.1016/j.childyouth.2020.104846

Alexander, K., Nordeck, C. D., Rosecrans, A., Harris, R., Collins, A., & Gryczynski, J. (2023). The effect of a non-congregate, integrated care shelter on health: A qualitative study. *Public Health Nursing, 40*(4), 487–496. https://doi.org/10.1111/phn.13197

Anggriana, Tyas Martika, & Dewi, Noviyanti Kartika (2016). Identifikasi permasalahan gelandangan dan pengemis di UPT Rehabilitasi sosial gelandangan dan pengemis. *Inquiry: Jurnal Ilmiah Psikologi, 7*(1). https://doi.org/10.51353/inquiry.v7i1.78

Aykanian, Amanda, Morton, Peggy, Trawver, Kathi, Victorson, Lane, Preskitt, Sarah, & Street, Kimberly (2020). Library-based field placements: Meeting the diverse needs of patrons, including those experiencing homelessness. *Journal of Social Work Education, 56*(Suppl. 1), S72–S80. https://doi.org/10.1080/10437797.2020.1723757

Baillergeau, E. (2014). Governing public nuisance: Collaboration and conflict regarding the presence of homeless people in public spaces of Montreal. *Critical Social Policy, 34*(3), 354–373. https://doi.org/10.1177/0261018314527716

Barton, S., Porter, H., Murphy, S., & Lysaght, R. (2022). Potential outcomes of work integration social enterprises for people who are homeless, at risk of homelessness, or transitioning out of homelessness. *Social Enterprise Journal, 18*(3), 409–433. https://doi.org/10.1108/SEJ-07-2021-0054

Baxter, Andrew J., Tweed, Emily J., Katikireddi, Srinivasa Vittal, & Thomson, Hilary (2019). Effects of housing first approaches on health and well-being of adults who are homeless or at risk of homelessness: Systematic review and meta-analysis of randomised controlled trials. *Journal of Epidemiology & Community Health, 73*(5), 379. https://doi.org/10.1136/jech-2018-210981

Bessel, Diane R. (2019). *Community-based strategies to address homelessness* (pp. 149–169). https://doi.org/10.1007/978-3-030-03727-7_7

Bhattacharya, P., & Priya, K. R. (2022). Stakeholders facilitating hope and empowerment amidst social suffering: A qualitative documentary analysis exploring lives of homeless women with mental illness. *International Journal of Social Psychiatry, 68*(4), 908–918. https://doi.org/10.1177/00207640211011186

Bingham, Brittany, Moniruzzaman, Akm, Patterson, Michelle, Distasio, Jino, Sareen, Jitender, O'Neil, John, & Somers, Julian M. (2019). Indigenous and non-indigenous people experiencing homelessness and mental illness in two Canadian cities: A retrospective analysis and implications for culturally informed action. *BMJ Open, 9*(4), e024748. https://doi.org/10.1136/bmjopen-2018-024748

Byrne, Thomas, Baggett, Travis, Land, Thomas, Bernson, Dana, Hood, Maria-Elena, Kennedy-Perez, Cheryl, Monterrey, Rodrigo, Smelson, David, Dones, Marc, Bharel, Monica, & Sartorius, Benn (2020). A classification model of homelessness using integrated administrative data: Implications for targeting interventions to improve the housing status, health and well-being of a highly vulnerable population. *PLoS One, 15*(8), e0237905. https://doi.org/10.1371/journal.pone.0237905

Calderón-Villarreal, Alhelí, Terry, Brendan, Friedman, Joseph, González-Olachea, Sara Alejandra, Chavez, Alfonso, Díaz López, Margarita, Pacheco Bufanda, Lilia, Martinez, Carlos, Medina Ponce, Stephanie Elizabeth, Cázares-Adame, Rebeca, Rochin Bochm, Paola Fernanda, Kayser, Georgia, Strathdee, Steffanie A., Muñoz Meléndez, Gabriela, Holmes, Seth M., Bojorquez, Ietza, Los Huertos, Marc, & Bourgois, Philippe (2022). Deported, homeless, and into the canal: Environmental structural violence in the binational Tijuana River. *Social Science & Medicine, 305*, 115044. https://doi.org/10.1016/j.socscimed.2022.115044

Chan, Dara V., Helfrich, Christine A., Hursh, Norman C., Sally Rogers, E., & Gopal, Sucharita (2014). Measuring community integration using Geographic Information Systems (GIS) and participatory mapping for people who were once homeless. *Health & Place, 27*, 92–101. https://doi.org/10.1016/j.healthplace.2013.12.011

Cheng, Li-Chen, & Yang, Yun-Sheng (2010). Homeless problems in Taiwan: Looking beyond legality toward social issues. *Housing Poverty, Homelessness, and the Transformation of Urban Governance in East Asian Cities, 1*(3), 165–173. https://doi.org/10.1016/j.ccs.2010.10.005

Crawford, Gemma, Connor, Elizabeth, McCausland, Kahlia, Reeves, Karina, & Blackford, Krysten (2022). Public health interventions to address housing and mental health amongst migrants from culturally and linguistically diverse backgrounds living in high-income countries: A scoping review. *International Journal of Environmental Research and Public Health, 19*(24), 16946. https://doi.org/10.3390/ijerph192416946

Dej, Erin, Gaetz, Stephen, & Schwan, Kaitlin (2020). Turning off the tap: A typology for homelessness prevention. *Journal of Primary Prevention, 41*(5), 397–412. https://doi.org/10.1007/s10935-020-00607-y

Deterding, N. M., & Waters, M. C. (2021). Flexible coding of in-depth interviews: A twenty-first-century approach. *Sociological Methods & Research, 50*(2), 708–739. https://doi.org/10.1177/0049124118799377

Edidin, J. P., Ganim, Z., Hunter, S. J., & Karnik, N. S. (2012). The mental and physical health of homeless youth: A literature review. *Child Psychiatry and Human Development, 43*(3), 354–375. https://doi.org/10.1007/s10578-011-0270-1

Fitri, Ifni Amanah (2019). Penanggulangan Gelandangan dan Pengemis di Indonesia (analisis program Desaku Menanti di Kota Malang, Kota Padang dan Jeneponto). *Share: Social Work Journal, 9*(1), 1. https://doi.org/10.24198/share.v9i1.19652

Fitzpatrick, Suzanne, & Stephens, Mark (2014). Welfare regimes, social values and homelessness: Comparing responses to marginalised groups in six European countries. *Housing Studies, 29*(2), 215–234. https://doi.org/10.1080/02673037.2014.848265

Fowler, Patrick J., Wright, Kenneth, Marcal, Katherine E., Ballard, Ellis, & Hovmand, Peter S. (2019). Capability traps impeding homeless services: A community-based system dynamics evaluation. *Journal of Social Service Research, 45*(3), 348–359. https://doi.org/10.1080/01488376.2018.1480560

Ganti, Mery, Yusuf, Husmiati, Wismayanti, Yanuar Farida, Setiawan, Hari Harjanto, Susantyo, Badrun, Nurhayu, Konita, Ita, Budiarti, Menik, & Sulubere, Muhammad Belanawane (2022). *The issues and social economic potentials of urban marginal groups in Indonesia* (pp. 246–259). https://doi.org/10.2991/978-2-494069-65-7_23

Harrison, R. L., Reilly, T. M., & Creswell, J. W. (2020). Methodological rigor in mixed methods: An application in management studies. *Journal of Mixed Methods Research, 14*(4), 473–495. https://doi.org/10.1177/1558689819900585

Huber, M. A., Brown, L. D., Metze, R. N., Stam, M., Van Regenmortel, T., & Abma, T. N. (2022). Exploring empowerment of participants and peer workers in a self- managed homeless shelter. *Journal of Social Work, 22*(1), 26–45. https://doi.org/10.1177/1468017320974602

Konduru, Laalithya (2019). Access clinic: A student-run clinic model to address gaps in the healthcare needs of the homeless population in adelaide. *Australian Journal of General Practice, 48*(12), 890–892. https://search.informit.org/doi/10.3316/informit.792950429214884. (Accessed 17 October 2023).

Kuntari, S., & Hikmawati, E. (2017). Melacak akar permasalahan gelandangan pengemis (gepeng). *Media Informasi Penelitian Kesejahteraan Sosial*, 11–26. https://doi.org/10.31105/mipks.v41i1.2272

Khayyatkhoshnevis, P., Choudhury, S., Latimer, E., & Mago, V. (2020). Smart city response to homelessness. *IEEE Access, 8*, 11380–11392. https://doi.org/10.1109/ACCESS.2020.2965557

Lastiwi, D. T., & Badruesham, N. (2017). *Library for the homeless: A case study of a shelter house and a school for the homeless in Indonesia and Malaysia*. https://library.ifla.org/id/eprint/2140.

Mago, Vijay K., Morden, Hilary K., Fritz, Charles, Wu, Tiankuang, Namazi, Sara, Geranmayeh, Parastoo, Chattopadhyay, Rakhi, & Dabbaghian, Vahid (2013). Analyzing the impact of social factors on homelessness: A fuzzy cognitive map approach. *BMC Medical Informatics and Decision Making, 13*(1), 94. https://doi.org/10.1186/1472-6947-13-94

Maryatun, Maryatun, & Muhammad Taftazani, Budi (2022). Policy for handling homeless beggars based on institutions for social functioning in need of social welfare services (PPKS): Study at social service homes for homeless beggars for homeless people mardi utomo Semarang. *Kebijakan: Jurnal Ilmu Administrasi, 13*(2). https://doi.org/10.23969/kebijakan.v13i2.5208. Juni 2022.

Merlindha, Astrini, & Hati, Getar (2017). Upaya rehabilitasi sosial dalam penanganan gelandangan dan pengemis di Provinsi DKI Jakarta. *Jurnal Ilmu Kesejahteraan Sosial, 16*(1). https://doi.org/10.7454/jurnalkessos.v16i1.67

Misje, Turid (2020). Social work and welfare bordering: The case of homeless EU migrants in Norway. *European Journal of Social Work, 23*(3), 401–413. https://doi.org/10.1080/13691457.2019.1682975

Mosley, J. E. (2021). Cross-sector collaboration to improve homeless services: Addressing capacity, innovation, and equity challenges. *The Annals of the American Academy of Political and Social Science, 693*(1), 246–263. https://doi.org/10.1177/0002716221994464

Murphy, Joseph F., & Tobin, Kerri (2012). Addressing the problems of homeless adolescents. *Journal of School Leadership, 22*(3), 633–663. https://doi.org/10.1177/105268461202200308. . (Accessed 3 June 2023)

Nasution, Lukman, Syamsuri, Abd Rasyid, & Ichsan, Reza Nurul (2021). Socialization of community participation in Bandar Khalifah village development planning Percut Sei Tuan district. *International Journal Of Community Service, 1*(2), 119–122. https://doi.org/10.51601/ijcs.v1i2.15

Ngo, A. N., & Turbow, D. J. (2019). Principal component analysis of morbidity and mortality among the United States homeless population: A systematic review and meta-analysis. *International Archives of Public Health and Community Medicine, 3*. https://doi.org/10.23937/2643-4512/1710025

O' Shaughnessy, Branagh, Manning, Rachel M., Greenwood, Ronni Michelle, Vargas-Moniz, Maria, Loubière, Sandrine, Spinnewijn, Freek, Gaboardi, Marta, Wolf, Judith R., Bokszczanin, Anna, Bernad, Roberto, Blid, Mats, Ornelas, Jose, & Group, The HOME-EU Consortium Study. (2021). Home as a base for a well-lived life: Comparing the capabilities of homeless service users in housing first and the staircase of transition in Europe. *Housing, Theory and Society, 38*(3), 343–364. https://doi.org/10.1080/14036096.2020.1762725

Phillips, Michael H., DeChillo, Neal, Kronenfeld, Daniel, & Middleton-Jeter, Verona (1988). Homeless families: Services make a difference. *Social Casework, 69*(1), 48–51. https://doi.org/10.1177/104438948806900108

Pierse, Nevil, Ombler, Jenny, White, Maddie, Aspinall, Clare, McMinn, Carole, Atatoa-Carr, Polly, Nelson, Julie, Hawkes, Kerry, Fraser, Brodie, Cook, Hera, & Howden-Chapman, Philippa (2019). Service usage by a New

Zealand Housing First cohort prior to being housed. *SSM - Population Health, 8*, 100432. https://doi.org/10.1016/j.ssmph.2019.100432

Rafferty, Yvonne (1995). The legal rights and educational problems of homeless children and youth. *Educational Evaluation and Policy Analysis, 17*(1), 39–61. https://doi.org/10.3102/01623737017001039

Rutenfrans-Stupar, Miranda, Van Der Plas, Bo, Den Haan, Rick, Regenmortel, Tine Van, & Schalk, René (2019). How is participation related to well-being of homeless people? An explorative qualitative study in a dutch homeless shelter facility. *Journal of Social Distress and the Homeless, 28*(1), 44–55. https://doi.org/10.1080/10530789.2018.1563267

Saewyc, Elizabeth M., Shankar, Sneha, Pearce, Lindsay A., & Smith, Annie (2021). Challenging the stereotypes: Unexpected features of sexual exploitation among homeless and street-involved boys in western Canada. *International Journal of Environmental Research and Public Health, 18*(11), 5898. https://doi.org/10.3390/ijerph18115898

Samari, Davood, & Groot, Shiloh (2023). Potentially exploring homelessness among refugees: A systematic review and meta-analysis. *Journal of Social Distress and the Homeless, 32*(1), 135–150. https://doi.org/10.1080/10530789.2021.1995935

Sim, J., & Waterfield, J. (2019). Focus group methodology: Some ethical challenges. *Quality and Quantity, 53*(6), 3003–3022. https://doi.org/10.1007/s11135-019-00914-5

Sinatra, R., & Lanctot, M. K. (2016). Providing homeless adults with advantage: A sustainable university degree program. *Education and Urban Society, 48*(8), 719–742. https://doi.org/10.1177/0013124514549832

Sisselman-Borgia, Amanda (2021). An adapted life skills empowerment program for homeless youth: Preliminary findings. *Child & Youth Services, 42*(1), 43–79. https://doi.org/10.1080/0145935X.2021.1884542

Somerville, Peter (2013). Understanding homelessness. *Housing, Theory and Society, 30*(4), 384–415. https://doi.org/10.1080/14036096.2012.756096

Stockemer, Daniel (2019). *Quantitative methods for the social sciences: A practical introduction with examples in SPSS and stata*. Springer International Publishing. https://doi.org/10.1007/978-3-319-99118-4

Suradi, Irmayani, Nyi R., Habibullah, Sugiyanto, Susantyo, Badrun, Mujiyadi, Benecdiktus, & Nainggolan, Togiaratua (2020). *Changes of poor family behavior through family development session* (pp. 22–26). https://doi.org/10.2991/assehr.k.200728.006

Temesváry, Zsolt, & Drilling, Matthias (2022). Beyond the state: Developments and trends in critical social work in Switzerland and Hungary. *International Social Work*. https://doi.org/10.1177/00208728211073792. . (Accessed 4 June 2023)

Yusuf, Husmiati, Setiawan, Hari Harjanto, Ganti, Mery, Wismayanti, Yanuar Farida, Susantyo, Badrun, Nurhayu, Konita, Ita, Budiarti, Menik, & Sulubere, Muhammad Belanawane (2022). *Access to basic needs for marginalized groups in Indonesia: A case study of the homeless and beggars* (pp. 237–245). https://doi.org/10.2991/978-2-494069-65-7_22

Zabkiewicz, Denise (2012). The vancouver at home study: Overview and methods of a housing first trial among individuals who are homeless and living with mental illness. *Journal of Clinical Trials, 02*(04). https://doi.org/10.4172/2167-0870.1000123

CHAPTER

22

Government policy transformation toward a reformed rental housing ecosystem as a mitigator of homelessness in postpandemic India

Ushnata Datta[1] *and Rewati Raman*[2]

[1]Mcgan's Ooty School of Architecture, Department of Architecture, Ooty, Tamil Nadu, India; [2]Indian Institute of Technology (BHU), Department of Architecture, Planning and Design, Varanasi, Uttar Pradesh, India

1. Introduction

Homelessness, extending beyond the lack of shelter to encompass exclusion and instability, has intensified in India following the pandemic. This crisis has exposed the housing sector's vulnerabilities, particularly in rental housing, challenging the country's social equity (Rogers & Power, 2020; Speak, 2019). In urban India, the pandemic's economic fallout has deepened the hardships of disadvantaged groups, spotlighting the inadequacies of the rental housing ecosystem—marked by high costs, regulatory hurdles, and informal markets (Mahadevia, 2013). This study arises from the pressing need to examine rental housing policies' role in mitigating homelessness in this new societal landscape.

This research investigates the transformation of India's rental housing policy in the postpandemic era as a strategy to alleviate homelessness. It scrutinizes existing policies, identifies gaps, and, informed by international benchmarks, suggests strategic reforms. The study's objective is to provide a comprehensive policy analysis to guide meaningful reforms in the rental housing sector. The study promises to enrich the dialog on homelessness and housing policy. It seeks to inform policymakers through a detailed analysis of the rental housing landscape and its intersection with homelessness, particularly in the postpandemic context. By integrating global best practices, the research aims to offer practical recommendations for policy reform, aiming to foster an inclusive and stable housing environment in India.

Homelessness to Hope
https://doi.org/10.1016/B978-0-443-14052-5.00022-7

2. Literature review

2.1 Overview of homelessness in India

Homelessness in India remains a critical issue, with the nation's rapid urbanization and economic expansion juxtaposed against persistent housing insecurity. Official figures report over 1.77 million homeless individuals (Chandramouli & General, 2011), a statistic considered to underestimate the true scope, as it fails to capture those in temporary or insecure dwellings (Dwivedi, 2021). This broader homelessness encompasses not only those on the streets but also those in inadequate housing, highlighting the multifaceted nature of this social challenge. The root causes of homelessness in India are diverse, involving economic factors like poverty, unemployment, and insufficient affordable housing, along with social marginalization and displacement from development projects (Dupont, 2008). Structural inequalities, land tenure complexities, and inaccessible housing markets aggravate the situation, disproportionately affecting vulnerable groups such as women, children, and the disabled (Tipple & Speak, 2009). Furthermore, environmental factors like natural disasters exacerbate housing instability, often leading to temporary or sustained homelessness (Gaillard et al., 2019). Addressing homelessness in India thus requires a holistic approach that integrates socioeconomic, institutional, and environmental strategies to create robust and inclusive housing policies.

2.1.1 *Impact of the COVID-19 pandemic on homelessness*

The COVID-19 pandemic exacerbated India's homelessness crisis, laying bare entrenched socioeconomic inequalities. Mandatory lockdowns, essential for public health, triggered an economic upheaval that disproportionately affected the homeless and those in precarious housing situations (Raju et al., 2021). The collapse of the informal sector, a significant employment source, heightened housing insecurity, leaving many without the financial means for shelter (Shekar & Mansoor, 2020). Homeless populations faced amplified health risks, as social distancing proved unfeasible in crowded shelters, increasing their susceptibility to the virus (Choudhari, 2020). The pandemic's strain on healthcare and social services further jeopardized this group's well-being, compounded by the digital divide that restricted access to critical online resources (Ian Litchfield et al., 2021). Consequently, the pandemic intensified the vulnerabilities of India's homeless, demanding urgent and inclusive policy interventions.

2.1.2 *Migrant workers and homelessness*

India's massive contingent of migrant laborers, the uncelebrated workhorses of its economic machinery, found themselves bearing the brunt of the pandemic's collateral damage. The abrupt imposition of lockdown measures, designed to arrest the virus's spread, left an estimated 140 million interstate and intrastate migrants stranded, bereft of income, sustenance, and, most critically, shelter (Bhagat, 2017; Ghosh et al., 2020). The resulting mass exodus of these migrants back to their rural homes constituted one of the largest peacetime migrations and starkly underscored their heightened vulnerability to homelessness (Khan & Arokkiaraj, 2021). Migrant workers, primarily engaged in construction, manufacturing, and informal service sectors, often subsist on daily wages and live in shared, makeshift accommodations or employer-provided housing. These living arrangements are typically characterized by substandard conditions and tenure insecurity (Kesar et al., 2021). The sudden loss of employment due to the lockdown measures meant the loss of these lodgings, effectively rendering a vast number of migrants homeless overnight.

For instance, a 2020 case study conducted by the Center for Policy Research in New Delhi provided empirical evidence of the vulnerabilities of migrant workers during the pandemic. The study documented that in the urban slums of Dharavi, Mumbai, many migrant workers who were employed in small manufacturing units lived in shared spaces with up to 10–15 individuals in a single room. When the lockdown was implemented, not only did they lose their jobs, but the enforced social distancing measures made their living conditions untenable, leaving them with no option but to return to their native places (Patel et al., 2020). Moreover, research by the Indian Institute for Human Settlements in Bengaluru highlighted the plight of migrant workers from the construction sector. The study revealed that most of these workers, without formal rental agreements, faced immediate eviction from their job-site living quarters upon the cessation of construction activities (Srivastava & Sutradhar, 2016). With limited savings and no immediate employment prospects, many were forced to seek shelter in overcrowded relief camps set up by local authorities, which posed significant health risks amidst the pandemic (Suresh et al., 2020).

The crisis illuminated the precariousness of migrant workers' living conditions, often disregarded due to their transient nature. It highlighted the crucial role of housing security as an aspect of employment and social protection for this group (Dubey et al., 2022). The mass reverse migration also had a ripple effect on rural areas, where the influx of returnee migrants amplified the existing strain on housing, food security, and public services. Many returnees found their ancestral homes in disrepair or occupied, thereby experiencing rural homelessness, a lesser-known yet equally pressing issue (Bhadra, 2021). In essence, the pandemic laid bare the extent of the housing crisis faced by migrant workers in India and underscored the urgent need for robust, inclusive, and flexible housing solutions that recognize their unique circumstances and needs.

2.2 Overview of the Indian rental housing policies and practices

Rental housing in urban India is a critical but troubled segment, vital for those financially precluded from homeownership or in need of temporary residence. The sector's growth is stunted by regulatory challenges and policy inconsistencies. Rent Control Acts designed to shield tenants have inadvertently disincentivized landlords due to rent caps and eviction difficulties, leading to a diminished rental stock and a reticent formal market (Mahadevia, 2013). Landlord-tenant conflicts, exacerbated by lengthy legal resolutions and perceived landlord vulnerabilities, further restrain the rental market's expansion (Gandhi et al., 2022). Predominantly informal, the sector is rife with unregulated agreements and inadequate living conditions, particularly impacting low-income and migrant groups who rely on the informality for its flexibility despite its inherent insecurities (Nijman, 2014). Historical policy oversight has resulted in a sector ill-equipped to address urban housing needs or homelessness effectively. India's rental housing requires a policy overhaul to revitalize its role in the housing ecosystem, ensuring it can adequately supplement homeownership and contribute to urban housing solutions.

2.2.1 *Rental housing demand and supply*

In India, burgeoning urbanization and internal migration fuel a pressing demand for rental housing, especially among urban poor, migrants, and young professionals seeking affordability and flexibility (Nijman, 2014). Despite such demand, outlined by the National Housing Bank of India, the supply, particularly of affordable rentals, is insufficient (National Housing Bank, 2022). Restrictive rent control laws aimed at tenant protection inadvertently dissuade landlords from renting, tightening the market (Mahadevia, 2013). Additionally, low rental yields, averaging 2%–3%, provide little incentive for property investment in the rental sector (Tiwari & Rao, 2016). Government housing policies, historically skewed toward homeownership, have neglected the rental sector, leaving a gap that private investment alone cannot fill. This imbalance has led to escalated rents and poor living conditions. Strategic policy reform is needed to bolster the rental market, offering incentives for landlords and integrating rental housing within broader housing programs.

2.2.2 *Key challenges in the Indian rental housing ecosystem*

2.2.2.1 Affordability

The affordability crisis is at the forefront of rental housing challenges. For instance, a case study from the slums of Mumbai highlighted that despite the squalid living conditions, the average rents could be as high as those in lower-middle-class neighborhoods (Desai & Yadav, 2007). The National Housing Bank of India (National Housing Bank, 2022) has quantified this burden, noting that a disproportionate amount of income is directed toward rent, leaving little for other necessities.

2.2.2.2 Security and discrimination

Tenant security is another challenge, with discrimination exacerbating the issue. In cities like Bengaluru, single individuals, especially women and those from marginalized communities, face significant barriers to accessing rental housing. Reports of rental refusals based on marital status, caste, or dietary preferences are not uncommon (Bhargava & Chilana, 2020; Oliveri, 2009; Peterman, 2018).

2.2.2.3 Living conditions

Living conditions in rental units, particularly within the informal sector, often fall below acceptable standards. An investigation into Delhi's informal settlements revealed that many tenants live in cramped quarters without proper ventilation or access to clean water (Bhan, 2019).

2.2.2.4 Regulatory complexities

The rental market is also plagued by regulatory complexities. Informal rental agreements dominate, leaving tenants without legal protection. A study conducted in Kolkata illustrated that disputes over eviction or rent increases could drag on for months or even years, largely due to the lack of standardized rental contracts (Anand & Rademacher, 2011).

Addressing these issues is critical to transforming India's rental housing ecosystem into an effective tool for mitigating homelessness. The provision of affordable, quality rental housing must be prioritized, alongside enhanced legal protections for tenants to heighten housing security and stability.

2.2.3 Rental housing and homelessness

The nexus between rental housing and homelessness is multifaceted and intricate. On the one hand, a robust and accessible rental housing market can significantly alleviate homelessness by providing affordable, flexible housing solutions to those unable or unprepared to venture into homeownership. This is particularly salient for migrant laborers, students, and single working individuals who require short-term economic accommodation (Nijman, 2014). However, on the other hand, an inefficient and unresponsive rental housing market can inadvertently contribute to housing instability and eventual homelessness. In India, high rents, fueled by the skewed supply-demand dynamics in the rental housing market, often result in overburdened tenants who struggle to balance rental expenses with other essential needs. The National Housing Bank in India (National Housing Bank, 2022) noted that escalating rents could precipitate housing instability, pushing vulnerable populations toward the brink of homelessness. Poor housing quality in the rental sector also plays a role in inducing homelessness. Substandard living conditions, lack of basic amenities, and overcrowding, common in low-cost rentals, can lead to health issues, psychological stress, and strained social relationships, making such housing untenable in the long run (Bhan, 2019). Furthermore, the prevalence of informal rental agreements and the associated lack of tenant protection can result in sudden evictions, catapulting tenants into homelessness. In the absence of a formal rental contract, tenants often have limited legal recourse in the event of disputes or unfair eviction, heightening their housing insecurity (Anand & Rademacher, 2011). Thus, while rental housing holds the potential to serve as a mitigator of homelessness, the present state of the rental housing ecosystem in India, marked by high rents, poor housing conditions, and weak tenant protection, often acts as a catalyst for homelessness. It underscores the critical need for comprehensive policy interventions to address these issues and reform the rental housing ecosystem.

2.3 Knowledge gaps

While the prevailing body of research offers valuable insights into the phenomena of homelessness and the rental housing landscape in India, it concurrently uncovers several areas of deficiency that underline the need for additional exploration. A pronounced knowledge gap exists concerning the potential interplay between rental housing policies and homelessness mitigation, particularly in the context of a postpandemic environment. Although the causes and consequences of homelessness in India are widely discussed (Alok & Vora, 2011; Bhan, 2019). The discourse often sidesteps the specific role of rental housing policies in shaping housing trajectories. Moreover, the bulk of the scholarship is concerned with the prepandemic state of rental housing, offering little illumination on how the realities of the COVID-19 pandemic may necessitate reshaping rental housing policies (Choudhari, 2020). With its far-reaching socioeconomic repercussions, the pandemic has markedly shifted the contours of homelessness and rental housing landscape. Yet, research tracing these shifts and advocating for policy adaptations remains scarce. Additionally, the potential of policy reform as a responsive measure to the escalated homelessness crisis in the aftermath of the COVID-19 pandemic remains underexplored. Most literature tends to underscore structural factors and direct interventions to address homelessness (Nijman, 2014). However, there is a dearth of scholarly attention on how policy transformations in the rental housing sector could provide a systemic solution to this persistent issue. Addressing these gaps, therefore, is of utmost importance. This study aims to contribute toward this end by situating rental housing policies at the center of the homelessness discourse and offering a nuanced understanding of the possibilities for mitigating homelessness in postpandemic India.

3. Assessment of existing government policies on rental housing

3.1 Timeline of rental housing policy evolution in India

India's rental housing policy has undergone significant evolution, marking the transition from a tenant-centric approach to a more balanced policy framework that addresses the needs of both tenants and landlords. Initially, post-independence policies like the Rent Control Act of 1948 aimed to shield tenants from exorbitant rents and unfair eviction, a response to the housing crisis of the era (Bertaud & Brueckner, 2005). However, such measures inadvertently discouraged property renting due to unattractive returns for landlords, skewing the market's supply-demand balance (D'souza, 2019). The 1990s introduced a paradigm shift with policies incentivizing rental housing through tax benefits and public-private partnerships. The National Housing and Habitat Policy of 1998 emphasized rental housing's significance for achieving "Shelter for All," acknowledging its role in urban housing provision (Planning Commission of India, 1998). The 2007 Policy further advanced this narrative, positioning rental housing as a key strategy for accommodating the urban poor and migrant workers (Ministry of Housing and Urban Poverty Alleviation, 2007). Recently, the Model Tenancy Act of 2021 marks a pivotal step, aiming to overhaul the rental market by balancing landlord and tenant rights, streamlining eviction and maintenance processes, and optimally utilizing vacant properties (Ministry of Housing and Urban Affairs, 2021). As reflected from the analysis in Table 22.1, this progression reflects a maturing understanding of the rental market's intricacies, emphasizing the need for a dynamic and responsive regulatory environment to tackle the challenges of housing shortages and homelessness effectively.

3.2 Critical analysis of the rental housing policies

India's rental housing policies have been characterized by a dichotomy, with the Rent Control Act of 1948 primarily focusing on tenant protection, and the more recent Model Tenancy Act 2020 seeks to balance the rights and responsibilities of tenants and landlords. Despite the intent behind these policies, their implications reveal a series of challenges and unintended consequences, often contributing to the issue of homelessness (See Table 22.2).

3.2.1 *The Rent Control Act, 1948*

The Rent Control Act was instituted shortly after India's independence, during acute housing shortage and widespread distress. The primary aim was to protect tenants from exploitative practices by landlords, such as arbitrary rent hikes and evictions (Bertaud & Brueckner, 2005). The Act capped rental rates, ensuring tenants' affordability, and provided robust eviction protection. While these measures offered considerable protection to existing tenants, they

TABLE 22.1 Timeline of rental housing policy evolution in India.

Year	Policy/legislation	Key features
1948	Rent Control Act	• Capped rental rates • Provided strong tenant protection against eviction
1990s–Early 2000s	Various State and Central Policies	• Tax benefits to landlords • Encouragement of public-private partnerships for rental housing construction
2007	National Urban Housing and Habitat Policy, 2007	• Emphasized rental housing as a solution for urban poor and migrant workers • Encouraged public and private sector participation in rental housing
2020	Draft Model Tenancy Act	• Aimed at balancing the rights and responsibilities of landlords and tenants • Encouraged growth of the formal rental housing market

TABLE 22.2 Comparative analysis of rental housing policies in India.

Policy	Focus	Key provisions	Impact	Challenges
Rent Control Act, 1948	Tenant Protection	Capped rental rates and robust eviction protection.	Provided affordability and security for existing tenants, contributed to shrinkage and degradation of rental housing stock.	Discouraged landlords from renting due to low returns, which led to the dilapidation of rental housing stock.
National Urban Housing and Habitat Policy, 2007	Holistic Approach	Advocated for rental housing, proposed fiscal incentives, public–private partnerships, and rental assistance.	Recognized rental housing as a viable strategy for urban housing but faced implementation challenges.	Limited financial resources, and difficulty in mobilizing private sector participation.
Model Tenancy Act, 2020	Balanced Approach	Capping security deposits, expedited dispute resolution, separate rent authority.	Aims to balance the rights and responsibilities of tenants and landlords, could potentially formalize the rental housing market.	Yet to be implemented nationwide, enforcement and awareness challenges.

inadvertently led to a shrinkage of the rental housing market. The Act has been widely criticized for its negative impacts on the housing sector. Firstly, capping rental rates disincentivized landlords from renting out their properties due to low returns, contributing to a supply-demand mismatch in rental housing (Alok & Vora, 2011; Dev, 2006; Dutta et al., 2022; More, 2021; Nath, 1984). Secondly, the Act led to a degradation of the rental housing stock. Since landlords received low rental income and tenants lacked ownership rights, neither party was incentivized to maintain and invest in property upkeep. As a result, many rental properties suffered from dilapidation and neglect.

3.2.2 *The National Urban Housing and Habitat Policy, 2007*

The National Urban Housing and Habitat Policy (NUHHP), 2007 marked a progressive shift in India's housing policy landscape. Recognizing the importance of rental housing in addressing the housing needs of urban poor and migrant workers, it proposed an array of measures to bolster the rental housing sector (Ministry of Housing and Urban Poverty Alleviation, 2007). These included fiscal incentives for rental housing development, public-private partnerships, and rental assistance for vulnerable groups. However, the policy's implementation witnessed numerous challenges. Limited financial resources and difficulties in mobilizing private sector participation in rental housing development were some of the key constraints in the execution of the NUHHP, 2007. Despite recognizing rental housing as a key strategy, the policy failed to articulate a comprehensive framework for rental housing development, thereby leading to insufficient implementation. Furthermore, the lack of a concrete action plan for rental assistance and weak regulations for tenant protection further undermined its impact on mitigating homelessness.

3.2.3 *The Model Tenancy Act, 2020*

The Draft Model Tenancy Act, though not yet implemented nationwide as of 2023, marks a critical step toward addressing the longstanding issues in India's rental housing sector. It proposes a balanced approach, striving to protect the interests of both tenants and landlords. Some of the key provisions include a cap on security deposits, the establishment of a separate rent authority, and provisions for expedited dispute resolution (Ministry of Housing and Urban Affairs, 2021). However, its potential impact on the rental housing market and homelessness remains to be seen. The Act's effectiveness will largely depend on how its provisions are translated into practice. This would require addressing challenges such as enforcement mechanisms, capacity building for dispute resolution, and ensuring tenant and landlord awareness about their rights and responsibilities.

3.2.4 *Implications for homelessness*

Rental housing policies have a direct bearing on homelessness. Policies that disincentivize rental housing or result in the deterioration of rental housing stock contribute to homelessness by reducing the availability of affordable and habitable housing. On the other hand, effective rental housing policies can potentially reduce homelessness by offering secure and affordable housing options to low-income households and vulnerable groups (Phang, 2018). In the Indian context, despite the various rental housing policies implemented, homelessness continues to be a pressing issue. This could be attributed to the policy deficiencies and the lack of a comprehensive approach to addressing homelessness by providing affordable rental housing. Therefore, it is crucial to devise and implement rental housing policies that address these gaps and effectively mitigate homelessness. The evolution of rental housing policies in India reflects the growing recognition of rental housing as a critical component of the housing strategy. However, the policies also illustrate the complexities and challenges associated with rental housing provision. These range from ensuring tenant protection and affordability, incentivizing landlords, and maintaining housing stock quality to effectively implementing policy measures. Each policy intervention has had its merits and shortcomings, thereby underlining the need for a balanced, comprehensive, and nuanced approach in formulating and implementing future rental housing policies.

3.3 Potential of rental housing as a mitigator of homelessness in India

Homelessness is a multidimensional issue with complex economic, social, and policy dimensions. In the Indian context, the absence of affordable and accessible housing options, especially rental housing, significantly contributes to homelessness. Despite this, rental housing as a potential mitigator of homelessness has often been overlooked in policy discussions. In theory, a well-functioning rental housing market can provide affordable housing options for low-income households, thereby reducing homelessness. On the other hand, an inefficient rental market can contribute to housing instability and homelessness (Phang, 2018). This section analyses the potential of rental housing as a mitigator of homelessness in India, evaluating current barriers and prospects.

3.3.1 *Rental housing: A vital housing option for vulnerable groups*

The importance of rental housing as a key option, particularly for vulnerable sections of the population, such as low-income households, migrant workers, and homeless individuals, cannot be overstated. It offers a flexible and affordable housing alternative for those who cannot afford homeownership or are in transient phases of their life, including students, young professionals, and migrant laborers (Mahadevia, 2013). A well-regulated and affordable rental housing market could conceivably mitigate homelessness by providing secure, dignified, affordable housing options to these vulnerable sections. It could also address "hidden homelessness," a scenario where individuals live in highly insecure and inadequate housing conditions, often sharing cramped spaces with others due to the unavailability of affordable housing options.

3.3.2 *Current barriers in rental housing market*

Despite its potential, the rental housing market in India is characterized by a myriad of challenges. Some of the key barriers include regulatory ambiguities, low rental yields, stringent rent control laws, affordability issues, and a widespread informal rental market (Mahadevia, 2013; Sethi & Mittal, 2020). Another significant challenge is the lack of adequate tenant protection, which often leaves tenants, especially those from vulnerable sections, at the mercy of landlords. In many cases, tenants face high rents, substandard living conditions, and the constant threat of eviction without legal recourse (Anand & Rademacher, 2011). The inadequate supply of rental housing for low-income households is another critical concern. While there is a growing trend of high-end rental housing in urban areas, the supply of affordable rental housing remains significantly constrained. This supply-demand mismatch further exacerbates the issue of homelessness among low-income households.

3.3.3 *Future prospects*

Tackling these challenges would require a comprehensive, multi-pronged approach that addresses supply and demand-side issues. On the supply side, policy measures are needed to incentivize landlords to rent out their properties. These could include rental income tax benefits, easing regulatory restrictions, and legal provisions to protect landlords' rights (Ministry of Housing and Urban Affairs, 2021). On the demand side, there is a need to enhance the affordability and accessibility of rental housing for low-income households and vulnerable groups. This could involve measures such as rental assistance programs, affordable housing

TABLE 22.3 Potential of rental housing as a mitigator of homelessness.

Aspect	Barriers/current practices	Prospects
Regulatory	Ambiguities, low rental yields, stringent rent control laws	Easing regulatory restrictions, protecting landlords' rights
Supply-side	Insufficient affordable rental housing, degradation of rental housing stock	Incentives for landlords, use of vacant properties, affordable housing construction
Demand-side	Affordability issues, lack of tenant protection	Rental assistance programs, strengthening tenant rights
Innovation	Traditional rental housing models	Alternative rental housing models, use of technology

construction by public-private partnerships, and strengthening legal protections for tenants (Ministry of Housing and Urban Poverty Alleviation, 2007). Moreover, fostering innovation in rental housing delivery could be a key strategy. This could include developing alternative rental housing models like community land trusts, cooperative housing, and shared housing. These models could provide affordable and secure housing options, thereby mitigating homelessness (Bates, 2022). Additionally, leveraging the existing housing stock, including vacant and underutilized properties, could help expand the supply of affordable rental housing. Policy interventions could aim to bring these properties into the rental market through fiscal incentives and regulatory reforms (See Table 22.3).

While there are significant challenges in the rental housing market, it holds considerable potential as a mitigator of homelessness in India. However, realizing this potential would require nuanced and comprehensive policy interventions that address existing barriers and leverage future prospects. This calls for a renewed focus on rental housing in policy discussions and urban housing strategies to effectively address homelessness in the country.

4. Case studies: International best practices

In order to understand potential effective strategies and tactics to improve India's rental housing sector, it is instructive to examine case studies of nations with successful policies and strategies in place. This section will delve into the rental housing frameworks of Germany, Singapore, and the United States to glean best practices and learnings.

4.1 Germany: A balanced rental housing market

Germany provides an interesting case study due to its high rate of rental housing and a balanced, well-regulated rental housing market. Approximately 60% of households in Germany reside in rented accommodation, reflecting one of the highest rates worldwide (Bates, 2022; Easthope, 2014; Kemeny, 2002; Kemp & Kofner, 2010; Kohl, 2017). This high prevalence of renting is supported by a comprehensive legal framework that ensures a fair balance of rights and obligations between landlords and tenants, providing robust protections for tenants without unduly infringing on the rights of landlords (Easthope, 2014; Kemeny, 2002). In contrast to many other countries where rental housing is often seen as a stepping stone to homeownership, in Germany, renting is widely accepted as a long-term housing solution. A strong social housing sector supports this cultural acceptance of renting, which provides affordable rental housing to low-income households (Bates, 2022; Kemp & Kofner, 2010). Furthermore, German local governments play a significant role in housing provision, with many acting as landlords in social housing initiatives. This public involvement in the rental housing sector helps to ensure a supply of affordable rental housing and prevents the exclusion of low-income households from the housing market (Kemeny, 2002). In terms of addressing homelessness, the German rental housing model offers a possible pathway. A well-regulated rental market, combined with strong tenant protections and a robust social housing sector, could significantly contribute to mitigating homelessness by providing secure, affordable housing options.

4.2 Singapore: Public rental scheme as a solution

Singapore's housing policy is unique in its emphasis on public rental schemes. The country's Housing and Development Board (HDB) provides heavily subsidized rental units to lower-income households who cannot purchase a flat and have no other viable housing options (Phang, 2018). These public rental schemes have significantly reduced homelessness in Singapore and ensured that low-income households have access to affordable housing. The programs are designed to be inclusive, ensuring that even the poorest segments of the population have access to decent, affordable housing. The success of Singapore's public rental schemes can be attributed to several factors, including a strong political commitment to housing provision, the effective use of state resources, and the central role of HDB in overseeing the housing sector (Addae-Dapaah & Juan, 2014; Phang, 2018). The Singapore model illustrates the potential of public rental schemes as a tool for addressing homelessness and providing affordable housing.

4.3 The United States of America: Housing Choice Vouchers

The United States offers a different approach to rental housing policy, strongly emphasizing rental subsidies. The Housing Choice Voucher Program, also known as Section 8, provides subsidies to eligible low-income families, the elderly, and people with disabilities to enable them to afford decent, safe, and sanitary housing in the private market (Williamson et al., 2009). As of 2023, the program assists over 2.2 million households in the United States, making it one of the country's key tools for preventing homelessness and ensuring housing affordability. The success of the Housing Choice Voucher Program illustrates the potential of rental subsidies as a policy tool for addressing housing affordability and homelessness. The program's flexibility, allowing recipients to choose where they live, and the fact that the subsidy adjusts with income and rental market conditions ensure that assisted households can afford their housing costs, thus reducing the risk of homelessness (Kathleen Moore, 2016; Seicshnaydre, 2016; Williamson et al., 2009).

TABLE 22.4 Key learnings from international case studies.

Country	Key learnings
Germany	Balanced rental market, strong tenant protections, the active role of local governments in housing provision
Singapore	Public rental schemes as a tool for reducing homelessness and providing affordable housing
United States of America	Value of rental subsidies as a tool for ensuring housing affordability and preventing homelessness

4.4 Key learnings from case studies

The case studies of Germany, Singapore, and the United States provide several key learnings for India's rental housing policies. From Germany, we learn the importance of a well-regulated rental market, strong tenant protections, and an active role of local governments in housing provision. From Singapore, we learn the potential of public rental schemes as a tool for reducing homelessness and providing affordable housing. And from the United States, we learn the value of rental subsidies as a flexible and effective tool for ensuring housing affordability and preventing homelessness. These international experiences underscore the potential of rental housing as a tool for addressing homelessness. They also highlight the importance of a comprehensive and nuanced approach to rental housing policy, considering the needs and rights of both landlords and tenants, the importance of affordability, and the potential of both public and private sector involvement in the housing sector (See Table 22.4).

As India continues to grapple with its housing challenges, these international experiences offer valuable insights for policymakers. They underline the potential of rental housing as a tool for addressing homelessness and provide practical examples of successful rental housing policies. The key is to adapt these strategies to the Indian context, taking into account the unique challenges and opportunities of the country's housing sector.

5. Recommendations for rental housing policy transformations for the homeless

In shaping India's rental housing policy recommendations, it's essential to distinguish between Western practices and the Global South's realities. Western nations typically integrate rental housing policies within a broader social welfare context, ensuring tenant protections through subsidies and legal frameworks, as seen in Germany's social housing programs (Jones & Murie, 2008). India, representative of the Global South, contends with rapid urbanization and fiscal limitations, resulting in a rental market dominated by informal arrangements and scant regulation. The lack of standardized contracts leads to disputes often settled outside legal channels (Bhan, 2019). Furthermore, the data-driven policymaking common in the West is constrained in India by data scarcity, affecting tailored policy development. Adapting Western insights to India's context involves crafting policies that navigate the nuances of its rental market, incorporating regulatory needs with the realities of informal tenancies. Beyond legislation, India needs innovative, culturally cognizant strategies that align with its socio-economic landscape to address homelessness effectively.

5.1 Potential policy changes for a reformed rental housing ecosystem

The complexities of homelessness and housing instability in India necessitate a nuanced, multifaceted approach to reforming the rental housing ecosystem. This reform should consider the broad array of factors contributing to housing instability and homelessness, such as eviction, unaffordability, unemployment, and other social and economic vulnerabilities. Here are potential policy changes that could be implemented to reform India's rental housing ecosystem.

5.1.1 Enhanced legal protection for tenants

One potential reform to consider is strengthening the legal protections afforded to tenants. The existing legal framework in many parts of India still does not offer sufficient protections against forced evictions, a significant contributor to homelessness. Enhancing tenants' legal rights could include prohibiting "no-cause" evictions, ensuring due process in eviction proceedings, and implementing rent controls or stabilization measures to prevent excessive rent increases. By enhancing tenant protections, policymakers can help maintain housing stability and prevent situations that may lead to homelessness.

5.1.2 Prioritizing affordable rental housing

A second potential reform is to prioritize affordable rental housing in housing policies. While the Model Tenancy Act represents a significant step toward encouraging the growth of the formal rental housing market, there is a need for policies that specifically target low-income and vulnerable groups. These could include the provision of subsidized rental housing, the development of social rental housing projects, or the use of housing vouchers to assist low-income households in accessing rental housing in the private market. Prioritizing affordable rental housing can help bridge the gap between the high cost of housing and the limited incomes of vulnerable households, thereby reducing housing insecurity and homelessness.

5.1.3 Improving housing quality

In addition to affordability, the quality of rental housing is another area where policy reforms could be beneficial. Poor housing conditions can contribute to health problems and instability, while high-quality, affordable housing provides a stable platform for employment, education, and other opportunities. Improving housing quality could involve stricter enforcement of building codes, subsidies or tax incentives for property improvements, or programs to upgrade informal housing settlements.

5.1.4 Facilitating access to support services

Finally, policy reforms could also improve access to support services for tenants, particularly those at risk of homelessness. This could include legal aid for tenants facing eviction, financial counseling or assistance for tenants struggling to pay rent, and social services for tenants dealing with issues such as mental health problems or substance abuse. Facilitating access to these services can help prevent homelessness and ensure that tenants have the resources they need to maintain stable housing. In sum, reforming the rental housing ecosystem in India requires a comprehensive, multi-pronged approach. Enhancing legal protections for tenants, prioritizing affordable rental housing, improving housing quality, and facilitating access to support services are potential policy changes that, if implemented, could significantly contribute to mitigating homelessness and improving housing stability in India. These policy recommendations are informed by international best practices and adapted to the specific challenges of the Indian context, ensuring their relevance and potential for impact.

5.2 Promoting the use of vacant housing stock and unsold housing inventories

The paradox of substantial vacant housing stock and unsold housing inventories coexisting with rampant homelessness and housing insecurity presents a significant, yet untapped, opportunity for mitigating the housing crisis in urban India. As per the 2011 Census, there were approximately 11 million vacant units in urban India (Chandramouli & General, 2011). These vacant houses could be harnessed to alleviate homelessness, but doing so necessitates policies and initiatives that encourage homeowners to rent out these properties.

5.2.1 *Tax incentives for homeowners*

One potential measure is the introduction of tax incentives for homeowners who rent out their vacant properties. The government could, for instance, offer property tax reductions or income tax deductions for rental income earned from these properties. Such incentives would lower the financial barriers to renting out properties and provide a tangible monetary benefit for homeowners, encouraging them to participate in the rental market.

5.2.2 *Streamlining the property letting process*

In addition to financial incentives, bureaucratic and procedural simplifications could be instrumental in mobilizing vacant housing stock. Often, the process of letting a property is fraught with complexity and inefficiency, dissuading many homeowners from renting out their properties. Streamlining the process of property letting through initiatives such as online platforms for registration and verification of rental agreements, expedited dispute resolution mechanisms, and straightforward procedures for eviction could incentivize homeowners to let their vacant properties.

5.2.3 *Adequate legal protection for landlords*

Enhancing legal protections for landlords is another avenue to explore. One of the concerns that homeowners often cite when hesitant to rent out their properties is the risk of problematic tenants and the subsequent difficulties in eviction. Strengthening landlord rights within the context of a balanced tenant–landlord relationship could reassure homeowners and make the prospect of renting out properties more attractive.

5.2.4 *Use of unsold housing inventories*

Furthermore, India's real estate sector has a substantial volume of unsold housing inventories. Policies encouraging real estate developers to rent out these unsold properties could help alleviate housing shortages. Mechanisms such as Real Estate Investment Trusts could be more actively promoted, allowing developers to put their unsold inventory into these trusts and earn rental income (Pricewaterhouse Coopers and the Urban Land Institute, 2023). By adopting measures to mobilize vacant and unsold housing stock, the government can tap into an existing resource to mitigate homelessness and enhance housing security. In doing so, it would be crucial to maintain a balance between incentivizing homeowners and protecting the rights of the tenants to ensure the sustainability and success of such initiatives.

5.3 Encouraging the development of rental housing for migrant workers

The COVID-19 pandemic unveiled the extent of the precarious housing conditions faced by migrant workers in urban India, thus underscoring an urgent need to provide secure and affordable housing for this vulnerable group. A substantial number of these workers reside in informal settlements or shared accommodations that lack basic amenities, exposing them to various health and safety risks (Sarkar & Mishra, 2020). This issue demands specific policy attention to create suitable rental housing options in urban centers, where the majority of migrant workers seek employment.

5.3.1 *Public–private partnerships for rental housing development*

One strategy to address the housing needs of migrant workers involves fostering public–private partnerships (PPPs) for rental housing development. Such partnerships can leverage the efficiency and resources of the private sector while ensuring affordability through government subsidies or incentives. This collaboration can facilitate large-scale, affordable rental housing projects targeting migrant workers. However, the design of these initiatives should consider the unique requirements of migrant workers, such as the need for flexible leasing arrangements and the provision of shared accommodation.

5.3.2 *Flexible housing models*

Given the transient nature of migrant work, housing solutions for migrant workers must incorporate flexibility. Traditional rental agreements may not suit migrant workers who often have to relocate based on employment opportunities. Consequently, policies should encourage flexible housing models, such as short-term leases, dormitory-style accommodations, and pay-per-use models, that cater to the specific housing patterns of migrant workers. Such models can be facilitated by PPPs or incentivized through fiscal measures, such as tax benefits for developers that provide flexible rental options.

5.3.3 *Legislation to protect migrant workers' housing rights*

The implementation of legislation that protects the housing rights of migrant workers is another necessary step. Despite their significant contribution to India's urban economy, migrant workers often face discrimination and exclusion in the housing market. Policies should ensure nondiscriminatory access to housing and provide legal safeguards against unfair rental practices. The development of rental housing for migrant workers is a critical yet complex task that requires a multifaceted approach. By facilitating public-private partnerships, encouraging flexible housing models, and legislating protections for migrant workers, policy changes could significantly alleviate the housing insecurity experienced by this demographic, thereby contributing to a more inclusive and equitable urban housing landscape.

5.4 Participation models for the private sector

Given its capacity to mobilize resources and execute large-scale projects, the private sector plays an essential role in expanding the supply of rental housing. However, for this potential to be realized, it is important to establish an enabling environment that encourages private investment in rental housing projects, particularly those aimed at providing affordable housing for low-income households. This could involve various participation models, ranging from tax incentives to innovative operational frameworks such as Build-Operate-Transfer.

5.4.1 Regulatory reforms and incentives

Regulatory reforms that provide fiscal incentives to private developers could be implemented to stimulate supply in the rental housing market. For instance, tax relief or subsidies for developers who construct affordable rental housing projects can make these projects more financially viable. Such incentives could help bridge the affordability gap and encourage the private sector to invest in rental housing, leading to an increased supply that caters to a wider demographic, particularly those at risk of homelessness.

5.4.2 Build-Operate-Transfer (BOT) model

In addition to fiscal incentives, innovative operation models can also help engage the private sector in providing rental housing. The BOT model is one such mechanism where a private entity is granted the right to finance, build, and operate a rental housing project for a certain period, after which the ownership is transferred back to the public sector. This approach allows private developers to recoup their investment and earn a return during the operation phase, thus making rental housing projects more attractive. Furthermore, the BOT model ensures that the projects revert to public ownership after a specified period, thus safeguarding the long-term availability of affordable rental housing (Pricewaterhouse Coopers and the Urban Land Institute, 2023).

5.4.3 Public–private partnerships

Public–Private Partnerships (PPP) can also play a significant role in leveraging the efficiency and resources of the private sector while ensuring affordability and accessibility. This model involves government and private companies collaborating to finance, build, and manage rental housing projects. The government's role in these partnerships is often to provide land or subsidies, thereby reducing the financial risks for private developers and encouraging their participation (Requena et al., 2021). In conclusion, to address the housing crisis and homelessness in India, it is crucial to involve the private sector more substantially. This could be achieved through a combination of regulatory reforms, innovative operation models, and fostering collaborations through public–private partnerships.

5.5 Fostering innovation in rental housing design and delivery

In the face of a burgeoning housing crisis, traditional approaches to housing provision are proving insufficient. Addressing this issue effectively will require fostering innovation in both rental housing design and delivery. This can encompass several dimensions, including novel construction techniques, shared housing models, digital platforms, and even social innovations that challenge traditional conceptions of what constitutes "housing."

5.5.1 *Innovative design approaches*

At the level of design, modular or prefabricated construction stands out as a promising approach. This involves constructing parts of homes (modules) in a factory setting before transporting them to the building site for assembly. Such a method reduces construction time, minimizes wastage of materials, and, importantly, can substantially lower construction costs, making the resulting homes more affordable. Similarly, shared housing models, wherein facilities such as kitchens and living areas are communal, can also contribute to affordability while fostering social cohesion. This can be particularly suitable for certain demographics, such as students, single professionals, or migrant workers who might value affordability and sociability over private space.

5.5.2 *Innovative delivery mechanisms*

In terms of delivery, digital platforms can play a transformative role. These platforms can match potential tenants with suitable rental housing options in a way that is efficient and transparent. Such platforms can also streamline administrative processes, like payment of rent or resolution of disputes, further enhancing the ease and accessibility of rental housing.

5.5.3 *Creating an enabling environment for innovation*

For these innovations to materialize, policymakers will need to create an enabling environment. This could involve, for example, funding for research and development into innovative housing design and construction techniques. It might also mean setting up "innovation labs" where novel ideas can be tested and refined. Furthermore, incentivizing the adoption of innovative practices, perhaps through tax breaks or priority in granting building permissions, could encourage the private sector to invest in innovative approaches (Kayanan, 2022; Yigitcanlar et al., 2020). In conclusion, fostering innovation in the design and delivery of rental housing could offer new ways to tackle the challenge of homelessness in India. This will require a concerted effort from policymakers, the private sector, and civil society to create an ecosystem that encourages and rewards innovative solutions.

5.6 Strengthen the role of the informal rental housing sector

The informal rental housing sector is a significant yet understudied facet of urban housing in India. Comprising a diverse array of living arrangements, from slum tenancies to rooms rented out by homeowners, this sector provides shelter for a large number of low-income households. Despite its importance, the informal rental sector tends to be overlooked in policy discussions, resulting in its potential being largely untapped. There is a pressing need to rectify this oversight, especially in the context of addressing homelessness in India.

5.6.1 Acknowledging the informal rental housing sector

The first step toward strengthening the informal rental housing sector is its acknowledgment in policy discussions. Policymakers need to recognize that the formal housing market alone cannot meet the housing needs of the burgeoning urban population, particularly those on low incomes. The informal rental sector can play a crucial role in plugging this gap, providing affordable and flexible housing options that are well-suited to the economic realities of many urban dwellers.

5.6.2 Legal recognition of informal rental agreements

Once recognized, measures can be taken to integrate the informal rental sector into the formal housing framework. One approach would be to provide legal recognition to informal rental agreements. This would not only protect the rights of tenants and landlords but also create a level of certainty and stability that is often lacking in informal rental arrangements. Legal recognition would also pave the way for tenants in the informal sector to access benefits like rental subsidies.

5.6.3 Upgradation of informal rental housing

Furthermore, the physical conditions of informal rental housing often leave much to be desired. Substandard housing not only has detrimental effects on the health and well-being of tenants but also reinforces their vulnerability and marginalization. Policymakers should consider promoting the upgradation of informal rental housing. This could involve providing subsidies or micro-credit facilities to landlords willing to invest in improving their properties. Such initiatives would enhance living conditions and potentially attract more landlords to the rental market, increasing the supply of affordable rental housing. In conclusion, the informal rental housing sector is critical in providing shelter for a significant segment of India's urban population. It is time that this sector is acknowledged and strengthened with policies that provide legal recognition, promote housing upgradation, and ultimately contribute toward mitigating homelessness in India.

5.7 Development of rent subsidy programs

Rent subsidies or housing vouchers are essential policy instruments in many developed countries to make housing affordable for low-income households. This form of rental assistance contributes to poverty reduction and prevents homelessness by enabling these households to access rental housing in the private market. With its acute housing affordability issue, India could benefit significantly from introducing similar rent subsidy programs.

5.7.1 The rationale for rent subsidy programs

The basic rationale for rent subsidy programs is straightforward: bridging the affordability gap. Housing is a fundamental need and a human right. However, many low-income households in Indian cities are unable to afford decent housing, leading to precarious living conditions and potential homelessness. By subsidizing rents, these programs enable households to access a broader range of housing options, thereby ensuring that even the most economically disadvantaged citizens have access to safe, adequate, and affordable housing.

5.7.2 *Potential models for India*

The design and implementation of rent subsidy programs can vary depending on a country's specific context and needs. In the case of India, policymakers could draw lessons from the Housing Choice Voucher program in the United States. Under this program, eligible households pay a portion of their income (typically 30%) toward rent, with the remainder covered by the government. A similar program, adapted to the Indian context, could provide a crucial safety net for low-income households, ensuring they have access to decent housing without compromising other essential needs.

5.7.3 *Implementation challenges*

While the benefits of rent subsidy programs are well-documented, their implementation in India would undoubtedly pose several challenges. These include determining eligibility criteria, establishing a reliable payment mechanism, ensuring sufficient funding, and preventing fraud and abuse. However, these challenges are not insurmountable and can be addressed through careful policy design and robust implementation mechanisms. International experiences in rent subsidy programs could offer valuable lessons for India in this regard. In conclusion, developing rent subsidy programs could be a critical step toward improving housing affordability for low-income households in India and mitigating the problem of homelessness. The success of such programs in other countries provides a strong rationale for India to consider adopting similar initiatives.

6. Future directions in policy, practice, and research on rental housing for mitigating homelessness

To effectively address homelessness, a holistic policy approach is imperative, one that transcends mere shelter provision to tackle poverty, social exclusion, and mental health issues. A National Rental Housing Strategy could serve as a cornerstone for such a policy, encompassing tenant rights protection, affordable housing creation, rental assistance, and incentives for utilizing vacant properties. Practically, policy translation into tangible actions demands robust funding, capable institutions, decisive governance, and collaboration across stakeholders, including pivotal roles for NGOs and CBOs in bridging policy-practice gaps. Research-wise, a dire need exists for in-depth studies that not only quantify homelessness but also qualitatively dissect the lived realities of affected individuals and the systemic barriers in accessing housing. Comprehensive efforts across policymaking, practical application, and research are essential to forge meaningful progress toward eradicating homelessness in India.

7. Conclusion

The role of rental housing in addressing the grave issue of homelessness in India cannot be overstated. As we have elucidated in this paper, a comprehensive and well-articulated approach to rental housing can significantly contribute to mitigating homelessness, thereby fulfilling India's commitment to "Housing for All." This approach not only necessitates the construction of additional affordable housing units but also requires innovative and multifaceted interventions. The case studies of Germany, Singapore, and the United States illustrate how diverse strategies in rental housing policies can be effectively deployed to create balanced and accessible rental markets. Adopting key learnings from these international best practices and aligning them with India's unique socioeconomic context could be a potent approach to housing policy formulation. Our analysis of the Indian rental housing landscape brings forth several areas of intervention, from fostering the development of rental housing for migrant workers to bolstering the role of the informal rental housing sector. Moreover, the private sector's participation, the promotion of innovative rental housing design and delivery, and the development of rent subsidy programs are other viable strategies that merit attention. Therefore, creating a dynamic, inclusive, and efficient rental housing market is instrumental in India's pursuit to eradicate homelessness. As India strides ahead on its path of economic growth, it is of paramount importance to ensure that no citizen is left without a secure and affordable place to call home.

References

Addae-Dapaah, Kwame, & Juan, Quah Shu (2014). Life satisfaction among elderly households in public rental housing in Singapore. *Health, 6*, 1057–1076.

Alok, Aditya, & Vora, Pankti (2011). Rent control in India-obstacles for urban reform. *NUJS Law Review, 4*, 83.

Anand, Nikhil, & Rademacher, Anne (2011). Housing in the urban age: Inequality and aspiration in Mumbai. *Antipode, 43*(5), 1748–1772. https://doi.org/10.1111/j.1467-8330.2011.00887.x

Bates, Lisa K. (2022). Housing for people, not for profit: Models of community-led housing. *Planning Theory & Practice, 23*(2), 267–302. https://doi.org/10.1080/14649357.2022.2057784

Bertaud, Alain, & Brueckner, Jan K. (2005). Analyzing building-height restrictions: Predicted impacts and welfare costs. *Regional Science and Urban Economics, 35*(2), 109–125. https://doi.org/10.1016/j.regsciurbeco.2004.02.004

Bhadra, Subhasis (2021). Vulnerabilities of the rural poor in India during pandemic COVID-19. *Asian Social Work and Policy Review, 15*(3), 221–233. https://doi.org/10.1111/aswp.12236

Bhagat, R. (2017). *Migration and urban transition in India: Implications for development.* https://doi.org/10.13140/RG.2.2.17888.17925

Bhan, Gautam (2019). Notes on a Southern urban practice. *Environment and Urbanization, 31*(2), 639–654. https://doi.org/10.1177/0956247818815792

Bhargava, Rashi, & Chilana, Richa (2020). A flat of my own: Singlehood and rental housing. *Urban spaces and gender in Asia* (pp. 95–107). Springer.

Chandramouli, C., & General, Registrar (2011). Census of India 2011. *Provisional population totals* (pp. 409–413). New Delhi: Government of India.

Choudhari, Ranjana (2020). COVID 19 pandemic: Mental health challenges of internal migrant workers of India. *Asian Journal of Psychiatry, 54*, 102254. https://doi.org/10.1016/j.ajp.2020.102254

D'souza, Renita (2019). Housing poverty in urban India: The failures of past and current strategies and the need for a new blueprint. *ORF Occasional Paper, 187*(1).

Desai, Gaurang, & Yadav, Madhura (2007). Housing tenure for the urban poor: A case study of Mumbai city. In *Proceedings of the Second Australasian Housing Researchers Conference*. Citeseer.

Dev, Satvik (2006). *Rent control laws in India: A critical analysis*. Available at: SSRN 926512.

Dubey, Shubhankar, Sahoo, Krushna Chandra, Dash, Girish Chandra, Sahay, Mili Roopchand, Mahapatra, Pranab, Bhattacharya, Debdutta, Del Barrio, Mariam Otmani, & Pati, Sanghamitra (2022). Housing-related challenges during COVID-19 pandemic among urban poor in. *Frontiers in Public Health, 10*, 1029394. https://doi.org/10.3389/fpubh.2022.1029394

Dupont, Véronique (2008). Slum demolitions in Delhi since the 1990s: An appraisal. *Economic and Political Weekly, 43*(28), 79–87. http://www.jstor.org/stable/40277717. (Accessed 30 June 2023).

Dutta, Arnab, Gandhi, Sahil, & Green, Richard K. (2022). Do urban regulations exacerbate rural-urban inequality? Evidence from rent control in India. In *Evidence from rent control in India (April 25, 2022).*

Dwivedi, Saurabh (2021). *Sustainable affordable housing for low income group in India.*

Easthope, Hazel (2014). Making a rental property home. *Housing Studies, 29*(5), 579–596.

Gaillard, J. C., Walters, Vicky, Rickerby, Megan, & Shi, Yu (2019). Persistent precarity and the disaster of everyday life: Homeless people's experiences of natural and other hazards. *International Journal of Disaster Risk Science, 10*(3), 332–342. https://doi.org/10.1007/s13753-019-00228-y

Gandhi, Sahil, Green, Richard K., & Patranabis, Shaonlee (2022). Insecure property rights and the housing market: Explaining India's housing vacancy paradox. *Journal of Urban Economics, 131*, 103490.

Ghosh, Somenath, Seth, Pallabi, & Tiwary, Harsha (2020). How does Covid-19 aggravate the multidimensional vulnerability of slums in India? A commentary. *Social Sciences & Humanities Open*, 2(1), 100068. https://doi.org/10.1016/j.ssaho.2020.100068

Ian Litchfield, David Shukla, & Sheila Greenfield. (2021). Impact of COVID-19 on the digital divide: A rapid review. *BMJ Open, 11*(10), e053440. https://doi.org/10.1136/bmjopen-2021-053440

Jones, Colin, & Murie, Alan (2008). *The right to buy: Analysis and evaluation of a housing policy*. John Wiley & Sons.

Kathleen Moore, M. (2016). Lists and lotteries: Rationing in the housing choice voucher program. *Housing Policy Debate, 26*(3), 474–487.

Kayanan, Carla M. (2022). A critique of innovation districts: Entrepreneurial living and the burden of shouldering urban development. *Environment and Planning A: Economy and Space, 54*(1), 50–66.

Kemeny, Jim (2002). *From public housing soc market.* Routledge.

Kemp, Peter A., & Kofner, Stefan (2010). Contrasting varieties of private renting: England and Germany. *International Journal of Housing Policy, 10*(4), 379–398.

Kesar, Surbhi, Abraham, Rosa, Lahoti, Rahul, Nath, Paaritosh, & Basole, Amit (2021). Pandemic, informality, and vulnerability: Impact of COVID-19 on livelihoods in India. *Canadian Journal of Development Studies/Revue Canadienne d'Études du Developpement, 42*(1–2), 145–164. https://doi.org/10.1080/02255189.2021.1890003

Khan, Asma, & Arokkiaraj, H. (2021). Challenges of reverse migration in India: A comparative study of internal and international migrant workers in the post-COVID economy. *Comparative Migration Studies, 9*(1), 49. https://doi.org/10.1186/s40878-021-00260-2

Kohl, Sebastian (2017). *Homeownership, renting and society: Historical and comparative perspectives*. Taylor & Francis.

Mahadevia, Darshini (2013). Institutionalising spaces for negotiations for the urban poor. *Inclusive urban planning: State of the urban poor report* (pp. 148–166).

Ministry of Housing and Urban Affairs. (2021). *The model tenancy act, 2021.* https://mohua.gov.in/upload/uploadfiles/files/Model-Tenancy-Act-English-02_06_2021.pdf.

Ministry of Housing and Urban Poverty Alleviation. (2007). *National urban housing and habitat policy 2007*. Government of India New Delhi: Ministry of Housing and Urban Poverty Alleviation.

More, Akshay Ramesh (2021). *Rental housing for urban poor assessing the effectiveness of Maharashtra rent control act of 1999*. Bangalore: National Law School of India University.

Nath, Shyam (1984). Impact of rent control on property tax base in India: An empirical analysis. *Economic and Political Weekly*, 805–810.

National Housing Bank. (2022). *Report on trend and progress of housing in India 2021.* https://nhb.org.in/publications/report-on-trend-and-progress-of-housing-in-india/.

Nijman, Jan (2014). India's urban future: Views from the slum. *American Behavioral Scientist, 59*(3), 406–423. https://doi.org/10.1177/0002764214550304

Oliveri, Rigel C. (2009). Discriminatory housing advertisements on-line: Lessons from Craigslist. *Indian Law Review, 43*, 1125.

Patel, Amit, Shah, Phoram, & Beauregard, Brian E. (2020). Measuring multiple housing deprivations in urban India using Slum Severity Index. *Habitat International, 101*, 102190.

Peterman, Danieli Evans (2018). Socioeconomic status discrimination. *Virginia Law Review, 104*(7), 1283–1357.

Phang, Sock-Yong (2018). *Policy innovations for affordable housing in Singapore: From colony to global city*. Springer.

Planning Commission of India. (1998). *National housing and habitat policy 1998.* Government of India: Ministry of Urban Development and Poverty Alleviation.

Pricewaterhouse Coopers and the Urban Land Institute. (2023). *Emerging trends in real Estate® Asia Pacific 2023.* Pricewaterhouse Coopers (PWC) and the Urban Land Institute (ULI). https://www.pwc.com/sg/en/publications/aprealestemerging.html.

Raju, Emmanuel, Dutta, Anwesha, & Ayeb-Karlsson, Sonja (2021). COVID-19 in India: Who are we leaving behind? *Progress in Disaster Science, 10*, 100163. https://doi.org/10.1016/j.pdisas.2021.100163

Requena, Alicia, Vanhuyse, Fedra, Agerström, Magnus, Nilsson, Astrid, & Arra, Venni (2021). *Public-private partnerships*.

Rogers, Dallas, & Power, Emma (2020). *Housing policy and the COVID-19 pandemic: The importance of housing research during this health emergency*. https://doi.org/10.1080/19491247.2020.1756599

Sarkar, Sudipta, & Mishra, Deepak K. (2020). Circular labour migration from rural India: A study of out-migration of male labour from West Bengal. *Journal of Asian and African Studies, 56*(6), 1403–1418. https://doi.org/10.1177/0021909620967044

Seicshnaydre, Stacy (2016). Missed opportunity: Furthering fair housing in the housing choice voucher program. *Law and Contemporary Problems, 79*, 173.

Sethi, Mahendra, & Mittal, Shilpi (2020). Improvised rental housing to make cities COVID safe in India. *Cities, 106*, 102922.

Shekar, K Chandra, & Mansoor, Kashif (2020). COVID-19: Lockdown impact on informal sector in India. *Transport, 13*(86.6), 13–14.

Speak, Suzanne (2019). The state of homelessness in developing countries. In *Annals in Expert Group Meeting on Affordable housing and social protection systems for all to address homelessness*. Newcastle University England.

Srivastava, Ravi, & Sutradhar, Rajib (2016). *Migrating out of poverty? A study of migrant construction sector workers in India*. Indian Institute for Human Development.

Suresh, Rajani, James, Justine, & RSj, Balraju (2020). Migrant workers at crossroads—The COVID-19 pandemic and the migrant experience in India. *Social Work in Public Health, 35*(7), 633–643.

Tipple, A. Graham, & Speak, Suzanne (2009). *The hidden millions: Homelessness in developing countries. Housing and society series/edited by Ray Forrest*. Routledge. xiii, 327 pp. https://ci.nii.ac.jp/ncid/BA90415197

Tiwari, Piyush, & Rao, Jyoti (2016). *Housing markets and housing policies in India*. ADBI Working Paper 565.

Williamson, Anne R., Smith, Marc T., & Strambi-Kramer, Marta (2009). Housing choice vouchers, the low-income housing tax credit, and the federal poverty deconcentration goal. *Urban Affairs Review, 45*(1), 119–132.

Yigitcanlar, Tan, Adu-McVie, Rosemary, & Erol, Isil (2020). How can contemporary innovation districts be classified? A systematic review of the literature. *Land Use Policy, 95*, 104595.

CHAPTER

23

Homelessness and solution-oriented pathways and recommendations

Surendra Kumar Yadawa

School of Liberal Arts, IMS Unison University, Dehradun, Uttarakhand, India

1. Introduction

Homelessness is a global humanitarian crisis having various dimensions: socioeconomic, psychological, governmental, personal and policy. The social phenomenon of homelessness has been present throughout recorded history, with concerns over the well-being of those who lead nomadic lives dating back to the Middle Ages. Each shift in the economic landscape has brought forth new groups of marginalized individuals, simultaneously reshaping the societal perception of these outcasts. However, only recently have homelessness and its associated problems, such as inadequate housing, poverty, crime, and victimization, been recognized as a critical issue, prompting the United Nations to urge all member states to collaborate in eradicating homelessness by 2030 (United Nations, 2016). Its rising trends pose a new challenge to communities across the globe. Globally, children, families, and marginalized adults face housing insecurity (Peressini, 2009, pp. 1–17), while most countries of the world are struggling to meet the extensive housing needs of the homeless population. Homelessness is viewed as neither discrete nor necessarily homogenous. Every nation has a unique socioeconomic structure, depending on the nature of government, the structure responses, and the economic structure. Examining the responses denotes unexpected effects of the policies, and the answers are exceptionally different. Despite policy efforts to reduce homelessness, trends in homelessness continue to be stubbornly high. The constant demand for housing assistance over service delivery, while prevention efforts remain inconsistently implemented in most countries. A response is needed to ensure the sustainable delivery of coordinated preventive efforts. Comprehension of the causes and nature of this menace is necessary for getting solution-oriented pathways (Pophaim & Peacock, 2021). This problem has caught nations: rich and poor, socialist and capitalist, and welfare and competitive market-based political systems. "Homelessness is one of the crudest manifestations of poverty, discrimination and inequality, affecting people of all ages, genders and

Homelessness to Hope
https://doi.org/10.1016/B978-0-443-14052-5.00023-9

backgrounds. According to UN-Habitat, 1.6 billion people live in inadequate housing conditions, with about 15 million forcefully evicted every year, which has noted an alarming rise in homelessness in the last 10 years" (United Nations, 2020a). In order to end homelessness altogether, firstly, we must acknowledge housing as a human right and build deeply affordable housing; secondly, adopt the Housing First principles plus additional service support, when implementing individualized choice-based supports; thirdly, hold the government responsible for their social policy choices in the past, present, and future, and fourthly, prevent homelessness from occurring in the first place by implementing systemic change while adopting various approaches. A better approach identifies capacities and constraints to solve homelessness sustainably. This study looks at homelessness, analytically, examines solution-oriented pathways, and offers recommendations.

1.1 Rationale of the study

Homelessness is a complex problem with various contributing factors, including poverty, a lack of affordable housing, mental illness, and substance abuse, and it significantly impacts individuals, families, and society. These consequences are both undeniable and, in many ways, immeasurable, spanning social, medical, behavioral, economic, and moral costs. The rationale for studying homelessness and solution-oriented pathways and recommendations is to better understand the causes and consequences of homelessness and to develop effective strategies for addressing it. These pathways can include a variety of services, such as housing assistance, employment assistance, and mental health and addiction treatment. The study focuses on how to prevent homelessness and help people who are experiencing homelessness to achieve stable housing. This study recommends policies and programs that are more effective in mitigating, preventing, and redressing homelessness.

1.2 Research objective

To identify causes and consequences of homelessness, analyze effective strategies for preventing and ending homelessness and identify solution-oriented pathways that can help to inform policymakers, practitioners, and the public about the best ways to address this issue.

1.3 Materials and methods

Different kinds of literature are being reviewed to gather a conception of the study and the factors which are responsible for homelessness. The study is purely based on secondary data and information gathered from various sources such as government reports, homeless service providers, and academic research on homelessness. The secondary data has been analyzed to identify trends and patterns in homelessness and solution-oriented pathways, while qualitative analysis delves into personal narratives. Recommendations encompass a holistic approach, including affordable housing initiatives, mental health services, addiction treatment, and job training programs.

1.4 Research questions

1. What are the most significant causes and consequences of homelessness, based on a review of the literature? What are the critical challenges faced by individuals or organizations while implementing a solution-oriented approach and how can they be resolved?
2. How can solution-oriented pathways be effectively integrated into existing problem-solving frameworks or methodologies and collaboration between government, nonprofit organizations, and other stakeholders can be enhanced to address homelessness?

2. Literature review

2.1 Homelessness: An overview

Homelessness is a global problem, even though countries define, measure, and report it differently. The problem of homelessness is becoming more widespread these days. Homelessness is the experience of living without a stable, safe, and permanent place to call home. It can happen to anyone, regardless of age, race, gender, or income. Every country on the globe is facing this problem, whether rich or poor, developing or developed, socialist or capitalist; even economically developed countries experience this issue severely. A large portion of the population lacks the resources to afford a house. Social exclusion is a defining feature of homelessness (Carpenter-Song et al., 2016). They lack the resources necessary to buy affordable lodging in a more affluent area. It Depends on how homelessness is defined within the given location and context. "Globally, 1.6 billion people worldwide live in inadequate housing conditions, with about 15 million forcefully evicted every year, according to UN-Habitat, which has noted an alarming rise in homelessness in the last 10 years. Young people are the age group with the highest risk of becoming homeless" (United Nations, 2020b). The collection of credible data is very crucial (Belanger et al., 2013).

The problem of homelessness has structural as well as social, economic, and personal dimensions (Mago et al., 2013). Countries like Finland and Norway are efficiently able to reduce this problem to near zero, and other countries from Europe, the United States, Canada, and developing countries are struggling to resolve this problem. Now, the question arises: why are other countries struggling with this problem? Indeed, there are some functional problems in governance, policymaking, and implementing those policies to make them feasible at the grass-roots level. Nonetheless, a few indicators are fairly similar across the globe.

Some political systems across the globe are offering social security, promoting social inclusion (Mashau & Mangoedi, 2015), good governance, and trying to make people's lives easier. It is supposed that the absence of affordable housing is the main reason behind this problem. Affordable housing is believed to be an important step in helping people come out of poverty. Additionally, individual difficulties also play an important role (Bassuk & Franklin, 1992). Other aspects, called "housing crisis," "family breakdown," "substance abuse," "mental health," and "youth to adult," play a crucial role in this problem (Chamberlain & Johnson, 2013; McCarty et al., 1991).

In fact, Housing First is the most discussed method in terms of combating homelessness. However, this single measure cannot resolve all aspects related to this complex problem. It needs a comprehensive policy to address this problem, in addition, it requires a multifaceted approach to get rid of homelessness. Providing affordable housing and employment opportunities, eradicating poverty, and removing other socio-psychological discrimination are supposed to be the right pathways to solutions for the primary requisites. Employment opportunities, medical support, psychological solutions, and old age care make the homeless more self-reliant, which will prevent social exclusion. Thus, Balanger the collection of credible data is very crucial (Belanger et al., 2013).

The individualistic approach looks at homelessness as a failure of an individual in the endeavor of his life. No doubt, personal irresponsibility occurs everywhere. If personal irresponsibility is the reason, then it should happen everywhere. The structural approach focuses on the shortage of resources—low income, housing prices, the impact of changing technology, globalization, the labor market, etc. Ironically, it does not care about human suffering alone. The "politics of compassion" prevail among contemporary social scientists. In this view, the new homeless are portrayed as victimized by various means like economic isolation, violence of intimate partner, and physical hardship. However, there are two problems this approach overlooks: the common economy and the politics of compassion.

The impacts of the welfare reforms and public expenditure being pursued by the governments, and other agencies together with the implications of its housing, homelessness, and other relevant policies. Walls (2017) in his memoir "The Glass Castle" tells the story of a family living in poverty and struggling with housing instability (Walls, 2017). It offers a personal perspective on the challenges faced by those without stable homes. Orwell (2021), in his first-hand account of living among the destitute in the two cities, provides a powerful exploration of poverty and homelessness. Orwell's account offers a unique perspective on the challenges faced by individuals on the margins of society (Orwell, 2021). Arnade (2019) encounters and interviews individuals living on the margins of society, including the homeless. He highlights the importance of dignity and human connection in addressing the underlying causes of homelessness (Arnade, 2019).

The conflict and violence in their households was the main reason many young people left home (Tyler & Schmitz, 2013). Violence by intimate partners leads to homelessness (Tutty et al., 2013). Desmond (2016) explores the devastating consequences of eviction and housing insecurity in America. He offers an in-depth examination of the challenges faced by low-income individuals and families, shedding light on the systemic issues that contribute to homelessness (Desmond, 2016). Padgett et al. (2016) investigate the Housing First approach, which prioritizes providing stable housing to individuals experiencing homelessness before addressing other needs. They examine the effectiveness of this approach and offer insights into how it can transform homeless service systems (Padgett et al., 2016). Beck and Twiss (2018) argue that neoliberal policies and piecemeal efforts to address the problem, they explore how government policies and practices have served to shape limited responses to the problem. They consider how a just, human-rights–based approach might be affected (Beck & Twiss, 2018). In almost every country on the globe, people can be seen sleeping on the streets, outskirts of the city or elsewhere. This is because homelessness affects people of all ages, sexes, races, and backgrounds, and is among the most obvious examples of poverty, discrimination, and inequality.

In brief, numerous factors contribute to the phenomenon, many of which are interconnected.

- When natural calamities destroy homes, leaving families without a roof over their heads becomes a compulsion.
- Domestic conflicts cause people to lose their homes and look for a place to stay but not everyone is self-sufficient to find adequate shelter.
- The absence of affordable housing is another major cause. Some people are unable to pay their rent or mortgage due to unemployment and poverty which can be caused by a crisis or physical or mental issues.
- Some homeless people, despite having a job, do not earn adequate to be able to cover their rent, sometimes as a result of an increase in local housing costs.
- Migration for search of job from rural to urban area (Fig. 23.1).

Homelessness is associated with a dramatic increase in mortality risk and a substantial reduction in life expectancy (2023). Individuals with a history of homelessness experience a disproportionately higher prevalence of various chronic health conditions (Sutherland & Rosenoff, 2021).

- Individuals experiencing homelessness face a substantially heightened risk of contracting infectious diseases. Homelessness places individuals at significantly greater risk of infectious diseases (Beijer et al., 2012). Individuals experiencing homelessness are

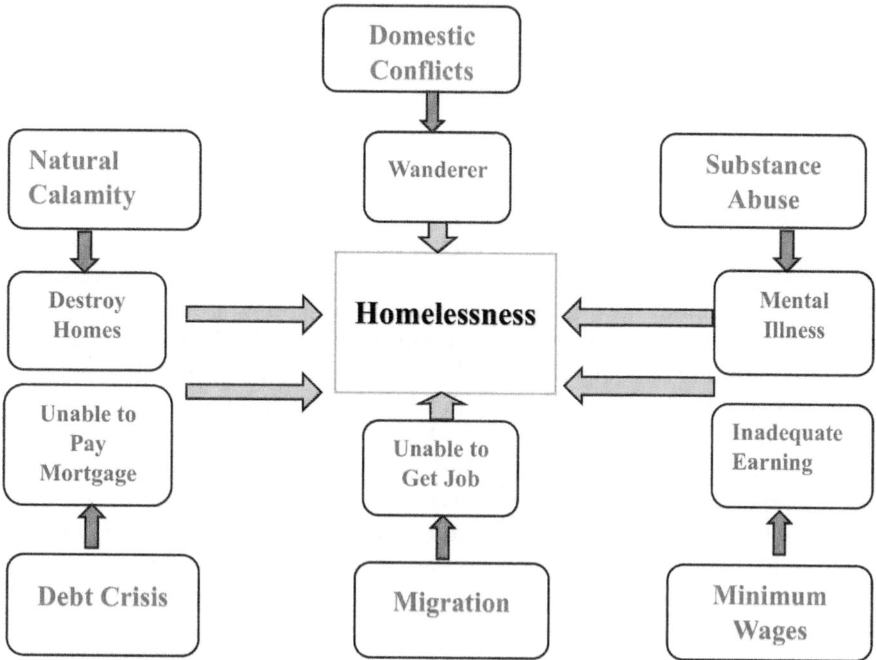

FIGURE 23.1 Causes of homelessness: Homelessness is the result of multiple factors.

disproportionately likely to experience a traumatic brain injury in their lifetime (Stubbs et al., 2020). Consequently, violence is the most common cause of traumatic brain injuries among people experiencing homelessness. Rates of violent victimization are alarmingly high among people experiencing homelessness (Hsu et al., 2020; Thompson & Tapp, 2021). Thus, experiencing homelessness dramatically increases women's risk of sexual victimization (Riley et al., 2020). Homelessness poses profound mental and emotional strains (Kilpatrick & Acierno, 2003). People experiencing homelessness are at an elevated risk of experiencing mental illness and substance use disorders.

It can harm prenatal development and young children's physical health (St. Martin et al., 2021), and can impact prenatal development (Clark et al., 2019). Compared to children in low-income families, children experiencing homelessness are twice as likely to be hospitalized (Weinreb et al., 1998). As adverse childhood experiences are strongly associated with negative health and social outcomes (Hughes et al., 2017), high rates of early trauma for children experiencing homelessness can fuel a range of adversity throughout the life course. Repeated exposure to stressful and traumatic events at a young age causes severe emotional distress, resulting in a high incidence of mental health challenges for children experiencing homelessness. Homelessness also disrupts children and youth's schooling, from elementary school to higher education, limiting opportunities to advance their education and economic standing.

Homeless program costs are often far greater than the cost of providing rental assistance and providing transitional or emergency housing to individuals and families who need longer-term housing support incurs very high costs (Spellman, 2010), compared to stably housed individuals, people experiencing homelessness have roughly three times as many visits to Emergency Departments each year (Coe et al., 2015). Even among patients who frequently visit Emergency Departments, those experiencing homelessness visit annually than frequent visitors with secure housing (Ku et al., 2014).

Individuals experiencing homelessness also have longer average hospital stays (Buck et al., 2012). Similar findings emerge for infants born during periods of housing instability that result in homelessness, who experience longer neonatal intensive care unit stays and more emergency department visits, resulting in higher healthcare spending (Clark et al., 2019).

While some studies indicate that enrollment in permanent supportive housing may not directly impact total annual healthcare costs, the Housing First approach has demonstrated a significant potential to reduce healthcare expenditures overall (Raven et al., 2020; Williams et al., 2023). These findings underscore the potential of supportive housing to act as a preventive measure, potentially leading to a sustainable decrease in overall healthcare utilization and associated costs. Permanent supportive housing can yield quantifiable cost savings across multiple systems, ultimately benefiting taxpayers. Furthermore, examinations of community courts indicate that this model is effective in reducing costs across various systems (Lee et al., 2013). Despite the need for more comprehensive cost–benefit studies, existing evidence suggests that problem-solving courts can effectively disrupt the revolving door between incarceration and homelessness, enabling individuals experiencing homelessness to access essential housing support and other services, thereby optimizing the allocation of limited public resources.

3. Solution-oriented pathways

Homelessness is a complex issue with multiple underlying causes so finding effective solutions requires a comprehensive and collaborative approach. There is no one-size-fits-all solution, as the most effective approach will vary based on the specific needs of each community and individual. Solution-oriented pathways are crucial in finding to assist in reducing and eventually preventing this global menace. Therefore, it becomes pertinent to identify the most prominent pathways among most homeless individuals. The pathways are generally associated with both the impact and consequences of homelessness and the risk factors for victimization. Reducing homelessness will not necessarily guarantee that society will become free of all other social challenges, such as poverty; unemployment; or the ongoing demand for adequate and affordable housing. The challenge of reducing and ultimately eradicating homelessness starts with the strengthening of the existing system and knowledge. It requires a concerted, intersectional effort by every single stakeholder involved in the eradication of this issue, which includes but is not limited to, researchers, community members, social welfare, governmental agencies, and civil society. Prevention and service interventions need a lot (Brown et al., 2016). Minnery and Greenhalgh (2007) advocate for a new policy. There are several key aspects to consider in this effort.

1. **Affordable Housing Solutions:** Building affordable housing and providing subsidies for housing affordability is a cornerstone of homelessness mitigation (Tsemberis et al., 2012). Governments, businesses, and NGOs all have a crucial role in providing affordable housing. Policies that prioritize affordable housing, investment in public housing programs, and housing strategies are essential in providing stable housing options.
2. **Support Services and Social Security:** Homelessness requires a comprehensive solution, including housing, support services, and addressing root causes like poverty and unemployment. Social security systems can significantly reduce the homelessness rate. Governments and nonprofit organizations should work together to offer tailored support services to help homeless individuals regain stability and independence.
3. **Changing Attitudes:** Changing societal attitudes toward homelessness is essential. School-level counseling and education and awareness campaigns can foster empathy and support, and help to change public perceptions. Homelessness is not a choice, and people experiencing it are individuals with unique needs. Treating them with dignity and respect is essential for facilitating their reintegration into society. Murphy (2022) underlines counseling at the school level (Murphy, 2022).
4. **Policies:** Government policies and resources can significantly impact homelessness, either exacerbating or preventing and alleviating it. Policies related to housing, social welfare, and income support can directly affect homelessness rates. Governments can also invest in homelessness prevention programs, emergency shelters, transitional housing, mental health services, substance abuse treatment, and support services. Policies related to housing, social welfare, and income support can directly affect homelessness rates (Anderson, 2001).
5. **Funding:** Adequate funding and resource allocation by the government are fundamental to the extent and effectiveness of homelessness interventions. Investment in homelessness prevention, emergency shelters, transitional housing, mental health services, and substance abuse treatment is necessary for addressing the issue effectively.

6. **Collaboration and Coordination:** Collaboration and coordination at all levels, involving stakeholders like community organizations, nonprofits, and service providers, are indispensable for tackling homelessness effectively. Comprehensive plans involving all relevant parties are crucial to create synergy in efforts. This includes sharing data and best practices and developing and implementing comprehensive plans to address the issue.
7. **Data and Research:** Data collection and research are crucial for informed policymaking, identifying service gaps, and allocating resources efficiently. Comprehensive data and national enumeration are essential for understanding homelessness and developing effective interventions. Credible data and comprehensive national enumeration are very crucial (Adams & Füss, 2010; Belanger et al., 2013).

4. Role of government

The nature of government and its approach to dealing with homelessness can vary significantly depending on the country and its political system. Governments often address homelessness, but specific policies and approaches may differ across different nations. Creating workable and effective policies requires clear and comprehensive definitions of homelessness, which is essential for policymakers to develop effective solutions (Minnery & Greenhalgh, 2007). Social welfare programs can include emergency shelters, transitional housing, subsidized housing, financial aid, food assistance, and healthcare services. In recent years, many governments have adopted Housing First approach to provide stable and permanent housing to homeless individuals, along with necessary support like rapid housing (Gallagher, 2023). Subsequently, Policy legislation can create laws to protect the rights of homeless individuals, such as antidiscrimination measures or laws to prevent the criminalization of homelessness while strengthening tenant protections, increasing affordable housing stock, and allocating funding specifically for homeless services. Tsemberis and Eisenberg (2000) find, "Clients with severe psychiatric disabilities and addictions are capable of obtaining and maintaining independent housing when provided with the opportunity and necessary support" (Tsemberis & Eisenberg, 2000). Homelessness can be reduced through a combination of short-term relief measures and long-term strategies to address the root causes. The specific nature and effectiveness of government responses to homelessness can vary and it's an ongoing challenge that requires an ongoing commitment to sustainable solutions (Fig. 23.2).

4.1 Strategies for combating homelessness

Homelessness is a complex issue with diverse causes, including socioeconomic conditions, housing market dynamics, mental health, substance abuse, and systemic inequalities. While the nature of government is a significant factor, it should be considered in conjunction with these other factors to develop holistic solutions to homelessness. There are a number of different approaches to addressing homelessness. Some governments focus on providing temporary shelter, while others focus on providing permanent housing (Hurtubise et al., 2009, pp. 1–16). Some governments focus on providing services to help people address the underlying causes of homelessness, such as mental illness and addiction. The best approach to addressing homelessness will vary depending on the specific circumstances of each

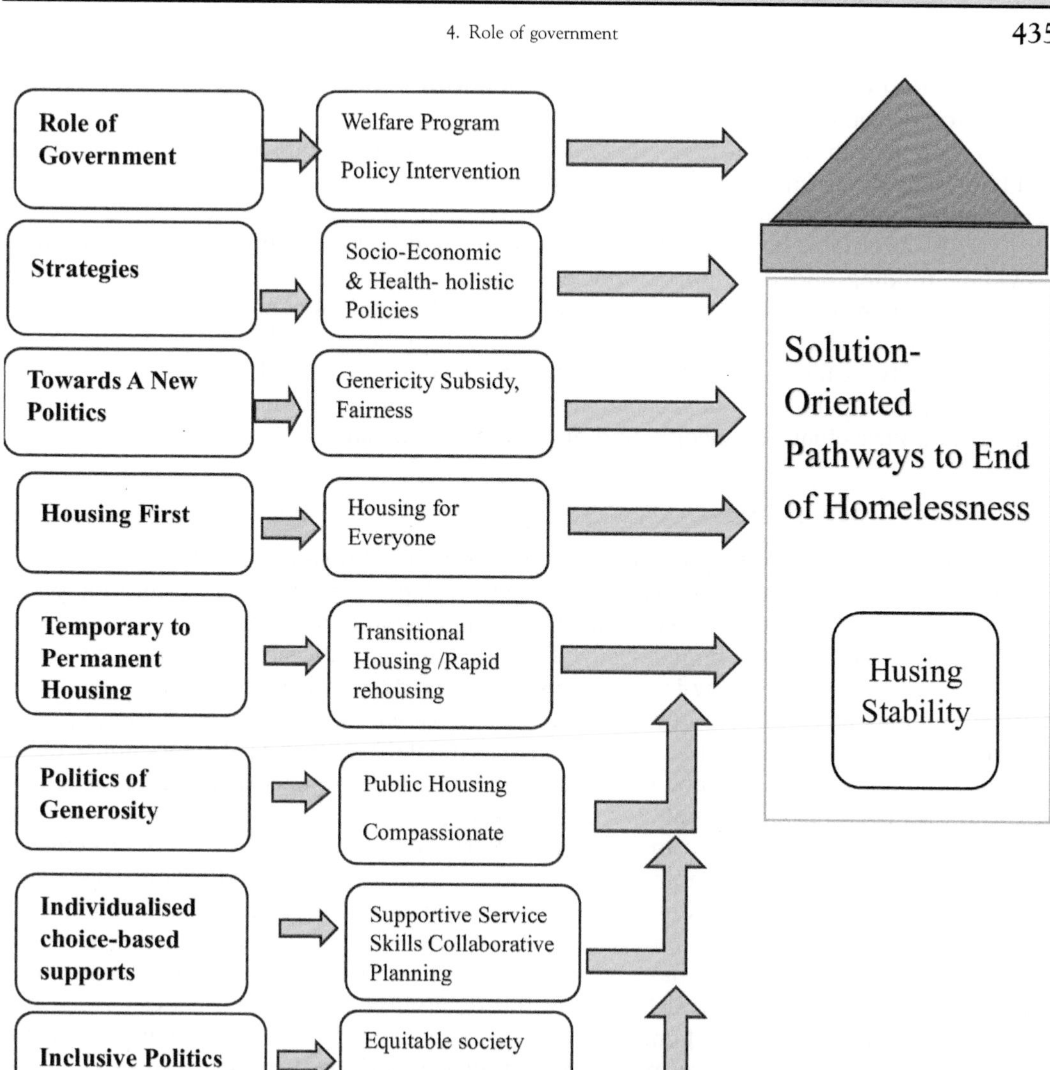

FIGURE 23.2 Solution-oriented pathways to end homelessness: All of the factors must exist simultaneously to end homelessness through solution-oriented pathways effectively.

community. However, all governments can play a role in making a difference. Working together we can create a world where everyone has a safe and stable place to call home. Psychiatric support has brought remarkable outcomes worldwide (Greenwood et al., 2013).

It's important to note that the nature of government is not the sole factor influencing homelessness. There are other social, economic, and individual factors that contribute to homelessness as well (Farrugia & Gerrard, 2016). However, a government that recognizes homelessness as a significant issue, implements evidence-based policies and provides the necessary resources and support that can have a positive impact on reducing homelessness and improving the lives of those affected.

Governments can alleviate homelessness by providing monetary support for programs that help individuals find and maintain stable housing, such as rapid rehousing and permanent supportive housing initiatives. They can also develop policies that make it easier for people to access affordable housing, such as rent control and making regulations. In addition to providing funding and developing policies, governments can also play a role in raising awareness of homelessness and advocating for the rights of people who are homeless. Importantly, the role of media is also important, the information they provide, in other words, seeks to move society in the right direction (Aitamurto & Varma, 2018). They can also work to change the public perception of homelessness so that people are more likely to support programs and policies that help people find and maintain housing.

4.2 Toward a new politics

Political systems across the globe play a crucial role in addressing homelessness through various policies, programs, and initiatives. Engaging in long-term planning, monitoring progress, and making adjustments based on feedback and outcomes are more likely to make sustainable progress in reducing homelessness. Despite potential criticisms of subsidy programs, rental assistance initiatives, low-income housing tax credits, and public housing solutions, these efforts play a critical role in expanding the accessibility of affordable housing options and providing essential supportive services, such as mental health and substance abuse treatment, job training, counseling, and case management, to individuals experiencing homelessness. Sullivan et al. (2000) advocate for mental add.

New Policies and legislation can include outreach and street-based programs to engage with individuals experiencing homelessness and by implementing antidiscrimination laws to protect homeless individuals, developing zoning regulations to encourage affordable housing construction, and adopting strategies to prevent evictions and promote housing stability. Social Welfare Programs should have robust social security, including programs such as unemployment benefits, welfare assistance, healthcare, and mental health services, which can help prevent individuals and families from falling into homelessness (Chambers et al., 2014). Funding and resources can prioritize and allocate sufficient resources toward homelessness prevention, affordable housing initiatives, and support services to have a better chance of reducing homelessness. Lastly, political will and leadership can lead to comprehensive strategies, policy reforms, and effective implementation.

4.3 Housing-led solutions

Housing First is an assistance approach serving as a platform to provide permanent housing to people experiencing homelessness. This approach believes that people need basic necessities like food and shelter before attending to anything such as getting a job and substance use. It is based on the understanding that client choice is valuable in housing selection and supportive service participation and that exercising that choice is likely to make a client more successful in improving their quality of life. The motive of Housing First is ending homelessness instead of tackling it. The basic idea is to offer homeless people permanent housing and needs-based support instead of temporary accommodations. Permanent housing means an independent rental flat with a rental contract. In Housing First, people do not have to earn their right to housing by proving their capability to manage their lives. Instead, they are provided with a stable home, making them self-sufficient. The Housing First Policy, which emphasizes collaboration between the government, local governments, and nonprofit organizations, has served as the foundation for Finland's national homelessness policy. Since its launch in 2008, has reduced the number of long-term homeless people in Finland by more than 35%, nearly eradicating rough sleeping in the country's capital (United Nations, 2020c).

Shelters have been transformed into assisted housing units, and investments have been made to offer affordable housing. To meet the varied requirements of individual tenants, new services and assistance techniques have been developed (Kaakinen & Turunen, 2021). Finland has minimized rough sleeping and sustainably housed a significant number of long-term. Finland is the only country in Europe where the number of homeless people has declined in recent years. The Housing First model can be replicated even though housing conditions may vary from country to country in Europe (Baker & Evans, 2016). Providing permanent homes for the homeless should be a target instead of a temporary solution. There is no quick fix to all life situations, but a solid foundation provides the foundation upon which to improve the welfare of the homeless through good governance (Baker & Evans, 2016). This new approach and its convincing results have been appreciated globally.

4.4 Uniqueness of housing first

Housing First is a unique approach to addressing homelessness that differs from traditional approaches in significant ways. It provides immediate and permanent housing to individuals experiencing homelessness, The philosophy behind Housing First is that stable housing is a basic human right and should be the primary focus. It gives autonomy to individuals by allowing them to choose where they want to live and offering them a range of housing options means client-oriented. This approach upholds that people are more likely to engage in supportive services and address their issues when they have a safe and stable place to call home.

Unlike traditional models that require individuals to complete treatment or meet certain behavioral requirements before being provided with housing, it separates housing from treatment and acknowledges that people are more likely to succeed in addressing their challenges when they have a stable living environment. It comprehensive, flexible, and ongoing support services to individuals in their own homes. These services are tailored to meet the specific needs and goals of each person and may include case management, mental health support, employment assistance, and substance abuse treatment.

The finish option is good but not finished (Kaakinen & Turunen, 2021). The focus is on helping individuals maintain their housing and improve their overall well-being. It recognizes housing stability is a crucial first step in addressing other challenges, such as mental health issues, substance abuse, or unemployment (Chen et al., 2004). By prioritizing housing stability, individuals have a stable platform where they can reach personal goals and improve their overall quality of life.

4.5 Temporary to permanent housing

Transitional housing or rapid rehousing is an approach that aims to provide temporary shelter and support services to individuals and families experiencing homelessness with the ultimate goal of helping them secure permanent housing. Such programs promise results in helping individuals and families exit homelessness and achieve long-term housing stability. By combining temporary shelter with comprehensive support services and assistance in securing permanent housing, these programs offer a holistic approach to addressing homelessness and promoting self-sufficiency.

Housing First is designed to help individuals who are experiencing homelessness, particularly those who face significant barriers to accessing and maintaining stable housing, particularly, effective in addressing chronic homelessness, which refers to individuals who have been continuously homeless (Tsemberis et al., 2012). Stable housing is a critical factor in improving the mental health of individuals experiencing homelessness.

It assists individuals struggling with substance abuse issues and often collaborates with substance abuse treatment providers to offer integrated services (Mallett et al., 2005). It has been widely implemented to address homelessness among military veterans. Many veterans experience challenges such as posttraumatic stress disorder (PTSD), substance abuse, and difficulties transitioning from military to civilian life (Hamilton et al., 2011). Women are more likely to face victimization compared to men (Dietz & Wright, 2005) (Fig. 23.3).

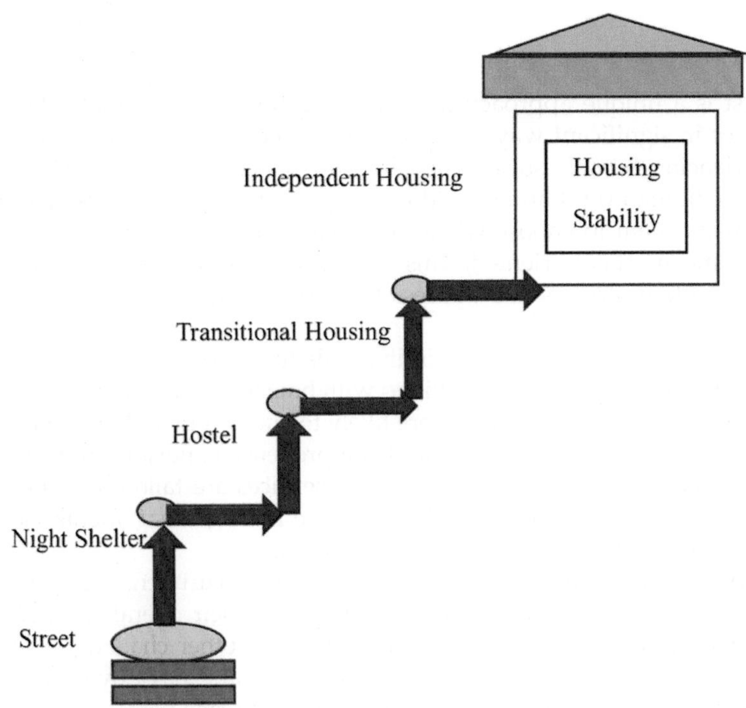

FIGURE 23.3 The housing first: The journey of hope with housing first.

It can also support homeless families with children and unaccompanied youth. Supportive services can be tailored to address the needs of families, such as parenting support, childcare assistance, and employment services. Moreover, it is a flexible approach that can be adapted to various populations and their specific needs. The focus is on addressing homelessness by providing housing as a fundamental right and offering ongoing support services to promote housing stability and individual well-being. "Housing First is an effective, evidence-based approach to ending youth homelessness" (Gaetz et al., 2021).

5. Politics of generosity

Political ecology reveals how economic and other forms of disparities are rationalized and normalized. This is evident in the way that housing is now seen as a commodity to be bought and sold, rather than a human right. This has led to skyrocketing housing prices and a growing homelessness crisis in many countries. Finland and Norway are two countries that have prioritized building deeply affordable public housing. This has resulted in significantly lower rates of homelessness than in other countries. In contrast, Canada, the United States, and many other countries have not prioritized affordable housing, and as a result, they have much higher rates of homelessness.

Private builders are not filling the demand for affordable housing because they are focused on profits. Public housing is essential for providing deeply affordable housing for people with very low incomes. Public housing should be rent-geared-to-income and committed to deep affordability. Housing First is a programmatic model that provides people with housing first, without any preconditions. This means that people do not have to be sober or employed before they can be housed. Once people are housed, they can then access other supports, such as mental health and addiction treatment, as needed.

The bigger issue is having governments commit to building deeply affordable housing and making policy choices that support this goal at all levels of governance. A politics of generosity requires a compassionate and comprehensive approach to homelessness. This includes adequate funding, collaboration, a Housing First strategy, supportive services, prevention efforts, long-term solutions, and cultivating public awareness and empathy.

6. Individualized choice-based supports

Individualized choice-based support (ICBS) is an approach that recognizes the unique needs and preferences of individuals experiencing homelessness. It focuses on providing tailored support and empowering individuals to make choices that align with their goals and aspirations. Christensen (2012) suggests individual pathways (Christensen, 2012). Couldrey (2010) argues that the impact of childhood abuse and trauma has been associated directly and indirectly with a pathway to homelessness (Couldrey, 2010). Here are some key elements of individualized choice-based support to end homelessness:

- Collaborative planning is essential for engaging individuals and identifying their goals, strengths, and needs. This process involves listening to their experiences, preferences, and aspirations and using this information to develop a personalized plan.
- Dedicated housing navigators are crucial for helping individuals find suitable housing options, understand eligibility criteria, and access resources while working closely.
- Housing options should be tailored to individual needs, including permanent supportive housing, affordable rentals, shared housing, and rapid rehousing programs. Empower individuals to choose housing support that meets their needs and preferences.
- Supportive services should be given to individuals, including mental health counseling, substance abuse treatment (Fountain et al., 2009), job training, life skills development, healthcare access, and income support assistance (Hwang, 2022).
- Recognize and address trauma, a significant factor in homelessness. Implement trauma-informed practices to create a safe and supportive environment.
- Incorporate peer support programs with mentors or advocates who have experienced homelessness. Peer support provides insights, empathy, and encouragement as individuals work toward housing stability (Fowler et al., 2009).
- Offering employment services, skills training, vocational training, and educational opportunities to help individuals secure stable employment and increase their income is vital.
- Foster partnerships with community organizations, service providers, and stakeholders for a coordinated approach. Engage the community in awareness, volunteerism, and public forums to reduce stigma, increase understanding, and garner support for ending homelessness.

This approach aims to empower individuals to come out of homelessness, increasing the likelihood of long-term success and stability.

7. Inclusive politics (social justice)

Inclusive politics with a focus on social justice can play a crucial role in addressing homelessness. By adopting policies and approaches that prioritize equity, fairness, and the protection of vulnerable populations, societies can work toward ending homelessness and ensuring housing security for all. By embracing inclusive politics with a focus on social justice, societies can work toward creating a more equitable and compassionate approach to homelessness. Such an approach recognizes the inherent dignity of all individuals, addresses systemic inequalities, and aims to ensure that everyone has a safe and stable place to call home.

A crucial step toward eradicating homelessness lies in significantly bolstering social investments. Governments must enhance income support to ensure adequate social and financial assistance for individuals experiencing homelessness. To ensure government accountability, we must advocate for stronger financial assistance programs, including increased welfare rates, improved disability support, higher shelter allowances, and expanded rent supplements. People must exercise their voting power to elect candidates who are genuinely committed to developing affordable housing and raising income support levels, recognizing the critical importance of elderly care (Crane et al., 2005).

The country, like Finland, has dramatically reduced homelessness by investing in affordable housing and building strong partnerships between all levels of government and the nonprofit sector. Finland already had more robust social programming so the addition of affordable housing resulted in a large and sustained reduction in the number of people experiencing homelessness. Nations need more investments in social housing and community-based mental health and social supports that are tailored to address homelessness. Unemployment and disability income supports are often not sufficient to meet the financial needs of people who rely on them. These income programs are not working to end homelessness, so a transformation is needed. This could include looking into the effectiveness of increasing the rates and/or investing in a universal basic income like the one briefly piloted.

Nordic countries spend substantially more on social housing and social protections for vulnerable people. Shifting to the Nordic model is the best way for other countries to move the needle on homelessness across the country. Homelessness could end if there existed the will to work on, and resources to invest in, the systemic and societal issues that are driving it. In addition to housing, we need programs that help people successfully make the transition. We could never end homelessness simply by having more emergency programs and services for people who are currently experiencing homelessness.

Social policy is primarily the responsibility of the government because it has the authority, power, and mandate, but the private sector and individuals also have a role to play (Horsell, 2006). How the government understands the issue of homelessness, such as through its ideology, impacts the policy responses it puts in place. Instead, it will shift the responsibility to individuals, families, and religious organizations to provide charity. Political mediation brings results (Cress & Snow, 2000).

8. Limitations of the study

This study on solution-oriented pathway offers a promising finding to address homelessness, it is pertinent to acknowledge its limitations the diversity of homelessness experiences, a complex interplay of factors, long-term sustainability challenges, and limited scope of solutions all contribute to the complexity of the issue. This study went through secondary resources, journals, and articles.

9. Challenges to solution-oriented pathways

When working on solution-oriented pathways, there are several challenges that need to be considered, including- One of the major challenges is the shortage of affordable housing (Knutagård, 2018). Housing costs, low vacancy rates, and limited subsidized housing programs make it difficult for individuals experiencing homelessness to find and maintain affordable housing (Caldera & Johansson, 2013). Many individuals experiencing homelessness have mental health problems, substance abuse issues, or both. These conditions can make it difficult to hold down a job, maintain relationships, and access housing. Hence, support services such as counseling, addiction treatment, and job training are crucial for individuals experiencing homelessness. However, these services are often limited and difficult to access. Homelessness is often stigmatized, leading to discrimination and social exclusion. Negative attitudes and misconceptions about homelessness can hinder efforts to implement effective solutions. Consequently, coordinating efforts among various agencies to address homelessness can be challenging due to differing priorities, bureaucratic hurdles, and limited communication channels then, Implementing comprehensive homelessness solutions requires significant financial resources, and securing sustainable funding streams to support affordable housing, support services, and preventive measures can be challenging. Moreover, many individuals experiencing homelessness have experienced trauma, and their safety is at risk while living on the streets. Young people are more at risk of developing substance abuse (Johnson & Chamberlain, 2008). Ensuring safe shelter options and trauma-informed care is essential, but it can be challenging to provide adequate support; otherwise, chances of victimization increase (Lee & Schreck, 2005) lastly, it is also important to recognize that homelessness is often a symptom of larger systemic challenges, such as poverty, inequality, and a lack of access to affordable housing and healthcare.

10. Recommendations

Solution-oriented pathways are focused on identifying and implementing effective solutions to address various challenges. Mangayi (2014) advocates for comprehensive solutions (Mangayi, 2014). Here are some major recommendations that take a solution-oriented pathway:

- A holistic approach that deals with mental health counseling, substance abuse treatment, vocational training, and case management is required to address homelessness.
- Public–private partnerships can be developed for better resolution.
- Microhousing, tiny homes, and other alternative shelter models should be cost-effective, provide transitional housing, and bridge the gap between homelessness and permanent housing.
- Vacant buildings can be converted into affordable housing units. This approach utilizes existing infrastructure to quickly increase the housing supply and reduce costs.
- Rent subsidies and housing vouchers expand rental assistance programs and housing vouchers to help individuals afford stable housing.
- Employment programs, job training, and placement for individuals experiencing homelessness can help them regain financial independence and housing stability.

11. Conclusions

Since solution-oriented pathways are combined with both the impacts and consequences of homelessness as well as the risk factors for victimization, determining which factors existed beforehand and which developed as a result of homelessness remains a challenge. Therefore, acquiring reliable data of this nature is crucial for gaining a deeper understanding of the intricate connection between homelessness's causes, victimization risk factors, and its effects and consequences.

Minimizing homelessness is not necessarily guaranteed that society will become free of all social challenges, such as poverty; unemployment; or the ongoing demand for affordable housing. The challenge of eradicating homelessness begins with the strengthening of the current body of knowledge and requires a concerted, intersectional effort by every single stakeholder involved in the eradication of this social issue, which includes but is not limited.

Solution-oriented pathways and recommendations offer a way forward in addressing the complex challenges we face in various domains. By focusing on practical and actionable strategies, can work toward tangible solutions and positive change.

In order to address homelessness effectively, it is essential to adopt a multidimensional approach that recognizes the interconnectedness of social, economic, and environmental factors. This requires collaboration and coordination among governments, businesses, communities, and individuals. By leveraging the strengths and expertise of each stakeholder, we can create comprehensive solutions that have a lasting impact.

Investing in outreach, education, training, and innovation is a very important pathway. By providing quality education and training opportunities, we can equip individuals with the skills and knowledge needed to thrive in the rapidly evolving global economy. Fostering entrepreneurship and supporting innovation can lead to breakthrough solutions and drive economic growth.

Furthermore, promoting social inclusion and equality is crucial. By addressing systemic discrimination, promoting gender equality, and ensuring equal access to housing, more inclusive societies can be built. This involves implementing policies that reduce inequalities, provide social protection, and empower marginalized groups.

Lastly, fostering global cooperation is essential in tackling global challenges. Strengthening international partnerships, sharing best practices, and pooling resources can achieve collective progress. Collaboration on issues such as climate change, poverty eradication, and humanitarian crises can lead to more effective and sustainable solutions.

Solution-oriented pathways and recommendations provide a roadmap for addressing complex challenges. By adopting a multidimensional approach, investing in education and innovation, promoting social inclusion and equality, and fostering global cooperation, can work toward a more sustainable, equitable, and prosperous future.

Future research on homelessness and solution-oriented pathways should prioritize rigorous evaluation of intervention programs, guided by a deeper understanding of root causes. Exploring technology's role (social media) in raising awareness and connecting individuals with resources is essential. The research should focus on identifying the most effective public policies while ensuring public–private partnerships for affordable housing to reduce homelessness.

References

Adams, Z., & Füss, R. (2010). Macroeconomic determinants of international housing markets. *Journal of Housing Economics, 19*(1), 38–50. https://doi.org/10.1016/j.jhe.2009.10.005

Aitamurto, T., & Varma, A. (2018). The Constructive Role of Journalism: Contentious metadiscourse on constructive journalism and solutions journalism. *Journalism Practice, 12*(6), 695–713. https://doi.org/10.1080/17512786.2018.1473041

Anderson, I. (2001). *Pathways through homelessness: Towards a dynamic analysis.* Urban Frontiers Program.

Arnade, C. (2019). *Dignity: Seeking respect in back row America.*

Baker, T., & Evans, J. (2016). 'Housing first' and the changing Terrains of homeless governance. *Geography Compass, 10*(1), 25–41. https://doi.org/10.1111/gec3.12257

Bassuk, E., & Franklin, D. (1992). Homelessness past and present: The case of the United States. *New England Journal of Public Policy, 8*(1).

Beck, E., & Twiss, P. C. (2018). *The homelessness industry: A critique of US social policy.* Lynne Rienner Publishers.

Beijer, U., Wolf, A., & Fazel, S. (2012). Prevalence of tuberculosis, hepatitis C virus, and HIV in homeless people: A systematic review and meta-analysis. *The Lancet Infectious Diseases, 12*(11), 859–870. https://doi.org/10.1016/S1473-3099(12)70177-9

Belanger, Y. D., Awosoga, O. A., & Head, G. (2013). *Homelessness, urban Aboriginal people, and the need for a national enumeration.*

Brown, R. T., Goodman, L., Guzman, D., Tieu, L., Ponath, C., & Kushel, M. B. (2016). Pathways to homelessness among older homeless adults: Results from the HOPE HOME study. *PLoS One, 11*(5). https://doi.org/10.1371/journal.pone.0155065

Buck, D. S., Brown, C. A., Mortensen, K., Riggs, J. W., & Franzini, L. (2012). Comparing homeless and domiciled patients' utilization of the Harris County, Texas public hospital system. *Journal of Health Care for the Poor and Underserved, 23*(4), 1660–1670. https://doi.org/10.1353/hpu.2012.0171

Caldera, A., & Johansson, A. (2013). The price responsiveness of housing supply in OECD countries. *Journal of Housing Economics, 22*(3), 231–249. https://doi.org/10.1016/j.jhe.2013.05.002

Carpenter-Song, E., Ferron, J., & Kobylenski, S. (2016). Social exclusion and survival for families facing homelessness in rural New England. *Journal of Social Distress and the Homeless, 26*(1), 41–52. https://doi.org/10.1080/10530789.2016.1138603

Chamberlain, C., & Johnson, G. (2013). Pathways into adult homelessness. *Journal of Sociology, 49*(1), 60–77. https://doi.org/10.1177/1440783311422458

Chambers, C., Chiu, S., Scott, A. N., Tolomiczenko, G., Redelmeier, D. A., Levinson, W., & Hwang, S. W. (2014). Factors associated with poor mental health status among homeless women with and without dependent children. *Community Mental Health Journal, 50*(5), 553–559. https://doi.org/10.1007/s10597-013-9605-7

Chen, X., Tyler, K. A., Whitbeck, L. B., & Hoyt, D. R. (2004). Early sexual abuse, street adversity, and drug use among female homeless and runaway adolescents in the Midwest. *Journal of Drug Issues, 34*(1), 1–21. https://doi.org/10.1177/002204260403400101

Christensen, Julia (2012). "They want a different life": Rural northern settlement dynamics and pathways to homelessness in Yellowknife and Inuvik, Northwest Territories. *Canadian Geographer/Le Géographe canadien, 56*(4), 419–438. https://doi.org/10.1111/j.1541-0064.2012.00439.x

Clark, R. E., Weinreb, L., Flahive, J. M., & Seifert, R. W. (2019). Infants exposed to homelessness: Health, health care use, and health spending from birth to age six. *Health Affairs (Project Hope), 38*(5), 721–728. https://doi.org/10.1377/hlthaff.2019.00090

Coe, A. B., Moczygemba, L. R., Harpe, S. E., & Gatewood, S. B. (2015). Homeless patients' use of urban emergency departments in the United States. *The Journal of Ambulatory Care Management, 38*(1).

Couldrey, C. (2010). *Violence within the lives of homeless people* (Doctoral dissertation).

Crane, M., Byrne, K., Fu, R., Lipmann, B., Mirabelli, F., Rota-Bartelink, A., Ryan, M., Shea, R., Watt, H., & Warnes, A. M. (2005). The causes of homelessness in later life: Findings from a 3-nation study. *Journals of Gerontology Series B: Psychological Sciences and Social Sciences, 60*(3), S152–S159. https://doi.org/10.1093/geronb/60.3.S152

Cress, D. M., & Snow, D. A. (2000). The outcomes of homeless mobilization: The influence of organization, disruption, political mediation, and framing. *American Journal of Sociology, 10*(4), 1063–1104. https://doi.org/10.1086/210399

Desmond, M. (2016). *Evicted: Poverty and profit in the American city.*

Dietz, T., & Wright, J. D. (2005). Victimization of the elderly homeless. *Care Management Journals, 6*(1), 15–21. https://doi.org/10.1891/cmaj.2005.6.1.15

Farrugia, D., & Gerrard, J. (2016). Academic knowledge and contemporary poverty: The politics of homelessness research. *Sociology, 50*(2), 267–284. https://doi.org/10.1177/0038038514564436

Fountain, Jane, Howes, Samantha, Marsden, John, Taylor, Colin, & Strang, John (2009). Drug and alcohol use and the link with homelessness: Results from a survey of homeless people in London. *Addiction Research and Theory, 11*(4), 245–256. https://doi.org/10.1080/1606635031000135631

Fowler, P. J., Toro, P. A., & Miles, B. W. (2009). Pathways to and from homelessness and associated psychosocial outcomes among adolescents leaving the foster care system. *American Journal of Public Health, 99*(8), 1453–1458. https://doi.org/10.2105/AJPH.2008.142547

Gaetz, S., Walter, H., & Story, C. (2021). *THIS is Housing First for Youth.*

Gallagher, K. (2023). *Rapid re-housing: Housing stability outcomes for families experiencing homelessness in San Francisco* (Doctoral dissertation).

Greenwood, R. M., Stefancic, A., & Tsemberis, S. (2013). Pathways housing first for homeless persons with psychiatric disabilities: Program innovation, research, and advocacy. *Journal of Social Issues, 69*(4), 645–663. https://doi.org/10.1111/josi.12034

Hamilton, A. B., Poza, I., & Washington, D. L. (2011). Homelessness and trauma go hand-in-hand": Pathways to homelessness among women veterans. *Women's Health Issues, 21.*

Horsell, C. (2006). Homelessness and social exclusion: A foucauldian perspective for social workers. *Australian Social Work, 59*(2), 213–225. https://doi.org/10.1080/03124070600651911

Hsu, H. T., Fulginiti, A., Petering, R., Barman-Adhikari, A., Maria, D. S., Shelton, J., Bender, K., Narendorf, S., & Ferguson, K. (2020). Firearm violence exposure and suicidal ideation among young adults experiencing homelessness. *Journal of Adolescent Health, 67*(2), 286–289. https://doi.org/10.1016/j.jadohealth.2020.02.018

Hughes, K., Bellis, M. A., Hardcastle, K. A., Sethi, D., Butchart, A., Mikton, C., Jones, L., & Dunne, M. P. (2017). The effect of multiple adverse childhood experiences on health: A systematic review and meta-analysis. *The Lancet Public Health, 2*(8), e356–e366. https://doi.org/10.1016/S2468-2667(17)30118-4

Hurtubise, R., Babin, P. O., & Grimard, C. (2009). *Shelters for the homeless: Learning from research. Finding home: Policy options for addressing homelessness in Canada.*

Hwang, S. W. (2022). Homelessness and health. *Canadian Medical Association Journal, 194*(6).

Johnson, G., & Chamberlain, C. (2008). Homelessness and substance abuse: Which comes first? *Australian Social Work, 61*(4), 342–356. https://doi.org/10.1080/03124070802428191

Kaakinen, J., & Turunen, S. (2021). Finnish but not yet finished—Successes and challenges of housing first in Finland. *European Journal of Homelessness, 3.*

Kilpatrick, D. G., & Acierno, R. (2003). Mental health needs of crime victims: Epidemiology and outcomes. *Journal of Traumatic Stress, 16*(2), 119–132. https://doi.org/10.1023/A:1022891005388

Knutagård, M. (2018). Homelessness and housing exclusion in Sweden. *European Journal of Homelessness, 12*(2), 103–119.

Ku, B. S., Fields, J. M., Santana, A., Wasserman, D., Borman, L., & Scott, K. C. (2014). The urban homeless: Super-users of the emergency department. *Population Health Management, 17*(6), 366–371. https://doi.org/10.1089/pop.2013.0118

Lee, B. A., & Schreck, C. J. (2005). Danger on the streets: Marginality and victimization among homeless people. *American Behavioral Scientist, 48*(8), 1055–1081. https://doi.org/10.1177/0002764204274200

Lee, C. G., Cheesman, F., Rottman, D., Swaner, R., Lambson, Rempel, M., & Curtis, R. (2013). *A community court grows in Brooklyn: A comprehensive evaluation of the Red Hook.* Community Justice Center Final Report.

Mago, V. K., Morden, H. K., Fritz, C., Wu, T., Namazi, S., Geranmayeh, P., Chattopadhyay, R., & Dabbaghian, V. (2013). Analyzing the impact of social factors on homelessness: A Fuzzy cognitive map approach. *BMC Medical Informatics and Decision Making, 13*(1). https://doi.org/10.1186/1472-6947-13-94

Mallett, S., Rosenthal, D., & Keys, D. (2005). Young people, drug use and family conflict: Pathways into homelessness. *Journal of Adolescence, 28*(2), 185–199. https://doi.org/10.1016/j.adolescence.2005.02.002

Mangayi, C. (2014). Poverty, marginalisation and the quest for collective wellbeing in the context of homelessness in the City of Tshwane. *Missionalia, 42*(3), 212–235. https://doi.org/10.7832/42-3-65

Mashau, T. D., & Mangoedi, L. (2015). Faith communities, social exclusion, homelessness and disability: Transforming the margins in the City of Tshwane. *HTS Teologiese Studies/Theological Studies, 71*(3). https://doi.org/10.4102/hts.v71i3.3088

McCarty, D., Argeriou, M., Huebner, R. B., & Lubran, B. (1991). Alcoholism, drug abuse, and the homeless. *American Psychologist, 46*(11), 1139–1148. https://doi.org/10.1037/0003-066X.46.11.1139

Minnery, J., & Greenhalgh, E. (2007). Approaches to homelessness policy in Europe, the United States, and Australia. *Journal of Social Issues, 63*(3), 641–655. https://doi.org/10.1111/j.1540-4560.2007.00528.x

Murphy, J. J. (2022). *Solution-focused counseling in schools.* John Wiley & Sons.

Orwell, G. (2021). *Down and out in Paris and London.* Oxford University Press.

Padgett, D., Henwood, B. F., & Tsemberis, S. J. (2016). *Housing First: Ending homelessness, transforming systems, and changing lives.* Oxford University Press.

Peressini, T. (2009). *Pathways into homelessness: Testing the heterogeneity hypothesis. Finding home policy options for addressing homelessness in Canada.*

Pophaim, J. P., & Peacock, R. (2021). Pathways into and out of homelessness: Towards a strategic approach to reducing homelessness. *Acta Criminologica: African Journal of Criminology & Victimology, 34*(2), 68–87.

Raven, M. C., Niedzwiecki, M. J., & Kushel, M. (2020). A randomized trial of permanent supportive housing for chronically homeless persons with high use of publicly funded services. *Health Services Research, 55*(2), 797–806. https://doi.org/10.1111/1475-6773.13553

Riley, E. D., Vittinghoff, E., Kagawa, R. M. C., Raven, M. C., Eagen, K. V., Cohee, A., Dilworth, S. E., & Shumway, M. (2020). Violence and emergency department use among community-recruited women who experience homelessness and housing instability. *Journal of Urban Health, 97*(1), 78–87. https://doi.org/10.1007/s11524-019-00404-x

Spellman, B. (2010). *Costs associated with first-time homelessness for families and individuals.* DIANE Publishing.

St. Martin, B. S., Spiegel, A. M., Sie, L., Leonard, S. A., Seidman, D., Girsen, A. I., Shaw, G. M., & El-Sayed, Y. Y. (2021). Homelessness in pregnancy: Perinatal outcomes. *Journal of Perinatology, 41*(12), 2742–2748. https://doi.org/10.1038/s41372-021-01187-3

Stubbs, J. L., Thornton, A. E., Sevick, J. M., Silverberg, N. D., Barr, A. M., Honer, W. G., & Panenka, W. J. (2020). Traumatic brain injury in homeless and marginally housed individuals: A systematic review and meta-analysis. *The Lancet Public Health, 5*(1), e19–e32. https://doi.org/10.1016/S2468-2667(19)30188-4

Sullivan, G., Burnam, A., & Koegel, P. (2000). Pathways to homelessness among the mentally ill. *Social Psychiatry and Psychiatric Epidemiology, 35*(10), 444–450. https://doi.org/10.1007/s001270050262

Sutherland, & Rosenoff, E. (2021). Health conditions among individuals with a history of homelessness research brief. *Health, 3.*

Thompson, A., & Tapp, S. N. (2021). *Criminal victimization.* NCJ.

Tsemberis, Sam, Kent, Douglas, & Respress, Christy (2012). Housing stability and recovery among chronically homeless persons with co-occurring disorders in Washington, DC. *American Journal of Public Health, 102*(1), 13–16. https://doi.org/10.2105/AJPH.2011.300320

Tsemberis, S., & Eisenberg, R. F. (2000). Pathways to housing: Supported housing for street-dwelling homeless individuals with psychiatric disabilities. *Psychiatric Services, 51*(4), 487–493. https://doi.org/10.1176/appi.ps.51.4.487

Tutty, L. M., Ogden, C., Giurgiu, B., & Weaver-Dunlop, G. (2013). I Built my house of hope: Abused women and pathways into homelessness. *Violence Against Women, 19*(12), 1498–1517. https://doi.org/10.1177/1077801213517514

Tyler, K. A., & Schmitz, R. M. (2013). Family histories and multiple transitions among homeless young adults: Pathways to homelessness. *Children and Youth Services Review, 35*(10), 1719–1726. https://doi.org/10.1016/j.childyouth.2013.07.014

United Nations. (2016). https://news.un.org/en/story/2016/03/523512. (Accessed 10 July 2023).

United Nations. (2020a). https://www.un.org/en/desa/homelessness-could-happen-anyone. (Accessed 12 May 2023).

United Nations. (2020b). *Everyone included – How to end homelessness.* https://www.un.org/tr/desa/everyone-included-%E2%80%93-how-end-homelessness. (Accessed 15 June 2023).

United Nations. (2020c). *First-ever united nations resolution on homelessness.* Department of Economic and Social Affairs Social Inclusion. https://www.un.org/en/desa/homelessness-could-happen-anyone. (Accessed 12 June 2023).

Walls, J. (2017). *The glass castle: A memoir.* Simon and Schuster.

Weinreb, L., Goldberg, R., Bassuk, E., & Perloff, J. (1998). Determinants of health and service use patterns in homeless and low- income housed children. *Pediatrics, 102*(3 I), 554–562. https://doi.org/10.1542/peds.102.3.554

Williams, J. L., Keaton, K., Phillips, R. W., Crossley, A. R., Glenn, J. M., & Gleason, V. L. (2023). Changes in health care utilization and associated costs after supportive housing placement by an urban community mental health center. *Community Mental Health Journal.* https://doi.org/10.1007/s10597-023-01146-6

CHAPTER

24

Social support system and rights for survival of homeless people

Vijay Yadav

Jawaharlal Nehru University, New Delhi, Delhi, India

1. Introduction

Homelessness is a rising problem in various urban cities around the world (Hwang & Dunn, 2005, pp. 19–41). According to the research findings, significant health issues, including physical and mental illness, communicable diseases, and chronic illness conditions, are faced by homeless people of Delhi. Homeless people are often socially isolated, with inadequate social care and social networks, and this shortage of social services leads to their ill health (Hwang et al., 2009). In this way, isolation from mainstream society not only create health issue but also raise the issue of homeless people's rights. UN declaration covers a broad aspect of Human rights related to social, economic, and political factors. In contrast, state or institution rights are compromised, then the question is raised of an individual's survival (Dwyer et al., 2015). The discussion upon the homeless people in the context of rights is new to social and political discourse (Eisenmann & Origanti, 2019). This chapter aims to discuss and explain homeless people's social support system and different rights in the context of health. Homeless people live in the excluded and dirty parts of urban areas, which center on different urban activities such as the high concentration of commercial activities, industries, population density and traffic congestion, etc. (Kumar, 2019). They have to bear climate extremity, air pollution, odors, noise, water pollution, sanitation problems, unhygienic living conditions, etc. These all things exclusively affect homeless people, which that they survive in open space for their livelihood (Harriss, 2013). Homeless people engaged in the informal sector such as daily wage labor, construction worker, rickshaw pulling, street vendors, rag picking, etc. These informal sector jobs required a lot of manual labor. If they didn't do work for a day, they would have to sleep without meals and shelter. Homeless people cannot satisfy their basic needs, such as food, clothing, shelter, health, education, etc. Yet they have to pay for everything like food, shelter, latrine, bathrooms, etc. Homeless-related regulations and guidelines are typically stringent and revolve around municipal ordinances, security, and criminalization. Factors that worsen the condition of

Homelessness to Hope
https://doi.org/10.1016/B978-0-443-14052-5.00024-0

homelessness are related to societal barriers such as perception, the right to a permanent address, public space, and personal property (Eisenmann & Origanti, 2019). The International Covenant on Economic, Social and Cultural Rights (ICESCR) of the United Nations is an important convention signed by India. The prominent Alma Ata Declaration was signed by the 134-nation government in September 1978, including an Indian delegate who proclaimed health a fundamental human right (Phadke, 2003, pp. 4567–4576). Although the Indian constitution does not consider the "right to health care" as a fundamental right, but the Supreme Court (SC) of India has interpreted in its judgment that Article 21 (Fundamental right to life) and Article 47 of Directive Principles of Indian states (one of the duties of respective state government on the health of citizens). It means that the right to life, including access to basic health care, should be considered a right to life with dignity. According to Lynch (2002), there has found a strong association between human rights and reduction in "poverty and homelessness." It can't be possible to construct the condition of social inclusion, participation, and empowerment without ensuring people's human rights (Hunt et al., 2003).

2. Study area

The national capital Delhi counts as one of the biggest cities in the world. It is located (28.61N; 77.23E) in India's northern part and has an extreme climate situation with a very hot summer and cold winter. Delhi accounts for 0.05% of India's geographical area but consists of 1.38% of the nation's population and 2.66% share of the total homeless population of India (Census, 2011). The homeless population has doubled, and growth has tripled from 2001 to 2011. Delhi is the center of policymaking bodies, administration, and medical facilities. But the homeless population lives without rights, security, privacy, and access to the medical facility.

3. Data and methodology

(a) Primary Data:

A field survey was conducted in nine Delhi districts through a key informant and detailed questionnaires that incorporate various social, economic, and political aspects of urban homeless people.

In this research, the working definition has two aspects; first, it focuses on absolute homeless people living either street or night shelter. The second is taking help from Hertzberg's categorization of homeless people. But the homeless person who lives less than 1 year in the city does not consider for data. Data were procured from night shelters (NS) and homeless persons' concentration zones from July 2018 to February 2019. The concentration zones mean public space where homeless persons are found easily, such as under flyover; foot overbridge; transport junction; religious places, and nearby public hospitals. There were visited 24 permanent and 15 temporaries (Porta Cabin) night shelters and 7 concentration zones to procure 505 homeless person data in Delhi in the extensive survey schedule. The plan of

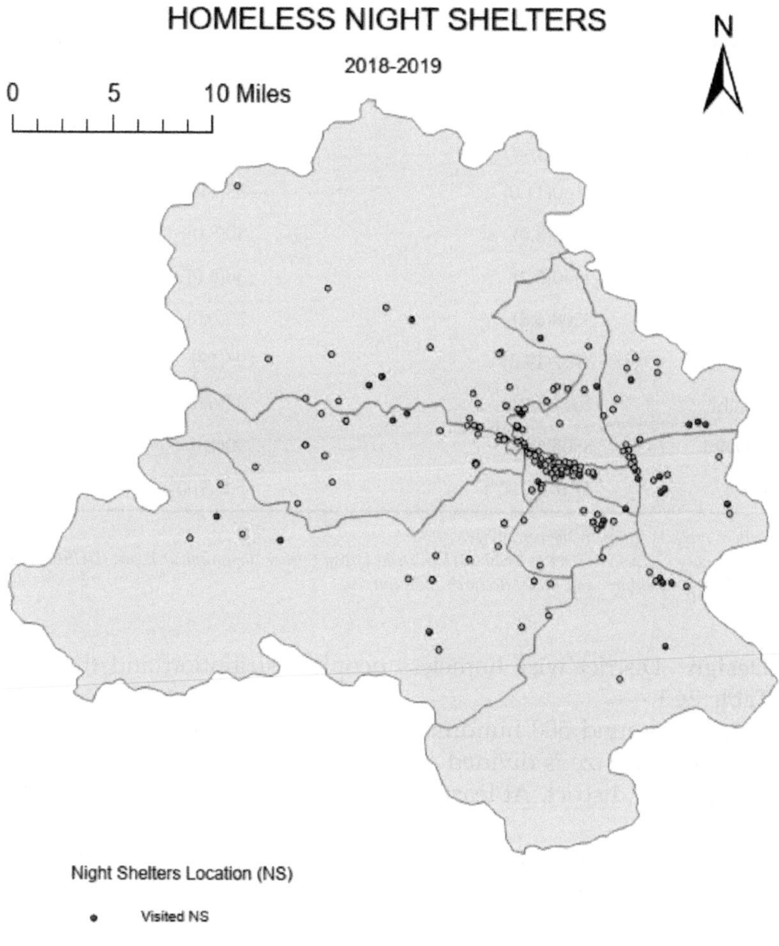

FIGURE 24.1 Distribution of night shelters across in Delhi: description of visited and nonvisited night shelters during field survey.

procurement of field survey data is covered from Delhi. It covers all nine districts: North, Central, South, North West, West, South West, North East, East, and New Delhi (Census, 2011). The distribution of the night shelter in Delhi and visited shelters during survey are given below Fig. 24.1.

Sampling Criteria—In the field survey process, Random Stratified Sampling Method was adopted for the survey. Some criteria were taken that could be recognized in the research sample from the homeless population in Delhi. The criteria are mentioned below.

1. Single male homeless persons
2. The age should be above 18 years
3. At least 1 year experience of homelessness condition

TABLE 24.1 Sample size across the district of Delhi.

Districts	Homeless population 2011	Representative sample size
NW Delhi	4903(10.4)	50(10)
North Delhi	8104(17.2)	85(17)
West Delhi	5160(11.0)	55(11)
SW Delhi	3755(8.0)	40(8.0)
NE Delhi	2440(5.2)	30(6.0)
East Delhi	3208(6.8)	30(6.0)
Central Delhi	8957(19.0)	95(19)
New Delhi	2044(4.3)	30(6.0)
South Delhi	8505(18.1)	90(18)
Total	47,076(100[a])	505(100[a])

[a]Numbers in brackets represent the percentage.
Based on Primary Census Abstract of Delhi 2011 & Delhi Urban Shelter Improvement Board (DUSIB) 2014 (https//:delhishelterboard.in), and Ph.D. Research Field Survey.

Sampling Design—District wise homeless people distribution and the sample size are given below. Table 24.1.

In the field survey, around 500 hundred homeless people surveyed in the summer and winter season. The sample size is divided according to the percentage of the homeless population in their respective district. At least 30 homeless persons are covered in each selected area.

There are used narratives to express the actual ground condition of homeless people and to analyze the data through tabulation, graphs, and correlation methods.

(b) Secondary Data

1. Census of India provides district-wise data of homeless people under primary census abstract (PCA) 2001 and 2011.
2. Central government (NITI aayog, NIUA, MHUPA) publications and reports.
3. Economic survey of Delhi 2017–18 and 2018–19; NGOs survey and reports.
4. Delhi urban shelter improvement board (DUSIB) provides running Night shelter's occupancy status data in Delhi.

4. Social support network

Family, relatives, and friends of a person are a necessary element of his social life. They are not only emotionally involved but also reliable financial support in their bad times. *In East Delhi, a night shelter is located nearby the Kalyanpuri JJ colony. A homeless person reported that he is a truck driver as a profession and lives in a night shelter due to his drinking habit. He had fractured his leg due to an accident in Sikandrabad, and he was admitted to a government hospital where is given primary care. Then his job providing agency is taken to Lal Bahadur Shastri Hospital (LBS) in Kalyanpuri for further treatment. The doctor of LBS hospital set plaster on his fractured leg, and then this procedure is going on for some time. He explained that after 2 months of treatment was no relief in his condition, but his leg injury was decayed. Then Doctor of LBS hospital told him that his leg would have to cut. He felt nervous to hear and think that he couldn't work without his leg. He immediately informs his family member. Family members were reached there and taken him to a private hospital in Noida. The private hospital assured him that there is no need to cut the leg, but they have to pay two lakhs of rupee for the treatment. The homeless person's family managed the money for his treatment. Now he is fine and continues his work.* (Source: Field Survey, Kalyanpuri Night Shelter East Delhi January 21, 2019)

These are the people who attached a person to the mainstream of society. The majority of homeless people have not to live without family but also reliable friends (Gupta & Ghosh, 2006; Letiecq et al., 1998) explained in his manuscript that the three kinds of social support networks are accessible for a person. His study was based on a single homeless mother. These social networks comprise social embeddedness, perceived social support, and enacted support networks. For the first time, these social support networks were identified by Barrera (1986). Social embeddedness depends upon, connections of the individual to others. It mainly depends on the size of social networks. Perceived social support is defined as the cognitive measure of being reliably connected to others. It is judged as the degree of support to get in the difficult time from their society. And enacted support addresses the actual assistance provided by one's social environment. This is measured by the amount and type of help which one receives from a specific individual or any institution. Homeless people are a vulnerable section of society (Mechanic & Tanner, 2007). They live in public places without basic amenities such as drinking water, shelter, electricity, and hygienic food. Their unstable life is a major restriction for making social networks and getting social support from social institutions. A person who lives without any fixed place or address; it means that is not trustworthy. A family could be the most reliable social support network for a human being.

4.1 Enacted social support networks

The purpose of discussion is to find the enacted social support network because it is represented that the degree of association to their family or "home" (Porteous, 1976). A homeless person who is deeply connected to their family and social ties represents that they want to change their homelessness condition (Shlay & Rossi, 1992). There are two variables to find the degree of association with their family: "medium of contact" and "Frequency of contact." The below table represent the connectivity of homeless people to their family.

4.1.1 Availability of enacted social support in Delhi

The government of Delhi provides community shelter for homeless people to sort out this problem temporarily. Yet, the facility of permanent community shelter is addressed only by 18% of homeless people in Delhi. There are around 82% of homeless people who sleep in open or different public places in Delhi. According to Table 24.2 around 68% of homeless persons live without a family member or a relative in Delhi. The district-wise table of homeless people who live without a family member or relative in Delhi is given below.

TABLE 24.2 Homeless persons reported to live without family.

	Family or relative live in Delhi	
Districts	**Yes**	**No**
NW Delhi	55.26	44.74
North Delhi	47.06	52.94
West Delhi	31.43	68.57
SW Delhi	25.00	75.00
NE Delhi	14.29	85.71
East Delhi	71.43	28.57
Central Delhi	12.99	87.01
New Delhi	5.56	94.44
South Delhi	26.79	73.21
Average	**32.20**	**67.80**

Based on Field Survey 2018–19.

There was taken data from 505 homeless people in which around 200 homeless persons were reported that they didn't contact their family last 1 year or more. There were reported around 300 homeless persons were connected to their families. Table 24.3 shows that those had not been contacted to their family around 41% of homeless people during the last year. It means they were dependent upon perceived social support networks such as local friends, job providing agency, etc. There had been connected to family members around 59% of homeless people in different mediums such as mobile, postage, and personal visit. Connectivity through post office was found insignificant due to the majority of them were uneducated. Mobile phone personal or friends was found a prominent medium of connectivity. It was followed by a personal visit like monthly or yearly. Data revealed that the majority of homeless people preferred to contact their families as per need and daily basis. District wise analysis, there was admitted less social support network in Central Delhi (45.45%), East Delhi (46.43%), and New Delhi (33.33%). The reason of less social support network in these three districts can be given by socio-demographic status. In Central Delhi, around 48% homeless people were unmarried and 50% were 35 plus age group. They were living in homelessness condition for a long duration (97% permanent nature). Central Delhi has been provided a high possibility of getting a job in the informal sector. So there was reported the highest number of unskilled labor that earned 100–300 Rs/day and spent their money to drink. Homeless people of New Delhi reported, around 60% were unmarried and separated or widow. In New Delhi, the majority of homeless people lived in their old age and depended upon food and shelter on nearby Gurudwara. New Delhi and East Delhi found majority of homeless people in their 35 plus age group and live as homelessness condition for a long duration. According to field survey, enacted support changes into perceived support in respect to long homelessness duration.

TABLE 24.3 Connectivity status of homeless person to their family.

	Contact with family		Medium of contact				Frequency of contact					
Districts	**Yes(a)**	**No(b)**	**Mobile**	**By post**	**Personally**	**NA**	**Daily**	**Weekly**	**Monthly**	**Yearly**	**As per need**	**NA**
NW Delhi	52.63	47.37	36.84	2.63	28.95	31.58	7.89	0.00	10.53	7.89	39.47	34.21
North Delhi	66.67	33.33	47.06	0.00	31.37	21.57	23.53	11.76	7.84	13.73	21.57	21.57
West Delhi	77.14	22.86	45.71	0.00	31.43	22.86	25.71	11.43	11.43	8.57	20.00	22.86
SW Delhi	85.00	15.00	45.00	0.00	40.00	15.00	10.00	10.00	0.00	35.00	30.00	15.00
NE Delhi	50.00	50.00	35.71	3.57	46.43	14.29	3.57	17.86	35.71	7.14	21.43	14.29
East Delhi	46.43	53.57	39.29	3.57	39.29	17.86	42.86	3.57	3.57	10.71	17.86	21.43
Central Delhi	45.45	54.55	44.16	2.60	15.58	37.66	9.09	14.29	9.09	9.09	20.78	37.66
New Delhi	33.33	66.67	38.89	0.00	0.00	61.11	16.67	5.56	0.00	0.00	22.22	55.56
South Delhi	75.00	25.00	41.07	0.00	33.93	25.00	10.71	14.29	3.57	28.57	17.86	25.00
Average	**59.07**	**40.97**	**41.53**	**1.37**	**29.66**	**27.44**	**16.67**	**9.86**	**9.08**	**13.41**	**23.47**	**27.51**
		N = *505*				*N* = *a*						*N* = *a*

'a' is based on the positive responses of respondents.
'N' stands total number of respondents.

Table 24.2 represents the susceptibility of homeless people in the Districts of Delhi. There has reported living without family around 80% of homeless people in three districts, such as New Delhi, Central Delhi, and North East Delhi. There are reported 50–80% of homeless people who live without relatives in four districts such as South West Delhi, South Delhi, West Delhi, and North Delhi, whereas the rest of the two districts are reported less than 50%. The majority of homeless people come to Delhi from nearby states such as Uttar Pradesh and Bihar. Table 24.3 presents that the homeless people of North-west and East Delhi have relatives in Delhi. These districts are found more local homeless populations. Family members and relatives are part of the emotional and active support of the social network (Belle, 1983). According to data, only 32% of homeless people responded that they have a family member or relative who lives in Delhi.

Rajesh (Changed name), 53 years old, is from Delhi. He has been living as homeless for the last 6 years. He has a son who lives with his wife in Palam. Earlier, he had also been lived with his son. He was a driver by profession and fractured his leg in a road accident 2 years ago. He had operated in Safdarjung hospital, and after treatment, he could not start his job due to having chronic pain. One year later of the treatment, he was facing a swelling problem in his leg. He had operated again in the Deen Dayal hospital. But the same swelling problem occurred after some time. Now, he is to live in a recovery shelter in south Delhi. He is continuing his treatment due to the regular support of his son. He lives in a recovery shelter because his son's home is not having much space. He said that his health problem is the main reason for his social, mental, and financial condition. (*Source*: Field Survey, South Delhi Recovery Shelter, January 2019)

Ravi (Changed name), 30 years old, is from Delhi. He has been living as a homeless for 4 years. In a road accident, he had lost his family except for his brother. He is an electrician by a profession. He was suffering from the breathing problem for the last 2 months. He consulted Deen Dayal hospital for the treatment. The hospital is 10 kilometers from his shelter, so he has to spend 100 rupees per visit. He did not also satisfy the treatment because of the time taking procedure, over-crowdedness, and unsupportive staff behavior of the hospital. Eventually, he decided to go for further treatment in a private hospital by financial support of a friend. (*Source*: Field Survey, West Delhi Night Shelter, January 2019)

4.2 Status of social support network

To show the homeless people's status of social support networks, three age groups is created, such as 1–4 years; 4–10 years; and more than 10 years. These categories are called teeters, acceptors, and accommodators, respectively (Hertzberg, 1992). The table is given below, which represents "family status" and "support during the illness."

The first two categories were represented a similar kind of situation of family status. In the third category (more than 10 years), 40.74% of homeless people didn't have a family member. So the homelessness condition is persistent for a long period creates solitude or less social support network (Eyrich et al., 2003). According to Table 24.4, Homeless people of Central Delhi, New Delhi, and North East Delhi has been living in Delhi for a long time as homeless condition. Therefore, most of them is living in solitude condition.

To examine support status during their illness period, first category (1–4 years) was taken less help from family due to depending upon saving and religious places as well as increase the duration of homeless; they were dependent more on family and friends with less saving. The last category was presented that they were dependent upon friends and religious places during their illness with no savings. Thus, the long duration of homelessness is extended the responsibility of state institutions.

Table 24.4 presents the reason for working during illness time, there are more than 58% of homeless person accepted that they have been working during their illness time due to different reason such as no support from anywhere; no saving; need money for family; fear to lose the job and drug addiction. In the district-wise analysis, three leading factors

TABLE 24.4 Social support status of homeless people.

	Family status		Support during illness period			
Homelessness duration	**Yes**	**No**	**Family member (enacted support)**	**Relatives or friends (perceived support)**	**Savings**	**Religious places/ begging**
1–4 years	80.88	19.12	13.94	43.82	13.94	28.29
4–10 years	80.82	19.18	20.55	49.32	9.59	19.18
More than 10 years	59.26	40.74	11.11	66.67	0.00	22.22

TABLE 24.5 Reason for working during illness across districts in Delhi.

Districts	Reason for working during illness time					
	No support from anywhere	No saving	Need money for family	Fear to lose the job	Drug addiction	NA
NW Delhi	15.79	0.00	31.58	0.00	15.79	36.84
North Delhi	15.69	1.96	25.49	5.88	19.61	31.37
West Delhi	2.86	0.00	40.00	2.86	14.29	40.00
SW Delhi	5.00	5.00	25.00	0.00	10.00	55.00
NE Delhi	17.86	0.00	14.29	0.00	25.00	42.86
East Delhi	21.43	3.57	21.43	3.57	10.71	39.29
Central Delhi	33.77	10.39	19.48	1.30	16.88	18.18
New Delhi	27.78	0.00	11.11	0.00	0.00	61.11
South Delhi	32.14	1.79	10.71	1.79	16.07	37.50

Based on Field Survey 2018–19.

are found in all districts, generally. These factors include the need for money for the family, no financial support from anywhere, and drug addiction. Homeless persons involve in informal sector jobs in which the majority of them are daily wage labor and rickshaw puller. So there is no certainty of wage. Data also revealed that factors like "fear to lose Table 24.5 the job" and "no saving" are not more relevant. The table is given below explains the reasons for working during illness.

There is found permanent (97.40%) and illiterate (45.5%) homeless population in Central Delhi. They were involved in nonskilled jobs such as daily wage laborers (50.65) and Thela puller (25.97). So they have a little possibility to get support from anywhere and no savings due to a little income. Many of them are married, so they work in Delhi to fulfill the money required for their family needs. South Delhi presents the same kind of pattern as central Delhi. In North East Delhi, most local homeless populations (50%) of Delhi reported their drug addiction problem and domestic problem.

4.3 Substance use a medium of social support creation

Homeless people make peer groups in a social and cultural context to the search for survival needs and improved by perceived needs such as alcohol and other substances (Snow & Anderson, 1987). *"I had been not come here to any cause of problem. I like to drink alcohol. At home, I couldn't have drunk due to parents. Therefore I had left home and come to Delhi." Sunil Kumar (Changed name), live in Phoolmandi Night Shelter, Delhi, November 10, 2018.*

Strauss described the homeless person is less socialized with a scarcity of lifetime activities, which would close him into contact with other people. Drinking provides a suitable environment for easy interaction with a friendly companion. And Grunberg articulated that-

"The purpose of obtaining liquor brings cohesion to a group of men. It allows a degree of intimacy between men, which might not be available otherwise. Thus the "bottle gang" appears to fulfill the needs of both drink and affiliate. It seems that eventually, the desire to affiliate and to drink become so intertwined that they are the most inseparable. The need for a few emotional relationships may very well assume the need to drink (Gurunberg, 1998)." Drug addiction helps to make social networks. According to Goodman (1991) it is created negative social support networks such as hurting, annoying, and draining by alcohol. According to the field survey, homeless people reported their money was stolen by other homeless people during the night, whereas the caretaker explained that they come into night shelter with an alcoholic condition, so they didn't have an idea of their belongings. The table explains the district wise condition of drug addiction is given below. Table 24.6 presents the involvement of homeless people in substance use. There are common phenomena of drinking among the homeless people in every district of Delhi. But there has been found a difference into use of drugs on some scale. In East Delhi, West Delhi, and North East Delhi are more associated with tobacco and Gutkha compared to other districts.

4.4 The role of support (care) by institution during illness period

They need extra care such as hygienic food, pure drinking water, proper rest, etc., during the illness period except medical treatment. Thus, it is necessary to relate the social support during the homeless people's illness to the range of economic, social, and welfare institutions of society (Whynes, 1990). The economy institution includes individual's savings and the job providing agency's support during illness time, whereas social institution includes family and relatives' help. And in the national capital, many religious institutions provide free

TABLE 24.6 Substance use among homeless persons.

	Substance use		Kind of substance			
Districts	**Yes**	**No**	**Alcohol**	**Tobacco and Gutkha**	**All**	**NA**
NW Delhi	73.68	26.32	42.11	15.79	15.79	26.32
North Delhi	76.47	23.53	52.94	13.73	9.80	23.53
West Delhi	88.57	11.43	60.00	28.57	2.86	8.57
SW Delhi	75.00	25.00	75.00	5.00	0.00	20.00
NE Delhi	71.43	28.57	50.00	28.57	7.14	14.29
East Delhi	53.57	46.43	39.29	28.57	10.71	21.43
Central Delhi	72.73	27.27	46.75	20.78	15.58	16.88
New Delhi	22.22	77.78	5.56	16.67	5.56	72.22
South Delhi	66.07	33.93	33.93	42.86	0.00	23.21

TABLE 24.7 District wise support during illness of homeless persons.

	Support during illness					
Districts	**Gov./NGO**	**Family member**	**Relatives or friends**	**Job providing agency**	**Savings**	**Religious places/ begging**
NW Delhi	0.00	10.53	50.00	10.53	21.05	7.89
North Delhi	0.00	13.73	41.18	11.76	17.65	15.69
West Delhi	0.00	11.43	54.29	2.86	14.29	17.14
SW Delhi	0.00	20.00	55.00	5.00	10.00	10.00
NE Delhi	0.00	25.00	25.00	7.14	14.29	28.57
East Delhi	0.00	50.00	14.29	10.71	10.71	14.29
Central Delhi	1.30	7.79	54.55	1.30	6.49	28.57
New Delhi	0.00	0.00	0.00	0.00	5.56	94.44
South Delhi	0.00	12.50	35.71	5.36	8.93	37.50

medical camp, food, medicine, and distribute other essential things under charity. The government and the NGOs work under welfare institutions for the better health of poor people. The data are taken from single homeless people. There are analyzed the support system of the homeless people during the illness period. The Table, which represents support during the illness period, is given below. Table 24.7 presents support during the illness of homeless people in Delhi. East Delhi and North East Delhi homeless people have more enacted social support than other districts during their illness. In these districts, homeless people's family or family members live from Delhi, so they provide not only financial support but also care and emotional support during their illness. There are seen a peer group of homeless people such as friends and relatives who help him financially during illness. The presence of family or relative or reliable friends in Delhi is a big support for a homeless person. The majority of homeless people live without treatment due to unavailable a reliable companion.

Veenit Sharma (Changed name), 55 years old, who is from Hapur district of Uttar Pradesh. He was educated at senior secondary level and worked as a shop assistant. He earned 1500 in a week. According to him, he got Tuberculosis (TB) due to living in a night shelter. He had to leave his job due to TB. His son had been got him treated for tuberculosis (TB) in a private hospital. He has not been working for the last 4 months regularly. Source – Fountain Chowk Temporary NS nearby Chandni Chowk Cent Delhi Field (Survey October 2018). According to 6.6 presents an insignificant role of government and NGOs to the care of homeless people during their illness. Consequently, the homeless people were preferred to leave the treatment during their illness due to many reasons. The table which presents the positive response of different indicators is given Table 24.8.

Table 24.7 data presents reported different reasons for homeless persons living without treatment during their health problems. There is around 27% of homeless people give

TABLE 24.8 Reasons for "no treatment or to live without treatment" among homeless.

Districts	Live without treatment (Yes)	Can't afford (Yes)	Rudeness of hospital staff (Yes)	No proper treatment (Yes)	No attendant (Yes)	Loss of wages (Yes)
NW Delhi	73.68	57.89	18.42	31.58	31.58	68.42
North Delhi	74.51	33.33	11.76	23.53	23.53	76.47
West Delhi	54.29	40.00	8.57	17.14	14.29	48.57
SW Delhi	100.00	70.00	0.00	50.00	30.00	95.00
NE Delhi	75.00	71.43	21.43	42.86	28.57	75.00
East Delhi	46.43	32.14	3.57	17.86	21.43	53.57
Central Delhi	59.74	37.66	6.49	18.18	18.18	59.74
New Delhi	83.33	66.67	0.00	11.11	66.67	50.00
South Delhi	91.07	48.21	5.36	23.21	28.57	91.07
Average	**73.12**	**50.82**	**8.40**	**26.16**	**29.20**	**68.65**

*Exclude the negative and NA response.
Based on Field Survey June to February 2018–19.

positive responses that they live without treatment during their health problems. Table 24.6 presents the reasons why the homeless people didn't get them treated during their illness period? There was 50.82% of homeless people reported that they had no money for treatment and 68.65% said they didn't want a loss of wages. There were other reasons responsible for living without treatments of homeless people: "no attendant" and "no proper treatment." Data revealed that most of the homeless population in the "36–60 years" age group was found to live without treatment because of a long experience of the health center's conditions. According to the respondent, there was around 29.20% of homeless people informed that due to "no attendant," and 26.16% admitted that "no proper treatment" was provided in public hospitals to the homeless people. There are given some narratives from different districts related to the responses such as *Live without treatment, can't afford, Rudeness of hospital staff, No proper treatment, and Loss of wages*, respectively.

Hari Parsad (Changed name), 40 years old, is from Sultanpur (Uttar Pradesh). He is married, and his family lives in his village. He is a rickshaw puller and earns around 250 rupees a day. He has been living in Delhi for 10 years. He is having to live as homeless last 3 years. He has a liver infection. Government hospital doctor's suggested some medical tests from outside. He did not get the test because he had no money. He could not afford a loss of wages, so he didn't go to another government hospital. He took some medicine from a medical store when he has pain. He knows very well that this is not good for his health in the long term, but he has not had any other choice. (Source – Sarai Kale Khan-2, Temporary Night Shelter South Delhi Field Survey January 2019)

Ashok Kumar (Changed name), 45 years old, is from Rajasthan. He works in a pantry shop and is got 5000 per month. He said that he was suffering from an unknown disease for the last 2 years. He had consulted in the government hospital, but they referred him to other government hospitals. He prescribed some medicine from that hospital. But he didn't feel any improvement. Lastly, he consulted from a private clinic, and it was suggested some test which was so costly. So when he feels unwell, he takes some medicine from a medical store. (Source – Govind Puri Night Shelter Nearby Machli Market South Delhi Field Survey January 2019)

Bitto Thapa (Changed name), 47 years old, is from Uttarakhand. He has been living as a homeless in Delhi for 25 years. He is uneducated but skilled labor. He was a carpenter and earned 6000 per month but, at present, unable to work. He is suffering from a hernia problem. He reported that doctors and staff behavior in a government hospital was not supportive, so he left the treatment. He has not enough money to go for the treatment in a private hospital. (Source – Gali Tel Mill, Nabi Karim Night Shelter Central Delhi Field Survey November 2018)

Manish Kumar (Changed name), 28 years old, is from Sitapur (Uttar Pradesh). He came for a job, and he has been living in Delhi for 2 years. He is unmarried, and his parents had died. He works as daily wage labor and earns 5000 per month. He was having dengue last year. He was admitted in Sanjay Gandhi Hospital; due to the unsatisfactory treatment, he moved into a private hospital with friends' support. In the private hospital, they were taken 15 days for the treatment and 5000 rupees. I managed money with the support of friends. He said that the government hospital was overcrowded and did not treat well. He said that my condition was so serious, so my friend suggested me to shift the private hospital. (Source – BVK D-4 Block Sultanpuri Night Shelter North West Delhi Field Survey December 2018)

Mukul Kumar (Changed name), 28 years old, is from Siwan (Bihar). he has been living in Delhi for 3 years, and he is having to live as homeless last 2 years. He says that he is not money for room rent. He is daily wage labor (Beldar) and earns around 300 rupees a day, but it is not regular. He has a large family, and he has to send money regularly. He said that he was fell ill with an infectious disease last year. He consulted a local private clinic for treatment but having spent 2000 rupees and a week's work, and there did not see any improvement in health conditions. Eventually, he had to go to the Safdarganj hospital. He was treated there. The treatment was good, but he had to lose 3 weeks of work. So he could not send money to home that month. (Source – Okhla Modi Mill behind TATA Indicom Night Shelter South Delhi Field Survey January 2019)

Majority of homeless people has been living without medical treatment for different reasons in the districts of Delhi during the last 1 year period. The reason "Loss of wages" which was reported the most prominent in all the districts. It depends on the possibility of getting a job. According to field survey revealed that central Delhi, South West Delhi, South Delhi, and North East Delhi had seen the majority of homeless people in low-income categories (100–200 Rs/day). Homeless persons reported that they didn't prefer to skip the job even for a day due to their low earnings. The districts of South West Delhi, South, and North East Delhi reported that the homeless people were unable to afford the treatment due to loss of wages, whereas in Central Delhi, West Delhi, and East Delhi reported comparatively less. North East and North West Delhi were admitted that unsupportive behavior of staff also. Homeless persons of South West Delhi and North East Delhi were reported that the public hospital was not helpful to cure their illness. Elderly homeless person of New Delhi was admitted that there was nobody present to go with them for treatment.

In the Mongolpuri Industrial area in North West Delhi, A 40 years old homeless person lived from Uttar Pradesh. The homeless person collided with a vehicle. He reached the night shelter, but nobody was available to help him. Finally, the Homeless person reached Sanjay Gandhi Hostel (SGH) with a fellow disable homeless person. SGH treated him the right-hand shoulder injury but after some time referred him to Ram Manohar Lohia (RML) or Safdarjung Hospital. RML and Safdarganj hospitals were distanced 16 and 23 km from Mangolpuri, respectively. He tried for further treatment in RML and, according to him, spent around 2000 Rupees for treatment or medicine, but he didn't get any satisfactory improvement in their right hand. He is unable to move his right hand due to a lack of proper treatment in a timely. He accepted that the negligence and blame to SGH doctors for his present condition. (Source: Field Survey, Mangol Puri Night Shelter West Delhi January 9, 2019)

5. Social, economic, and political rights of homeless people

Homelessness is one of the extreme conditions of socioeconomic marginalization and represents a high level of isolation and rejection of basic rights (Tipple & Speak, 2006). Homelessness reduction is highly related to the value of human rights conditions in that country. It is observed that respect for human rights guarantees to facilitate conditions of social, political, and economic rights (Lynch, 2005). The UN international covenant on economic, social, and cultural rights (ICESCR) and Alma Ata declaration signed in September 1978 by 134 countries, including the Indian representative, has also declared health as a fundamental human right are such significant conventions to which India is a signatory (Phadke, 2003, pp. 4567–4576). Yet the Indian constitution has not included the right to health care as fundamental rights. There is a judgment of Supreme Court of India which has interrelated Article 21 and Article 47 of the directive principle (which mentions improving health of the citizen as one of the duties of the government) to mean the right to life to life with dignity, including access to basic health care.

5.1 Economic rights

In the Economic rights encircled, such as the rights to work, equal pay for work of equal value, safe and healthy working conditions, and a fair wage, etc. According to field survey data, many homeless people have accepted homelessness as a lifestyle to save money for their families. Homeless people avoid revealing or hesitate to accept that they have a bank account. But when asked about the mode of transaction of money to their family, their response was "through bank account." According to data, around 22% of the homeless people have bank account, whereas only six persons reported (1.2%) that they have health insurance also. The table which presents the age group wise status of the bank account and the health insurance is given below Table 24.9.

TABLE 24.9 Bank account and health insurance among homeless across age-groups.

Age groups	Having bank A/c			Having health insurance		
	Yes	No	Not response	Yes	No	Not response
18–25 years	22.5	77.5	0.0	2.5	95.0	2.5
26–35 years	32.7	67.3	0.0	1.9	96.3	1.9
35–60 years	27.5	70.8	1.7	0.6	95.5	4.0
More than 60 years	7.7	92.3	0.0	0.0	100.0	0.0
Average	22.6	77.0	0.4	1.2	96.7	2.1

Based on Field Survey 2018–19.

According to the field survey, there was around 22.6% of homeless people used a bank account to transfer money to their families. The account of homeless people is mainly opened under Jan Dhan Yojna. Many of them reported their account is not functioning, and some of them explained, the account has been opened before becoming homeless. Health insurance is a key indicator of determining the accessibility of health care in western countries like a lack of health insurance has been a hurdle for most of the shelterless persons in the United States (Stark, 1992, pp. 151–164), and although Canada has adopted a mechanism of universal health insurance (Hwang, 2000).In India, health insurance can't be considered an indicator of health care accessibility in the context of homeless people. Data revealed that there was only 1.2% of homeless persons having health insurance but were not functioning.

5.2 Social rights

There are many indicators to explain the social condition of a person. The constitution of India promotes every citizen to pursue their social rights such as health, education, Shelter, adequate standard of living, social security, etc.

5.2.1 Basic amenities

There are many things to determine the health of human beings, such as shelter, food, drinking water, living environment, etc. (Harvey, 2003) Hygienic Food and clean drinking water are a fundamental need for human life. The table presents the district-wise utilization of basic amenities by homeless people is given below Table 24.9.

Health represents a compound of the physical, social, mental, and economic status of a person. There is a fundamental problem to access basic amenities for a homeless person who lives in a community shelter or on the street. Table 24.10 presents district-wise use of basic amenities. Homeless persons are transient in nature. They move from one place to another place as their requirement. So they don't make their food due to shelterless. The majority of homeless people depend on their food on street hotels (thela), religious places, charity, and begging. According to field the survey, homeless people (86.4%) depend on street hotels and religious places for their meals. The source of water for drinking and bathing is

TABLE 24.10 Basic amenities among homeless people.

	Source of food		Source of drinking water			Defecation		Bath	
Basic amenities	**Self-cooked**	**Others**	**Community tap**	**NS water**	**Tanker**	**Free public toilet**	**Free public toilet and ODF**	**NS**	**Community tap**
NW Delhi	18.4	81.6	0.0	100	0.0	100	0.0	100	0.0
North Delhi	11.8	88.2	17.6	82.4	0.0	100	0.0	100	0.0
West Delhi	14.3	85.7	0.0	100	0.0	100	0.0	100	0.0
SW Delhi	10.0	90.0	0.0	100	0.0	0.0	100	0.0	100
NE Delhi	3.6	96.4	60.7	39.3	0.0	35.7	60.7	39.3	60.7
East Delhi	50.0	50.0	78.6	21.4	0.0	0.0	92.9	0.0	100
Central Delhi	5.2	94.8	20.8	74.0	5.2	75.3	11.7	72.7	27.3
New Delhi	0.0	100	0.0	100	0.0	100	0.0	100	0.0
South Delhi	8.9	91.1	91.1	8.9	0.0	8.9	91.1	8.9	91.1
Average	**13.6**	**86.4**	**29.9**	**69.6**	**0.6**	**57.8**	**39.6**	**57.9**	**42.1**

night shelter and community tap. Every permanent night shelter has been a toilet and washroom facility within the premises. But these facilities are inadequate compared to occupants such as Lahori Gate night shelter with occupancy of 350 homeless persons rather than only six toilets and four bathrooms; a similar condition is found in Mori gate night shelter. The homeless people who live on the streets or temporary night shelters depending community tap for drinking water. In South Delhi, East Delhi, and southeast Delhi are dominant in the share of the homeless population who live on the streets and temporary night shelters. The homeless person prefers to live on streets in summer because night shelter charges money to stay of Rs 10 per night. These districts are more associated with free public toilets and open defecation facility (ODF).

5.2.2 *Education*

Education is one of the key indicators of the human development index (UNDP, First HDI Report, 1990). Education increases the skill of understanding and empowers people to form a healthy society. Age is a dominant determinant of health status (Whynes, 1990). In the context of homeless people, around 32% were illiterate. Illiteracy was reported high in all age groups. The graph which represents the age group wise education level is given below Fig. 24.2.

Fig. 6.1 presents the age group wise education level of homeless people in Delhi. The majority of homeless people were found illiterate in every age group due to social status and poverty (Field Survey 2018–19). According to the graph, the number of homeless people is decreasing as well as the attainment of a higher level of education. Many of them attained lower- than middle-class education, which gradually decreased from primary to a higher

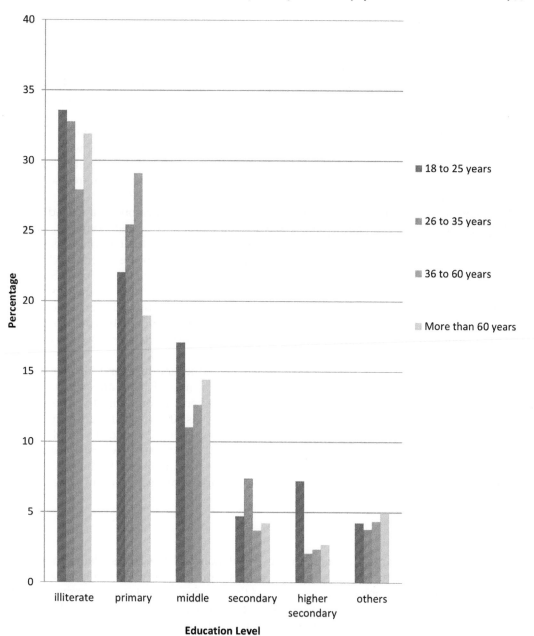

FIGURE 24.2 Age group-wise education level among homeless people homeless people's education level age group-wise. *Based on Field Survey for Ph.D. Research 2018–19.*

education level. Education provides the opportunity to become a more productive person of a civilized society by acquiring skills compared to an illiterate or less educated person. And it is prepared a person how to tackle challenges and overcome difficulties. The relationship between education and work status is given below Fig. 24.3.

Fig. 24.2 presents, educated homeless workers were more skilled in comparison to a less-educated homeless worker. It means uneducated homeless workers have to struggle or be unable to develop the skill in the homeless duration. The explanation is based on field survey experience; there are many reasons such as shyness, lack of communication, hesitation, fear, and unwillingness. Data revealed that the majority of illiterate laborers were involved in a job that did not require any skill, such as rickshaw puller (45.60%) and daily wage labor (38.06%). On the other side, educated labor involved in a job which was required some kind of skill such as painter (37.50%), hotel worker (35.14%), and guard (30%).

Data revealed the involvement of higher educated homeless persons in a skilled or non-skilled job that they were earning more money than less uneducated homeless workers. The homeless people are involved in a different informal job, so their income depends on their labor, job nature, and skill. According to an informant, labor means a person who didn't have any skill or new in that area, just go and wait for a job in the labor Chowk. Skilled laborers work on a contract basis, and they have been in touch with their contractors. Further explained that it is not very important of education in these works but being an educated person makes it easy to keep the record of their daily wages. Sometimes, job providers demand educated labor and offer comparatively good wages. Field survey data revealed the degree of association between education and income. The table is given below.

Table 24.11 presents a positive association between education level and income. There is around one third in Rs.6000 to 9000 income group, and most of them are skilled workers. Therefore, education level is related to the homeless person's income directly. The majority of the homeless population is illiterate and earns hardly Rs.3000 to 6000. According to field survey, 51.8% of homeless people were listed in the group of Rs.3000 to 6000. But the survey revealed the majority of the homeless population was uneducated. As earlier discussed, many young homeless people (18–25 age-group) were also uneducated. Illiteracy is also a dominant cause of homelessness whenever basic education has become a fundamental right in India.

5.3 Political rights

India is the second-largest country in the population and the largest democracy in the world. The equal opportunity of the right to vote is a prior need for securing a real democracy. The right to vote empowers the people, and they are free to elect such a government that is committed to improving their conditions. A person must be having citizenship of that country to cast their vote. In the context of poor people like the homeless, they migrate to the city for different structural causes such as failure of agriculture policies, deforestation, landlessness, and large developmental project (hydroelectric project, construction of dam) (Dasgupta & Bisht, 2010) and they have become the part of the city. Their lifestyle and living conditions don't allow them to keep belonging (Barrett, 2010). There are many of them had been lost their identity documents (Hwang et al., 2009). So the citizenship card is a prior need to use their voting rights. There are issued different types of cards by the government, such as

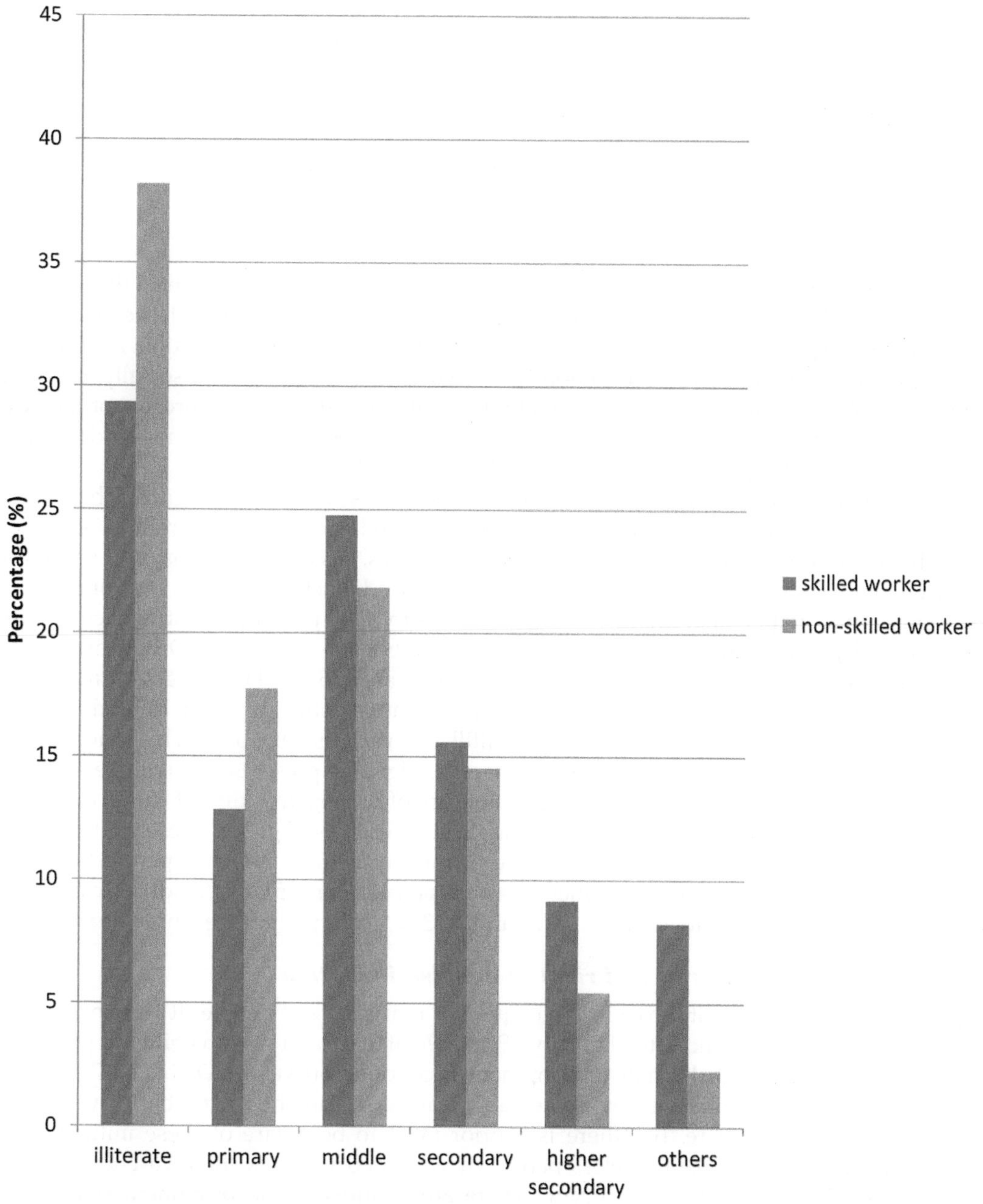

45
40
35
30
25
20
15
10
5
0
Percentage (%)
skilled worker
non-skilled worker
illiterate
primary
middle
secondary
higher secondary
others

FIGURE 24.3 Education and work status among homeless persons association between education level and work status. *Based on Field Survey for Ph.D. Research 2018–19.*

TABLE 24.11 Education and income level of homeless people.

Correlations	Education level	Monthly income
Education level	1	0.256[a]
Monthly income	0.256[a]	1

[a]*Correlation is significant at the 0.01 level (2-tailed).* N = *505.*
Based on Field Survey 2018–19.

Aadhar card, Voter card, Ration card, BPL card, Driving license, PAN card, etc., that ensured citizenship. These cards provide not only voting rights but also ensured accessibility to different welfare programs or public institutions related to healthcare, education, and finance, etc. There is the government's accountability to secure the participation of socially, economically, and politically marginalized sections to be involved in the forming process of the government to get their interest ensured. The graph which presents the citizenship document status among the homeless people is given below Fig. 24.4.

Identity includes not only self-realization but also one's image as recognized by others. The word "Identity" is a comprehensive meaning which related or differed from its respected research area. But there are discussed only social identities. According to Maslow's social human well-being concept, it presents third hierarchy of need. Each person has two kinds of social identity that are personal identity and public identity. The personal identity encircled the family, friends, and neighborhood where a person can behave informally. On the other side, Public identity includes certain types of documents that are provided by the government. The government provides different kinds of identity documents such as Voter Id card, Aadhar Card, Ration Card, Driving license, etc. to fulfill the various purposes. These documents also determine the citizenship status of that person. Fig. 24.4 presents the average status of citizenship cards in which there is made one category of having any kind of cards, and other "No" category clubbed two variables "no" and "lost" variables. Fig. 24.4 presents the district wise citizenship status of homeless people. There is around 30% of homeless people those live without citizenship card in Delhi in which 17.4% "Voter Id" card, 51.7% "Aadhar" card, 1.3% "Ration" card, 19.3% "no citizenship" card, and 10.3% "lost" their citizenship card.

5.3.1 *Right to social security and right to nondiscrimination*

Social security plays a vital role in the context of marginalized people. It is mentioned in the Constitution of India under the Directive Principle of State as a responsibility to provide a favorable environment for human development (Economic Survey of Delhi, 2017–18). So, there are taken many initiatives by center and state government to combat poverty and ensure access to health care. But there is a prior need to be aware of these initiatives by the marginalized section like homeless people. It was tried to check the awareness level of homeless people during the field survey. There are included some prominent schemes of the center and the state government, such as the free mobile health scheme, Mohalla clinic, National Urban Health Mission (NUHM), and Ayushman Bharat. The table which explained the awareness level of the health program is given Table 24.11.

Data revealed that the awareness level of state government health schemes is very high compared to the central government health schemes. Homeless people live their whole life

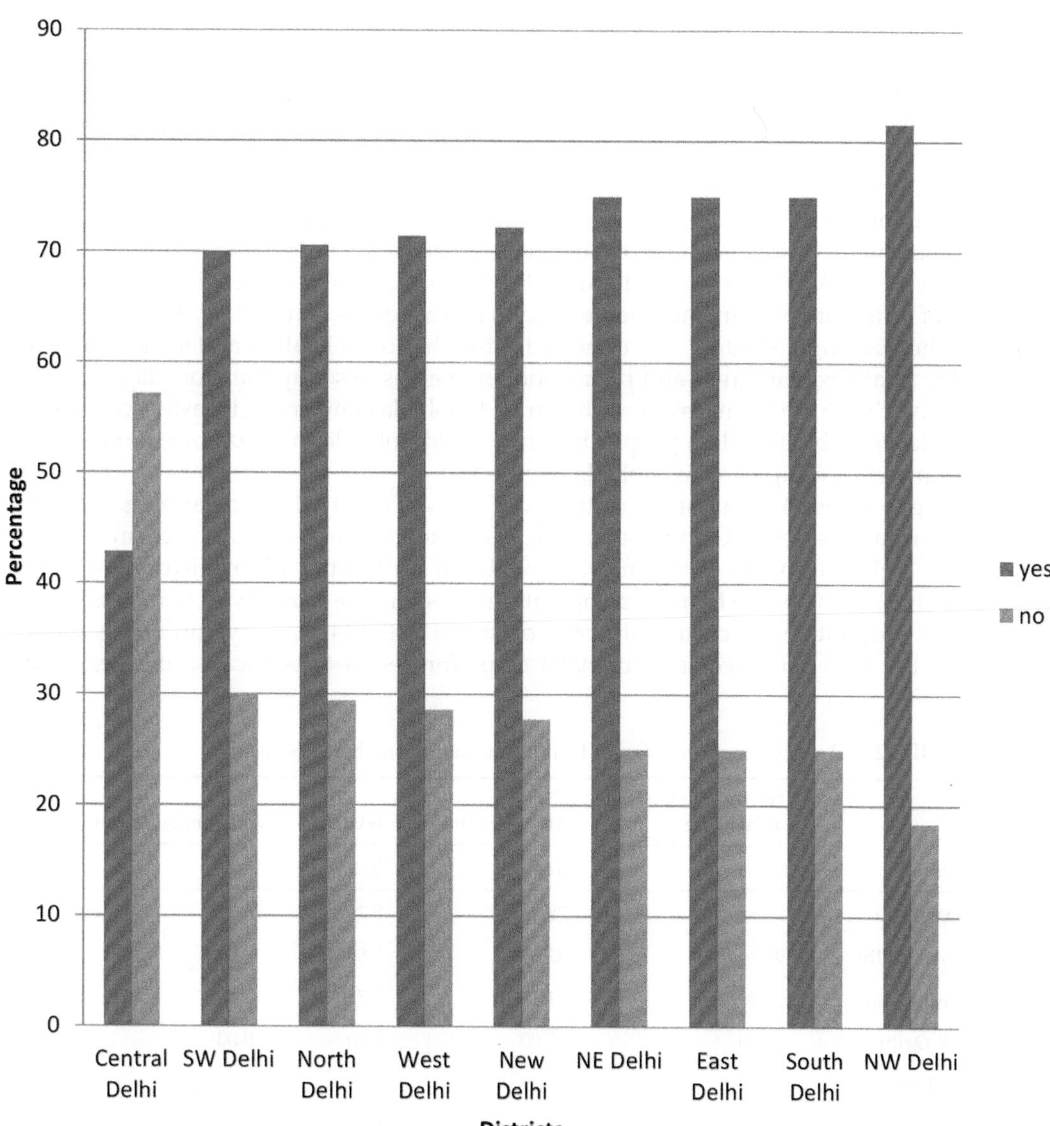

FIGURE 24.4 District-wise citizenship status of homeless people: citizenship status of homeless people in Delhi. *Based on Field Survey for Ph.D. Research 2018–19.*

in isolation due to many reasons such as poverty, illiteracy, fear, drug addiction, mental and physical problems. They make a distance from mainstream society due to their poor conditions. The majorities of homeless people is illiterate or have a primary level of education, so they don't understand and access newspapers (Report, 2012). Homeless people are illiterate and communicate their regional language due to the lack the knowledge of a prevalent language of the city; they fear or hesitate to contact other people. So they are like to live in their peer group like other people. Drug addiction is a prominent reason for their isolation. And their lifestyle also creates many health problems. The entire factors create a distance from relevant information about government initiative. Political leaders are not taken any interest because many of them lost or did not have citizenship cards, so they do not become part of the vote bank (Walters & Gaillard, 2014). Thus the homeless people are left behind the access of different government schemes due to unawareness, particularly the central government schemes. Night shelters are connected through mobile health van facility. There are targeted the localities of migrant and poor population areas to set up the mohalla clinic from 2015 (Lahariya, 2017). So the mobile health van and mohalla clinic make it easily accessible to health care facilities by homeless people, but they could not able to utilize medical test facilities without a citizenship card in Mohalla clinics also.

According to Table 24.12, all of the districts of Delhi are reported less awareness about the central government scheme. On the other side, the state government health schemes are reported in a high percentage except for Central Delhi. In Central Delhi district has found the largest share (42%) of the homeless population in Delhi. The majority of homeless people have been living without a citizenship card or stolen it (57.14%). They can access mobile health van, but they can't access to mohalla clinic for health tests because the health test

TABLE 24.12 Awareness of different health schemes among homeless person.

	Free mobile health van scheme	Mohalla clinic	NUHM	Ayushman Bharat
Districts	**Yes**	**Yes**	**Yes**	**Yes**
NW Delhi	86.84	78.95	5.26	5.26
North Delhi	92.16	82.35	5.88	0.00
West Delhi	94.29	91.43	5.71	11.43
SW Delhi	95.00	95.00	0.00	0.00
NE Delhi	96.43	89.29	0.00	0.00
East Delhi	57.14	64.29	0.00	0.00
Central Delhi	54.55	41.56	1.30	1.30
New Delhi	100.00	72.22	5.56	11.11
South Delhi	98.21	85.71	1.79	3.57
Average	**86.07**	**77.87**	**2.83**	**3.63**

Based on Field Survey 2018–19.

facility is provided through legal Identity proof. They live as homelessness due to many problems such as financial, health, domestic tension, drug addiction, etc. Homeless persons who suffer from a health problem depend upon others such as family, public hospitals, NGO, Charitable trust, friends, or relative (Field Survey 2018–19) due to "no saving" and expensive or time taking medical services in Delhi. Many healthcare providers are in the private sector that is shared more than 50% of medical facilities in Delhi (Economic Survey of Delhi 2018–19). So the government health centers have faced overcrowdedness due to inexpensive treatment (Economic Survey of Delhi 2017–18). A single homeless person who lives without a citizenship card and a small saving or "no saving," there is a high possibility to live without treatment or unable to afford due to expensive or overcrowded health center.

6. Conclusion

Every year thousands of people experience chronic homelessness in Delhi (Census, 2011). According to literature and field survey data revealed that social support networks of individuals who experience homelessness are found the much lesser comparison to housed people (Solarz & Bogat, 1990; Stovall & Flaherty, 1994). And among the homeless people are found a high frequency of sickness. Generally, they are fallen to sick five times more than general people (Martel & Dupuis, 2006). There had been living around 67% of homeless people without any family members in Delhi. There had been lived for more than 10 years as homelessness, found 40% as solitude condition with "no saving." Two-third of homeless people were used drugs regularly and encircled with a negative support system. There were admitted by 73% of homeless people that they skipped their treatment due to unable to afford, loss of wages, and nobody for care.

One of the important objectives of the national urban health mission (NUHM) is that connect with health insurance to urban poor, but an insignificant number of homeless people access health benefits through the health insurance scheme. Homeless people who are struggling for their survival without any voice (Mander, 2009). A homeless person cannot complain or express their circumstances because there is nobody to hear the pain of a homeless person (Matter et al., 2009). They live gross or partially violated their rights such as the right to life and security, the right to education, the right to health care, the right to social security, the right to freedom of expression, the right to vote, the right to privacy, the right to nondiscrimination, the right to participation. These rights are also accepted by the International Covenant on Economic, Social, and Cultural Rights (ICESCR). They have been living on a road around 82% of the homeless population in Delhi due to the existing night shelter facility, only for 18% of homeless people (DUSIB, Delhi Government). So there is left a major population of homeless people from their basic rights, such as shelter, pure water, electricity, etc. Illiteracy is found on a large scale among homeless people. There were illiterate around 32%, and if included primary and middle, then it reached 70%. Education has affected not only their income but also their existence. There were admitted that stolen or lost and not having and unable to make citizenship card by 30% of homeless people. They are unable to access advanced health care facilities or emergency, the government welfare schemes, night shelter in some cases, and unable to raise their voices for injustice.

References

Barrera, M. (1986). Distinctions between social support concepts, measures, and models. *American Journal of Community Psychology, 14*(4), 413–445. https://doi.org/10.1007/BF00922627

Barrett, A. (2010). The new homelessness revisited. *Annual Review Sociology, 36*, 501–521.

Belle, D. E. (1983). The impact of poverty on social networks and supports. *Marriage & Family Review, 5*(4), 89–103. https://doi.org/10.1300/J002v05n04_06

Census. (2011). *Primary census abstract of Delhi.* Office of the Registrar General & Census Commissioner. New Delhi, Government of India.

Dasgupta, R., & Bisht, R. (2010). The missing mission in health. *Economic and Political Weekly, 45*(6), 16–18.

Dwyer, P., Bowpitt, G., Sundin, E., & Weinstein, M. (2015). Rights, responsibilities and refusals: Homelessness policy and the exclusion of single homeless people with complex needs. *Critical Social Policy, 35*(1), 3–23. https://doi.org/10.1177/0261018314546311

Eisenmann, A., & Origanti, F. (2019). Homeless rights: A call for change. *Journal of Social Distress and the Homeless.* https://doi.org/10.1080/10530789.2019.1705519

Eyrich, K. M., Pollio, D. E., & North, C. S. (2003). An exploration of alienation and replacement theories of social support in homelessness. *Social Work Research, 27*(4), 222–231. https://doi.org/10.1093/swr/27.4.222

Goodman, L. A. (1991). The relationship between social support and family homelessness: A comparison study of homeless and housed mothers. *Journal of Community Psychology, 19*(4), 321–332. https://doi.org/10.1002/1520-6629(199110)19:4<321::AID-JCOP2290190404>3.0.CO;2-8

Gupta, S., & Ghosh, S. (2006). Homelessness in the context of the Delhi master plan 2001: Some results from the census. *Social Change, 36*(2), 57–82. https://doi.org/10.1177/004908570603600205

Gurunberg, J. (1998). Homelessness as a lifestyle". *Journal of Social Distress and the Homeless, 7*, 241–261.

Harriss, W. (2013). Multiple shocks and slum household economies in South India. *Economy and Society, 42*(3), 400–431.

Harvey, D. (2003). The right to the city. *International Journal of Urban and Regional Research, 27*(4), 939–941. https://doi.org/10.1111/j.0309-1317.2003.00492.x

Hertzberg, E. L. (1992). The homeless in the United States: Conditions, typology and interventions. *International Social Work, 35*(2), 149–161. https://doi.org/10.1177/002087289203500205

Hunt, P., Nowak, M., & Osmani, S. (2003). *Human rights and poverty reduction: A conceptual framework, office of the high commissioner for human rights.*

Hwang, S. W., & Dunn, J. R. (2005). *Homeless people* (pp. 19–41). Springer Science and Business Media LLC. https://doi.org/10.1007/0-387-25822-1_2

Hwang, S. W. (2000). Mortality among men using homeless shelters in Toronto, Ontario. *JAMA, 283*(16), 2152. https://doi.org/10.1001/jama.283.16.2152

Hwang, S. W., Kirst, M. J., Chiu, S., Tolomiczenko, G., Kiss, A., Cowan, L., & Levinson, W. (2009). Multidimensional social support and the health of homeless individuals. *Journal of Urban Health, 86*(5), 791–803. https://doi.org/10.1007/s11524-009-9388-x

Kumar, A. (2019). Negotiating income and identity in cities: A case of ethnic migrants living beyond slum in India. *Asian Ethnicity, 20*(3), 298–311. https://doi.org/10.1080/14631369.2019.1575717

Lahariya, C. (2017). Mohalla Clinics of Delhi, India: Could these become platform to strengthen primary healthcare? *Journal of Family Medicine and Primary Care, 6*(1), 1. https://doi.org/10.4103/jfmpc.jfmpc_29_17

Letiecq, B. L., Anderson, E. A., & Koblinsky, S. A. (1998). Social support of homeless and housed mothers: A comparison of temporary and permanent housing arrangements. *Family Relations, 47*(4), 415–421. https://doi.org/10.2307/585272

Lynch, P. (2005). Homelessness, human rights and social inclusion. *Alternative Law Journal, 30*(3), 116–119. https://doi.org/10.1177/1037969X0503000304

Mander, H. (2009). *Living rough: Surviving city streets a study of homeless population in Delhi, Chennai, Patna and Madurai for the planning commission of India.*

Martel, J. P., & Dupuis, G. (2006). Quality of work life: Theoretical and methodological problems, and presentation of a new model and measuring instrument. *Social Indicators Research, 77*(2), 333–368. https://doi.org/10.1007/s11205-004-5368-4

Matter, R., Kline, S., Cook, K. F., & Amtmann, D. (2009). Measuring pain in the context of homelessness. *Quality of Life Research, 18*(7), 863–872. https://doi.org/10.1007/s11136-009-9507-x

Mechanic, D., & Tanner, J. (2007). Vulnerable people, groups, and populations: Societal view. *Health Affairs, 26*(5), 1220–1230. https://doi.org/10.1377/hlthaff.26.5.1220

Phadke, S. (2003). *Thirty years on: Womens studies reflects women movements.* EPW 38.

Porteous, J. D. (1976). Home: The territorial core. *Geographical Review, 66*(4), 383. https://doi.org/10.2307/213649

Report. (2012). *Non-profit organization working for vulnerable sections of society.* Indo-Global Social Service Society (IGSS).

Shlay, A. B., & Rossi, P. H. (1992). Social science research and contemporary studies of homelessness. *Annual Review of Sociology, 18*, 129–160. https://doi.org/10.1146/annurev.so.18.080192.001021

Snow, D. A., & Anderson, L. (1987). Identity work among the homeless: The verbal construction and avowal of personal identities. *American Journal of Sociology, 92*(6), 1336–1371. https://doi.org/10.1086/228668

Solarz, A., & Bogat, G. A. (1990). When social support fails: The homeless. *Journal of Community Psychology, 18*(1), 79–96. https://doi.org/10.1002/1520-6629(199001)18:1<79::AID-JCOP2290180112>3.0.CO;2-B

Stark, L. R. (1992). *Barriers to health care for homeless people.* John Hopkins University Press.

Stovall, J., & Flaherty, J. (1994). Homeless women, disaffiliation and social agencies. *International Journal of Social Psychiatry, 40*(2), 135–140. https://doi.org/10.1177/002076409404000205

Tipple, A. G., & Speak, S. E. (2006). Who is homeless in developing countries? Differentiating between inadequately housed and homeless people. *International Development Planning Review, 28*(1), 57–84. https://doi.org/10.3828/idpr.28.1.3

Walters, V., & Gaillard, J. C. (2014). Disaster risk at the margins: Homelessness, vulnerability and hazards. *Habitat International, 44*, 211–219. https://doi.org/10.1016/j.habitatint.2014.06.006

Whynes, D. K. (1990). Research note: Reported health problems and the socio-economic characteristics of the single homeless. *British Journal of Social Work, 20*(4), 355–364.

CHAPTER

25

Trends, associated factors, and the policy necessities with a special focus on the Quality of Life (QoL) of the homeless populations

Kasturi Shukla

Symbiosis Institute of Health Sciences, Symbiosis International (Deemed University), Pune, Maharashtra, India

1. Introduction

Homelessness is a global scenario and is a global issue irrespective of whether it is a developed, underdeveloped or a developing country. It is a wide term that includes unavailability of shelters for sleeping, living in an emergency shelter, and staying in temporary accommodations that are set up by different organizations or being episodically homeless and living with friends or relatives (Gadermann et al., 2021). The Housing Act, 1996 passed by the European Union (EU) has defined a homeless person as "someone who lacks accommodation, cannot access accommodation, or resides in a vehicle or building which is unsuitable for occupation" (Barker & Maguire, 2017; Bennett et al., 2005). A recent study defined homeless populations as "lacking adequate housing or living in housing below a minimum adequacy standard" (Song et al., 2022). However, there is a lack of a cornerstone definition that can be used for policy making and planning support programs for these populations.

Sustainable Development Goals (SDGs) set by United Nations have set a relevant context for countries across the globe to pave way toward superordinate goals (UN SDG, 2015). However, a major shortcoming of SDG is noncoverage of homelessness which is a global crisis irrespective of the development stage of countries (Casey & Stazen, 2021). Global organizations play a large role in streamlining the efforts and providing a conceptual framework for critical issues that have a global prevalence. In this chapter, we have tried to explore some

Homelessness to Hope
https://doi.org/10.1016/B978-0-443-14052-5.00025-2

definitions, models and theories given by various organizations that have defined homelessness thereby paving a path to develop strategies for its management. This needs a higher emphasis after the COVID-19 pandemic that has resulted in mass homelessness at a global level. Moreover, the homeless populations are essentially different than other populations due to a high preponderance of complex morbidities and multimorbid conditions (Barker & Maguire, 2017). Another major challenge in framing policy decisions is the lack of an outcome variable that will support the authorities to evaluate the success of the programs for homeless populations.

Homelessness can be an outcome of mental illness, domestic violence, chronic illness, substance abuse, social exclusion, disturbed family and relations, and socioeconomic factors. Services for homeless populations vary widely based on national and local situations, policies, and socioeconomic conditions (Fowler et al., 2019). Collaborative efforts are required to combine homelessness, health, and law enforcement together to develop strategies for creating a strong support system for such populations. These populations also suffer from higher morbidities and lower life expectancy levels. Almost all outcome indicators like health, education, safety, employment, and overall Quality of Life (QoL) has been found to be low in homeless populations (Fowler et al., 2019). Moreover, multiple studies have revealed that higher communicable and noncommunicable diseases and higher hospitalizations rates are prevalent in homeless populations (Gadermann et al., 2021). Hence, it is a complex vicious circle of individual, interpersonal, and socioeconomic factors that need to be explored in depth.

Different governments across the world attach varied importance to tackle homelessness although it may not be on their official agenda. In addition to the formal and informal policies, partnerships go a long way in setting up systems to reduce home instability among these populations. In some instances, slowly moving to temporary housing and then permanent housing may be preferred, but where situations permit, it is preferable to provide permanent shelters to control morbidities and various other types of threats. The overall impact of these strategies can be assessed by evaluating how this improved the QoL of these populations in totality. Literature is scant in the context of QoL as majority of the research focusses on functional status and very few studies have reported the subjective QoL (Gadermann et al., 2021). Specific populations like women, children, adolescents and those suffering from certain diseases are poorly represented in research. The research on these subjects is too sparse and the literature is highly fragmented. Furthermore, the linkages of homelessness with SDGs and outcome like QoL are not available for certain populations and locations. This chapter intends to present an exploratory review of the recent literature to identify the definitions of homelessness and related trends. The chapter also attempts to throw some light on the policy initiatives of the United Nations and the lessons that can be learned from the different countries in this context. Furthermore, the chapter attempts to review the literature to analyze the impact of homelessness on various health aspects and the linkages of QoL across diverse homeless populations. Toward the end, the chapter proposes certain policy necessities and future directions of research for these populations.

2. Limitations of the study

This is an exploratory review and no primary data was collected which may pose some limitations. Due to the drastic changes brought upon by the COVID-19 pandemic it is possible that most of the earlier literature may not be relatable or comparable now. Therefore, only the recent literature that reported QoL in homeless populations from 2021 till 2023, which is the post-COVID-19 lockdown period, has been included. However, the literature review for definitions, policies, trends, and interventions was not limited to this period. Hence, the study poses many strengths by analyzing QoL as an outcome in various categories of homeless populations.

3. Materials and methods

This chapter is an exploratory literature review of the past studies that have reported QoL in homeless populations. The first part of the study was conducted through a literature search for article review of the impact of homelessness on QoL (See Fig. 25.1). The online platforms used were Google Scholar and Scopus and the studies published from 2021 until June 2023 (time of writing this chapter) were screened and included in the study. The keywords included: homeless, quality of life, impact, and policy. All the studies that were returned through search and matched the objectives of the study were included to extract the relevant information. Around 18 studies were found through online search that were conducted on homeless populations, out of which the 10 studies that reported QoL as an outcome were included in the review. The remaining 8 studies that were done on past homeless populations

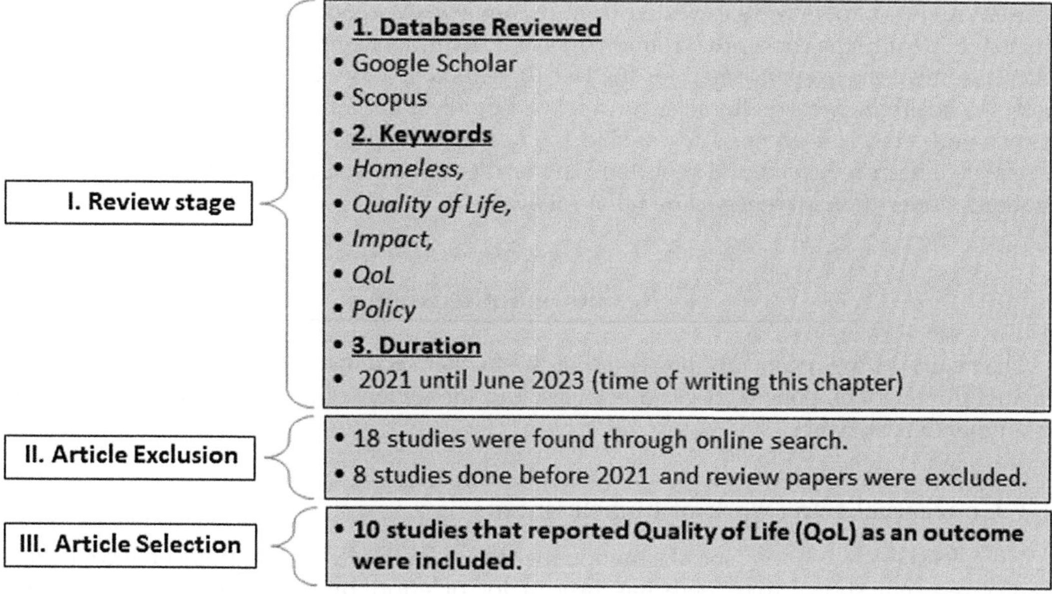

FIGURE 25.1 Methodological flow diagram for article selection that have reported the impact of homelessness on QoL.

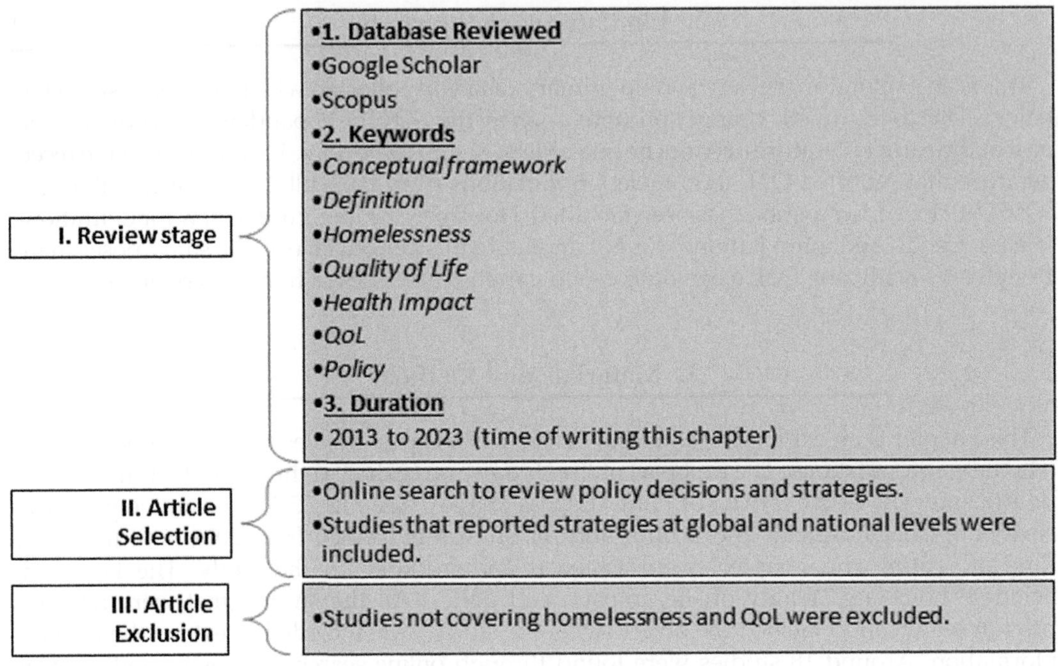

FIGURE 25.2 Methodological flow diagram for the review process to identify the conceptual frameworks of homelessness.

or were review papers were excluded from the review. A second literature search was conducted to identify the conceptual frameworks and definitions of homelessness (See Fig. 25.2). Literature review was performed for the last 10 years to explore the definitions of homelessness, the health impacts on homeless and related policy interventions. Further, the policy decisions and strategies adopted to cater to the homeless populations at global and national levels have been reviewed and reported. The results of the review have been compiled and reported under the different sections that follow.

4. Results and discussion

The results of the review are presented in the further sections to explain the relevant conceptual frameworks, policies at various levels, and the review of the studies that report QoL as an outcome variable.

4.1 Conceptual framework of homelessness

Homelessness is a serious social situation that affects both low- and high-income countries. Earlier research shows that there has been a lot of effort by different groups to define homelessness. It is widely agreed that it is an extreme manifestation of poverty

(Pumariega et al., 2022). Housing was declared as a basic right in 1991 under the Universal Declaration of Human Rights, however, it is still a human rights crisis (Fowler et al., 2019). Homelessness is also defined as a condition of extreme social exclusion and social inequality. As per some authors "Homelessness is a state of deprivation and exists when people do not have access to safe, stable, and appropriate places to live" (Pumariega et al., 2022; Song et al., 2022). Homelessness continues to be on the rise in multiple countries and is largely a global phenomenon but a globally acceptable definition is crucial (Fowler et al., 2019). In 2015, the United Nations adopted Sustainable Development Goals (SDGs) as the 2030 agenda after much deliberations and negotiations (UN SDG, 2015). The 17 SDGs ranged from ending hunger, ending poverty, healthy lives, quality education, gender equality and so on, however, it is striking that it does not cover homelessness as any of the goals. Goal 1 "No Poverty" and Goal-2 "Zero Hunger" apparently seem unachievable without fighting against the cause of homelessness. However, combatting homelessness is inevitable to achieve further goals like good health and hygiene (Goal 3), clean water and sanitation (Goal 6). Similarly, Goal 10 "Reducing inequalities," Goal 11 "Sustainable cities," Goal 16 "Peace, justice and strong institutions" are the goals that necessitate reducing homeless populations. The Goal 17 "Partnership for the goals" indicates the essentiality of collaboration for any social issue at national and global level. Some studies have classified homelessness as chronic, episodic, or transitional (Song et al., 2022). However, still the vacuum of a standard definition remains that creates difficulty in setting priorities for the homeless populations (Fornaro et al., 2022).

The Institute of Global Homelessness (IGH) addressed this gap in 2015 by proposing a framework to define homelessness (Busch-Geertsema et al., 2016; Casey & Stazen, 2021). As shown in Table 25.1, IGH classified the homeless populations into three broad categories under the *"Global framework for Homelessness"*.

TABLE 25.1 Global framework for homelessness.

Category 1—people without accommodation(a) people sleeping in the streets or in other open spaces (such as parks, railway embankments, under bridges, on pavement, on river banks, in forests, etc.) (b) people sleeping in public roofed spaces or buildings not intended for human habitation (such as bus and railway stations, taxi ranks, derelict buildings, public buildings, etc.) (c) people sleeping in their cars, rickshaws, open fishing boats and other forms of transport.(d) "Pavement dwellers"—individuals or households who live on the street in a regular spot, usually with some form of makeshift cover.
Category 2—people living in temporary or crisis accommodation(a) people staying in night shelters (where occupants have to renegotiate their accommodation nightly.(b) People living in homeless hostels and other types of temporary accommodation for homeless people (where occupants have a designated bed or room).(c) Women and children living in refuges for those fleeing domestic violence.(d) People living in camps provided for "internally displaced people," i.e., those who have fled their homes as a result of armed conflict, natural or human-made disasters, human rights violations, development projects, etc. but have not crossed international borders.(e) People living in camps or reception centers/temporary accommodation for asylum seekers, refugees and other immigrants.
Category 3—people living in severely inadequate and/or insecure accommodation.(a) People sharing with friends and relatives on a temporary basis.(b) People living under threat of violence.(c) People living in cheap hotels, bed and breakfasts and similar(d) people squatting in conventional housing(e) people living in conventional housing that is unfit for human habitation.(f) People living in trailers, caravans and tents(g) people living in extremely overcrowded conditions(h) people living in nonconventional buildings and temporary structures, including those living in slums/informal settlements.

Reproduced from IGH (n.d.).

The framework of these definitions was piloted in 13 "Vanguard Cities" comprising the cohort of communities on six continents comprised of Adelaide, Australia; Brussels, Belgium; Bengaluru, India; Chicago, USA; Edmonton, Canada; Glasgow, Scotland; Manchester, England; Little Rock, USA; Montevideo, Uruguay; Rijeka, Croatia; Santiago, Chile; Sydney, Australia; and Tshwane, South Africa. These countries are now using this framework to define and classify the homeless populations (Casey & Stazen, 2021). As per the United Nations, around 2% of the world's population (approx. 1.6 million) is homeless and surprisingly this is independent of being a resident of a developed, underdeveloped or a developing country (Filipenco, 2023; Pumariega et al., 2022). Italy, Spain, Nigeria, and Syria are some of the countries with the highest number of homeless populations (Filipenco, 2023). On the other hand, Iceland, Sweden, Finland, and Japan are the countries with the lowest number of homeless populations (Filipenco, 2023; Pumariega et al., 2022). Lifetime prevalence estimates from high income countries like USA (4.2%) and Europe (4.9%). However, the prevalence data regarding homelessness from low- and middle-income countries (LMICs) is sparse (Fornaro et al., 2022). Article 25 of the Universal Declaration of Human Rights set the goal at the inception of the United Nations (U.N.): "Everyone has the right to a standard of living adequate for the health and well-being of himself and of his family, including food, clothing, housing and medical care and necessary social services, and the right to security in the event of unemployment, sickness, disability, widowhood, old age or other lack of livelihood in circumstances beyond his control." The U.N. Convention on the Rights of the Child, under Article 26 (Right to benefit from Social Security) and Article 27 (Right to Adequate Standard of Living) emphasize the similar social benefit for the children (United Nations, 1948). Globally, the only three nations that did not ratify this Convention are Somalia, South Sudan, and the United States (Pumariega et al., 2022). The United Nations Human Settlements Program (UN-Habitat) is a mandate of the United Nations General Assembly to promote sustainable towns and cities among 90 countries of the world (UN-Habitat, 2023). UN-Habitat is the focal point for all urbanization and human settlement matters within the UN system. The Institute of Global Homelessness (IGH) has partnered with "UN-Habitat" (IGH, n.d.). In order to further promote partnership and collaboration to tackle the challenge of homelessness, The UN formed the NGO Working Group to End Homelessness (WGEH) in 2017 which comprises approximately 30 organizations. Recently, in December 2021, The United Nations General Assembly had, for the first time ever, passed the resolution for homeless populations (IGH, n.d.).

4.2 Causes of homelessness

Situation of homelessness is a complex vicious cycle of individual, interpersonal, and socioeconomic factors (Fowler et al., 2019). Having a home, a permanent residential place, is a critical indicator of a person's life and their QoL. Recent data shows that around 1.77 million Indians are homeless (Deokar, 2022; Reuters, 2016). The 2011 Indian census shows that around 65 million people stay in urban slums (Reuters, 2016). Uncertainty in housing and evictions are major contributors to challenges in homelessness. In 2022, under "Housing for all" project, the Government of India, envisaged to construct 20 million residential settings with toilets, water and electricity. However, under the slum-free "city-beautification" initiatives, in 2017 the government authorities have demolished 53,700 dwellings, leading to eviction of around 260,000 people (Deokar, 2022). In 2016, in the EU, poor families (whose earning was 60% lesser than the median national income) expended >40% of their income on rent. Similarly, in the United States more than 80% of the households below the federal poverty line spent at least 30% of their incomes on rent. Eviction is a major challenge in EU and around 1 million US households suffered evictions. Sub-Saharan Africa shares more than half of the entire global extreme poor population. It is projected that by 2030, Sub-Saharan Africa will have nearly 9 out of 10 extreme poor people (Pumariega et al., 2022).

Extreme poverty is the predominant factor for homelessness in developing countries. Studies have shown that homelessness is directly and indirectly linked to the increased morbidity and multiple poor outcomes are associated with gender, malnutrition and increased vulnerability (Lewer et al., 2019; Racette et al., 2021). Homelessness can be a result of poverty, chronic and/or stigmatized illness, poor government support, natural disaster, and traumatic experiences like domestic violence or physical and mental trauma (Pumariega et al., 2022). Studies also report that the major reasons for homelessness are conflicts, natural calamities, affordability issues, chronic unemployment, and high housing costs (Filipenco, 2023). Homelessness is linked with terror, fear, loneliness, shame and a severe negative impact on the concerned family (Pumariega et al., 2022). Mental illness and addiction of any type are critical risk factors for homelessness (Fowler et al., 2019). Mental illness is also linked to pessimistic thoughts and negative feelings that reduces the ability of the homeless people to cope up with the situation (Mejia-Lancheros et al., 2021). Some of the major drivers of homelessness are female-headed households, step parents or teen parents or substance addicts, having parents who were physically violent, and runaway youth, particularly for those who identify as lesbian, gay, bisexual, or transgender LGBT (Pumariega et al., 2022). Disturbed family life, social conflict and abuse are predisposing factors for homelessness and an estimated 20%–40% of them are identified as LGBT (Embleton et al., 2016; Pumariega et al., 2022). Runaway youth are further at risk of posttraumatic stress leading to mental disorders, substance abuse, unsafe sexual behaviors, and HIV infections resulting in LGBT homeless youth at seven times higher risk of becoming victims of violent crimes (Pumariega et al., 2022). Homelessness has been reported to have an extreme impact on physical and mental health as it is closely linked with extreme poverty and poor education or illiteracy (Fowler et al., 2019). Another close linkage with homelessness is belonging to minority populations, sex, race, and ethnicities.

4.3 Linkages with quality of life as the outcome

The World Health Organization Quality of Life (WHOQOL) group defines QoL as "a multidimensional concept that encompasses individual's goals, expectations, standards, and concerns within their cultural context and value systems" (WHO, 1998). QoL is a complex concept with overlapping constructs pertaining to objective and subjective well-being; wherein, objective QoL includes "aspects of the physical environment and social functioning" and subjective QoL assesses the "individual preferences, opinions, and life satisfaction" (Gentil et al., 2019). Subjective QoL further consists of five domains: daily life and social relations, housing, neighborhood, personal relationships, spare-time activities; and autonomy

(Gentil et al., 2019). Another landmark study that was done to identify the various parameters that are important in defining subjective QoL of homeless populations has identified the six themes namely, health care, living conditions, financial situation, employment situation, relationships, and recreational and leisure activities (Palepu et al., 2012). One of the essential components of research among homeless populations is developing interventions to improve the subjective QoL. At times enhanced objective parameters (e.g., temporary housing) may not be the only solution, hence, the subjective side should not be ignored (e.g., quality of housing, safety, etc.). Unlike objective QoL, the subjective QoL varies with the expectations of the respondents and the discrepancy between their expectations and their actual want/need (Gadermann et al., 2021). Sometimes the reference point for this comparison can be internal (past condition of self) or may have an external referent (comparison with others). On the basis of past research, social support and satisfaction with social life are the factors that are reported to be very critical by homeless respondents in longitudinal research.

In order to frame suitable policies and support programs it becomes inevitable to study the impact of the various factors and analyze the objective and subjective outcomes through QoL assessments (Kaltsidis et al., 2021). There are multiple studies on homeless populations that have not reported QoL as an outcome variable but have indirectly measured one or the other QoL constructs. Homelessness is a major challenge for people with mental disorders and the programs aimed at such populations should be linked to enhancement of QoL (Sullivan et al., 2000). A study from Taiwan reported the prevalence of metabolic syndrome (MetS) and its association with marital status, hyperlipidemia, and cholesterol levels (Gu et al., 2021). Although women are an understudied population in regard to homelessness, they are known to be higher sufferers of lifetime trauma experiences, that correlates with their specific needs and health requirements (Gentil et al., 2019; Maroko et al., 2021; Pumariega et al., 2022; Racette et al., 2021; van Rüth et al., 2021). A review study that analyzed the impact of poverty and homelessness in children and adolescents stated that extreme poverty is the strongest determinant of homelessness in children in developing countries (Pumariega et al., 2022). The study reported the psychosocial effects of poverty, hunger, and homelessness on children and youth and assessed their impact on psychopathology and mental health. A cohort study was conducted on the homeless population across Wales who died during COVID-19 pandemic either due to infection or other reasons (Song et al., 2022). The study reported that poor health seeking behavior and poor health outcomes are prevalent in the homeless population and better integrated services are needed for them. Many of these factors, like psychosocial effects on children, psychopathology, effect of poor health seeking behaviors, and trauma, require QoL assessment for a better comparison of the outcome (Kaltsidis et al., 2021; Sullivan et al., 2000). However, there are a few studies that have assessed and reported QoL as an outcome for the variables associated with homelessness. Past research in this area has highlighted that health facilities, living conditions, financial support, ability to earn livelihood, relationships, effect of disability on physical and mental health; and recreational and leisure activities are the essential determinants of QoL in homeless populations (Gentil et al., 2019; Guillén et al., 2021; Krabbenborg et al., 2017; Magee et al., 2019; Palepu et al., 2012; van der Laan et al., 2017).

We have reviewed the recent studies published between 2021 and 2023 that focus on reporting QoL as an outcome for the homeless population. The compilation of these studies is presented in Table 25.2.

TABLE 25.2 Summary of studies reporting QoL as an outcome in homeless population.

S. No.	Study variables	Linkages reported with QoL	Instrument	Study population	Results/ Observations	References
1.	Sociodemographic variables, social support received, and social support rating scale (SSRS)	Social support received and QoL	EQ-Visual Analog Scale	Chronically homeless patients with schizophrenia in Hunan, China.	Homeless patients could be assisted by providing accommodation, family intervention, medical services, and other comprehensive measures.	Chen et al. (2022)
2.	Sociodemographic, psychosocial, drug use, and oral health-related factors	Dental pain negatively affected QoL	Oral Impact on Daily Performance scale (OIDP)	Individuals aged $\geq$18 attending a public homeless shelter in a Brazilian state capital.	Dental pain negatively affected QoL.	do Carmo Matias Freire et al. (2022)
3.	Dental caries, periodontal disease, and the consequences of untreated caries	Sociodemographic, oral health, and QoL	Oral Health-Related Quality of Life (OHRQoL)	Homeless population in Brazil.	The negative impact on the OHRQoL of homeless persons was associated with low educational level, presence of decayed teeth, gingival bleeding, and dental calculus.	Bernardino et al. (2021)
4.	Dermatologic diagnoses, actinic keratosis/skin cancer and infection	Skin-associated health-related quality of life (HRQL)	Dermatology Life Quality Index (DLQI)	Homeless in Salt Lake City, Utah.	Lower HRQL, were associated with nonwhite participants, poorer skin health, those in committed relationships, and rash or infection diagnoses.	Truong et al. (2021)
5.	Housing, health, substance use, and social support	Association of Subjective QoL (SQoL) with study variables	Quality of Life in Homeless- or-Unstably-Housed Individuals (QoLHI) Instrument	Homeless or vulnerably housed from three Canadian cities.	Important to address poverty, employment, housing challenges, mental and physical health problems, substance use and increase social supports.	Gadermann et al. (2021)

(*Continued*)

TABLE 25.2 Summary of studies reporting QoL as an outcome in homeless population.—cont'd

S. No.	Study variables	Linkages reported with QoL	Instrument	Study population	Results/ Observations	References
6.	Case management, financial/housing assistance, and mental health services	Analyze the trauma histories and need assessment for care provision	Cluster analysis, No QoL scale used	Adult homeless women in the Midwestern metropolitan area, USA.	Designing care systems to improve QoL and housing outcomes.	Racette et al. (2021)
7.	Sociodemography, health service use, clinical variables	Typology on the changes in QoL over a 12-month period	Satisfaction with Life Domains Scale (SLDS) (French version)	Presently or formerly homeless in Canada.	Positive change in QOL was associated with fewer needs variables. Housing stability and better service use was linked to high QoL.	Kaltsidis et al. (2021)
8.	Absolute homelessness or one in last year; a diagnosed mental disorder with or without co-occurring substance or alcohol use disorder	Linkage of resilience with generic and mental Health Related QoL (HRQoL)	Generic QoL measured using the validated global 20-item of the Lehman's 20-item QoL interview	Those who suffer homelessness and severe mental disorders in 5 cities across Canada (Toronto, Moncton, Montreal, Winnipeg, and Vancouver).	Higher resilience levels are positively associated with higher long-term global and mental HRQoL in homeless adults with mental illness.	Mejia-Lancheros et al. (2021)
9.	Sociodemographic variables, clinical history, and symptoms	Self-care and QoL	Qualitative self-report using grounded theory approach	Homeless population in Portugal suffering from Onychomycosis (OM).	Subjects had self-care difficulties, were worried about the condition of their feet; their QoL was compromised.	Silva-Neves et al. (2021)
10.	Sociodemographic variables, health insurance, and chronic alcohol consumption	Linkage of sociodemographic variables with QoL	EQ-5D-5L questionnaire	Hamburg survey of homeless individuals.	Higher age linked with lower HRQoL.	van Rüth et al. (2021)

Author contribution.

The QoL studies on homeless populations have used a variety of scales like the Satisfaction with Life Domains Scale (SLDS) (Kaltsidis et al., 2021) and EQ-5D (van Rüth et al., 2021). Another study attempted to develop the Quality of Life for Homeless and Hard-to-House Individuals (QoLHHI) instrument that is aimed at evaluating QoL in homeless persons (Gadermann et al., 2021; Palepu et al., 2012). Poor access and unaffordable health services are closely linked to higher morbidity in homeless populations. Further, the trends and policies linked to homeless populations necessitate the assessment of QoL.

5. Challenges and solutions

5.1 Health challenges and impact

People experiencing homelessness (PEH) are reported to suffer a high rate of all-cause hospitalization, mortality due to external causes, hepatitis-C injection among those using injectable drugs, and limitations in activities of daily living (ADL) (Barker & Maguire, 2017; Fornaro et al., 2022). These populations are also more prone to metabolic and cardiovascular diseases and respiratory illnesses (Gadermann et al., 2021). High-density lipoprotein (HDL) deficiency, elevated triglycerides, and elevated total cholesterol (TC) were the most predisposing factors for MetS (Gu et al., 2021). One of the studies reported poor health-related QoL (HRQOL) among homeless populations due to high prevalence of dermatologic issues (Truong et al., 2021). Around 82% prevalence of onychomycosis (OM), fungal nail infection, was found in homeless populations and was a chronic issue (Silva-Neves et al., 2021). It was found to negatively impact the QoL of patients as they were worried and had non-negligent therapeutic regimen management styles.

Homeless populations in temporary accommodations are 8 times and those sleeping rough are 11 times more prone to mental illness (Barker & Maguire, 2017). PEH are 6times more prone to mortality rate and about 15 times more likely to die from either accidents or intentional self-harm. It has been reported that severe mental disorders like schizophrenia, and bipolar disorders persist in the homeless populations (Pumariega et al., 2022). Even in developed nations like England and Wales, mortality in homeless men as well as women was 30% higher than remaining populations (Song et al., 2022). Homeless people were also more prone to depression, serious conditions (like psychosis), personality disorders, and suicidal thoughts which all together negatively impacted their Quality of Life (QoL) (Gentil et al., 2019). Stressful conditions, uncertainty, social isolation, and loss of autonomy and power are critical impacts of homelessness. The personal and social trauma is often linked with conditions that existed prior to homelessness like violence at home, serious sickness, negative childhood experiences, and so on (Mejia-Lancheros et al., 2021). There are studies that argue whether the health status of the homeless is worse than the underprivileged housed general populations (Lewer et al., 2019). The comparative study reported that homeless people were 3 times more likely to suffer a chronic illness and diseases like asthma, COPD, epilepsy, and heart problems. Further, homeless populations have reported worse QoL and suffer very high anxiety levels as compared to the most disadvantaged housed people (Lewer et al., 2019).

In homeless children, low birth weight, malnutrition, ear infections, and a higher exposure to environmental toxins and chronic illness (e.g., asthma) are crucial health issues (Pumariega et al., 2022). Homeless children are likely to be more violent, aggressive and poor learning outcomes are found commonly among these children (D'Sa et al., 2021; Pumariega et al., 2022). The psychological impact of homelessness on children could be analyzed under various categories, like "transgenerational," "new-onset homelessness," and hidden homelessness (D'Sa et al., 2021).

Women are an understudied population in the context of homelessness. Women usually try to stay in hidden housing, which is one of the reasons for poor representation of women in such studies (van Rüth et al., 2021). The determining factors for wellness and QoL for women are very different from other populations. A study done in Manhattan, New York reported the need for public toilets and the necessity to cater to the menstrual hygiene management (MHM) products for the homeless population (Maroko et al., 2021). The study reported that the MHM products were more available in upper socioeconomic localities. Further, the public toilets with MHM products and disposal facilities were difficult to access and did not provide free MHM products to the homeless. This has a tremendous impact on the QoL of homeless but the linkage to QoL was not the objective of the study. A cluster analysis of a female homeless population has identified five clusters of women based on the experiences: emotional abuse, physical abuse, nonsexual polytrauma, polytrauma, and no/minimal trauma (Racette et al., 2021). More than half of women who are homeless have reported experiences of sexual trauma. This has further consequences as women fare worse than men like psychiatric disorders, suicidal ideation, and suicidal attempts. In homeless women, appearance of psychological symptoms was more pronounced in individuals with nil to low amounts of traumatic experiences in contrast to those who experienced greater trauma. There may be many reasons for this as homelessness itself may be the biggest trauma in comparison to other issues. There is a huge possibility that the women in homeless situations would be experiencing more psychopathology but were underreporting it due to low trust, poor empathy and confidentiality issues (Racette et al., 2021). It is recommended that female staff should be the first line of care providers for homeless women populations who have suffered polytrauma. These respondents may at times be nonresponsive or hostile due to which the female staff needs to take time for rapport building, enhance trust and reduce hostility for providing care and making observations of interviewing them. Furthermore, there may be a higher consciousness and body awareness among female homeless populations which is indicated by higher frequency of physician visits (Hajek et al., 2021). As per a study conducted in eight different countries, prevalence of schizophrenia was higher among women, young people and those who are chronically homeless; with less than half of them not receiving any treatment (Pumariega et al., 2022). Another study that reported cluster analysis classified the homeless populations into four clusters comprising women of higher age with one episode of homelessness; those having a high prevalence of mental health Disorders (MHDs) and functional disability; individuals with milder MHDs but in temporary housing, and lastly women who are severely suffering from personality disorders, substance abuse, MHDs and suicidal ideation (Gentil et al., 2019). This study reported that higher QoL was associated with older women; whereas lower QoL was linked to MHDs, substance use and high functional disability. Homelessness and mental health programs, addressing issues of trauma, stigma and discrimination, and community integration must be intertwined but these are in nascency (Schreiter et al., 2021).

5.2 Successful models

Quality of Life in the homeless populations is very complex and multifactorial (Flike & Aronowitz, 2022). Past studies have shown that stable housing contributes more to QoL enhancement rather than housing status improvement (Kaltsidis et al., 2021). Housing stability is an important factor and a critical outcome for programs linked to housing assistance provided to homeless populations (Palimaru et al., 2023). Further, better QoL was associated with increased health services and reduced needs of homeless people. Social support is critical for maintaining Quality of Life as having professionals, like, advocates is known to improve QoL in homelessness (Babayan et al., 2021). Enabling resources like health insurance are also important for maintaining the QoL in such populations (Hajek et al., 2021).

Though there are no standard solutions for managing homelessness, there are some success stories that need due consideration. Finland is one of the known success stories for fighting homelessness across the world (Greater Change, 2023). In 2008, the Finnish government declared the "Housing First policy" to bring an end to homelessness. Finland has seen a 75% decline in homelessness over the last 30 years and is the only country in the European Union (EU) to achieve any decline in homelessness in the last 10 years. Working on these lines, Finland predicts to eradicate homelessness by 2027. Another Case study is Japan which has witnessed a 12% drop in homelessness rate since 2018 as a result of many strategies and policies. In contrast to Finland, Japan creates a social prejudice for the homeless populations and the urban infrastructure supports these populations to prevent them from sleeping on the streets and in open areas (Greater Change, 2023). Due to this invisibility of the homeless, there is a projected drop in the homelessness rates as the majority of the homeless are out of sight. Therefore, these populations are responsible themselves to manage their stay and extensive social support makes this further possible. During COVID-19 pandemic many countries had launched programs to mitigate homelessness. Like, in The United Kingdom, the "everyone in" strategy helped around 90% homeless populations to move into temporary accommodations (Casey & Stazen, 2021).

A French study reported the long term effects of the *"Housing First" approach* which is an independent housing program for the homeless with severe mental disorders (Lemoine et al., 2019). The approach was found to be very successful due to a recovery-oriented approach as homeless patients suffering from schizophrenia or bipolar disorders reported a higher recovery, better self-esteem and improved mental health symptoms. One of the major factors that works in long term improvement of QoL is stable housing instead of temporary housing provisioning (Kaltsidis et al., 2021). Further, the models inclined toward health improvement are more successful due to the higher incidence of mental health issues among homeless populations. Frequency of physician visits for homeless populations with an increased focus on those suffering from mental and health issues, women, elderly and children are necessary to build a model that has a higher focus to enhance welfare and overall care models (Hajek et al., 2021). Research from some countries shows that predisposing characteristics like age and gender and enabling resources like health insurance are highly correlated with better utilization of health services and thereby better health outcomes (Hajek et al., 2021). Besides these factors, utilizing peer groups for training and behavior changes is much more effective in terms of impact like reduction of drug/alcohol abuse/use, increasing mental and physical health, and increasing social support (Barker & Maguire, 2017). A very high "shared experience" is the key success factor behind peer group based interventions which is otherwise missing. Further, systems that can enhance the prevention of homelessness proactively are much more ahead in tackling the related issues rather than managing homelessness once it has set in (Fowler et al., 2019). Hence, proactivity, prevention and focus on specific populations while utilizing the peer support is the key to developing successful models.

5.3 Quality of life paradox

QoL paradox is an important finding in homeless populations indicating a higher QoL when it is expected to be low (Lewer et al., 2019). In some instances, except for high anxiety levels, homeless populations have not reported a lot of health problems, possibly due to a higher threshold and a low expectation level for good health. Another reason can be a higher acceptance or normalization of pain and illness and adjustments with the situations (Lewer et al., 2019; van Rüth et al., 2021). There is a range of studies that report mixed results like reporting of higher QoL by older homeless women, those experiencing a first homeless episode, fewer days of homelessness (Gentil et al., 2019). One of the critical reasons is a lower expectation among homeless populations as compared to the remaining population. Many homeless people tend to adopt resilience strategies to cope up with their social, health and housing challenges like "stay strong and thankful," "looking to live," "hope to move forward," "do not give up," and remain "optimistic" and "confident" (Mejia-Lancheros et al., 2021). Hence resilience is a life and health protective factor as it motivates participation in meaningful activities and setting life goals despite adversities, thereby improving overall QoL. Studies have also found that better coping ability despite the challenges of homelessness may be due to higher resilience (Mejia-Lancheros et al., 2021). Further, across socioeconomic groups in the general population there is a "slope" in health outcomes whereas a "cliff" was observed in homeless populations.

5.4 Policy recommendations

Poverty, homelessness, and hunger are all interrelated through systemic factors like employment, economy, social support, immigration and so on. Underdeveloped and developing countries are facing bigger challenges due to added complexities like resource limitations, unmanageable population size, corruption and political disinterest (Pumariega et al., 2022). Both government and nongovernment bodies need to address poverty, employment, housing challenges, mental and physical health problems, and provide social support (Gadermann et al., 2021). There is an increased income inequality and a smaller population has more control of social wealth (Pumariega et al., 2022). This divide has increased further after COVID-19 pandemic as wealthier classes have better access to facilities, technological support, and job security. One of the unique studies that conducted focus group discussion on the homeless population in Canada has identified the following six most critical life areas that affect the QoL of youth and adults namely, health care, living conditions, financial situation, employment situation, relationships, and recreational and leisure activities (Palepu et al., 2012).

Health is indeed the most critical priority as the majority of these populations were reportedly addicted to alcohol, tobacco and drugs, suffering from HIV and did not have any association with healthcare providers (Pumariega et al., 2022; Racette et al., 2021). Studies have strongly reported that mental disorders and infectious diseases (e.g., HIV, and hepatitis B and C infections), cardiovascular diseases and premature deaths are highly prevalent among homeless (Hajek et al., 2021). Among the female population, the services need to be tailor-made for those who have significant trauma histories. In-depth diagnosis, counseling and removing the treatment barriers is necessary to improve the systems of care and contribute to enhancing the quality of life and housing outcomes (Pumariega et al., 2022; Racette et al., 2021). Moreover, regular monitoring and surveillance regarding the health status of homeless people is required to bring in higher accountability and reduce poor health in these populations (Lewer et al., 2019). Secondly, living conditions like temporary shelters need better monitoring as these spaces are mostly cramped up with many people in the premises indulging in drug consumption, alcohol abuse, physical and mental abuse. This creates a tremendous trauma that can have a much more impact than homelessness has on these people. Rehabilitation services for addicted populations and deaddiction programs are related areas that needs to be scaled up. Moreover, financial support through government programs for daily needs is crucial for social acceptance and maintaining a basic life. Relationships were reported as another crucial aspect that determines QoL as many wish to connect again with their children or family. Legal job, self-respect, social respect, food, and autonomy were some more determinants that are reported by various studies and need to be addressed (Gentil et al., 2019; Palepu et al., 2012). Media representation of homeless people was a very important finding in one study which reported that these people have an objection to being misrepresented in the media like "being shown to eat from dumpsters" which is very demeaning and not generalizable (Palepu et al., 2012).

Another study identified Intentional Peer Support (IPS) to play a critical role in supporting PEH (Barker & Maguire, 2017). Peers in this context are defined as "those individuals or groups who have 'shared experiences' of homelessness and can provide various support for someone who is 'new' to the experience or entering recovery." IPS being intentional denotes that here the peer support is fostered and not coincidental making peers themselves as the intervention. Hence, these peers can better engage and empathize with those who are new to certain experiences, emotions and social situations that are a part and parcel of homelessness. This can be expanded to cover health seeking and medication behaviors, coping skills and even reduce hospitalizations. Nine areas in which IPS has an impact are (See Fig. 25.3): overall QOL, social support, physical and mental health, addiction/drug and alcohol use, life skills, homelessness, criminality, employment/finances, and attendance/interest (Barker & Maguire, 2017).

These areas of IPS should be utilized to develop policy interventions and implemented through fostered peer group systems. Evaluating the success of these interventions may get tedious due to the wide spectrum of recovery. QoL assessment and improvements is a key method to evaluate the outcome of mental health programs and initiatives for homeless populations which is often found missing (Kaltsidis et al., 2021). QoL assessment in homeless populations has been largely restricted to mental illness (Kaltsidis et al., 2021) and infections (Lewer et al., 2019). However, very few studies have focused on categorical analysis for groups with short-term, medium-term, and long-term shelters. There are almost nil studies on the homeless population in comparison to their housing trajectories and QoL evaluation. There is a need for such studies that characterize the homeless women and children to understand their specific needs and identify factors that are strong determinants of their health and QoL. Subjective self-reported measures are recommended for such studies. Longitudinal studies are necessary to avoid selection bias and make suitable interpretations (van Rüth et al., 2021).

Data availability from low- and middle-income countries is a big knowledge gap. Gelberg–Andersen Behavioral Model for Vulnerable Populations is a tool for health service

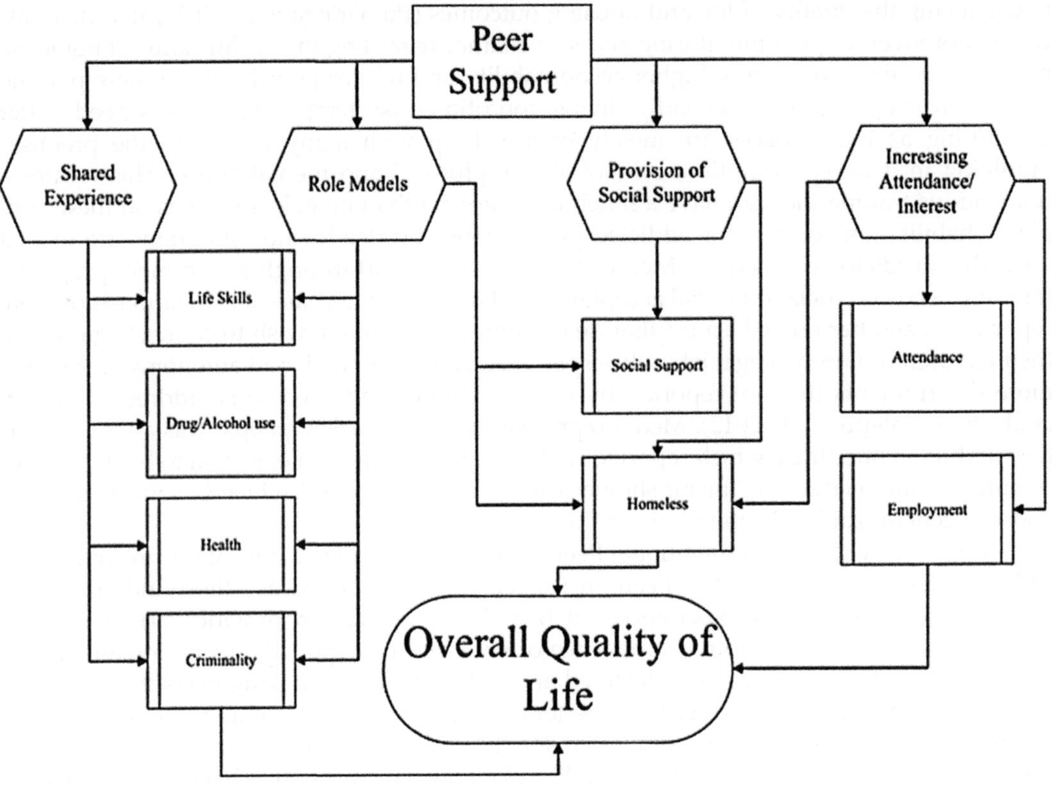

FIGURE 25.3 Impact of peer support on QoL of homeless (Barker & Maguire, 2017).

evaluation. In this model, sociodemographic variables are the predisposing factors, clinical variables are the needs, and enabling factors are the service-related variables (Kaltsidis et al., 2021). These models have a high relevance in developing and underdeveloped countries for application in similar research studies. By far the fundamental factors that decide the success of any program is trust and safety and targeted interventions for age groups and gender (Schreiter et al., 2021). Hence, the health workers who work in related projects need to be trained in rapport building and trust building while dealing with the affected populations.

6. Recommendations

Though the United Nations has launched many programs like UN-Habitat and formed the NGO Working Group to End Homelessness (WGEH), still there are policy gaps pertaining to homelessness. The SDGs do not cover homelessness directly in any of the goals. Globally, as the countries are preparing their priorities based on SDGs, it is necessary that SDG or the UN goals should attach certain priorities for the homeless populations. There is a strong need to conduct regular surveys for keeping an account of the number of homeless populations. Unless the volumes are known, policy formulation remains a challenge. The accounting of the homeless population should be done on the basis of some conceptual framework, like the one formulated by IGH, as there are multiple categories of homeless. Generically defining homeless people as one group cannot guide suitable policy definition and priority setting. Homeless populations may also be classified as chronic, episodic, and transitional for policy decisions and further research. Health needs of homeless people are very pronounced due to poor hygiene, malnutrition, bad food quality, and severe mental disorders. Review suggests that in this population metabolic conditions, all types of infections, malnutrition, Hepatitis-C, HIV infections, and mental disorders are very common. Accidents, epilepsy, respiratory illness like asthma and COPD, cardiovascular conditions, sexual trauma, and mental health disorders are very frequent. Children, youth, and women are the populations that need to be on high priority for policy recommendations. Like some of the studies, conducting cluster analysis and performing analysis in smaller well-defined homogeneous clusters is a recommended method to identify the necessary tailor-made initiatives. As women, children and LGBT may be hidden communities, snowball sampling and similar techniques may be used for case identification. Detailed research on facilities provided to them and its impact on QoL needs to be conducted. Recommendations and policy priorities should be set based on the domains that have been reported to affect QoL among the homeless populations, like Health/health care: conditions of Living, income and financial conditions, legal job/employment, relationships, and leisure activities. Intentional Peer support (IPS) has been known to improve the experience and recovery and creates a positive impact in major dimensions like improving physical and mental health, life skills and keeping a check on substance use. Therefore, creating peer groups in these settings may support the people who are experiencing homelessness for the first time or are at a new location. As we found very limited studies that reported QoL, it is recommended that QoL assessment and planning initiatives that can lead to QoL improvements is critical to evaluate the outcome of health programs for these populations. Subjective self-reported measures are recommended for such studies. Longitudinal studies are necessary for assessing the variations in QoL among permanent and temporary homeless populations.

7. Conclusion

One of the major limitations of research in this area is the lack of a standard definition and the representation of specific populations in research. Firstly, countries need to define their context specific definition of homelessness which may be based on the framework given by IGH. The situation of homelessness in every country is unique, hence, each country needs a tailor-made program in this regard. Women and children have unique needs and face the major brunt of homelessness and specific studies for these population groups are required from different parts of the world. In research studies women are underrepresented and need to be given more representation to make available relevant statistics. As the majority of homeless populations are known to suffer from mental illness, infections, metabolic conditions, respiratory and cardiovascular conditions, health provisioning is a critical mandate to be implemented for these populations. In future studies it is recommended that QoL assessment should be monitored in the homeless population to evaluate the success of the policy measures and assess the outcomes of intervention. Future research must also focus on longitudinal and self-reported measures to arrive at accurate interpretations. The governing bodies need an acute focus on creating employment opportunities and initiatives to control substance abuse and stable housing policies as these are the essential focus areas.

Another area where future research should be conducted is the further development of the complex model of utilizing peer support to address the challenges and behavior change required among homeless populations. As seen in the chapter, only one study has been conducted on the peer support and how it is crucial due to a high shared experience among these groups. Furthermore, psychosocial or personality-related determinants be included in future research. Different personality types and personality related factors may push people toward certain behaviors. Hence, personality related research can gather relevant insights to find ways to address the challenges. Health seeking behavior is also an extremely compromised factor among homeless populations due to multiple reasons. Therefore, factors like substance use related disorders, health literacy, lack of transportation, cognitive functioning or perceived discrimination in health care settings, needs to be evaluated for improving health outcomes. In addition to longitudinal QoL research, several factors and especially the rarely examined variables need to be considered in future research. Some of these can be housing status, perceived health, frequency of primary care, specialized ambulatory care and community-based service use, as well as service satisfaction.

References

Babayan, M., Futrell, M., Stover, B., & Hagopian, A. (2021). Advocates make a difference in duration of homelessness and quality of life. *Social Work in Public Health, 36*(3), 354–366. https://doi.org/10.1080/19371918.2021.1897055

Barker, S. L., & Maguire, N. (2017). Experts by experience: Peer support and its use with the homeless. *Community Mental Health Journal, 53*(5), 598–612. https://doi.org/10.1007/s10597-017-0102-2

Bennett, A., Beresford, P., Betts, C., Brady, G., Clelland, D., Cummings, J., & Sanders, A. (2005). ODPM: Housing, planning. *Local Government and the Regions Committee, 1*(3).

Bernardino, R. M. P., Silva, A. M., Costa, J. F., Silvas, M. V. B., dos Santos, I. T., Neta, N. B. D., Júnior, R. R. P., & Mendes, R. F. (2021). Factors associated with oral health-related quality of life in homeless persons: A cross-sectional study. *Brazilian Oral Research, 35*, 1–12. https://doi.org/10.1590/1807-3107bor-2021.vol35.0107

Busch-Geertsema, V., Culhane, D., & Fitzpatrick, S. (2016). Developing a global framework for conceptualising and measuring homelessness. *Habitat International, 55*, 124–132. https://doi.org/10.1016/j.habitatint.2016.03.004

Casey, L., & Stazen, L. (2021). Seeing homelessness through the sustainable development goals. *European Journal of Homelessness, 15*(3), 63–67.

Chen, J., Song, H., Li, S., Teng, Z., Su, Y., Chen, J., & Huang, J. (2022). Social support and quality of life among chronically homeless patients with schizophrenia. *Frontiers in Psychiatry, 13*. https://doi.org/10.3389/fpsyt.2022.928960

do Carmo Matias Freire, M., de Campos Lawder, J. A., de Souza, J. B., & de Matos, M. A. (2022). Dental pain in adult and elderly homeless people: Prevalence, associated factors, and impact on the quality of life in Midwest Brazil. *Journal of Public Health Dentistry, 82*(2), 211–219. https://doi.org/10.1111/jphd.12452

D'Sa, S., Foley, D., Hannon, J., Strashun, S., Murphy, A. M., & O'Gorman, C. (2021). The psychological impact of childhood homelessness—a literature review. *Irish Journal of Medical Science, 190*(1), 411–417. https://doi.org/10.1007/s11845-020-02256-w

Deokar, P. (2022). *Homelessness in India: Causes and remedies*. https://leaglesamiksha.com/2022/06/19/homelessness-in-india-causes-and-remedies/. (Accessed 19 June 2023).

Embleton, L., Lee, H., Gunn, J., Ayuku, D., & Braitstein, P. (2016). Causes of child and youth homelessness in developed and developing countries: A systematic review and meta-analysis. *JAMA Pediatrics, 170*(5), 435–444. https://doi.org/10.1001/jamapediatrics.2016.0156

Filipenco, D. (2023). *Homelessness statistics in the world: Causes and facts.* https://www.developmentaid.org/news-stream/post/157797/homelessness-statistics-in-the-world. (Accessed 15 June 2023).

Flike, K., & Aronowitz, T. (2022). Factors that influence quality of life in people experiencing homelessness: A systematic mixed studies review. *Journal of the American Psychiatric Nurses Association, 28*(2), 128–153. https://doi.org/10.1177/1078390320985286

Fornaro, M., Dragioti, E., De Prisco, M., Billeci, M., Mondin, A. M., Calati, R., Smith, L., Hatcher, S., Kaluzienski, M., Fiedorowicz, J. G., Solmi, M., de Bartolomeis, A., & Carvalho, A. F. (2022). Homelessness and health-related outcomes: An umbrella review of observational studies and randomized controlled trials. *BMC Medicine, 20*(1). https://doi.org/10.1186/s12916-022-02423-z

Fowler, P. J., Hovmand, P. S., Marcal, K. E., & Das, S. (2019). Solving homelessness from a complex systems perspective: Insights for prevention responses. *Annual Review of Public Health, 40*, 465–486. https://doi.org/10.1146/annurev-publhealth-040617-013553

Gadermann, A. M., Hubley, A. M., Russell, L. B., Thomson, K. C., Norena, M., Rossa-Roccor, V., Hwang, S. W., Aubry, T., Karim, M. E., Farrell, S., & Palepu, A. (2021). Understanding subjective quality of life in homeless and vulnerably housed individuals: The role of housing, health, substance use, and social support. *SSM - Mental Health, 1.* https://doi.org/10.1016/j.ssmmh.2021.100021

Gentil, L., Grenier, G., Bamvita, J. M., Dorvil, H., & Fleury, M. J. (2019). Profiles of quality of life in a homeless population. *Frontiers in Psychiatry, 10.* https://doi.org/10.3389/fpsyt.2019.00010

Greater Change. (2023). *Which country handles homelessness the best?.* https://www.greaterchange.co.uk/post/which-country-handles-homelessness-the-best#:~:text=In%20conclusion%2C%20while%20it%20may,that%20handles%20homelessness%20the%20best. (Accessed 15 June 2023).

Gu, M., Lu, C. J., Lee, T. S., Chen, M., Liu, C. K., & Chen, C. L. (2021). Prevalence and risk factors of metabolic syndrome among the homeless in Taipei city: A cross-sectional study. *International Journal of Environmental Research and Public Health, 18*(4), 1–10. https://doi.org/10.3390/ijerph18041716

Guillén, A. I., Panadero, S., & Vázquez, J. J. (2021). Disability, health, and quality of life among homeless women: A follow-up study. *American Journal of Orthopsychiatry, 91*(4), 569–577. https://doi.org/10.1037/ort0000559

Hajek, A., Bertram, F., Heinrich, F., van Rüth, V., Ondruschka, B., Kretzler, B., Schüler, C., Püschel, K., & König, H. H. (2021). Determinants of health care use among homeless individuals: Evidence from the Hamburg survey of homeless individuals. *BMC Health Services Research, 21*(1). https://doi.org/10.1186/s12913-021-06314-6

IGH. Global advocacy. IGH (Institute of Global Homelessness). https://ighomelessness.org/advocating-for-change/#:~:text=UN%20NGO%20Working%20Group%20to%20End%20Homelessness%20(WGEH)&text=One%20of%20the%20Working%20Group's,and%20inclusive%20asks%20and%20statements. (Accessed 15 June 2023).

Kaltsidis, G., Grenier, G., Cao, Z., L'Espérance, N., & Fleury, M. J. (2021). Typology of changes in quality of life over 12 months among currently or formerly homeless individuals using different housing services in Quebec, Canada. *Health and Quality of Life Outcomes, 19*(1). https://doi.org/10.1186/s12955-021-01768-y

Krabbenborg, M. A. M., Boersma, S. N., van der Veld, W. M., Vollebergh, W. A. M., & Wolf, J. R. L. M. (2017). Self-determination in relation to quality of life in homeless young adults: Direct and indirect effects through psychological distress and social support. *The Journal of Positive Psychology, 12*(2), 130–140. https://doi.org/10.1080/17439760.2016.1163404

Lemoine, C., Loubiere, S., Tinland, A., Boucekine, M., Girard, V., & Auquier, P. (2019). Long-term effects of a housing support intervention in homeless people with severe mental illness. *The European Journal of Public Health, 29*(Suppl. ment_4). https://doi.org/10.1093/eurpub/ckz185.086

Lewer, D., Aldridge, R. W., Menezes, D., Sawyer, C., Zaninotto, P., Dedicoat, M., Ahmed, I., Luchenski, S., Hayward, A., & Story, A. (2019). Health-related quality of life and prevalence of six chronic diseases in homeless and housed people: A cross-sectional study in London and Birmingham, England. *BMJ Open, 9*(4), e025192. https://doi.org/10.1136/bmjopen-2018-025192

Magee, C., Norena, M., Hubley, A. M., Palepu, A., Hwang, S. W., Nisenbaum, R., Karim, M. E., & Gadermann, A. (2019). Longitudinal associations between perceived quality of living spaces and health-related quality of life among homeless and vulnerably housed individuals living in three canadian cities. *International Journal of Environmental Research and Public Health, 16*(23). https://doi.org/10.3390/ijerph16234808

Maroko, A. R., Hopper, K., Gruer, C., Jaffe, M., Zhen, E., Sommer, M., & Kidd, S. A. (2021). Public restrooms, periods, and people experiencing homelessness: An assessment of public toilets in high needs areas of Manhattan, New York. *PLoS One, 16*(6), e0252946. https://doi.org/10.1371/journal.pone.0252946

Mejia-Lancheros, C., Woodhall-Melnik, J., Wang, R., Hwang, S. W., Stergiopoulos, V., & Durbin, A. (2021). Associations of resilience with quality of life levels in adults experiencing homelessness and mental illness: A longitudinal study. *Health and Quality of Life Outcomes, 19*(1). https://doi.org/10.1186/s12955-021-01713-z

Palepu, A., Hubley, A. M., Russell, L. B., Gadermann, A. M., & Chinni, M. (2012). Quality of life themes in Canadian adults and street youth who are homeless or hard-to-house: A multi-site focus group study. *Health and Quality of Life Outcomes, 10*. https://doi.org/10.1186/1477-7525-10-93

Palimaru, A. I., McDonald, K., Garvey, R., D'Amico, E. J., Tucker, J. S., & Greco, G. (2023). The association between housing stability and perceived quality of life among emerging adults with a history of homelessness. *Health and Social Care in the Community, 2023*, 1–16. https://doi.org/10.1155/2023/2402610

Pumariega, A. J., Gogineni, R. R., & Benton, T. (2022). Poverty, homelessness, hunger in children, and adolescents: Psychosocial perspectives. *World Social Psychiatry, 4*(2).

Racette, E. H., Fowler, C. A., Faith, L. A., Geis, B. D., & Rempfer, M. V. (2021). Characteristics of trauma among women experiencing homelessness: An exploratory cluster analysis. *North American Journal of Psychology, 23*(4), 569–582. http://najp.us/.

Reuters. (2016). *India's home-for-all by 2022 plan ignores slum dwellers and the homeless, says UN*. https://www.indiatoday.in/india/story/indias-home-for-all-by-2022-plan-ignores-slum-dwellers-and-the-homeless-says-un-319967-2016-04-26. (Accessed 19 June 2023).

Schreiter, S., Speerforck, S., Schomerus, G., & Gutwinski, S. (2021). Homelessness: Care for the most vulnerable - a narrative review of risk factors, health needs, stigma, and intervention strategies. *Current Opinion in Psychiatry, 34*(4), 400–404. https://doi.org/10.1097/YCO.0000000000000715

Silva-Neves, V., Hugo, V., Alves, P., Amado, J. C., Pais-Vieira, C., Sousa, F., Cerqueira, F., Pinto, E., & Pais-Vieira, M. (2021). Quality of life and therapeutic regimen management in onychomycosis patients and in vitro study of antiseptic solutions. *Scientific Reports, 11*(1). https://doi.org/10.1038/s41598-021-92111-4

Song, J., Grey, C. N. B., & Davies, A. R. (2022). Creating an e-cohort of individuals with lived experience of homelessness and subsequent mortality in Wales, UK. *Journal of Public Health, 44*(4), 805–809. https://doi.org/10.1093/pubmed/fdab180

Sullivan, G., Burnam, A., Koegel, P., & Hollenberg, J. (2000). Quality of life of homeless persons with mental illness: Results from the course-of-homelessness study. *Psychiatric Services, 51*(9), 1135–1141. https://doi.org/10.1176/appi.ps.51.9.1135

Truong, A., Secrest, A. M., Zhang, M., Forbes, B. R., Laggis, C. W., McFadden, M., Gardner, L. J., Powell, D. L., & Lewis, B. K. H. (2021). A survey of dermatologic health-related quality of life and resource access in patients experiencing homelessness. *Journal of the American Academy of Dermatology, 85*(3), 775–778. https://doi.org/10.1016/j.jaad.2020.08.085

UN SDG. (2015). *United Nations Sustainable Development Goals (SDGs). Transforming our world: The 2030 agenda for sustainable development [A/RES/70/1]*. https://sustainabledevelopment.un.org/content/documents/21252030%20Agenda%20for%20Sustainable%20Development%20web.pdf. (Accessed 15 June 2023).

UN-Habitat. (2023). https://unhabitat.org/about-us. (Accessed 15 June 2023).

United Nations. (1948). *The universal declaration of human rights. General assembly resolution 217 A, adopted on December 10th, 1948*. https://www.un.org/en/about-us/universal-declaration-of-human-rights#. (Accessed 23 May 2022).

van der Laan, J., Boersma, S. N., van Straaten, B., Rodenburg, G., van de Mheen, D., & Wolf, J. R. L. M. (2017). Personal goals and factors related to QoL in Dutch homeless people: What is the role of goal-related self-efficacy? *Health and Social Care in the Community, 25*(3), 1265–1275. https://doi.org/10.1111/hsc.12429

van Rüth, V., König, H. H., Bertram, F., Schmiedel, P., Ondruschka, B., Püschel, K., Heinrich, F., & Hajek, A. (2021). Determinants of health-related quality of life among homeless individuals during the COVID-19 pandemic. *Public Health, 194*, 60–66. https://doi.org/10.1016/j.puhe.2021.02.026

WHO. (1998). *Programme on mental health. Division of mental health and prevention of substance abuse*. World Health Organization.

Index

'*Note:* Page numbers followed by "f" indicate figures and "t" indicate tables.'

N

S

T